AF698513

Essentials of Behavioral and Social Science Statistics

Essentials of Behavioral and Social Science Statistics

Third Edition

K. Paul Nesselroade, Jr.
Asbury University
USA

Laurence G. Grimm†

WILEY

Copyright © 2025 by John Wiley & Sons, Inc. All rights reserved, including rights for text and data mining and training of artificial technologies or similar technologies.

Published by John Wiley & Sons, Inc., Hoboken, New Jersey.
Published simultaneously in Canada.

No part of this publication may be reproduced, stored in a retrieval system, or transmitted in any form or by any means, electronic, mechanical, photocopying, recording, scanning, or otherwise, except as permitted under Section 107 or 108 of the 1976 United States Copyright Act, without either the prior written permission of the Publisher, or authorization through payment of the appropriate per-copy fee to the Copyright Clearance Center, Inc., 222 Rosewood Drive, Danvers, MA 01923, (978) 750-8400, fax (978) 750-4470, or on the web at www.copyright.com. Requests to the Publisher for permission should be addressed to the Permissions Department, John Wiley & Sons, Inc., 111 River Street, Hoboken, NJ 07030, (201) 748-6011, fax (201) 748-6008, or online at http://www.wiley.com/go/permission.

Trademarks: Wiley and the Wiley logo are trademarks or registered trademarks of John Wiley & Sons, Inc. and/or its affiliates in the United States and other countries and may not be used without written permission. All other trademarks are the property of their respective owners. John Wiley & Sons, Inc. is not associated with any product or vendor mentioned in this book.

Limit of Liability/Disclaimer of Warranty: While the publisher and author have used their best efforts in preparing this book, they make no representations or warranties with respect to the accuracy or completeness of the contents of this book and specifically disclaim any implied warranties of merchantability or fitness for a particular purpose. No warranty may be created or extended by sales representatives or written sales materials. The advice and strategies contained herein may not be suitable for your situation. You should consult with a professional where appropriate. Further, readers should be aware that websites listed in this work may have changed or disappeared between when this work was written and when it is read. Neither the publisher nor authors shall be liable for any loss of profit or any other commercial damages, including but not limited to special, incidental, consequential, or other damages.

For general information on our other products and services or for technical support, please contact our Customer Care Department within the United States at (800) 762-2974, outside the United States at (317) 572-3993 or fax (317) 572-4002.

Wiley also publishes its books in a variety of electronic formats. Some content that appears in print may not be available in electronic formats. For more information about Wiley products, visit our web site at www.wiley.com.

Library of Congress Cataloging-in-Publication Data
Names: Nesselroade, K. Paul, Jr., author. | Grimm, Laurence G., author. | John Wiley & Sons, publisher.
Title: Essentials of behavioral and social science statistics / K. Paul Nesselroade, Jr., Laurence G. Grimm.
Description: Third edition. | Hoboken, New Jersey: Wiley, [2025] | Second edition published in 2019 as Statistical applications for the behavioral and social sciences. | Includes bibliographical references and index.
Identifiers: LCCN 2024017994 (print) | LCCN 2024017995 (ebook) | ISBN 9781394184101 (hardback) | ISBN 9781394184125 (adobe pdf) | ISBN 9781394184118 (epub)
Subjects: LCSH: Social sciences–Statistical methods.
Classification: LCC HA29 .N435 2025 (print) | LCC HA29 (ebook) | DDC 300.72/7–dc23/eng/20240514
LC record available at https://lccn.loc.gov/2024017994
LC ebook record available at https://lccn.loc.gov/2024017995

Cover Design: Wiley
Cover Image: © Arthimedes/Shutterstock

Set in 9.5/12.5pt STIXTwoText by Straive, Pondicherry, India

MWEP363446150525

To my Cheryl – you will always be the best thing that happened to me.

Contents

Preface

This textbook is an outgrowth of our combined 45+ years of experience teaching undergraduate statistics for social and behavioral science students, an experience that has impressed us with the dread students face when entering the course and the frustration they voice in trying to understand statistics. The dread is most likely a result of the unimaginative way mathematics is taught in the American grade school system. Unfortunately, there is nothing a statistics instructor can do about that. However, there is something the instructor of an undergraduate statistics course can (and must) do to combat this frustration. To be sure, most students will not find a statistics course as engaging as a course in social psychology or child development, but it need not rival the forced reading of the unending pages of terms and conditions associated with approving a new software program!

This book has been written with the typical student in mind – one who may dislike math and may also have little confidence in their ability to "deal with numbers." Even the student with only a basic familiarity with algebra will be able to understand the flow of this textbook. A knowledge of algebraic derivations and proofs is not needed to master the material in this text.

1 Goals of the Text

The primary goal of this book is to teach students the conceptual foundations of statistical analyses, particularly inferential statistics. Throughout the chapters on hypothesis testing, the surface mechanics of computing a test statistic are always related to the underlying sampling distribution of relevance. In this way, students learn *why* the formula for a test statistic looks as it does, and they gain an appreciation of the *statistical* meaning of each analysis.

Emphasis on the conceptual underpinnings of hypothesis testing distinguishes this textbook from those that offer a "cookbook" approach. In addition, this text places heavy emphasis on the research context of the statistical analysis under discussion. As a result, students will feel "connected" to the research activities of social and behavioral scientists and come to view formulas as tools to answer questions about human behavior.

Furthermore, students are exposed to a degree of arithmetic operations but also, they are relieved of endless mind-numbing calculations that are now so easily performed with the aid of statistical processing software. Yes, learning to "crunch the numbers" is important and has not been fully removed from this text. However, much of the more extensive calculations have been removed, presuming that students will be given the opportunity to use software programs such as Microsoft® Excel and SPSS® to generate the needed numbers for interpretation. It is the authors' contention that exposure to the number-crunching aspects of simple versions of descriptive and inferential concepts is important to develop an intimate familiarity with the quantitative work that is being done. However, once a student has this familiarity, then the introduction of a computer software program becomes a welcome timesaving aid, and not a method of obscuring what is going on. For this reason, Appendix C has been included at the end of the text. It contains both a general description of how to input data into both Microsoft® Excel and SPSS® as well as a set of brief tutorials with visual aids to help students learn how to navigate these programs and run various inferential tests.

2 Organization and Flexibility

The text has 16 chapters organized into five parts. After an introductory chapter, the parts can be described as follows: (i) Descriptive Statistics, (ii) Inferential Statistics: Theoretical Basis, (iii) Inferential Statistics: z Test, t Tests, and Power, (iv) Inferential Statistics: Analyses of Variance, and (v) Inferential Statistics: Bivariate Data and Chi-Square Tests. The breadth of coverage of topic areas makes this book suitable for a semester course, a two-quarter course, or a one-quarter course. If students have had exposure to research design, Chapter 1 may be skipped or used as a quick summary of research concepts.

Because earlier chapters build the conceptual foundation for later chapters, there is only a modest amount of leeway in assigning chapters out of sequence. Nonetheless, the chapters covering chi-square and other nonparametric tests may be assigned before the chapter on one-way ANOVA. The chapters covering two-way and repeated-measures ANOVA can be omitted without hampering the students' understanding of subsequent chapters. The chapters covering linear correlation and regression treat these procedures in the context of inferential statistics. Consequently, it is not recommended that they be presented immediately after the section on descriptive statistics. The chapter on probability can be left out if there is limited time or a desired lack of emphasis on the theoretical underpinnings of inferential statistical tests. Finally, the chapter on power may be omitted without sacrificing the students' understanding of hypothesis testing.

3 Student Aids in the Text

Because most students approach statistics with considerable angst, we have included several pedagogical features in the text to enhance learning and foster motivation:

1) The application of many formulas is illustrated in step-by-step computational procedures so that students can master the sequential process of arriving at the correct answer to sample problems.
2) Boxes that highlight the topic under discussion using published and unpublished research are presented in each chapter. The material is selected for its interest value to students.
3) One series of boxes addresses, head-on, a major issue in the social and behavioral sciences, the so-called replication crisis. These boxes bring students into this larger discussion, help them understand the underlying issues, and empower them to think critically about their own and others' research.
4) Spotlights present biographical sketches of some of the luminaries in the field of statistics. Interesting aspects of the person and their times are provided to bring the material to life.
5) Recognizing that many instructors want their students to be able to communicate the outcomes of statistical analyses in written form, most chapters contain a section informing the students how to present statistical findings in sentence form.
6) Each chapter ends with an extended summary, a presentation of the key formulas and key terms, and numerous exercises and work problems for concept checks and practice. Many of the questions and exercises are based on published research findings of high interest to students, so that students not only receive practice in data analyses but also increase their knowledge of the content of psychology.
7) At the end of Parts 3, 4, and 5 there are section reviews. These reviews recap the similarities and differences between the analytical tools covered in the preceding chapters. Furthermore, these reviews include questions designed to develop students' diagnostic skills regarding which previously covered statistical tool is appropriate for a given research context. These reviews are cumulative as the text advances.
8) A glossary of terms is provided at the end of the text.
9) A list of selected formulas can be found in the back of the text for easy reference.
10) A list of selected symbols is provided in the back matter.

Appendix A contains helpful tables for determining various critical values needed for determining probability and testing null hypotheses. (Although the tables are incomplete, they will provide the appropriate values for almost all the exercises and work problems. However, students may need to reference tables online to find the critical values needed to answer a few questions.)

It is our experience that students overwhelmingly prefer that *all* the answers to exercises and work problems be provided, and so they are, in Appendix B. In addition, for many of the computation problems, the answers are provided along with the interim steps, thereby allowing students to locate the source of potential errors in the use of formulas. Most of the chapters also include work problems with short data sets that can be used with any statistical software program. The answers to these problems are also provided in Appendix B.

Appendix C presents brief instructions for basic data entry procedures for Microsoft® Excel and SPSS® and a brief step-by-step instruction concerning how to run the various statistical analyses presented in this text. This resource further supports student's ability to use these software products for statistical calculation purposes.

Acknowledgments

Many people at John Wiley & Sons, including Monica Chandrasekar and Isabella Proietti, have contributed to the creation of this textbook.

We are grateful to the literary executor of the late Sir Ronald A. Fisher, F. R. S., to Dr. Frank Yates, F. R. S., and to Longman Group Ltd., London, for permission to reprint Tables III, IV, and VII from their book *Statistical Tables for Biological, Agricultural, and Medical Research* (6th edition, 1974).

A special thanks is also extended to Emma Miller for the graphic design work, Jezri Leong and Samantha Aquino for proofreading the chapters, Della Nevitt for her help with the PowerPoint slides and quizzes, and Daniel Nesselroade for consultation and wording advice.

Our largest debt of gratitude goes to the reviewers of the manuscript. For many years they remained anonymous, yet we came to know many through their styles of criticism, their preferences for how to teach statistics, and their thoroughness. Some went beyond the call of duty for time spent on the manuscript – to each of you, a special thanks.

Judy Britt performed an accuracy check on the entire manuscript, worked all the statistics problems, arranged the index, and continually amazed me with her "eagle eye." Her diligence is appreciated beyond words.

To the student, as well as the instructor, please send any suggestions or comments that you think ought to be considered for the next edition to Paul Nesselroade, Asbury University, Psychology Department, 1 Macklem Drive, Wilmore, KY 40390.

About the Companion Website

This book is accompanied by a companion website:

https://www.wiley.com/go/Essentials_of_Behavioral_and_Social_Science_Statistics3e

The Instructor Companion Site includes:

1) **PowerPoint slides** for each chapter of the book for instructors
2) **Kick Start Quizzes** for each chapter of the book for instructors

Introduction

1

Basic Concepts in Research

1.1 The Scientific Method

This is a textbook about statistics. Simply defined, statistics are the mathematical tools used to analyze and interpret data gathered for scientific study. It is paramount to remember that statistical analyses and interpretations do not exist in a vacuum. They occur within the larger scientific research process. Both *how to analyze* and *how to interpret* the data are quite dependent upon the surrounding research context. While the subject of statistics can be singled out and studied in isolation (as this textbook demonstrates), it is inextricably linked to the larger scientific enterprise. As such, it is appropriate to first review the basic features of the practice of science.

The scientific method can be conceptualized as a three-step recursive process. Each step can be summarized as follows:

Theory. Theories are attempts to explain and organize collections of data observed about a topic (or "phenomenon") under scrutiny by appealing to general principles and relationships that are independent of the topic itself. Take, for example, a line of research on the endurance of friendships. In theorizing *why* some acquaintances lead to enduring friendships while others do not, one could propose that personalities are a bit like magnets; similar ones repel one another, while dissimilar ones are drawn together (i.e. opposites attract). This theory, then, appeals to the concepts of "magnets," "personality," "similarity," and "dissimilarity."

Not all theories can be considered "scientific." For a theory to qualify as "scientific," it must be testable. By testable we mean: Is it potentially falsifiable? Can it be placed into jeopardy and potentially observed to be untrue? If it cannot, it remains a theory, but it is not considered to be properly "scientific." Using testability as a criterion, for example, the theory that each of our choices (past and

Essentials of Behavioral and Social Science Statistics, Third Edition.
K. Paul Nesselroade, Jr. and Laurence G. Grimm.
© 2025 John Wiley & Sons, Inc. Published 2025 by John Wiley & Sons, Inc.
Companion website: https://www.wiley.com/go/Essentials_of_Behavioral_and_Social_Science_Statistics3e

future) are fully predetermined by some combination of our current environment, our DNA, and behavioral conditioning from our previous experiences, could hardly be considered "scientific." While many people believe it to be true, how exactly would we go about putting it in jeopardy and testing it?

Hypothesis. In the light of any scientific theory, it should be possible to generate predictions about the data one expects to observe – this is a hypothesis. Sticking with our magnetic theory of friendships, one hypothesis might be as follows: If we measure the personalities of incoming university students who are randomly assigned living arrangements in a dorm, we might expect to find that those with quite discrepant personality profiles are more likely to be friends at the end of the semester than those with similar profiles.

Because it is possible to find evidence that would not support this hypothesis, we can say that this theory is "testable." However, we cannot stop at hypothesizing. To say anything meaningful, we must complete the research process by going out and doing the work, setting up the study, gathering the participants, and carefully collecting and analyzing the data. This leads us to the third step.

Observation. The gathering of scientific observations is done by careful and systematic measurements of events occurring in the world by using our five senses, often with the aid of various scientific tools and instruments. In our example, we would want to meticulously measure our incoming freshman's personalities as well as the nature of their friendships at the end of the semester. These observations, then, would be organized and interpreted. Ultimately, what is concluded will be brought back in contact with the stated theory. Observations will either support the theory, fail to support it, or, perhaps, partially support it. The circle is made complete as we compare our findings with our original theoretical proposition.

In our example, we should not be too confident that supporting data will be found – previous research suggests we will probably be disappointed (e.g. Back et al. 2008). And that is an important point – if supporting data are not found, so much the worse for the theory. We may need to think differently about why some friendships begin and endure while others do not. As would be expected, accurate theories will be supported by our observations. Supporting observations can both affirm a theory and lead to clearer and more refined articulations of that theory. More precise theories, in turn, lead to new hypotheses, and the cycle starts over again. The process is circular and recursive, with each cycle spiraling toward a more accurate understanding of the topic at hand.

The specific role of statistical analysis is found in the interpretation of our numerically represented observations. What do the numbers mean and *not* mean? How certain are we that our conclusions are accurate? On what do we base our sense of certainty? These are often not easy determinations to make. The central purpose of this text is to dissect and explain how this part of the research process works. The remainder of this introductory chapter will lay out an overview of the research enterprise.

1.2 The Goals of the Researcher

Scientific researchers set out with earnest intention to study carefully, logically, and objectively a particular topic of interest. Depending upon what is already known about the topic, what one *wants* to learn about the topic, and what one realistically *can* learn about the topic, researchers adopt different "goals" for their projects. Often, the initial goal for a researcher is that of **description**. Scientific description is the process of defining, identifying, classifying, categorizing, and organizing the topic of interest. Explicit delineation of topic boundaries is crucial. What exactly constitutes the topic? How many forms can it take? How frequently are these various forms found?

For example, if we were interested in studying the various ways in which people take vacations, we would first have to define what a vacation *is*. Is an afternoon day trip to a community park a vacation? What about an extra day tacked onto a work-related business trip? It is *not* a requirement for all researchers to agree on the same definition for a topic to be studied. However, it is imperative that the readers know explicitly what we, the researchers conducting the present study, mean by "vacation" when we use that term. In other words, concepts must be operationally defined. An **operational definition** is a precise description of the concrete measurement of that concept, as it will be used in a given research project.

Another related issue would be to decide how many ways "vacation" can take place. For example, someone might suggest that there are fundamentally two different *kinds* of vacations: one kind is designed around relaxation and focuses on bodily rest, while the other kind features action and excitement. Another researcher may come along and suggest that there is a third kind of vacationing – one that combines the two, incorporating dedicated time to both bodily rest *and* being active. For this reason, it is crucial for investigators to clearly indicate their particular interpretation of the concept in question. For example, one researcher might find that only 20% of vacations are of the relaxation variety, while another, using a different operational definition, might find a different percentage. A proper interpretation of any statistical statement first requires an understanding of variable operationalization. Finally, it should be noted that the statistical needs associated with meeting the goal of "description" are usually not too sophisticated.

Another aim of the researcher would be one of **correlation** (or **prediction** or **association** – these are all analogous terms). Correlation involves a description of the degree of relationship between the topic of interest and other variables. For example, in our study of vacations, we might be interested to see if there was a relationship between the age of the vacationer and the type of vacation chosen. Here, we would be measuring two variables ("vacationer age" and "type of vacation chosen"). As we will learn later in the text, mathematical procedures applied to these measurements can determine if a relationship exists, and if so, the strength of that relationship.

It is critical to realize that research designed to show correlations does not allow us to draw causal conclusions. For example, if we find that older individuals, more so than younger ones, prefer to take vacations centered on rest, we could not justifiably conclude that age *causes* people to want to take restful vacations. It could very well be, for example, that older people grew up in a time when vacations were generally understood to be restful in nature, and they formed their vacationing expectations and habits accordingly. Another possibility might be that there are not as many exciting vacationing experiences geared toward an older audience compared with those available to a younger crowd. If the set of vacation options were different, then perhaps the numbers of older vacationers choosing active vacations would increase. Understand this clearly: one of the most frequently observed critical thinking errors is the tendency to impose a causal interpretation on data that were gathered correlationally. Interpretation of data from correlational studies must refrain from using causal language. In our example, it would only be correct to conclude that age and vacationing style are related, such that older folks are more likely to rest on vacation when compared with those who are younger.

The most ambitious goal of the researcher would be one of scientific **understanding**. Here, the researcher draws a cause-and-effect relationship between the topic of interest and some other variable. There is a myriad of subtle and yet important issues related to making "cause-and-effect" statements. As such, it is impossible to fully treat the subject of causation in this text. At a minimum, one must realize that even when causal statements are warranted by the research process, there are limits to these causal claims. See Section 1.6 in this chapter for more on this topic.

To gain even this limited claim of causality, we will need to engage in a specific form of research investigation termed an **experiment**. This is a technical term reserved for a research project in which each of the following three features is scrupulously maintained: (i) a high degree of control over the presumed *causal* variable, (ii) careful measurement of the presumed *effected* variable, and (iii) complete control over all other variables. An experimental study of "vacationing" would require looking at the role of something that can be manipulated, like disposable income. Perhaps we could give (thanks to a generous research grant!) some randomly selected participants $2000 and ask them to choose between two different but equally expensive vacation packages: one centered around resting and the other around doing. We could then grant other randomly selected participants $6000 and pose the same question. If different rates of vacation preferences emerged between the groups, the researchers would then be justified in claiming that their findings suggested the following: the amount of disposable income influenced (or, in some sense, *caused*) the type of vacation people chose to experience. However, before further discussion of experiments, we need to step back and survey the different ways scientific researchers think about variables.

1.3 Types of Variables

In a general sense, a variable is anything that can assume different values – anything that can *vary*. In the research process, variables are employed in different types of roles. The following paragraphs will introduce some of the major roles played by variables, as well as some of the related terminology.

1.3.1 Independent and Dependent Variables

As previously noted, when designing an experiment, the researcher attempts to examine how one variable influences another variable. The **independent variable** is the one imagined to play the *causative* role in a cause–effect relationship. It is selected and manipulated by the experimenter to observe its effect on another variable. By **manipulation**, we mean the controlled presentation of that variable. The nature of this manipulation can be either quantitative or qualitative. A **quantitative independent variable** means that participants are exposed to different *amounts* of the variable. Suppose a cognitive psychologist believes ambient noise levels influence performance on a memory task. Three groups of participants are used, with each group being exposed to different decibel levels of static noise: 20 (low), 40 (medium), and 60 (high). In this example, the three levels of the independent variable are distinguished from each other quantitatively.

A **qualitative independent variable** means that participants are exposed to different *kinds* of treatments or the *presence or absence* of the independent variable. Using the memory recall example, suppose one group is exposed to static noise while a second group hears music played at the same decibel level. The groups do not differ in the volume of sound, but rather by the *kind* of sound used. A related example would be for one group to experience static noise while the other works in absolute quiet. Here, the groups differ by the *presence or absence* of the independent variable. (Note: It is also possible to think of this "on" or "off" difference in quantitative terms.)

When a condition is marked by the absence of the independent variable, it is oftentimes referred to as the **control group**. Conversely, when a condition is marked by the presence of the independent variable, it is oftentimes referred to as the **experimental group**. Many research efforts are described in brief as "control-versus-experimental" studies. Of course, more than one experimental condition can be involved. A three-condition study might have a control condition and two different experimental conditions (e.g. no noise, static noise, and music) and so on.

Finally, note that when the procedure involves direct interaction with the participant, the term **treatment** is often used instead of "independent variable." For example, if instead of ambient noise levels, we manipulated the tone of voice used by the experimenter when giving instructions to the participants (e.g. stern vs. soothing), then we might say the participants are getting different *treatments*. (These two terms are somewhat overlapping and interchangeable.)

The **dependent variable** designates what is thought to play the *effected* role in a cause–effect relationship. Continuing with our memory recall and ambient noise example, either the time taken to finish the recall task or the number of recall errors (or both) could serve as a dependent variable.

Although the unfolding of an experiment may produce a variety of participant responses, only dependent variables are being carefully observed and measured. For example, suppose we are interested in teaching people self-control techniques to improve pain tolerance, operationally defined as the amount of time participants can keep their arm submerged in a bucket of ice water. As such, "pain tolerance" serves as the dependent variable in this experiment, and "self-control techniques" serve as the independent variable (or treatment). Participants learn one of two different mental imaging techniques designed to control their reaction to pain, and then they are measured to see how long they can keep one arm submerged in ice water. While their arm is submerged, an array of other behavioral responses may occur, such as fidgeting, vocalizations, increased breathing rate, and so on. These other behaviors are not dependent variables because the researcher is not systematically observing or measuring these responses. If the researcher decided to measure both the amount of time participants keep their hands submerged *and* the number of breaths per minute, then the study would contain two dependent variables.

■ **Question.** *A psychologist is interested in determining which of two sales techniques is more effective in influencing participants to purchase a more expensive car. In the "positive condition," participants are told about the many positive attributes of Car A compared with Car B, even though Car A costs $5000 more. In the "negative condition," another group of participants is told about all the undesirable attributes of Car B compared with Car A (the pricing remains the same in both conditions). After the sales pitches, each participant is asked to state which of the two cars they would most likely purchase. Identify the independent and dependent variables.*

Solution. The independent variable can be identified by asking, "What is being manipulated by the experimenter?" It should be clear that the "sales technique" is being manipulated. To determine the dependent variable, we can ask ourselves, "What particular behavior is being measured?" This variable is identified at the end of the example: "Each participant is asked to state which of the two cars they would most likely purchase." More simply, one might call this dependent variable something like "car preference." ■

As stated earlier, two defining features of an experiment are the careful control (manipulation) of the presumed causal ("independent") variable and the careful measurement of the effected ("dependent") variable. There exists, however, a third defining feature in any experiment: the control of all other variables. With any experiment, there will be several other variables that could influence the outcome. For example, in the aforementioned pain tolerance study, one could imagine several variables of interest like the age, ethnicity, and health status of the participants, as well as differences in the experimental environment (e.g. the room temperature, the presence of others encouraging the participant, and so on). Any variable present in an experiment other than the independent and dependent variables is referred to as an **extraneous variable.** Extraneous variables must be *controlled* for accurate conclusions to be drawn from the results of the study.

1.4 Controlling Extraneous Variables

1.4.1 Holding Constant and Balancing

When the experimental situation is not controlled, there is the possibility that variables other than the independent variable will be bringing about change in the dependent variable. This potential interference by an uncontrolled variable creates confusion regarding what caused the change. An uncontrolled extraneous variable is called a **confounding variable**. The presence of confounds in an experiment threatens the researcher's ability to draw meaningful conclusions from the study. Furthermore, there are oftentimes no statistical techniques that can be used to salvage a study with confounds, especially when the confounding variable has not been measured. Statistical analysis rests fully and firmly on sound research methodology. If the methodology is poor and confounding variables are present, the quality of the study suffers. It becomes impossible to figure out how much the confounding variable, as opposed to the independent variable, caused the change in the dependent variable. As the effect of the independent variable becomes ambiguous, the study becomes meaningless.

Holding constant an extraneous variable involves treating the variable like a constant; simply not letting it vary. Consider the previous research example comparing sales techniques. Presumably, one factor that influences car purchases has to do with the visual presentation of the vehicle. How the car is displayed, then, is an extraneous variable in our example. If this were uncontrolled, it could confound the study. For example, what if both Car A and Car B were displayed on presentation ramps for the "positive" sales condition, but Car B was moved off the ramp and placed blandly amid a row of other cars for the "negative" sales condition? If it were discovered that people in the "negative" condition opted for Car A more so than those in the "positive" condition, is there any clear way of knowing that the increased preference for Car A was due to the different sales technique (the actual independent variable) and not the differing presentations (the confounding variable)? "Sales technique" is confounded with "presentation style." Imagine, however, if "presentation" were controlled by holding it constant for both cars and across both sales conditions. This would effectively remove it as a competing explanation. In this way, the confounding variable is neutralized by holding presentation style constant across conditions.

Consider another example (see Table 1.1). A simple psychotherapy study contrasts two therapeutic approaches to depression. One group of depressed participants receives cognitive behavioral therapy, while a second receives psychodynamic therapy, each from a specifically trained therapist (Diagram A, Table 1.1). Suppose, at the end of the study, those who received cognitive behavioral therapy were, on average, less depressed than those in the other group.

What is the most obvious confound? It is possible that the cognitive behavioral therapist had some personal qualities (e.g. warmth, understanding) that were responsible for the better results as opposed to the therapy style itself. If the therapists had switched roles, perhaps the psychodynamic approach would have

Table 1.1 Diagram A depicts "therapist" confounded with "treatment." Diagram B depicts a redesign in which therapist effects are spread equally across treatment conditions. Diagram C depicts a redesign in which the therapist is held constant.

Diagram A			
Cognitive behavioral		Psychodynamic	
Therapist 1		*Therapist 2*	
30 depressed clients		30 depressed clients	
Diagram B			
Cognitive behavioral		Psychodynamic	
Therapist 1	*Therapist 2*	*Therapist 1*	*Therapist 2*
15 depressed clients	15 depressed clients	15 depressed clients	15 depressed clients
Diagram C			
Cognitive behavioral		Psychodynamic	
Therapist 1		*Therapist 1*	
30 depressed clients		30 depressed clients	

appeared superior. In this example, "therapist" is confounded with "treatment type." Any difference between the two groups of participants in terms of their recovery from depression cannot be unambiguously attributed to the relative effectiveness of the therapy styles. How could we control this extraneous variable? One solution (described in Diagram B, Table 1.1) would be to have half the participants in both conditions treated by each therapist. This technique to control extraneous variables is similar to "holding constant." It is called **balancing**, and it works by representing two forms of an extraneous variable equally in both conditions. The effect of each therapist would now be equally represented across the conditions, thereby removing any bias toward one form of therapy.[1] Since the effects of different therapists would be balanced across the conditions, any difference in the improvement of depression could be attributed to the treatment rather than the therapist. Another solution (Diagram C, Table 1.1) would be to hold the troublesome variable constant by using the same therapist for both conditions.[2]

One downside to holding variables constant is the interpretive limitations that follow. For example, if we decided to control the variable "biological sex" by only using females, the findings should be applied only to females, for how could one know that the same results would extend to biological males when males were never studied? Suppose only biological females were used in the imagery and pain tolerance study; if one imagery technique was found to result in higher pain tolerance than the other, could it be fairly said that this finding also pertained to biological males? When a variable is controlled by holding it constant, the interpretation of the findings regarding that variable is limited, and any extrapolation to other forms of that variable is subject to debate. We will have more to say about generalizing the findings of a study when we introduce the concept of "external validity" presented later in this chapter.

1.4.2 Randomization

Another method for controlling extraneous variables is a technique known as randomization. A host of potentially confounding variables associated with (i) differences between the various participants in a study, (ii) differences between the experimental settings, and (iii) even differences in the stimuli used can be controlled by careful use of randomization. It is important to analyze each of these potential problem areas one at a time, for each illustrates a way in which the technique of randomization can be applied to control the effect of extraneous variables. First, consider the numerous differences that naturally occur between different participants in a study.

1 It would be important to ensure that each therapist was equally proficient in conducting both forms of therapy.

2 It would be important to ensure that the therapist was equally proficient in both forms of therapy.

Table 1.2 Some common participant variables.

Physical attributes	Demographics	Psychological
Height	Income	Intelligence
Weight	Family size	Need for approval
Handedness	Ethnicity	Mental discipline
Running speed	Occupation	Degree of past trauma
Strength	Education	Intro/Extroversion
Sex	Religion	Emotional maturity

Participant variables are characteristics of the participant that are fixed before the experiment has even begun. This term "fixed" applies to their a priori existence and not to their inability to be changed (though sometimes they truly cannot be altered). One way to categorize participant variables is to consider them in three distinct groups: physical attributes, demographics, and psychological traits. Table 1.2 lists many common participant variables in these respective categories. Obviously, an exhaustive listing of participant variables is impossible.

Participant variables will create difficulties in an experiment when they are not adequately controlled. Most problematically, they can become confounding variables when they are proportionately linked to differing levels of the independent variable. This thorny dilemma, however, can be elegantly overcome by using the tool of **random assignment**: the designation of participants into various conditions within a study such that each participant is equally likely to be assigned to a given condition. Random assignment is an extremely powerful tool that can control innumerable participant variables all at the same time. The following examples illustrate this indirectly through the nonuse of random assignment, thereby confounding the experiment.

► **Example 1.1** An educational psychologist is interested in evaluating a psychological technique that is expected to help college students become more efficient in completing arithmetic problems. Efficiency, the dependent variable, is defined as the time it takes students to complete one page of simple math problems. The technique, or independent variable, is the use of self-motivating statements. A self-motivating statement is a phrase the student learns to repeat if they become distracted from solving arithmetic problems. The experimental group is trained to repeat these self-motivating phrases, while a second group of students, the control group, is not.

The experiment is conducted in a typical classroom setting. To save time, the experimenter assigns the first 20 students to arrive for class to the experimental group. All later arriving students are assigned to the control group and are asked to wait in the hall until the self-motivational training has been completed. These students are brought in later when it is time for everyone to work on the math problems. The results of the study showed that students trained to use self-motivational techniques, on average, worked more efficiently than those in the control group.

While it is impossible to know, it is certainly plausible to consider that a participant variable (e.g. a psychological or demographic trait) was responsible for at least some of the differences in performance. For example, suppose it is true that organized people tend to be more punctual (thus overrepresented in the training group) and more attentive than people in general. This participant variable existed prior to the study. It is also an extraneous variable that might be confounding our study since it was not controlled. While one could control it by holding it constant (e.g. only use punctual students), it could also be controlled by randomization – by arbitrarily assigning participants to the two conditions. In this example, a better methodology for placing students into the two groups would have been to flip a coin to assign each to a condition. In fact, employing a random group-assignment strategy, like coin tossing, eliminates the prospects of overrepresenting any specific participant variable in either group. Potentially important confounds like age, intelligence quotient (IQ), motivation, biological sex, and so on would all be neutralized. Notice how all participant variables can be controlled with one well-executed randomizing technique. ◄

It is critical to realize that a variable is controlled even if it is not perfectly balanced between the two conditions. For example, suppose one group contained a couple more biological male students than the other; this would not mean that random assignment had failed to control for biological sex. If an imbalance occurs through a truly random process (e.g. flipping a coin), then the resulting disparity cannot confound the study, rather it is just introducing what is called *error*. This is not as worrisome as it seems, for it is not the sort of error that invalidates a study (like a mathematical mistake). Rather, it is this kind of error that statistical methods are well suited to handle, namely, a feature termed *random error*.

Random error will be addressed in more detail later in the text, but a brief comment here will highlight the issue of group size. Because randomization does not perfectly balance a particular participant variable across conditions, it is important for large groups to help dilute the problem.

A thought experiment will help to explain. Imagine someone flipping a coin 10 times and noting how many heads are found. Now imagine them repetitively flipping a coin 10 times. On average, they will get 5 heads for each series of 10 flips, but oftentimes they will get a few more or a few less. However, occasionally, they may get as few as 1 or 2 heads or as many as 8 or 9. Now, imagine them flipping a coin 1000 times. Furthermore, imagine this being done repetitively in sets of 1000 (thankfully, we are just imagining this). On average, about 500 heads will be found for each set of 1000 flips. A few more or a few less would likely be found each time, but the rough percentage of both outcomes would hover around 50%. Very rarely would less than 450 heads or more than 550 heads be recorded. Therefore, it would be exceedingly infrequent to find the sort of large percentage differences between heads and tails that was seen in the 10-coin-flip scenario. This illustrates how larger sample sizes minimize error. The probabilistic benefits attached to large groups will be revisited in more depth later in the text.

Consider another example where randomization should have been introduced.

► **Example 1.2** Some of the earliest experimental demonstrations of the development of ulcers were conducted by Brady (1958) and Porter et al. (1958). Pairs of monkeys were exposed to electric shocks, which could only be avoided if one of the monkeys pressed a lever. The monkey responsible for working the lever was labeled the "executive" monkey. If the executive monkey failed to respond in time, not only they but also the "control" monkey would receive the electric shock. Therefore, while the numbers of shocks received by both monkeys were the same, the executive monkey was always responsible for managing the onset of the electric shock. Within two months, all the executive monkeys either had died or had become too incapacitated to continue in the study. Autopsies performed on the executives showed extensive gastric lesions. In contrast, the stomach linings of the control monkeys were discovered to be free of any significant lesions. Brady suggested that the stress associated with having responsibility and control led to the development of fatal ulcers in the executive monkeys.

Other researchers, however, could not replicate these findings. Ultimately, a confounding variable was uncovered. When designating which monkey would become the executive, Brady selected the one who learned the fastest how to press a lever to avoid the shock. Subsequent research revealed that emotional members of an animal population learn avoidance responses more quickly (Sines et al. 1963) and are also more susceptible to developing ulcers (Sines 1959). Without realizing it, Brady had assigned the more emotional monkey to the executive position, thereby introducing a confound and rendering the study uninterpretable. ◄

In the previous examples, if participants had been arbitrarily assigned to conditions, participant variables would have been controlled. However, in both cases, the researchers applied a rule to help create the two groups. (Example 1.1: First students to arrive. Example 1.2: First monkeys to learn how to avoid a shock.) This is always dangerous because it introduces a systematic approach to creating the groups (see Box 1.1 for more on this important topic).

As noted earlier, the technique of randomization can also be used to control differences related to the setting in which an experiment takes place. For example, suppose a drug manufacturer wanted to see if their newest product to control

Box 1.1 Is the Scientific Method Broken? The Wallpaper Effect

The public has recently been informed about a troubling discovery in the world of social and medical science investigation – the nonreproducibility of many scientific findings. Titles like "Scientific Regress" (Wilson 2016), "Does Social Science Have a Replication Problem?" (Tucker 2016), "Dozens of Major Cancer Studies Can't Be Replicated (Haelle 2021), and "Over Half of Psychology Studies Fail Reproducibility Test" (Baker 2015) seem to be popping up all over the place. The titles are unnerving, and the issues that are raised are both real and serious. Briefly stated, an alarming amount of published research, when attempts at replication are later performed, does not generate the same findings that were stated in the original publication. Upon learning this, it is quite natural for people to wonder if there is something fundamentally wrong with the scientific method or the conduction of scientific research. There are many aspects to this problem. This box attempts to analyze one aspect.

Many suspect that a leading reason for replication failure is the inability of the replicating researchers to reproduce accurately, fully, and completely the precise conditions under which the original study was performed. Even though the replicators faithfully follow every aspect of the methodological situation as recorded by the original researchers, it is proposed that there are other features of the original study that perhaps went unnoticed or were not deemed important to note in the initial write-up of the experiment, but nonetheless have influenced the recorded outcomes. Perhaps an extraneous variable thought to be negligible or to have been controlled was not controlled and yet *was* influential. These differences are colloquially referred to as "wallpaper effects" – a tongue-in-cheek way of suggesting that the original experimental situation must have been influenced by the color of the wallpaper in the room. If this is the case, then the incongruent findings between the original study and the replicated study may not be in genuine conflict; both may be accurate reflections of reality. The different outcomes could be explained by the effect of some, yet-to-be-identified other variable responsible for causing the difference. Unfortunately, because the influential variable is not identified, the findings of both studies are inconclusive and best left as mysteries, to be explained later when further light can be shed on the topic.

Look for other boxes in this series (Is the Scientific Method Broken?) scattered throughout the textbook. Each one addresses a different aspect of the replication problem.

blood pressure worked better than the current leading medication. When testing this, the manufacturer would be wise to not only control participant variables but also the setting differences through a randomizing procedure. For example, if the pharmaceutical company chose to test their new drug on patients in small private hospitals and compared the results with outcomes of patients from large public hospitals who were treated with the leading medication, might a critic of the research properly suggest that the level of care may be different in these two types of hospitals, thereby confounding the study? Randomly assigning not only the participants but also the study setting will solve this problem.

Random assignment can also be used to control the use of different stimuli in a study. For example, in the first illustration, students solved math problems and were timed for speed. We could design the study to hold the math problems constant and have students in both groups work to solve the same ones. However, suppose there was a concern that frustrated students might cheat by looking at their neighbor's work. To address this, the selection of the problems themselves could be controlled by randomly assigning them from a pool of acceptable problems to the individual pages being handed out to the students, thereby making each set of problems different. We might be concerned that some problems are easier to solve than others, but even if this were true, we must remember that the ratio of easy to hard problems will tend to even out as the number of problems on the page and the number of students solving them grows. Furthermore, any aggregate differences in difficulty between the pages of problems would not generate confounding error but rather random error, the kind of error statistical analysis can handle.

There are other techniques that can be used to control extraneous variables, but these are the ones most often employed. To learn more about controlling extraneous variables, a textbook on research methodology can be consulted (e.g. Marczyk et al. 2005; Morling 2020). Failure to control all extraneous variables coupled with a failure to recognize or acknowledge this lack of control can lead to the making of false claims. The box below is the first in a series of boxes that address the current conversation surrounding the erosion of public trust in science.

1.5 Validity Issues

1.5.1 Internal Validity

Much of this chapter has focused on how well conclusions can be made regarding the effect of an independent variable on changes to a dependent variable. As we might imagine, some studies do a better job of this than others. We use the term **internal validity** to capture this idea. That is, to what degree can one unambiguously attribute changes in the dependent variable to the manipulation of the independent variable? As our certainty grows, internal validity goes up. As we realize more and more competing explanations for changes in the dependent variable, our internal validity goes down (Shavelson 1996). We can think of internal validity as a measure of how logically tight and tidy are the inner workings of the study. Various kinds of reasons can account for a lack of internal validity. For example, real-world constraints may limit an experimenter's ability to control all extraneous variables, properly manipulate an independent variable, or carefully measure a dependent variable. However, some challenges to internal validity can be solved with careful methodological thinking and experimental execution (see Section 1.4).

Furthermore, we can often learn from our mishaps. In fact, frequently, the critique of an initial study can reveal an unconsidered rival explanation for the findings, and this, in turn, will open the door to a new and fruitful second study. What was initially found to be a confounding variable can either be controlled or used as an independent or dependent variable in a follow-up study. For example, for the mental imagery and pain tolerance study, let us introduce "anxiety" as a variable. Evidence from past research suggests that highly anxious people tolerate pain poorly (e.g. Cimpean and David 2019). Suppose that in our study we unintentionally varied anxiety systematically with the imagery conditions. For example, perhaps a close analysis of the experiment reveals that we had participants in the second imagery condition waiting longer in the reception room before their pain tolerance was measured, compared with those in the other condition. This delay in time could have allowed more anxiety to build. When a reviewer alerts us to this potential confound, not only could we rerun the study correcting this confound, but we might also decide to add anxiety as a new independent or dependent variable to explore.

When real-world constraints do not allow for full control of all variables, a researcher may still choose to move forward with a study, even though the internal validity may be compromised. For example, a professor who wanted to see if a review game was an effective teaching tool in the classroom might use the game in one of the course sections but leave another section unchanged and then compare test performances between them. Please note that students were not randomly assigned to the classes, rather they signed up for the class they wanted or that best fit into their schedule. This opens the possibility that one class may represent a different type of student than the other. Imagine that one class meets at 8:00 in the morning and the other at 2:00 in the afternoon. These classes, although they are both composed of students from the same institution and will have many other similarities, may represent quite different subgroups of students, namely, those who prefer morning classes and those who prefer afternoon classes. Furthermore, during the experiment, each class may experience other unique events that only occur in that class (e.g. a fire drill disruption, problems with the classroom technology). These other events are not the independent variable and now compete with the new teaching tool to explain any difference in the outcomes between the classes. These are not ideal experimental situations; however, oftentimes they are the only viable method available to a researcher. Such designs are called **quasi-experiments** because they share many of the same characteristics with true experiments and yet the participants are not randomly assigned to conditions and/or idiosyncratic differences between the conditions (like fire alarm disruptions) may be present.

1.5.2 External Validity

A researcher can also analyze the extent to which a study's findings can be justifiably generalized, thereby reaching beyond the limited context of the study. After all, no one would be interested in learning about the relationship between mental imagery and pain tolerance if it is believed that the imagery findings only pertain to the participants involved in the study and to the act of holding one's hand in ice-cold water. Obviously, the value of the study emerges when we consider the findings in a more generalized context – namely, people in general and pain tolerance in general. The degree to which a study legitimately applies to broader categories is critically important. Formally, "**External validity** asks the question of generalizability: To what other *populations, settings, treatment variables, and measurement variables* can this effect be generalized" (Campbell and Stanley 1963, p. 5). The external validity of a study can be very difficult to judge and is often subject to intense professional debate. Short of running innumerable slight variations of a study, there is no way to determine with certainty the generalizability of a finding. Thankfully, there are ways to think about these issues that do not require extreme measures.

Let us look at the problem of different populations first. To the degree the variables in the study tap into our shared human experience, the problem of different populations fades. When dealing with our common human nature, one human is like all others. However, many studies seek to investigate *particular* groups of people, for example, athletes, the elderly, cancer patients, and so on. These studies are not exploring human commonality, but rather human particularity. In these situations, the method used to select participants is of greater importance, and **random sampling** is the "gold standard." Random sampling occurs when the means of selecting participants for a study is such that each participant in a given population has an equal chance of being included in the sample. (Note that this is different from *random assignment*, a method used to assign already selected participants into the various conditions of a study. Random assignment pertains to *internal validity,* not external validity.) A *biased sample* may occur whenever each member of a population does *not* have an equal chance of being selected for the sample.

Before moving on, let us clarify the new terms just introduced: populations, sampling, and samples. A **population** is defined as "every member of a given group." For example, imagine we want to study the effectiveness of a new teaching tool designed to help adults learn how to read English. We can label our population, then, as *adults who are engaged in learning how to read English.* Of course, we cannot include every person in this population in our study; rather, we will select a subset of individuals from that population to investigate. The process of selecting participants is called **sampling**; the group of individuals, once selected, is called a **sample**.

Within any given population, there will be a host of participant variables distinguishing one participant from another. Just as *random assignment* creates groups within a study that are roughly similar across all participant variables, *random sampling* creates a group of participants that roughly captures the blend of participant variables found in that population. This is important because it allows the researcher to claim that findings obtained from a sample should also pertain to the population from which the sample came. This gives us good reason to presume our findings are externally valid and applicable to other participants in that population. Conversely, if a sample is not randomly gathered, certain participant variables may be overrepresented while others may be underrepresented. When this is the case, confidence regarding the external validity of the findings falters. For example, if we tested our new teaching technique only on new immigrants coming from countries using the Roman alphabet (the same letters used in English), we should be cautious about concluding that our findings will pertain to immigrants coming from places where non-Roman alphabets (e.g. Chinese characters) are used. Our sample does not "represent" adequately this larger population. Logically, if this were the case, we should realize that the population we have sampled from is not *adults learning how to read English*, but rather *adults learning how to read English who came from countries using the Roman alphabet.* Clearly, this is a much smaller population, but the external validity of our findings necessarily becomes more suspect when we extrapolate to people outside the population from which we sampled.

It is important to realize that even though random sampling is the ideal sampling technique, it is rarely employed. Take the example we used above: How could one select a truly random sample of *adults learning how to read English*? While the population can be easily described in the abstract, it is functionally impossible to gain access to it for sampling purposes. Thankfully, other more-or-less adequate sampling procedures can be used for social science research. The specifics of these techniques are typically treated with greater detail in methodology textbooks.

In a similar vein, there is the issue of sample size. How large of a sample is needed to ensure representation of the population from which it is drawn? This can be a complex determination, one that involves assumptions about the presumed variability within the population in question. As with the issue of nonrandom sampling styles noted above, this topic is also customarily discussed with greater precision in textbooks focused on research methodology.

Let us now return to the concept of external validity regarding other "...settings, treatment variables, and measurement variables." Determining the external validity of findings in these situations is a matter of reasoned argumentation. Sometimes the findings of a study will naturally prompt researchers to think of other similar settings and variables for which these findings might apply. A good example is a study by Baddeley and Longman (1978) comparing mass practice with distributed practice for learning the new skill of typing. The basic question was this: How is practice time best spent if one is learning to type with a new arrangement of keys? One group practiced in mass (i.e. long practice sessions within a short window of time), while another group distributed their practice (i.e. shorter practice sessions spread out over a longer stretch of time). Both groups, however, spent the same amount of time typing on the new keyboards. The findings showed clear evidence that those with distributed practice sessions typed more quickly and made fewer errors. Many theorists suspected that Baddeley and Longman demonstrated a more general principle regarding the relationship between mass and distributed practice (to learn more, see Box 1.2). The external validity for Baddeley and Longman's findings appears to be high.

The external validity of a particular finding is best judged by examining the body of existing research to see if a similar finding has occurred with participants from different populations using different experimental procedures and different variables. Findings that hold up under a wide range of circumstances (like the "distributed over mass practice" finding) are termed *robust*. Oftentimes, a researcher, who has uncovered an interesting finding, will go on to conduct a series of related studies in which they systematically alter populations, settings,

Box 1.2 A Strategy for Studying Statistics: Distributed over Mass Practice

Baddeley and Longman's (1978) finding that "distributed practice is more effective than mass practice" is actually a very old one, first uncovered by one of the pioneers of psychology, Hermann Ebbinghaus (1885), when conducting his famous studies on memory. It is also one of the more *robust* findings of psychology shown to apply to the acquisition of a variety of both neuromuscular skills like archery (Lashley 1915) and cognitive abilities like face recognition (Mammarella et al. 2002). Given the demonstrated *external validity* of this finding, the implications for studying material in this text should be clear. It is far better to spend a small but meaningful amount of time each day studying statistics than it is to set aside a few large blocks of time to cram just before a test. The challenge for the student is to discipline themselves to daily review recently covered material, run through a few practice problems, and read a bit ahead in the textbook. Research from numerous studies suggests this will be the best way to use the time one allocates to learning this material.

and related variables to establish the robustness of their finding. This is referred to as conducting a *line of research.*

Internal and external validity are two among many types of validity, each being important to the research process. A fuller treatment of validity can be found in methodology textbooks.

1.6 Causality and Correlation

Earlier (in Section 1.2), the limited way in which researchers think about causality was noted. Let us develop this a bit more here. To say that "*X*" *causes* "*Y*" is *not* to suggest that *X* is the necessary and sufficient reason for *Y* coming about. After all, there may be many other reasons why *Y* has occurred. When we say, "*X* causes *Y*," we are merely saying that if we place *X* in this particular situation, we will get more *Y* than if we had not placed *X* there. This understanding of causality is indeed helpful, but it is clearly limited. Saying that the use of a mental imagery technique increases one's pain tolerance is not to say that it is necessary for any amount of pain toleration, nor it is to say that this is the only means of changing pain toleration. Rather, the statement is merely saying that if one uses a certain form of mental imagery (*X*), one can expect to experience more pain tolerance (*Y*) than if one had not. Finally, causality statements in the social and behavioral sciences do not imply that something is now known about the more profound question, *why is reality such* that if we place *X* in this situation, we get more *Y*?

Nonetheless, the term *cause*, understood modestly, can be appropriately used in the behavioral and social sciences. However, other phrases can be used to designate a limited understanding of causality. For example, some causal relationships might be described in phrases like "*X* increases the probability that *Y* will occur" or "*X* tends to bring about a change in the occurrence of *Y*." In fact, many different action verbs can be found in the social and behavioral science literature.

Scientists highly value discovering causal relations because they help to bring about a deeper sense of understanding. Recall that this is the highest goal of the researcher (see Section 1.2). It is critical to see, however, that *causality* is not the only way in which a relationship between two variables can be described. If two different variables tend to change reliably in association with each other, they are said to *covary*. Variables that covary are *correlated* variables. This more limited description of relationship meets the second goal of the researcher: correlation. Height and foot size, for example, tend to covary. Taller people usually have larger feet, while smaller feet tend to be attached to shorter people. When data are gathered nonexperimentally (i.e. without manipulation), two variables may be observed to covary. And yet, this covariation does not imply a causal relationship between the variables, let alone the exact nature of a causal relationship.

Please note this important point: *how the data are collected determines the type of interpretation allowed.* Causal relations can *only* be established in an *experimental* setting when a manipulated variable is observed to influence the occurrence of another variable. Methods of gathering data without manipulation are called **correlational designs**. (These designs, as well as experiments and quasi-experiments, comprise the three most common designs that employ statistical analysis.) A prototypical example of a correlational design would be the *survey*. There is no independent variable manipulated, and no causal statement can be made. Data are gathered simply as they present themselves to the researcher. In this way, it is correct to say that data gathered from "correlational *designs* do not imply causation." (This topic will be covered in much greater detail in Chapter 14.)

Research on clinical depression demonstrates how variables can be found to covary without understanding the causal relation between them. Depression has many characteristics, and the reasons why people become depressed are numerous. The phenomenon is not fully understood. One view holds that people are depressed due to negative thinking. They are pessimistic, self-critical and do not praise themselves when they do something well. This perspective suggests that these cognitions play a causative role in depression. However, it is also possible that when people become depressed, they are more likely to think in a negative fashion; depression causes negative thinking. Furthermore, it could be that neither variable causes the other. They may covary because some third variable (e.g. "loss of control") causes both depression and negative thinking. In Figure 1.1, the question marks above the various arrows reflect this interpretive problem.

Unfortunately, some important research questions are, for ethical, logical, or logistical reasons, not amenable to experimentation. The relationship between negative thinking and depression is a good example. Even if we discovered a way researchers could manipulate "depression," it would seem to be unethical to do so. Likewise, is it ethical to manipulate the amount of negative thinking a participant experiences? Some questions, it appears, are restricted to correlational analysis.

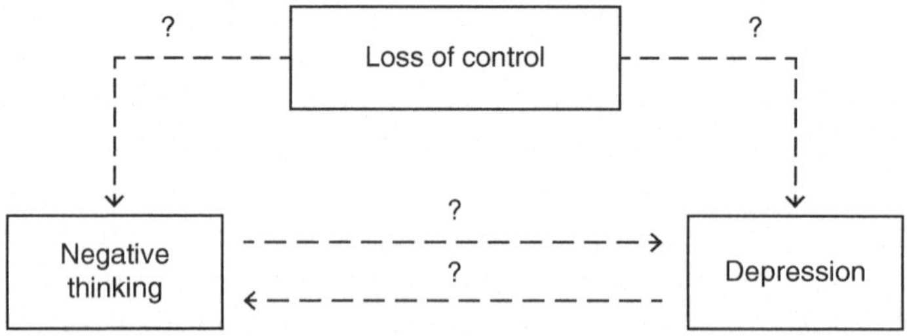

Figure 1.1 "Negative thinking" and "depression" are correlated, but which one causes the other? Might a third variable, like "loss of control," cause them both?

1.7 The Organization of the Textbook

The critical role of statistical analysis may not be obvious from this brief overview of the scientific method. Jones (2015), in fact, suggests there are as many as nine discrete steps to the research process; only one, data analysis, involves statistics directly. Nonetheless, it is crucially important. There is simply no way to make sense of gathered data without proper and careful statistical analysis. Additionally, statistical knowledge informs at least two other steps in the research process, namely, establishing the research design and collecting the data. Just as a woodworker needs to know what their tools can and cannot do before they design a table, so the researcher needs to know what their statistical "tools" can and cannot do before they design a study. Similarly, when gathering data, awareness of the capabilities of statistical tools influences which variables are measured and how. Knowledge of research methodology and statistical analysis will always go together.

Once the research plan has been set and the data have been carefully and properly gathered, the researcher will use statistical analysis to help make sense of the observations. **Descriptive statistics** are techniques designed to depict and summarize data in an abbreviated form. Frequency counts, distribution shapes, measures of centrality and variability, and standardized scores are all statistical tools that can be used to help describe the basic features of a dataset and are found in Part 1 of the text. More sophisticated concepts, found in Parts 2–5, allow the researcher to test hypotheses, specify differences, and quantify covariation between variables. These are the tools of **inferential statistics**, because they grant the researcher the ability to make inferences about the characteristics of populations from withdrawn samples.

The concepts found in Part 1 (descriptive statistics) are foundational and must be mastered before advancing to inferential statistics. Part 2 is composed of two chapters designed to introduce some important theoretical underpinnings for inferential statistical analysis, namely, elementary probability theory and the basics of hypothesis testing and sampling distributions. Parts 3–5 cover four different groupings of inferential statistical analyses. Part 3 deals with the basic set of z and t tests as well as the important concepts of *decision errors* and *statistical power*. Part 4 examines three basic analysis of variance tests including the very important interpretive concept of *interactions*. Part 5 introduces bivariate data and the fundamentals of bivariate data analysis. It also includes one chapter devoted to nonparametric statistical analysis, chi-square tests in particular. It is possible to rearrange the order of progression through the last three parts; however, some concepts needed for full comprehension of material in Parts 4 and 5 are covered in Part 3.

Each chapter will conclude with a summary, a collection of relevant formulas and keywords introduced in that chapter, and a series of exercises and work problems to be used by students to practice their skills and for self-assessment. Each part (starting with Part 3) will conclude with a broader summary of the material introduced in the preceding part, including a few questions designed to challenge our ability to select the appropriate test for a given research situation from all previously covered tests. This feature is included to address a fundamental problem with textbook learning: the tendency for end-of-chapter exercises to be solved by making use of only concepts found within that given chapter. By incorporating all the previously introduced concepts and tests, the questions at the end of each part will assess our ability to make sense of real-world problems through the application of an increasingly wider assortment of concepts and statistical tools.

The textbook has many addendums at the end. There are three appendices containing reference tables (A), answers to exercises and work problems (B), and tutorial information for Microsoft® Excel and SPSS® (C), two commonly used statistical software packages which can aid in answering calculation and tabulation questions. There is also a glossary with definitions for all key terms, a list of references, a list of selected formulas, a list of selected symbols, and an index of keywords, terms, and concepts.

Summary

This chapter presented an overview of the basic concepts in research. The scientific method, broadly speaking, is a three-step cyclical process where theories lead to hypotheses, then to observations, and then back to theories. Statistical analyses are used to interpret what observations mean for theories.

Researchers develop methods of studying topics of interest that are designed to answer a particular type of question. Some studies are designed to provide a better description of the basic features of the target of study. These are called descriptive studies. Other research methods, termed correlational studies, explore the strength of the relationship between a variable of interest and other variables. Another subset of scientific investigations, termed experiments, is marked by the controlled manipulation of one variable. These designs allow researchers to gain a better understanding of what *causes* another variable to present itself or change.

Basic research terminology involves familiarity with different types of variables. Independent variables are the hypothesized causal variables within experiments and are manipulated by the experimenter. Dependent variables measure the hypothesized effect of the independent variable. All other variables are extraneous and must be controlled. Holding constant, balancing, and randomization are some of the mechanisms employed to control them.

The degree to which a study controls all extraneous variables reflects internal validity. Quite separately, external validity describes the extent to which the findings of a study are generalizable to other populations, settings, and similar variables.

When researchers are interested in measuring the degree of relationship between two variables, they use correlational designs. These methods do not involve variable manipulation, but rather the careful measurement of two variables as they present themselves to the researcher.

Although the role of statistical analysis seems limited in the grand sequence of the research process, it is vitally important, such that its absence would render most empirical studies meaningless. The remainder of the text will examine various descriptive and inferential statistical concepts and tools designed to help the researcher make sense out of gathered data.

Key Terms

Theory
Hypothesis
Observation
Description
Operational definition
Correlation (prediction, association)
Understanding
Experiment
Independent variable
Manipulation
Quantitative independent variable
Qualitative independent variable
Control group
Experimental group
Treatment
Dependent variable
Extraneous variable
Confounding variable
Holding constant
Balancing
Participant variable
Random assignment
Internal validity
Quasi-experiment
External validity
Random sampling
Population
Sampling
Sample
Correlational designs
Descriptive statistics
Inferential statistics

Exercises

1 A researcher has just begun to study the "tiny house" trend in home construction. Think of two researchable questions that line up with each of the three different goals of the researcher: description, correlation, and understanding.

2 Imagine an experiment investigating the effectiveness of different rewards used by parents to "potty train" their children. Identify a quantitative independent variable and the different amounts of that variable to be used. Then identify a qualitative independent variable and the different kinds of that variable to be used. Also, describe how to use the technique of "holding constant" to control at least one extraneous variable.

3 For each scenario below, design a correlational study and then an experiment. (Hint: Some research questions may not be amenable to experimentation.)
  a Is there a relationship between pain and anxiety?
  b Is there a relationship between exercise and resting heart rate?
  c Is there a relationship between the need for achievement and the hours worked per week?
  d Do children who attend preschool daycare show better social skills in first grade?

4 Identify each of the following phrases as a theory, hypothesis, or observation.
  a 20 out of 25 basketball players improved their free-throw shooting percentage.
  b More positive reinforcement by the professor will improve test scores.
  c Humans seek to maximize gains and minimize costs.
  d "Chickens" come before "eggs."
  e More people choose Car A over Car B.
  f When giving names in a circle, people will not recall names given just prior to theirs.

5 Which of the following descriptions best operationally defines the concept of "vacation?"
  a An extended period of recreation, especially when spent away from home.
  b Release from obligation, business, or ordinary activity.
  c A stretch of time set aside to relax or travel for pleasure.
  d The number of days in a year a person spends not working and away from home.

6 Identify the independent and dependent variables in the following four studies.
  a A psychologist is interested in the effects of vitamin E on physical endurance. One group of participants receives 20 units of vitamin E, another 60 units, and a third gets a placebo. Endurance is assessed by the length of time participants can ride a stationary bicycle.
  b A teacher evaluates the effectiveness of different educational programs on reading speed and comprehension.
  c A psychologist hypothesizes that attitude change is greatest when people are prompted to believe their counter-attitudinal actions were made voluntarily as opposed to being forced.
  d An industrial psychologist hypothesizes that the amount of natural light will increase productivity. For 15 days a month, the blinds are drawn, and only indoor lighting is used. For the other 15 days in the month, the shades are left open. Productivity is measured by the number of widgets made.

7 Identify a confound affecting internal validity in the following four studies. (Hint: Some studies may not be confounded.)

a An independent marketing company has been hired to assess people's preference for Morgan's root beer versus Rose's root beer. To prevent bias, all test cans are covered with paper, with letter *A* placed on the Morgan cans and letter *B* placed on the Rose cans. The order of experience is counterbalanced such that half of the participants try Morgan's beer first and the other half try Rose's beer first.

b Researchers believe self-serving biases extend to objects associated with the self. Participants are asked to evaluate several objects in terms of usefulness and attractiveness. Half of the participants are given one of the objects, prior to evaluation, ostensibly as a gift for participating in the study. The target object is evaluated by both owners and nonowners. The target object is evaluated more favorably when it is owned by the evaluator.

c A particular stretch of highway is noted for an excessive number of traffic fatalities. The city council decides to reduce the speed limit, since evidence from national statistics clearly shows that traffic fatalities are correlated with speed limits. To make sure that the proper speed limit is observed, radar units are positioned every 5 miles along the highway. Not only did accidents significantly decrease, but motorists followed the speed limit better. Obviously, changing the speed limit has led to a decrease in accidents.

d Evidence shows that our reactions to pain are, in part, due to psychological factors. For procedures involving moderate discomfort, a dentist decides to offer patients the use of headphones playing soothing music. Since some patients may prefer novocaine, that option is also made available. Patients cannot both listen to music and use novocaine. At the end of the study, the dentist found that those patients who chose the headphones reported less anxiety and pain than those who opted for novocaine.

8 "Controlling extraneous variables" is to "generalizability" as:

a Independent is to dependent variable.

b Internal is to external validity.

c Random sampling is to random assignment.

d Population is to sample.

9 A professor claims to not care if submitted papers have a title page. Suppose we obtain several graded papers from our classmates and compare them. If our results show the average grade for title page papers to be higher than those without a title page, can we conclude the professor does, in fact, prefer title pages? What other interpretation could be made?

10 Which of the following descriptions represents "random assignment?"

- **a** Flipping a coin right before each participant is due to arrive at the lab.
- **b** Looking at a student's ID card and putting numbers that end in an odd number in one group and those that end in an even number in the other.
- **c** Assigning students who signed up in class to the experimental condition and those who signed up online to the control.
- **d** Asking each student to choose which color they prefer – those that choose "red" go to one condition and those that choose "blue" go to the other.
- **e** Asking participants their age. Even ages go to one group, and odd to the other.

11 Which of the following procedures represent the "random sampling" of university students?

- **a** Standing outside the cafeteria asking every third person if they will participate in a study.
- **b** Asking the registrar for the name of every twentieth person, going alphabetically through the university enrollment list.
- **c** Sending out an email to the entire campus asking for volunteers.
- **d** Getting professors to make an appeal for participants at the start of class.
- **e** Using the university directory to assign student names to slips of paper and then blindly drawing them out of a basket.

Part 1

Descriptive Statistics

2

The Nature, Scales, and Display of Measurements

2.1 The Nature of Measurement

One of Plato's most important assertions about thinking was that our best theories should "...carve nature at its joints" (Phaedrus, 265e). By this, it was meant that our *ideas* about what things exist should be properly descriptive of reality, that is, of what really *does* exist. As any butcher knows, finding the joints is key to proper meat categorization. Trying to dismantle an animal by cutting where there is no joint will only result in a frustrating (and messy) outcome. In the same way, it is important for researchers to have a good sense of what it is they want to measure and to understand where it is jointed and distinct from other forms or entities. This is one of the fundamental issues associated with measurement – namely, the clear identity of what is to be measured. Philosophers use the term *ontology* to mean the nature of being or the identification of the kinds of things that are in existence. The question of ontology asks, "what is really there?"

Some of the concepts used by behavioral and social scientists are rather clear-cut. For instance, a person's age is not particularly debatable. It seems well-jointed to stick with Plato's terminology. Other commonly used notions by social science researchers, however, are more problematic. One example is the idea of intelligence. Most of us feel rather comfortable with this concept, using it in casual conversation with little difficulty. And yet, where it is jointed with other ideas such as knowledge-set, wisdom, mental discipline, and attention is not particularly clear. (Each of these other concepts, by the way, is similarly hard to define.) Notice that the underlying issue is not one of operational definition (a concept presented in Chapter 1). A claim about what intelligence *is* differs from a description of *how* it can be measured. Even if intelligence is a real entity, there can be different ways researchers may choose to measure it.

A second issue related to measurement has to do with the contours and dimensions of the form being measured. Age, our example of a simple concept, seems

Essentials of Behavioral and Social Science Statistics, Third Edition.
K. Paul Nesselroade, Jr. and Laurence G. Grimm.
© 2025 John Wiley & Sons, Inc. Published 2025 by John Wiley & Sons, Inc.
Companion website: https://www.wiley.com/go/Essentials_of_Behavioral_and_Social_Science_Statistics3e

rather one-dimensional. It grows or changes in only one way; namely, it gets larger as time passes. But what about the concept of intelligence? Is it simple and unidimensional like age? As one grows in intelligence, is it right to simply make a single measurement grow larger to represent what is happening? What if intelligence has multiple dimensions and irregular edges? What if intelligence is less like one's height and more like the coastline of Britain? (see Figure 2.1a) Measurement strategies which force a complex form into a simple numbering system may misrepresent, perhaps grossly, the true nature of that form.

A third issue related to measurement is the degree of sensitivity used by the measurement system. Here, we can see that even our example of a simple, well-defined variable has a little trouble. Typically, once a person reaches about age 2, the finer measurement unit of months is dropped in preference for years. However, what is lost by identifying two 14-year-olds, born 300 days apart, with the same number? Perhaps not much. But let us turn back to the measure of Britain's coastline. The coastline paradox (Mandelbrot, 1967) shows that the claimed length of a coastline depends on the method used to measure it. If Britain's coastline is measured in units of 100 km, then its length is approximately 2800 km. However, if it is measured in units of 50 km, the total length swells to 3400 km (see Figure 2.1b and c). Surprisingly, Post and Eisen (2000) showed that as the unit of measure decreases from kilometers to meters to millimeters and so on, the true length of the British coastline approaches infinity!

If measuring physical entities such as the British coastline is this hard, we should not be surprised by persistent challenges to the measurement of abstract entities like privilege, nationalism, and intelligence found in the social sciences. In the field of psychology, the question of how to measure mental states and

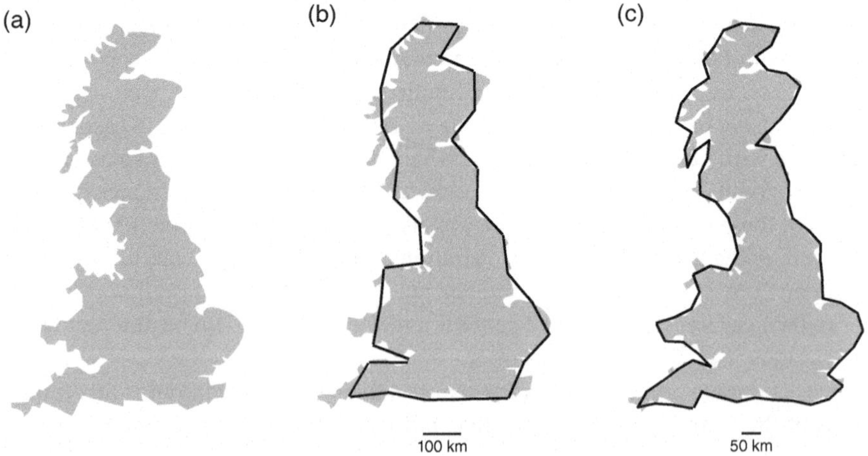

Figure 2.1 (a) A map of Britain. (b) Britain's coastline as measured in units of 100 km. (c) Britain's coastline as measured in units of 50 km.

psychological constructs, an activity termed *psychometrics*, dominated the work of many psychologists in the late 1800s and early 1900s. Much has been learned since then, to be sure, but challenges with definition and measurement still sit at the very foundation of all sciences, particularly the behavioral and social variety. It is important to note that the darkest chapters in the history of psychology were the early misguided attempts to measure intelligence (e.g. Jackson and Weidman 2006; Turiel 2020; Zenderland 1998). A historical analysis shows that each of the three issues noted above played a role in this tragic history. For more on the history of statistics, See Box 2.1.

2.2 Scales of Measurement

In addition to the philosophical issues of measurement, there are practical issues regarding *how* numbers are used. In practice, **measurement** is the assignment of numbers to attributes, objects, or events according to predetermined rules. A proper understanding of the different sets of rules, or scales of measurement, is required to make sense of what a particular number means within a given context. The four most frequently used measurement scales for social and behavioral scientists are introduced below.

2.2.1 Nominal Scales

Our first scale conveys no quantitative information. A **nominal scale** uses numbers merely to distinguish one type of thing from another type of thing or one event from another event. For instance, the numbers assigned to the members of a soccer team do not carry any quantitative value; they merely distinguish one player from the others on the team. Using a "1" for biological males and a "2" for biological females on a spreadsheet is not meant to suggest that females are somehow *more* than males. Or consider dividing people into three groups by having them line up and count off: one, two, three, one, two, three, and so on. The numbers used in these examples do not represent quantitative differences between groups, but rather *qualitative* ones. Since no quantitative information is being communicated, we are free to exchange one number for any other currently unused number. For instance, a midfielder who does not like the number 8 can exchange it for 6, provided it is not currently being used. In fact, in principle, we could even use letters or symbols instead of numbers. This scale uses numbers, "in name only," thus the name, "nominal."

2.2.2 Ordinal Scales

As with the nominal scale, ordinal scales categorize things; however, ordinal measures additionally reflect a quantitative relationship between the various categories. Stated more succinctly, an **ordinal scale** consists of a set of categories organized quantitatively. The degree of quantitative information communicated, however, is limited to the *relative position* of one event compared to others. For instance, noting that "Jim is less reliable than Pam," "Audis are faster than Accords," and "Muhammad Ali was the greatest of all time" are all measurements of relative position.

It may also help to think of the commonly used notion of *ranking*. Rankings reflect more or less of something, but not *how much* more or less of something. The difference between the winner and runner-up in a pie-baking contest may not be the same amount as the difference between the fourth- and fifth-place finishers. In other words, the quantitative intervals between adjacent ranks are not held constant over the entire range of the scale. Consider house numbers on a neighborhood street. They only mark the relative distance a house is from a given point, typically the center of town. If we are standing at a house marked 105 Maple, we know that 121 Maple is closer to us than 173 Maple, but we do not know how much closer. Also, as is typical with house numbering, an ordinal scale system can start at any number (it need not be 1) and not every number needs to be used (105 Maple may be next door to 111 Maple). The only feature that is needed for the system to work is for the numbers to increase as the houses move farther away from a given point.

2.2.3 Interval Scales

The next step-up in quantitative information is the interval scale. An **interval scale** consists of a set of quantitatively ordered categories for which all the intervals between the categories are held constant (or "conserved"). A good example is the Fahrenheit temperature scale. The amount of heat needed to add to a room to move it from 80 to 85 °F would be the same amount of heat needed to move it from 90 to 95 °F. At each point along the scale, the heat difference between adjoining degrees is the same. However, interval scales do not possess a *true* zero point. That is, the value "zero" is just an arbitrary point on the scale and not the absence of quantity. A room that is measured at 0 °F is not devoid of heat; it could be made colder. It is also important to realize that any given interval scale does not have a monopoly on how the variable is to be measured. The degree of heat in a room, for instance, can be measured using a Celsius scale. Even though the Celsius scale possesses larger intervals and has a different zero point, both have conserved intervals across the length of the scale.

The distinction between ordinal and interval scales is not as straightforward as it may appear, especially in the behavioral and social sciences. In these areas of study, many scales have been constructed to measure various psychological concepts – like intelligence, pain, extroversion, and self-esteem. The problem with these measures is the difficulty in determining if the intervals are conserved.

For instance, is the intelligence difference between individuals with intelligence quotients (IQs) of 100 and 105 the same as the intelligence difference between individuals with IQs of 45 and 50? The numerical distance between the pairs of scores is the same, but is the difference in the *amount of intelligence* the same? This is not an easy question to answer. Yet, it is a critical question; many statistical tests require data to be measured on a scale with equal intervals across the entire continuum. This is a necessary scale feature for the important statistical concept of an "average" to be meaningful. This thorny issue is discussed in Spotlight 2.1 and in Chapter 3 (see Box 3.1).

Spotlight 2.1 Rensis Likert

Use the scale below to respond to the following statement: I enjoy studying statistics.

O	O	O	O	O
1	2	3	4	5
Strongly disagree	Disagree	Neither disagree nor agree	Agree	Strongly agree

If we have ever had to respond to a question in this manner, we can thank the social scientist Rensis Likert. Likert, born in 1903 in Cheyenne, Wyoming, first began studying civil engineering in 1922, but by his senior year he had been enticed into studying the social sciences (Kish 1990). Over the next 10 years, he studied a variety of disciplines, including sociology at the University of Michigan and theology at Union Theological Seminary, culminating with a PhD in psychology from Columbia University. In his dissertation, he presented a measurement tool he developed to assess how people felt about various topics related to international affairs. This tool, which asked respondents to place their attitude on a 5-point "favor to disfavor" scale with a neutral midpoint, was much simpler compared with the then popular but cumbersome method developed by another psychologist, Leon Thurstone (Croasmun and Ostrom 2011). Since Likert's scale yielded similar results to Thurstone's and yet was much easier to use, Likert's procedure soon became the method of choice for measuring personal attitudes on almost any topic across the behavioral and social sciences. However, there was a price to be paid for the scales' simplicity; it was unclear if these sequential categories from 1 to 5 were to be best understood as an ordinal scale or an interval scale. The debate still goes on today and will be discussed more in Box 3.1.

In 1939, then working for the US Department of Agriculture (USDA), Likert began to use his scale to measure farmers' reactions and attitudes toward a variety of President Roosevelt's New Deal programs that were sponsored by the USDA (Kish 1990). Gaining information about people's perceptions is extremely important in a democratic system where those who are governing are beholden to the people they govern. Likert realized this, and over time his measurement efforts gathered a lot of interest and attracted many professionals from a variety of other disciplines who were likewise interested in assessing the personal attitudes of people regarding all sorts of policies, products, and perspectives. As the United States entered World War II, his research and measurement team became involved in the construction of several national surveys designed to measure Americans' attitudes toward a variety of war-related government efforts like the selling of war bonds, the instituting of price controls, and rationing.

After World War II, Likert and several colleagues, eager to apply their measurement methods to new problems in a postwar world, started what is now called the Institute for Social Research at the University of Michigan (see https://isr.umich.edu/). Despite the never-ending challenge to secure funds for research support, Likert's unfailing optimism that evaluation research would become increasingly recognized as needed was soon shown to be accurate. His institute grew rapidly, becoming the largest university organization of its kind in just a few years.

In addition to Likert's measurement insights, he was also a theorist. For instance, his interest in organizational leadership led him to propose a theory of management-style development. He argued that management style culturally evolved through a set of four stages, from "exploitable authoritarianism" through "benevolent authoritarianism" and "consultative management," finally arriving at "participative management" (Hall 1972). Likert was a prolific writer and thinker, authoring several books and over 100 published articles. Looking over the totality of his life, it is easy to conclude that his influence on numerous fields of study has been unquestionably significant and yet, perhaps ironically, a bit hard to measure.

2.2.4 Ratio Scales

A **ratio scale** possesses all the properties of an interval scale, with the addition of an absolute zero point. For instance, one could argue that the Kelvin scale of heat measurement is a ratio scale based on a theoretical state of no energy: 0 °K. A measure of length is a ratio scale since there is an absolute zero point (i.e. a point of no length). With an absolute zero point, a value can be compared to any other value on that scale; for instance, on a ratio scale, a value of 5 is half of the value of 10. In addition, a given number also represents an absolute amount of something.

Could *intelligence* be measured on a ratio scale? No, because an IQ of 0 is meaningless. Furthermore, it would require us to claim that someone with an IQ of 100 is twice as smart as someone with an IQ of 50. However, because the concept of *weight* has an absolute zero point, a person weighing 200 lbs. is indeed twice as heavy as another weighing 100 lbs. *Time* can also have an absolute zero point. A reaction time of two seconds is twice as slow as that of one second. Difference scores using interval scales are also ratio measurements. For instance, the increase in heat in a room from 60 to 65 °F is half the size of an increase from 60 to 70 °F. Even though the Fahrenheit scale is interval, measures of temperature *differences* are ratio (i.e. a difference of zero means no difference).

The ratio scale offers the researcher much more quantitative specificity. However, the price of specificity is the loss of numeric freedom. If "A" is deemed to be 5, and if "B" is twice the size of "A," then "B" must be assigned the number 10. With other scales, there is more flexibility. The movement from nominal scales through ordinal, interval, and finally ratio scales can be understood as a movement from high to low flexibility as well as from low to high quantitative specificity. These two properties of a scale move in opposite directions.

In the behavioral and social sciences, many concepts are measured by an ordinal or interval scale. For instance, measures such as achievement, aptitude, or intelligence do not have an absolute zero point. However, studies that investigate performance often use a ratio scale – the number of correct answers or the amount of time needed to complete a task. These are variables measured on scales with constant intervals, where a "zero" marks the absence of quantity.

2.3 Types of Variables and Their Features

2.3.1 Discrete Variables

A measurement concept related to the sensitivity issue noted earlier in the chapter concerns how many different values can be assigned when a measurement is taken. A **discontinuous** (or **discrete**) **variable** can be assigned only a finite number of values. No meaningful values exist between any two adjacent values. For instance, an undergraduate student is a freshman, sophomore, junior, or senior; an adult is single, married, divorced, or widowed; and a roll of a typical die yields a one, two, three, four, five, or six; there are no "in-between" possibilities. One cannot claim to be halfway married or roll a die and hope to get a 4.5. It is permissible, however, to present statistical features of discrete data sets as fractions. For instance, it may be true that the average American family size is 3.13 persons, even though a .13[1] person is not possible. We just need to keep in mind a correct interpretation of this statistic. In this instance, it means that for every 100 American families, there are, on average, 313 people.

1 For text uniformity, a "0" will not be used in front of a fractional value that is less than one, except on occasion.

2.3.2 Continuous Variables

A **continuous variable**, theoretically, has an infinite number of points between any two numbers. Unlike discrete variables, which jump from one value to the next, continuous variables flow. Although 7 and 8 cm may be adjacent options on a ruler, there are an infinite number of values between them. Even if millimeters are marked on the ruler, there are an infinite number of values between, say, 7.3 and 7.4 cm. When a variable is continuous, any assigned number to a measure is an approximation. Even though we could measure someone's reaction time down to the tenth of a second, this measurement could still be refined with a more precise instrument. Therefore, one can always theoretically increase the precision of measurements for continuous variables.

Detecting the continuousness of a variable is not as simple as looking at how it is reported. Age is often recorded in whole-number years, but the underlying dimension is clearly continuous. This is also true for many psychological measures. Suppose a psychologist administers an anxiety questionnaire which scores respondents from 0 to 63. Although the measuring tool may only allow for the assignment of whole numbers, the underlying concept is arguably continuous.

2.3.3 Midpoints and Real Limits

If a variable is continuous, any assigned number is an approximation. When someone weighs 195 pounds, it does not mean that the person is exactly that weight. A person who weighs 195.1 and another who weighs 194.8 might both be listed as weighing 195. The number 195 is located at the **midpoint** of an interval of weights, that is, at the balance point of the interval. The upper and lower boundaries of the interval are called the **real limits**. The upper real limit of the number is one-half the unit of measurement above the number, and the lower real limit is one-half the unit of measurement below the number. If the unit of measurement is 1, the real limits for the number 13 are 12.5 and 13.5. If the unit of measurement is .1, then the real limits for 13.0 are 12.95 and 13.05. Figure 2.2 graphically illustrates the concept of upper and lower limits for numbers with different units of measurement.

■ **Question** *The following data set contains the average temperature and amount of rainfall in several cities for the month of March. For each number, specify the upper and lower real limit.*

Place	Temperature (°C)	Rainfall (cm)
Acapulco	31	0.3
Chicago	6	6.6
Honolulu	25	7.9
Orlando	74	8.6

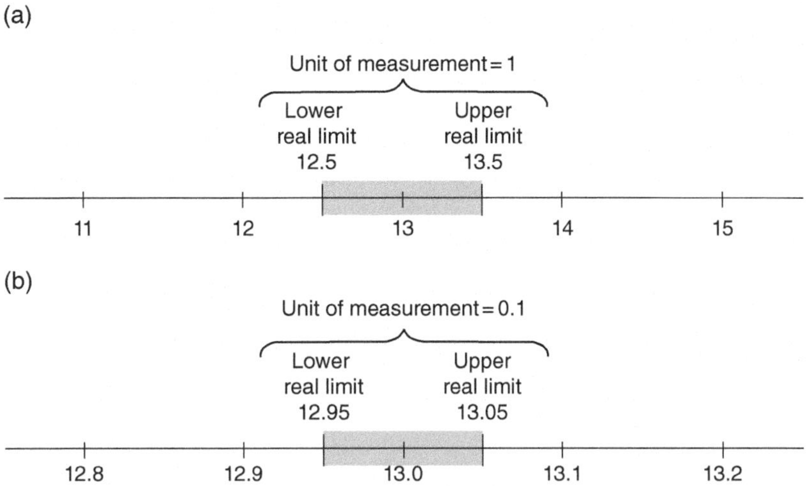

Figure 2.2 The upper and lower limits of a score of 13. (a) The unit of measurement is 1. The real limits are 13 + .5 = 13.5 and 13 − .5 = 12.5. (b) The unit of measurement is .1. The real limits are 13 + .05 = 13.05 and 13 − .05 = 12.95.

Solution The unit of measurement appears to be 1 °C for temperature. Since the boundaries of a number are one-half the unit of measurement, the upper and lower real limits of 1 °C are .5 °C above and below the number used to report temperature. Therefore, the upper and lower real limits for the average temperature in Chicago are 6.5 and 5.5 °C, respectively.

Since the unit of measurement for rainfall is .1 cm, the upper and lower real limits are specified as .05 cm. When establishing the real limits, it helps to think of the decimal place that is one notch greater in precision. If the scale uses whole numbers, the limits will be to the first decimal place. If the scale uses the first decimal place (e.g. rainfall), the limits will be to the second decimal place. For example, the upper and lower real limits for the March monthly rainfall of Honolulu are 7.95 and 7.85, respectively. The number 7.9 is the midpoint between 7.85 and 7.95. Table 2.1 presents the upper and lower real limits for the temperature and rainfall data of this problem. ■

In most research situations, a list (or *distribution*) of numbers, called **raw** (or **original**) **scores**, will be obtained. Unorganized data are hard to interpret. However, if the distribution is presented in a tabular or graphical form, summaries and important features of the data set can be communicated. The remainder of the chapter explores numerous ways data can be presented in tables or graphs.

Table 2.1 The midpoint, upper, and lower real limits for average temperatures and amount of rainfall for several cities in the month of March.

Temperature			Rainfall		
Lower limit	**Midpoint**	**Upper limit**	**Lower limit**	**Midpoint**	**Upper limit**
Acapulco					
30.5	31	31.5	0.25	0.3	0.35
Chicago					
5.5	6	6.5	6.55	6.6	6.65
Honolulu					
24.5	25	25.5	7.85	7.9	7.95
Orlando					
23.5	24	24.5	8.55	8.6	8.65

2.4 Using Tables to Organize Data

2.4.1 Simple Frequency Distributions

Table 2.2 presents scores from 90 participants who completed a questionnaire measuring their "need for achievement." A quick glance tells us very little about the scores. Indeed, even a longer look only tells us that the values seem to be between 1 and about 30. At a minimum, it would be nice to know how many participants received each score.

Table 2.2 Unorganized raw data.

15	8	20	16	12	18	14	22	17	5
19	15	18	29	6	13	16	19	10	24
15	3	26	30	13	17	7	16	23	25
1	15	18	14	5	27	16	20	14	6
24	14	20	25	21	15	17	8	23	21
17	14	10	13	18	16	21	9	11	22
15	12	9	16	20	11	13	22	17	13
9	22	16	12	19	17	14	10	19	18
11	16	12	18	13	17	15	14	15	28

Table 2.3 The simple frequency distribution constructed from the unorganized data given in Table 2.2.

X	*f*	*X*	*f*	*X*	*f*
30	1	20	4	10	3
29	1	19	4	9	3
28	1	18	6	8	2
27	1	17	7	7	1
26	1	16	8	6	2
25	2	15	8	5	2
24	2	14	7	4	0
23	2	13	6	3	1
22	4	12	4	2	0
21	3	11	3	1	1

Table 2.4 A grouped frequency distribution based on the raw data from Table 2.2.

Lower limit	Class interval	Upper limit	Midpoint	*f*
29.5	30–32	32.5	31	1
26.5	27–29	29.5	28	3
23.5	24–26	26.5	25	5
20.5	21–23	23.5	22	9
17.5	18–20	20.5	19	14
14.5	15–17	17.5	16	23
11.5	12–14	14.5	13	17
8.5	9–11	11.5	10	9
5.5	6–8	8.5	7	5
2.5	3–5	5.5	4	3
−0.5	0–2	2.5	1	1

A **simple frequency distribution** can accomplish this by systematically listing all the *possible* scores as well as the frequency with which each score appears. Table 2.3 shows a simple frequency distribution for the scores found in Table 2.2. Note that all possible scores are listed, typically in descending order, under the heading *X*. (The letter *X* typically represents the concept "scores.") The number of participants that received each score is placed correspondingly under the heading *f* (for "frequency"). Adding up the scores under *f* will tell us the number of participants in the study. If the variable being measured is continuous, recall that *X* values are midpoints of intervals.

2.4.2 Grouped Frequency Distributions

Some data sets cover a wide range of possible scores, making the resulting frequency distributions long and cumbersome. In these situations, researchers are often willing to exchange the loss of some information to create a table that is easier to understand. A **grouped frequency distribution** indicates the number of scores that fall into each of several ranges of scores (see Table 2.4).

Instead of noting a frequency count for each score, the viewer learns how many scores are found within a given range. **Class intervals** are groups of equal-sized ranges, determined by the researcher based on the desired trade-off between information loss and simplicity. Class intervals, typically organized in descending order, cover the full range of scores with no gaps or overlaps. Each raw score belongs to exactly one interval. Table 2.4 features class intervals of 3 units.

Class intervals have midpoints, and when depicting continuous variables, they also have upper and lower real limits. An interval of, say, 20–25, would have a midpoint of 23, a lower limit of 19.5, and an upper limit of 25.5. In Table 2.4, the midpoints, lower limits, and upper limits for each interval from the "need for achievement" data are represented. Rarely are the midpoints and real limits presented in published research. They are included here for educational purposes.

Box 2.1 Some Notes on the History of Statistics

Although ancient civilizations like the Egyptians and Chinese used tabulation and other simple statistics to keep track of tax collections, government expenditures, and the availability of soldiers, the modern use of statistics arguably began with the Englishman John Graunt (1620–1674). Graunt tabulated information on death rates in his hometown of London and noted that the frequency of certain diseases, suicides, and accidents occurred with remarkable regularity from year-to-year. (This realization, by the way, helped to develop the establishment of insurance companies.) Graunt also found the birth rate of biological males to be slightly greater than that of biological females. However, due to the greater male mortality rate (occupational accidents and wars), the number of men and women at the marriageable age was about equal. Graunt believed this arrangement was nature's way of assuring monogamy (Campbell 2001).

Most early uses of statistics focused on simple descriptions of data, but starting around the seventeenth century, advances in statistics began to take place, mostly springing from mathematicians' interest in the "laws of chance" as they apply to gambling. The French mathematician Blaise Pascal (1623–1662) was asked the following question by Chevalier de Méré, a professional gambler: "In what proportion should two players of equal skill divide the stakes remaining on the gambling table if they are forced to stop playing the game?" Pascal and Pierre Fermat (1602–1665), another French mathematician, arrived at the same answer, albeit by different proofs. It was the correspondence between these two over the practical issue of how to divvy up gambling stakes that initiated modern probability theory (Hald 2003).

Interestingly, the work of Pascal and Fermat was anticipated a century earlier by the Italian mathematician and gambler Girolamo Cardano (1501–1576). His volume, *The Book on Games of Chance*, was published posthumously in 1663 (after the Pascal and Fermat dialogs). In it, Cardano introduced many of the basic concepts of probability theory (as well as offered some tips on how to cheat when gambling). Cardano also practiced astrology. Indeed, by using astrology charts and horoscopes, he even predicted the exact date of his death. Several years later, upon arriving at that date and finding himself in perfect health, he allegedly decided to drink poison to ensure the accuracy of his prediction (Gliozzi 2008).

Several advances in the field of statistics have occurred in the last two centuries. Many of the chapters in contemporary statistics textbooks are based on advances from the period between 1850 and 1930. During this time, noted polymath Sir Francis Galton (1822–1911) formalized a method for making predictions of one variable with knowledge of a second related variable (regression analysis) (see also Spotlight 15.1). William Gosset (1876–1937) ushered in the era of modern statistics by developing tools that allow one to generalize based on only a few observations (the t test) (see also Spotlight 9.1). Sir Ronald Fisher (1890–1962) extended this work to allow for the comparison of several treatment variables on a dependent variable (the F test) (see also Spotlight 11.1). Contemporary statisticians are continuing to make advances in statistics, each allowing researchers to ask increasingly complex questions about the mysteries of human behavior.

2.4.3 Conventional Rules for Establishing Class Intervals

A grouped frequency distribution sacrifices some information by collapsing numbers into a set of intervals, but it is assumed that this information loss is inconsequential and perhaps even beneficial, drawing our attention to the forest instead of each particular tree. Table 2.4 uses 11 intervals. As we view the frequency column of the table, we can now easily see that just a few people received scores at the extreme ends of the distribution. Most of the scores are in the middle, with the greatest number of scores in the interval 15–17. If too few or too many intervals are used, it can be difficult to see how the numbers are concentrated. The use of about 10 class intervals is customary; however, the needs of the researcher vary by situation. The proper number of intervals to use should be determined by what best illustrates a meaningful distribution of the scores.

Common interval sizes, symbolized by i, are $i = 3$, $i = 5$, $i = 10$, or $i =$ some multiple of 10. There are no fixed rules for constructing a grouped frequency distribution. However, the following additional guidelines will be helpful:

1) Select a suitable interval size – one that leads to about 10 class intervals.
2) Some graphs of continuous measures require the use of the interval midpoint. Whole-number midpoints make graphs easier to read. Combine the number of intervals and interval width in such a way that the midpoint is a whole number. Using an odd number i will accomplish this.
3) The first number of the interval should be a multiple of i. For instance, if the interval width is 10, then the first number of the interval should be a multiple of 10. This guideline is sometimes violated when the interval width is 5. Instead of starting intervals on a multiple of 5 (e.g. 20–24), we may opt to create intervals where the *midpoint* is a multiple of 5 (e.g. 23–27).

2.4.4 Cumulative Frequency Distributions

A **cumulative frequency distribution** adds a column with a running tally of all the scores up through each class interval. Table 2.5 includes a cumulative frequency column, (*Cum f*), to the grouped frequency data presented in Table 2.4. It is customary to start accumulating the scores from the bottom of the distribution. For instance, the cumulative frequency for the 15–17 interval is 58. This is

Table 2.5 A cumulative frequency column added to the grouped frequency data in Table 2.4.

Class interval	*f*	*Cum f*
30–32	1	90
27–29	3	89
24–26	5	86
21–23	9	81
18–20	14	72
15–17	23	58
12–14	17	35
9–11	9	18
6–8	5	9
3–5	3	4
0–2	1	1

the sum of frequencies found at that interval plus all preceding intervals (1 + 3 + 5 + 9 + 17 + 23 = 58). The number at the top of the *Cum f* column provides the total number of scores.

2.5 Using Graphs to Display Data

One of the best ways to display data trends is to summarize them in the form of a graphic. In this section, we will learn about some of the most used graphic displays in the behavioral and social sciences. We will also learn how to view graphs with a healthy degree of skepticism. Information presented in graphical form can be technically correct and yet entirely misleading.

2.5.1 The Axes of a Graph

The typical graph has two axes. The horizontal axis is called the *X* axis, or **abscissa**. The vertical axis is called the *Y* axis, or **ordinate**. Larger numbers are to the right on the abscissa and upward on the ordinate. Smaller numbers progress to the left on the *X* axis and downward on the *Y* axis. Figure 2.3a shows the positive sides of the *X* and *Y* axes of a graph. Figure 2.3b reflects a broader perspective. Typically, the *X* axis reflects the possible values (*X*'s), either quantitative or qualitative, and the *Y* axis reflects frequency. As these figures show, the two axes do not need to be scaled identically.

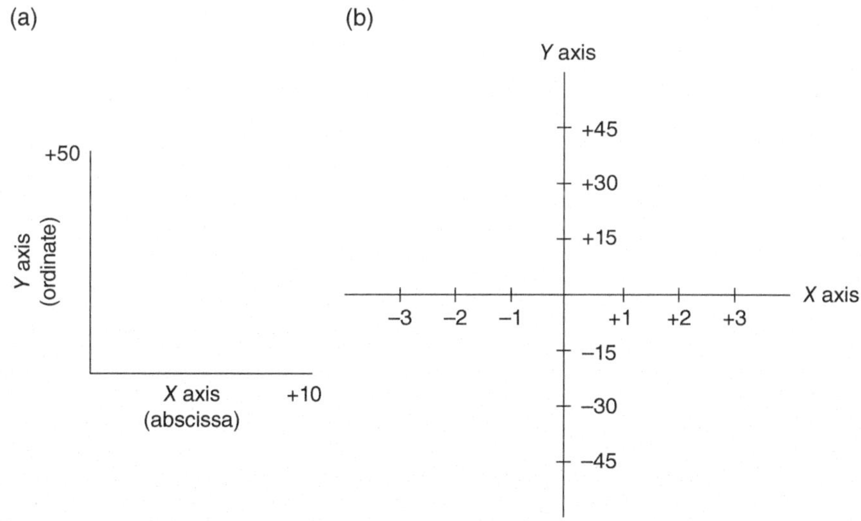

Figure 2.3 (a) The positive sides of the two axes. (b) A graph with a broader perspective.

2.5.2 The Frequency Polygon

A **frequency polygon** plots the number of scores in each of the intervals of a frequency distribution. The interval width may be 1, as in a simple frequency distribution, or greater than 1, as in a grouped frequency distribution. Figure 2.4 is a frequency polygon drawn from the grouped frequency distribution in Table 2.4. Note in Figures 2.4 and 2.5 that the first and last points of the graph do not meet the horizontal axis. Whether or not to draw the frequency polygon to meet the X axis is a matter of personal preference.

As we view Figure 2.4, note that the X axis marks the midpoints of the class intervals. The Y axis is labeled "Frequency" and presents equally spaced numbers that specify the frequency of scores. A *single* point on the graph indicates both the midpoint of a class interval (X axis) and the number of scores found in that interval (Y axis).

The intersection of the X and Y axes, called the *origin*, usually represents the 0 point for each of the variables. However, sometimes the first number of a class interval is some distance from 0, or the smallest frequency count for any interval is much greater than 0. Should this situation arise, the X and/or Y axes can be truncated (i.e. shortened where not needed) with broken lines. Figure 2.5 shows a

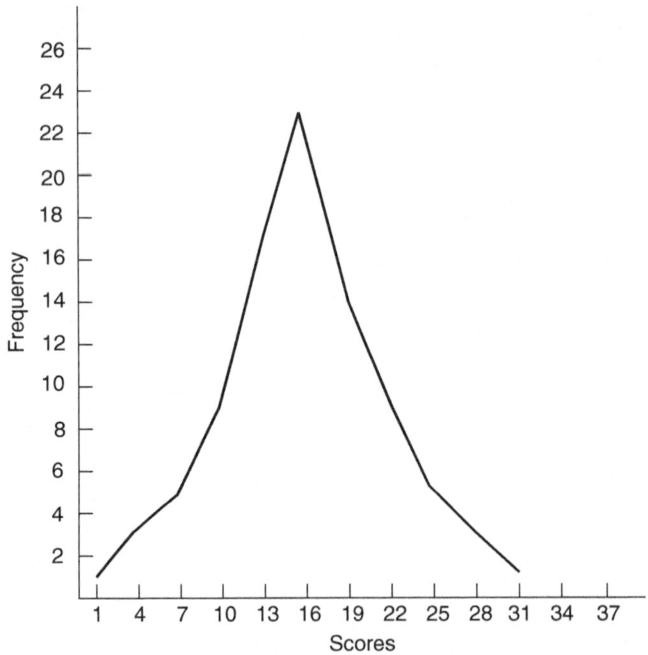

Figure 2.4 A frequency polygon of the data in Table 2.4 using interval midpoints.

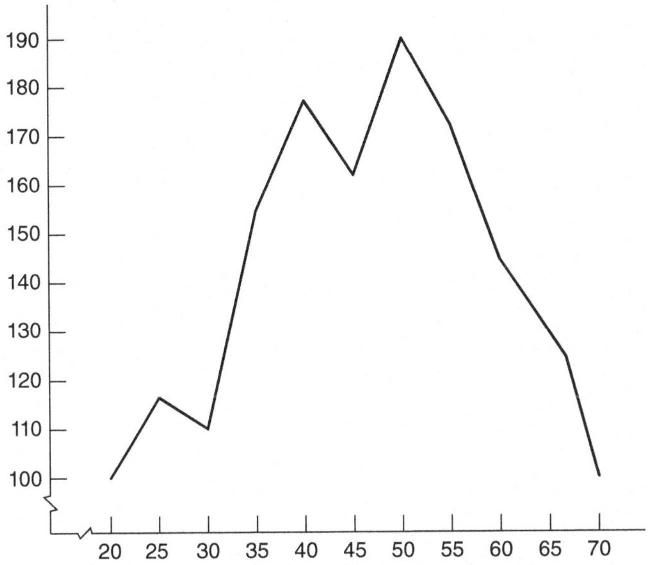

Figure 2.5 The *X* and *Y* axes are broken between the origin and the lowest scores.

frequency polygon in which the data points are far from the origin. Note the truncation marks at the base of the axes. Truncating axes are an acceptable practice, provided they are clearly indicated.

The frequency polygon is a useful graphic for depicting the overall concentration of numbers. It is easy to construct, and it is possible to compare two or more similarly scaled distributions on the same graph. However, many suggest frequencies are easier to read when using a different type of graphic display – a histogram.

2.5.3 The Histogram

The **histogram** is a graph of vertical bars with shared borders in which the height of each bar corresponds to the frequency of scores for a given class interval (see Figure 2.6). The width of the bar spans the width of the class interval, including the real limits and, thus, the shared borders. The bars of a histogram are typically colored in contrast with the background.

The frequency polygon and the histogram are related. If we were to place a point at the midpoint of the top of each bar of the histogram, erase the bars, and connect the data points, we would have a frequency polygon. A frequency polygon has been superimposed on the histogram depicted in Figure 2.6 so that we can directly compare these two ways to display data graphically.

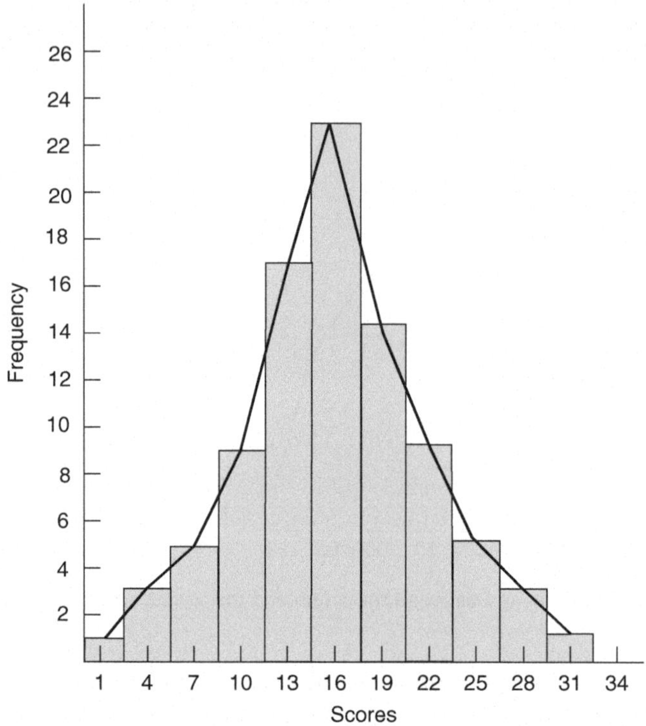

Figure 2.6 A frequency polygon superimposed onto a histogram based on the data in Table 2.4.

2.5.4 The Bar Graph

A **bar graph** is used to represent the frequency of scores associated with categories. It looks like a histogram, except the bars do not share a common border. Since the categories represented on the X axis are discrete in nature, they do not have real limits. Gaps between the bars clearly communicate this. For example, in Figure 2.7, the scale used on the X axis is nominal. Psychology, history, and so on are names of different majors. Figure 2.7 shows hypothetical data depicting the number of undergraduate majors in each of several university programs.

2.5.5 Graphs Can Be Misleading

Suppose a researcher compares two different methods for enhancing learning. As it turns out, Method B produces a modest three-point improvement for those who use it, whereas Method A does not register any change in learning. Figure 2.8 labels pretest and posttest scores on the X axis. The data points above the pretest

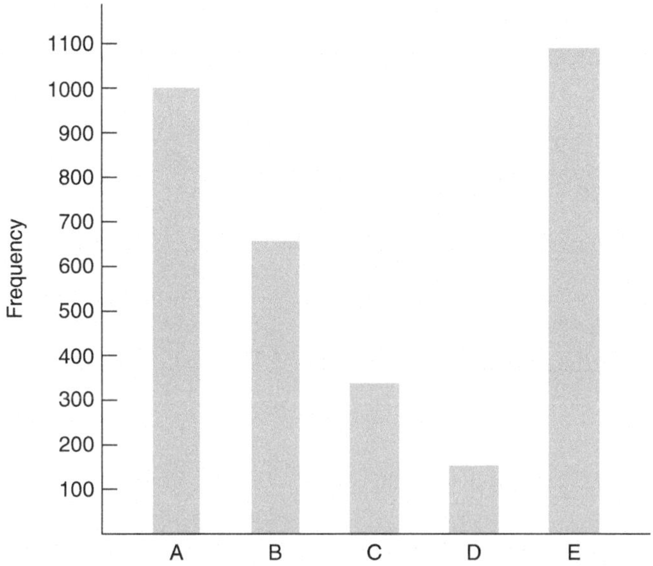

Figure 2.7 The number of undergraduates majoring in psychology (A), sociology (B), history (C), biology (D), and business (E).

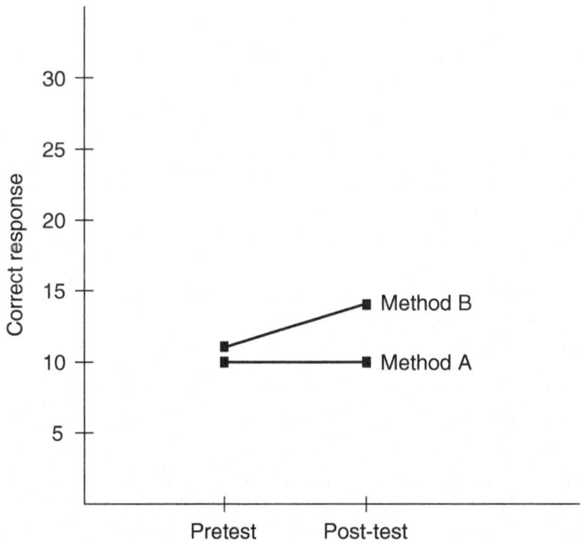

Figure 2.8 A graph that indicates little difference between the two methods. Note the scaling of the *Y* axis; it appears to start at zero.

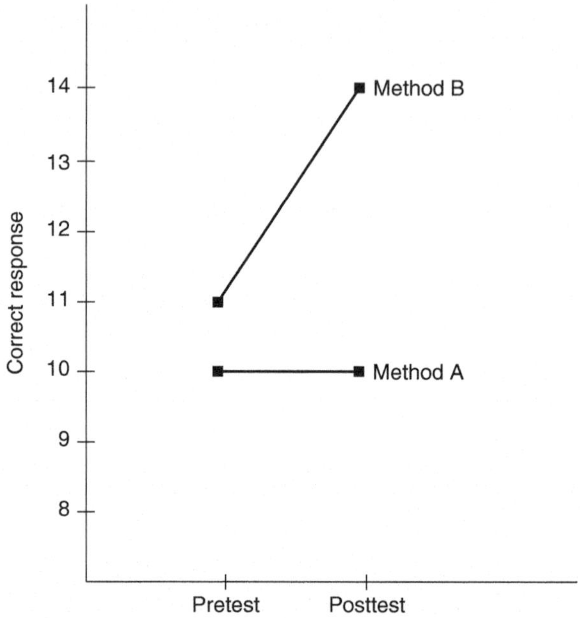

Figure 2.9 The data points of Figure 2.8 rescaled to create the impression of a large difference.

indicate the average number of correct responses for each method *before* participants are administered the training. The data points above the posttest represent the average number of correct responses *after* training. Note that the line for Method *A* is parallel with the *X* axis, indicating no change in performance due to training. The line for Method *B* rises *slightly*, reflecting the modest increase of three points. Since the lines show little divergence, it appears that the two methods have similar effects on performance.

Now examine Figure 2.9. The same data are graphed, but now it looks like Method *B* is vastly superior to Method *A*. Why? Notice the *Y* axis. The ordinate has been altered such that an increase of three points spans a much greater distance. The two graphs are both *technically* accurate. However, the second graph is very misleading; the *Y* axis has been truncated, but without including the broken axis line. Furthermore, the highest value on the *Y* axis is now 14 instead of 30. This leaves the impression that those using Method *B* virtually topped out in terms of performance. Together, these techniques serve to amplify the differences between the data lines. However, this amplification seems to misrepresent the difference between the two methods. Viewers should always pay close attention to not only the data lines but also the scaling of the axes. See Box 2.2 to learn about the role of data misrepresentation in the replication crisis.

Box 2.2 Is the Scientific Method Broken? Misrepresenting Data/Findings

In Box 1.1, we started a series asking whether the scientific method is broken. One of the reasons for the replication crisis might be the misrepresentation of scientific data and findings.

Data misrepresentation can occur in several ways. One concerns how science writers interpret scientific findings for the public. Since most people get their scientific information from the media, those who interpret science for the public bear the responsibility to accurately convey what scientific investigators have found. However, some interpreters of science are not sufficiently familiar with the scientific process and the subtleties of doing and interpreting research. Furthermore, there is no getting around the fact that there is a financial incentive to write eye-catching headlines. This situation can lead to oversimplified descriptions of findings presented to the public. One example is from a team of psychologists who, in 2013, reported no cognitive improvement for preschoolers briefly exposed to a music enrichment experience (Mehr et al. 2013). It was a limited study designed only to see if positive effects could be found in young children with a brief exposure to music. Great lengths were taken by the authors to clarify the limits of the study. Nonetheless, headlines soon appeared like this one from the *Times of London*, "Academic benefits of music a myth" (Devlin 2013), clearly overstating the study's modest conclusions. Indeed, another study the following year found improved neuroplasticity and language development in children from underserved communities after exposure to community music classes (Kraus et al. 2014). Some of the public's distrust of science comes from the intentionally provocative way many popular interpreters choose to report scientific findings.

Another form of data misrepresentation involves the researchers themselves, either through data collection or interpretation. Assuming the purest of motives, for the moment, researchers can unintentionally bias participant responses through the ordering of questions (which question comes first, then second, and so on), the limited number of response options available, or even the specific wording of the questions. How much of the voting population is pro-gun, pro-choice, pro-environment, or pro-peace, for instance, depends on the words that are used in the question. Collecting and interpreting data is rarely a straightforward process.

Some researchers also selectively report findings, only publishing relationships that stand out, failing to put them into a broader context. This is like the claim of a coin-flipper to control which side comes up. The evidence offered is a series of nine "heads" in a row. Impressive. However, we might be less impressed if we learned that this string of heads was mined out of the middle of 4000-coin flips. Or consider the reporting of positive COVID tests for a given period of time. Should they be presented as a total or in terms of a rate, given the number of tests administered? (Sometimes higher positivity rates corresponded with higher testing numbers, not necessarily a greater number of cases.) Context matters. (This topic will be explored more in Box 8.1.) Unfortunately, several scientific articles, many of which misrepresented findings unintentionally, are retracted by journals every year. The website retractionwatch.com is an example of one service that monitors retractions.

Finally, there is the issue of academic fraud (e.g. Carey 2016). Science, we must remember, is not practiced by purely objective robots or angels but by people, people who possess the same frailties, temptations, and pressures common to us all. Science is also a cultural enterprise, with its own hierarchy of authority, internal rewarding structure, and value system – a value system that places a premium on new findings, new ideas, and numerous publications. Researchers who do not make original discoveries, propose interesting innovative theories, or generate numerous publications often find themselves out of a job. Given this reality, we should not be surprised to learn that just as the enterprise of professional sports, financial investment, politics, and virtually all other human communities deal with different cheating scandals, this practice can and does take place within the world of scientific investigation. Thankfully, just as in these other professions, there are correcting mechanisms designed to ferret out falsehoods and eventually get to the truth. Nonetheless, when the public finds out that a headline is wrong, a journal article is retracted, or a scientist is a fraud, we should not be surprised to learn that for some people it feels as if science is broken.

2.6 The Shape of Things to Come

Graphs allow us to look at the "forest" of a data *set* instead of the "trees" of individual data *points*. As data sets grow, the ability to describe efficiently, yet accurately, the features of a distribution becomes more important. It is interesting to learn that most data sets, whatever the size, can be described well using just three characteristics: the distribution *shape*, a measure of *centrality*, and a measure of *variability*. Centrality and variability will be the subject matter of the next two chapters, but several terms and concepts related to distribution shape can be presented now.

2.6.1 Bell-Shaped Distributions

If we were to keep adding numbers to our set of "need for achievement" values (see Table 2.2) and then graph them, using either a histogram or a frequency distribution, we would likely see a bell-shaped curve start to emerge. In fact, with the

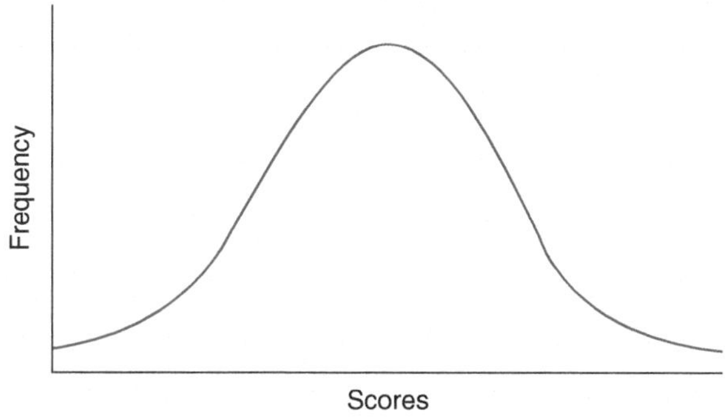

Figure 2.10 One type of bell-shaped curve is the normal curve. Note that it is symmetrical, and most of the scores are found in the middle of the distribution.

existing 90 scores, this shape is already forming (see Figure 2.6). Many naturally occurring phenomena, when measured and graphed, take this common shape. It is so common, in fact, we could call it *normal*. A **normal distribution** (or *normal curve*) is a symmetrical, bell-shaped curve escalating gradually at first and then more sharply, eventually inflecting, and then tapering to a peak (see Figure 2.10 for one example). We will learn much more about normal curves in later chapters. Of course, frequency distributions and histograms, even ones with large numbers, do not perfectly take this shape, but a smooth-lined representation of the data is certainly invited by the viewer. It makes sense, theorists argue, to begin to think of many data sets in terms of being normally distributed. This shape will become exceptionally important to us as we learn about descriptive and inferential statistics.

2.6.2 Skewed Distributions

Not all data sets approximate a normal curve. One non-normal shape finds most of the scores near one end. This is called a **skewed distribution**. Figures 2.11 and 2.12 show positively and negatively skewed distributions, respectively. The type of skewness depends on the direction of the elongated tail. If it runs to the larger positive numbers, it is a **positively skewed distribution**; if it runs toward the smaller or negative numbers, it is a **negatively skewed distribution**.

2.6.3 Kurtosis

Kurtosis refers to the quality of the peak of the curve. The sharpness of the peak reflects the relative concentration of the scores. In Figure 2.13, most of the scores are found very close to the middle of the distribution. When the shape of the curve is relatively narrow and possesses an accentuated peak, the distribution is labeled **leptokurtic**. In Figure 2.14, most of the scores are spread out and widely

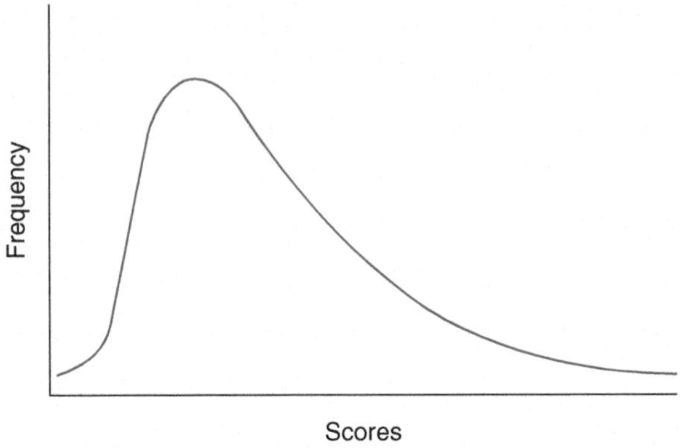

Figure 2.11 A positively skewed distribution. Here, most of the scores are found in the lower end of the distribution, and the elongated tail is pointing toward larger positive numbers.

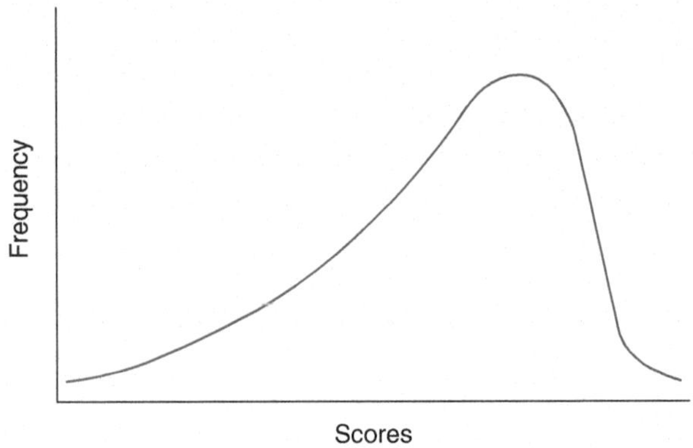

Figure 2.12 A negatively skewed distribution. Here, most of the scores are found in the upper end of the distribution, and the elongated tail is pointing toward smaller or negative numbers.

dispersed. When the shape of the curve is relatively broad and possesses a muted peak, the distribution is labeled **platykurtic**.

The curves drawn to illustrate the concepts of skewness and kurtosis do not exhaust the myriad of ways in which scores can be patterned. A distribution can be shaped like a *J*, a box, a *U*, or even an *M*; in fact, distributions can assume practically any shape. And these shapes matter. In Chapters 3 and 4, we will learn that different statistical concepts developed to describe data sets should be used for different shaped distributions.

Figure 2.13 An example of a leptokurtic distribution.

Figure 2.14 An example of a platykurtic distribution.

Summary

There are many challenging issues related to measurement, including philosophical concerns related to the definition, dimensions, and measurement sensitivity of concepts. Additionally, there are practical issues related to measurement. This chapter introduced us to the four most useful scales of measurement for social and behavioral scientists.

A nominal scale merely distinguishes one type of thing from another. Nominal numbers are not quantitative, but rather qualitative and categorical.

Ordinal measures also categorize things, but in quantitative relation to one another. An ordinal scale is used to identify the relative position of an event as compared to others. Terms like "more" and "less," as well as the concept of "ranking," are helpful when considering the ordinal scale.

More specific descriptions of quantity are found in the interval scale. Here, all intervals between units on the scale are held constant. The amount of quantity needed to go from a five to a six is the same amount needed for a one-unit movement at any other position on the scale. This additional feature is necessary for many statistical techniques since it allows averages to be found.

A ratio scale possesses all the properties of an interval scale with the addition of an absolute zero point. This allows ratio statements to be made. For example, on a ratio scale, 5 is half of 10, and 20 is one-third of 60.

A discontinuous or discrete variable can be assigned only a finite number of values. No meaningful values exist between any two adjacent values. A continuous variable, on the other hand, has an infinite number of possible values. The level of precision of measurement for continuous events is only limited by the sensitivity of the measuring tool. The upper and lower boundaries for a unit interval of a continuous variable are called the real limits.

Using tables and graphs to organize data allows us to view a summary of the raw scores of the study. A simple frequency distribution lists all the possible scores and the frequency with which each score appears. A grouped frequency distribution indicates the number of scores that fall within each of several class intervals. Class intervals are sets of equal-sized ranges used to organize and simplify the data. A cumulative frequency distribution includes a column that shows the accumulated number of scores for a given interval as well as all preceding intervals.

The typical graph has two axes: a horizontal axis called the X axis or abscissa, and a vertical axis called the Y axis or ordinate. Larger positive numbers are to the right on the abscissa and upward on the ordinate. A frequency polygon uses class interval midpoints to plot the number of scores in each interval of a frequency distribution. A histogram represents score frequency by using the real limits. A bar graph is used to represent the frequency counts of categorical data. A bar graph looks like a histogram, except the bars do not share a common border. Viewers of graphs should take note of the scaling on the axes to avoid misunderstanding the presented data.

Statisticians use terms to describe important features of the shape of distributions. A normal curve is a symmetrical, bell-shaped line escalating gradually at first and then more sharply, inflecting at some point and then tapering to a peak. A distribution that has scores that bunch at one end of the distribution is skewed, positively if the elongated tail points toward the larger numbers and negatively if it points toward the smaller or negative numbers. Kurtosis is a term that refers to the quality of the peak of a distribution. A leptokurtic distribution features a narrow width and accentuated peak, while a platykurtic distribution features a wide width and muted peak.

Microsoft® Excel and SPSS®

Information regarding how to use Microsoft® Excel and SPSS® (two common statistical software packages) to assist in finding descriptive statistics can be found in Appendix C. While sophisticated, both programs are easy to use with the help of tutorials.

Key Terms

Measurement
Nominal scale
Ordinal scale
Interval scale
Ratio scale
Discontinuous (discrete) variable
Continuous variable
Midpoint
Real limits
Raw (original) scores
Simple frequency distribution
Grouped frequency distribution
Class intervals
Cumulative frequency distribution
Abscissa
Ordinate
Frequency polygon
Histogram
Bar graph
Normal distribution
Skewed distribution
Positively skewed distribution
Negatively skewed distribution
Kurtosis
Leptokurtic
Platykurtic

Exercises

1 Think of different ways to measure these concepts using as many different scales as possible.
   a Academic proficiency
   b Athletic prowess
   c Creativity
   d Daily food consumption

2 Think of two variables that may be normally distributed; defend the rationale.

3 Think of two variables that may be negatively skewed; defend the rationale.

4 Think of two variables that may be positively skewed; defend the rationale.

5 Indicate which scale of measurement (nominal, ordinal, interval, or ratio) best applies.
- **a** Abortion attitudes (for/against)
- **b** Ratings of popularity
- **c** Time tolerating pain
- **d** Heart rate under stress
- **e** A measure of need for approval
- **f** Amount of weight lifted
- **g** Identifying political affiliation
- **h** Identifying diagnostic categories
- **i** Mood disorder vs. no mood disorder
- **j** A list of tennis players, best to worst

Work Problems

6 For each class interval, specify the width, the real limits, and the midpoint.
- **a** 1–3
- **b** 6–10
- **c** −4 to −8
- **d** −2 to +2

7 The data in the table below are from a midterm exam. Set up frequency distributions with:
- **a** $i = 1$ (simple frequency distribution)
- **b** $i = 3$
- **c** $i = 10$
- **d** $i = 20$
- **e** Include *cum f* columns for each distribution

Midterm examination scores

41	98	63	90	70	60	45	43	78
67	56	54	78	87	43	90	81	81
77	80	79	80	81	66	75	88	84
49	63	78	79	80	92	89	84	77

8 For each frequency distribution in Problem 7, specify the real limits of each class interval.

9 Construct histograms for b and c frequency distributions of Problem 7.

10 Based on the histograms of Problem 9, draw frequency polygons.

11 The amount of sugar per serving in breakfast cereal might be misrepresented on the side of a cereal box. Draw two different bar graphs, both using the data in the table below: one graph faithfully representing the relationship between the cereals and the other misrepresenting the relationship in such a way as to suggest Cereal A is far superior in terms of sugar content.

Cereal type	**A**	**B**	**C**
Sugar in grams/serving	12	14	15

3

Measures of Central Tendency

3.1 Describing a Distribution of Scores

When social and behavioral scientists collect data, they gather up the numerically represented observations (or "scores") associated with a given variable. As we learned in Chapter 2, viewing a list of numbers can be rather uninformative. For example, suppose someone has used a tracking device to record the distance walked each day during a 10-week period. If we were to ask them about the data they gathered, we might find it rather cumbersome and not particularly helpful if they responded by reciting a list of 70 numbers. Although a lot of information would be communicated, it would be hard to make much sense out of it. In Chapter 2, we learned ways in which a distribution of scores could be *visually* displayed by using various tables and graphs. In addition to these visual displays, a variety of *descriptive statistics* have been developed that are designed to communicate quickly, efficiently, and *numerically* the basic features of a distribution.

What are the most relevant aspects of a distribution to communicate? In Section 2.6, we learned that there are three key features of any dataset: the shape of the distribution, a measure of central tendency, and a measure of variability. Some concepts related to distribution shape were presented in Chapter 2, and more will be presented as we go along. This chapter will focus on measures of centrality, while Chapter 4 will focus on measures of variability. **Central tendency** measures are statistical indices designed to identify the "center" or "middle" of a distribution, while *variability measures* describe the degree to which scores are dispersed around this center or middle point.

Essentials of Behavioral and Social Science Statistics, Third Edition.
K. Paul Nesselroade, Jr. and Laurence G. Grimm.
© 2025 John Wiley & Sons, Inc. Published 2025 by John Wiley & Sons, Inc.
Companion website: https://www.wiley.com/go/Essentials_of_Behavioral_and_Social_Science_Statistics3e

3.2 Parameters and Statistics

In Chapter 1, we introduced the terms *population* and *sample*. Recall that a population is every member of a given group, and a sample is a subset of a population. Deciding whether a collection of scores is one or the other can easily change based on one's perspective. If a researcher is only interested in how a particular class is performing, then a complete set of test scores from that class constitutes a population. However, if they are interested in students in general, then this set of test scores should be considered a sample, a subset of this larger population. **Parameters** are numerical values describing the distribution features of populations, while **statistics** are numerical descriptions of samples. In one case, a test score average of 78 represents a parameter; in the other, that same number represents a statistic. Unfortunately, the term "statistic" can still be confusing, as it can refer to several concepts. For instance, it can be colloquially referenced as a number (e.g. "Bob got divorced, now he is just another statistic"), it can be a value that summarizes a feature of a sample of scores, or, if stated in its plural form (*statistics*), it can refer to a general field of study.

Most studies involve the use of samples since researchers are usually interested in generalizing their findings to people not in the study. The distinction between populations and samples is very important. To help distinguish them, different symbols are used in formulas; this is important, as some formulas are constructed differently depending on how a dataset is defined. Wrong statistical values can result if we misunderstand the nature of a dataset.

3.3 The Rounding Rule

Before going further, a brief digression is in order. Since the rest of the text involves making calculations, we should reach some agreement regarding precision. Given the nature of most statistical data, rounding to two decimal places seems appropriate. To do this, compute values to the *third* decimal place and then round to the *second*. For example, 27.534 would become 27.53. If the third decimal place is a 5, always round up. So, 34.785 becomes 34.79. When hand calculating, sometimes intermediate values are found. For consistency, these can also be rounded to two places. However, computers may not round these intermediate values. This might result in slightly different final values than those derived from hand calculations. These minor discrepancies, however, should not concern us.

3.4 The Mean

The **mean**, colloquially referred to as the "average," is the most frequently used measure of central tendency. As a descriptive measure, the mean has advantages and disadvantages that will be discussed later. Formula 3.1 shows how to find the mean for a population.

Population mean

$$\mu = \frac{\Sigma X}{N} \quad (3.1)$$

---

where

μ = (pronounced "mew") the symbol for the mean of a population
X = a score in the distribution
N = the total number of scores in the population (or population size)
Σ = (pronounced "sigma") a notation that directs one to sum up a set of scores. Thus, $\Sigma X = X_1 + X_2 + X_3 \ldots X_n$

The formula for the *sample* mean functions identically to the population formula, only with different symbols. As different formulas are introduced in the text, we will see that Greek letters represent features of a population, while Romanized letters represent features of samples.

Sample mean

$$M = \frac{\Sigma X}{n} \quad (3.2)$$

---

where

M = the symbol for the mean of a sample
n = the total number of scores in the sample (or sample size)

■ **Question** *What is the mean of this population of scores?*

5, 8, 10, 11, 12

Solution $\mu = \frac{46}{5} = \mathbf{9.20}$

Notice that the answer of 9.2 would be the same whether the set of scores is considered a population or a sample. Although the designation is theoretically important, it does not impact the calculation of this statistic. ■

The common practice in many statistics books is to use $\bar{X}$ (pronounced "X bar") to represent the sample mean. This is the more traditional symbol. However, most recently published manuscripts in the social and behavioral sciences report sample means using an M. Since students are much more likely to encounter the M symbol in their readings and to use it in their professional writing, this will be the symbol used in this textbook. If we see the $\bar{X}$ symbol elsewhere, keep in mind that it also stands for the sample mean.

There are three measures of central tendency discussed in this chapter: the mean, median, and mode. Each one is designed to communicate where scores tend to be centered in the distribution. However, each measure approaches the concept of "centrality" or "middleness" differently. In what way does the mean reflect centrality? Or stated in other words, what does the mean *mean*?

Each raw score in a distribution can be thought of as being "off" of some middle point or deviating from the center by a certain amount, even if that amount is zero. The mean is the value where the sum of those raw score deviations equals zero. To clarify, let us tackle this in a different way. A **deviation score** (sometimes called an **error score**) is the distance a raw score is from the mean ($X - M$) and let us symbolize it as x (pronounced "little x"). Therefore, $x = X - M$. So, if the mean of a distribution is 10, a raw score of 12 has a deviation (or error, or little x) score of 2.

In Table 3.1, the deviation score for each raw score is listed in the fourth column. Note that a raw score has a negative deviation score when it falls below the mean and a positive deviation score when it falls above the mean. The mean is the number where *the sum of all the deviation scores equals 0*; this is how the mean defines the middle or the center of a distribution. Stated mathematically, $\Sigma(X - M) = \Sigma x = 0$. In Table 3.1, both distributions have identical scores except for Participant 5. A score of 30, instead of 10, is obtained by Participant 5 in Distribution B. Consequently, the M of Distribution B (10) is greater than the M of Distribution A (6). However, the deviation scores still sum to 0. In a manner of speaking, the mean has been adjusted so that the Σx is still = 0. It is precisely in this sense that the mean is the center of a distribution. For every distribution, no

Table 3.1 Deviation scores always sum to zero.

Distribution *A*				Distribution *B*			
Participant	Score	*M*	*X* − *M*(*x*)	Participant	Score	*M*	*X* − *M*(*x*)
P_1	2	6	−4	P_1	2	10	−8
P_2	4	6	−2	P_2	4	10	−6
P_3	6	6	0	P_3	6	10	−4
P_4	8	6	+2	P_4	8	10	−2
P_5	**10**	6	+4	P_5	**30**	10	+20
			$\Sigma x = 0$				$\Sigma x = 0$

matter what its shape or number of raw scores, *the sum of the deviation scores from the mean always equals 0.*

3.4.1 The Properties of the Mean

The mean is an attractive measure of centrality not only because it incorporates every value in a dataset but also because it includes each score's interval distance away from the center. As we will see, the other measures of centrality cannot do this. As a result, it is the most frequently used measure of centrality. However, this ability is a double-edged sword. In Table 3.1, we saw that replacing Participant 5's raw score of 10 with the quite discrepant value of 30 shifted the mean tremendously – from 6 all the way to 10. This highlights a problem with the mean; it is sensitive to extreme scores. This problem is most evident with small datasets.

To illustrate this, consider Congressman Aster Windblows. The politician has promised to bring prosperity to the district. During the next election, Windblows would like to convince the constituents that this promise has been kept. The definition of prosperity Windblows uses is the mean income of families living in this exceptionally small district. When the congressman was first elected, the mean income was \$50 000. Two years later, one couple moved into the district with a yearly income of \$250 000. Everyone else's income remained the same. Look what happens to the average family income when the mean is used as the measure of central tendency.

Family income (beginning of term)	Family income (end of term)
\$44 000	\$44 000
\$48 000	\$48 000
\$50 000	\$50 000
\$52 000	\$52 000
\$56 000	\$56 000
	\$250 000
$\mu = \$50\,000$	$\mu = \$83\,333$

Congressman Windblows can honestly report that the average family income has dramatically increased during his term in office. The outsized influence of extreme scores on small datasets is one reason why an index of central tendency that is less sensitive to extreme scores is necessary.

Furthermore, many distributions possess more than one extreme score. Skewed distributions, in fact, can feature a moderate percentage of scores trailing well off to one side. If we are interested in communicating where scores are most concentrated, and the existence of extreme scores would generate a misleading impression, then a different measure of centrality is needed.

3.5 The Median

We have learned that the mean defines centrality as the point in a distribution where the $\Sigma x = 0$. However, *centrality* can be thought of in different ways. The **median** defines centrality as the value which splits the bottom half of the scores, quantitatively speaking, from the top half. The median is not algebraically defined but is rather typically found through a counting procedure.

3.5.1 Finding the Median When Given an Odd Number of Scores

■ **Question** *What is the median of this distribution?*

1, 4, 6, 8, 40, 42, 43, 45, 47

Solution 40

The median is 40 because the same number of scores (four) fall above 40 as fall below 40. To find the median, first position the scores in ascending (or descending) order, and then identify the midpoint of the distribution. ■

Note that in the abovementioned example, the values of the scores surrounding the median are irrelevant. The median is based strictly on the ranking of scores. We could say that the median is "rank sensitive," whereas the mean is "valuc sensitive." The median takes care of the problem of extreme scores; they only count as one score, no matter their distance from the middle of the distribution. But the downside comes for numbers measured on an interval or ratio scale. Their exact position carries important quantitative information, and yet this information is not factored into the median. Only *relative* position matters. In this example above, a median of 40 might seem to misrepresent "centrality." The next set of questions will further emphasize this point.

■ **Question** *What are the mean and median of this sample distribution?*

2, 4, 8, 9, 10, 15, 22

Solution $M = 10$ median $= 9$ ■

■ **Question** *What are the mean and median of this sample distribution?*

2, 4, 8, 9, 10, 15, 22, 26, 30

Solution $M = 14$ median $= 10$ ■

Look closely at the two previous distributions. They are identical except for two scores added to the high end of the second list. This has greatly influenced the mean, changing it from 10 to 14. The median, however, was only shifted one value to the right, from 9 to 10. The two added values do not seem to be extreme. It may be problematic that their exact values (26, 30) are not being factored into this measure of central tendency.

3.5.2 Finding the Median When Given an Even Number of Scores

In the examples provided so far, it was easy to identify the median because there were an odd number of scores in the distribution. But what do we do in the datasets below?

4, 6, 9, 10, 11, 12

1, 2, 4, 6, 8, 11, 14, 18

The median will fall between two scores anytime there is an even number of scores; it will typically be a value that does not occur in the distribution. The median may even be a number that does not seem to make sense in terms of what is being measured, like a fraction. For instance, imagine the median number of traffic tickets handed out each month for a given city equals 207.5. (What is the meaning of one-half of a traffic ticket?) Remember that statistical concepts convey a feature of an entire distribution; it is not a requirement that the statistic itself make sense as a score in that distribution. The following examples will show us how to calculate medians when there is an even number of scores in the distribution.

■ **Question** *What is the median of this distribution?*

3, 9, 15, 16, 19, 22

↑

Median

Solution 15.50

The median of a distribution having an even number of values is the mean of the middle two numbers, provided there is not a string of identical numbers in the middle. ■

Finding the Median When There Are Identical Scores in the Middle of the Distribution

■ **Question** *What is the median of the following distribution?*

7, 7, 7, 8, 8, 8, 9, 9, 10, 10

↑

Median

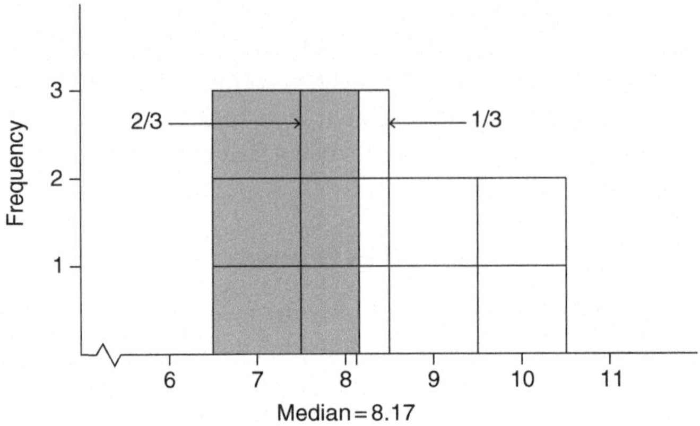

Figure 3.1 A visual representation of finding the median when identical scores are in the middle.

Solution

If discrete, the median = 8.00; if continuous, the median = 8.17.

It may not be immediately obvious why the type of variable matters and why one answer would be 8.17. Recall from Chapter 2 that discrete variables can take on only a finite number of values. No meaningful values exist between any two adjacent values. In situations like this, since the same number is found for the two values in the middle, the resulting median is simply that number.

For continuous variables, however, recall that every number is at the midpoint of an interval. (A person's claimed height of 72 in. merely means their height is somewhere between 71.5 and 72.5 in.) In our example, there are 10 values; therefore, the median is located such that there are 5 values on each side. Coming up from the bottom, the three 7's get us to within two values of the median. But we will need two more values to get us there. But those three 8's cannot be separated since they are stacked on top of each other. (Look at Figure 3.1.) We will need to split those three boxes identically so that a total of two of them fall to the lower half of the distribution and the remaining parts of the boxes fall to the upper half. If we drew a line down through the three boxes such that 2/3rds of each box were to the left and 1/3 to the right, that would "do the trick." What value is 2/3rds of the way from 7.5 to 8.5? Adding .67 to 7.5 gives us 8.17. ■

■ **Question** *What is the median of this distribution?*

7, 7, 7, 8, 8, 8, 8, 8, 9, 9, 10, 10

↑

Median

Solution

If discrete, the median = 8.00; if continuous, the median = 8.10.

Here is another example where, if the measure is continuous, we will need to split up the 8's. Three of them need to go to the lower half of the distribution and two of them to the higher half. The fairest way to get three of the five 8's would be to get 60% or .6 of each one. Since the 8's actually start at 7.5, the answer would be (7.5 + .60) 8.10. ■

3.6 The Mode

The third and final measure of centrality is the mode. Recall that the mean defines centrality as the point where all the deviation scores sum to 0 ($\Sigma x = 0$). The median defines centrality as the value which splits the bottom half of the scores from the top half. The **mode**, in turn, defines centrality as the most typical or most frequent score in the distribution. It is the easiest of all three measures to determine. All that is needed is to observe which score occurs most often.

■ **Question** *What is the mode of this distribution?*

100, 101, 105, 105, 107, 108

Solution 105 ■

Some distributions may have two scores that are most typical. Consider the frequency distribution in Table 3.2. Here, the distribution has two modes: 40 and 34. This is known as a **bimodal** distribution ("bi," meaning two). A distribution with a single mode is termed **unimodal** ("uni," meaning one). The graph of a bimodal

Table 3.2 A distribution with two modes: 40 and 34.

X	*f*
43	1
42	4
40	6
39	3
37	2
34	6
30	1

distribution has two distinct humps. The humps do not have to be the same exact height to be considered bimodal. In the rare case where all scores occur with the same frequency, there is no mode.

3.7 Distribution Shape and Measures of Central Tendency

Chapter 2 introduced us to the notion that distributions can assume different shapes. In fact, distributions can take on a wide variety of shapes, and there are names for some of them (platykurtic, leptokurtic, positively skewed, negatively skewed, normal, etc.). The particular shape of the distribution has implications for the relative position of the mean, median, and mode. If the distribution is symmetrical, then all three measures of central tendency will be identical. Figure 3.2 depicts this fact. Note that symmetry is what is important here, not kurtosis (peakedness).

There is one exception to this rule: A symmetrical bimodal distribution, as shown in Figure 3.3, has identical values for the mean and median, but not the mode. Also, note that in a bimodal distribution, the mean and median are not located amidst highly frequent scores. In this instance, the mean and median do not reflect where scores tend to bunch.

If a distribution is skewed, then the mean, median, and mode will all be different. Figure 3.4 shows two distributions: (a) negatively skewed and (b) positively skewed. In a negatively skewed distribution, the mean is to the left of the median. The scores in the elongated tail pull both the mean and the median to the left, but because the mean is based on deviation scores, it is pulled even more so. This same rationale holds for positively skewed distributions.

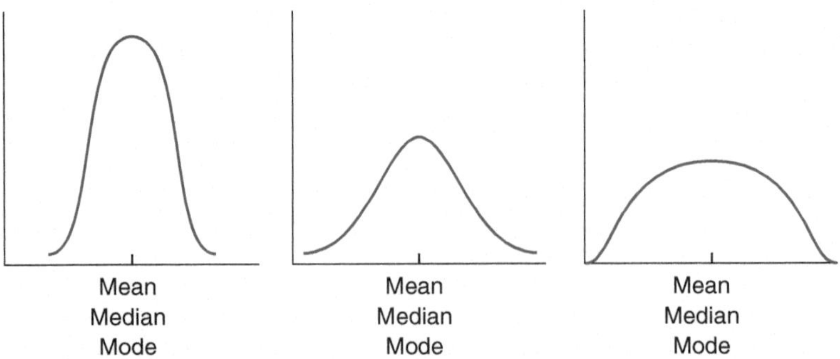

Figure 3.2 Three symmetrical distributions, each with the same mean, median, and mode.

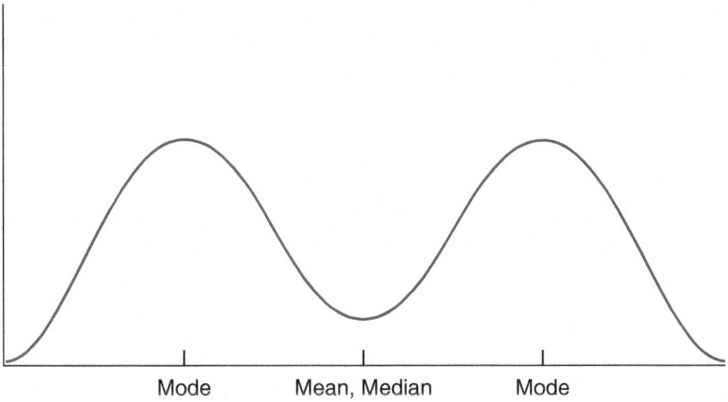

Figure 3.3 A symmetrical, bimodal distribution. The mean and median are the same but not where the scores tend to bunch.

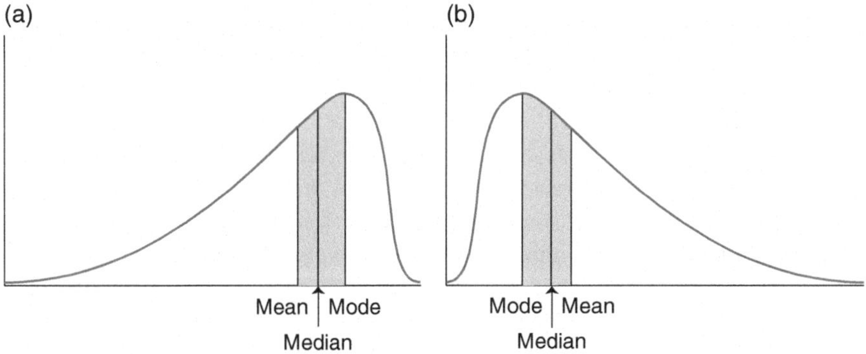

Figure 3.4 (a) The mean is to the left of the median in a negatively skewed distribution. (b) The mean is to the right of the median in a positively skewed distribution.

3.8 When to Use the Mean, Median, and Mode

Using the mean as an index of central tendency has several advantages. First, as mentioned earlier, the mean considers all scores in a distribution, including each score's interval distance from the center. For this reason, the mean usually captures centrality well. Second, as the size of a random sample grows, the sample mean becomes an increasingly more accurate estimate of the population mean. This will make more sense when we learn about sampling distributions (Chapter 7). Third, because of the preceding two reasons, the mean is used in many statistical formulas.

However, there are two disadvantages to using the mean. First, means have dubious meanings for data gathered from either nominal or ordinal scales. As a result, many statistical procedures which make use of the mean cannot be applied to data measured on these scales. (See Box 3.1 for more information about measures of central tendency and Likert scale data.) Second, the mean is sensitive to extreme scores; if present, they can produce a mean value that does not seem to be well centered. Sometimes, it only takes one extreme value, if the distribution set is small, to distort the mean. But even for more numerous distribution sets, if they are tremendously skewed (either positively or negatively), the mean may not appear to capture "the center" very well (see Figure 3.4). Medians are less influenced than means by extreme scores. For this reason, medians are oftentimes preferred as measures of centrality for skewed distributions.

Box 3.1 The Central Tendency of Likert Scales: The Great Debate

In Chapter 2, readers were introduced to a critical difference between ordinal scales and interval or ratio scales – namely, the nature of the relationship between numerical values. Ordinal scales are a quantitatively organized series of categories and, as such, make no assumptions about the quantitative distance between these categories. Interval and ratio scales hold the intervals constant throughout the measure. In this chapter, we learned that the concept of a deviation score is necessary to find a mean. A mean is defined as the point where the deviation scores sum to zero. Deviation scores, however, cannot be found for numbers on an ordinal scale; the intervals are not held constant. (It would be like suggesting first- and third-place finishers in a pie bake-off are equidistant from second. We have no reason to presume that.) Readers were also made aware (see Spotlight 2.1) of the ambiguity that surrounds how to interpret numbers generated by a Likert scale – scales that typically offer 5–11 options ranging from strongly disagree to strongly agree (usually with a neutral point in the middle) that are used to measure the amount of agreement people have with a given statement. The great debate question is this: Is it appropriate to generate means for Likert-scale data?

It is a very important question because if we decide the answer is "no," then we must restrict ourselves from using statistical tools that make use of the concept of the mean when analyzing Likert-scale data. We may not realize this yet, but most statistical tests utilize the mean concept. A few tests (called nonparametrics) do not. (For example, only Chapter 16 in this text is devoted to nonparametric tests.) Furthermore, nonparametrics are oftentimes less powerful tests (a concept we will explore in Chapter 10) and, therefore, less likely to help us find meaningful differences between groups. The stakes of this debate are indeed high.

The conservative approach is to argue that Likert scales have no way to determine constancy between values and should therefore be considered ordinal. As a result, data gathered using Likert scales must be analyzed nonparametrically, end of discussion. Others argue that the line between ordinal and interval is rather vague, some even calling it "fuzzy" (e.g. Abelson 1995). If, they argue, the data from a Likert scale take on the shape of a normal distribution and if there are a good number of options for the respondent to choose from, then the data can be considered "normal" or "sufficiently close" to normal and analyzed with more standard statistical techniques.

We do not aim to settle the debate here but merely raise it as an important issue. Perhaps it will be a good class discussion topic. As we think about this issue, keep in mind some recommendations made by Karen Grace-Martin, a specialist in data analysis. They are paraphrased below (Grace-Martin 2008):

1) Realize the difference between a Likert-type item and a Likert scale. A Likert scale can be made up of many Likert-type items. Collectively, these items attempt to provide a measure of the attitude in question. Many times, however, the term "Likert scale" is used to refer to a single item.
2) Proceed with caution. Look at the particulars of the Likert-scale data in question. Would treating it as interval data influence conclusions? The fact that everyone else is treating it as interval data is not sufficient justification in and of itself.
3) At a minimum, insist that (i) the item has at least nine points, (ii) the underlying concept be continuous, and (iii) there be some indication that the intervals between the values are approximately equal. Also, make sure the other statistical assumptions for the test are met.
4) When possible, run the nonparametric equivalent to the test. If the same results are found, confidence in the conclusions increases.
5) If Likert data are used in a parametric procedure, make sure the results are particularly strong before making a claim.
6) Consider the consequences of reporting inaccurate results. Is the analysis going to be published? Will it be used by others to make decisions?

The goal here is not to argue for a particular position but rather to help us develop an appreciation for the more subtle and yet important debatable issues related to data analysis.

Another problematic situation for the mean occurs when scores in a distribution are truncated. For instance, recall the study of pain tolerance using the technique of holding one's hand in a bucket of freezing water for as long as possible. Suppose one participant keeps their hand in the bucket for more than 300 seconds, well past all other participants. If the researcher, needing to move on to the next participant, records 300 seconds and asks the participant to remove their hand, the resulting mean value that includes that score will be artificially low. That score was truncated. The median, however, would be the same value whether that score was cut off at 300 or not. Since the median is not influenced by the value of extreme scores, it should be used when the distribution is skewed, truncated, or has inexact upper or lower cutoff scores.

The median is also the central tendency measure of choice for data drawn from an ordinal scale. Neither the ordinal scale nor the concept of the median makes any assumptions about the uniformity of the intervals between values.

The mode ignores all numbers in a distribution except the one score that occurs most often. For this reason, it is the least used measure of central tendency. The mode of a sample, for instance, is never used to infer the mode of a population. It does, however, have its place. The mode is used when one wants to convey the most typical score in a distribution, such as when students want to know what score on an exam was received by most students. The mode is the preferred measure to use when working with nominal data. Recall from Chapter 2 that a nominal scale merely distinguishes one kind of thing from another. Suppose we ask students at our university to name their favorite leisure activity. We report that 12% prefer gaming, 50% report tending to social media accounts, and 38% point to sporting activities. These three activities are categories on a nominal scale. There is no way to compute a mean or median, but the mode can be declared to be tending to social media accounts. Oftentimes, when the data are in the form of *how many* (i.e. a nominal scale), the mode is the appropriate measure of central tendency.

It should now be clear that each measure of central tendency has advantages and disadvantages. Bear in mind that when using descriptive statistics, the goal is to communicate the features of a distribution in as accurate a manner as possible. Further, there is no rule against reporting two or even all three measures of central tendency.

Summary

Descriptive statistics are numerical indices that summarize and communicate basic characteristics of a distribution. Statistical values that describe the characteristics of a population are called parameters; those that describe the characteristics of a sample are called statistics.

Values that communicate where scores center in the distribution are called measures of central tendency. The mean is the most important and most often used measure of central tendency. Not only can it be used as a descriptive index of central tendency, but the mean is frequently used in formulas designed to test experimental hypotheses. The degree to which a score deviates from the mean is $X - M$. This deviation amount can be called a deviation score (or error score). The mean is the value at which all the deviation scores sum to 0.

The mean has several advantages. First, it considers not only all the scores in a distribution but also their precise distance from the middle. Second, as distribution size increases, the mean becomes a stable measure of central tendency, serving as a good estimate of the population mean. Third, it is used in many statistical formulas. However, it can only be used on interval or ratio data. Also, being sensitive to extreme scores, it may not be a good measure of centrality for skewed distributions or if the data has been truncated or limited by the nature of the measuring instrument.

The median is the value which splits the bottom half of scores from the top. Since the median is less affected by extreme scores, it should be used when the distribution is skewed, truncated, or has scale-limited upper or lower cutoff scores. The median is also the appropriate measure for central tendency when the values in the distribution come from an ordinal scale.

The mode is defined as the most typical or most frequent score. It is the least used measure of central tendency. The mode ignores all the numbers in a distribution except the one value that occurs most often. On the other hand, the mode is the only measure of central tendency to use when evaluating scores measured on a nominal scale.

Distribution shape has implications for the location of the mean, median, and mode. If the distribution is unimodal and symmetrical, then all three measures of centrality will be identical. With skewed distributions, both the mean and median are pulled to the side of the elongated tail; the mean, however, is pulled farther.

Microsoft® Excel and SPSS®

Information regarding how to use Microsoft® Excel and SPSS® to assist in calculating various descriptive statistics can be found in Appendix C.

Key Formulas

Population mean, μ

$$\mu = \frac{\Sigma X}{N} \quad (3.1)$$

Sample mean, M

$$M = \frac{\Sigma X}{n} \quad (3.2)$$

Key Terms

Central tendency
Parameters
Statistics
Mean
Deviation score (error score)
Median
Mode
Bimodal
Unimodal

Exercises

1 Think of two examples where a mean is typically used but may not be the best measure of central tendency. Defend these examples.

2 Think of two examples of variables that may be distributed in such a way that they might take on a bimodal shape. Defend these choices.

3 Given a distribution in which the mean = median, what is almost certainly true about the shape of the distribution?

4 If a population mean = 15, how will the mean change if it is later claimed to be a sample?

Work Problems

5 Find the deviation scores for the following raw scores in a distribution that has a mean of 10.
a 12
b 9
c 0
d 10
e −1
f 9.5

6 Identify the mean, median, and mode of these six distributions.
a 3, 3, 4, 5, 6, 8, 8, 8, 9
b 2, 4, 4, 4, 6, 7, 7
c 7, 7, 8, 9, 10, 10, 10
d 1, 1, 3, 4, 4, 5, 9
e 1, 4, 6, 7, 8, 8

f 9, 11, 6, 8, 12, 15, 3, 5, 5

7 What is the median of this distribution? 4, 5, 7, 7, 7, 7, 9, 10

8 What is the median of this distribution? 13, 16, 18, 19, 19, 19, 22, 26

9 What is the likely shape of each of these distributions?
a $M = 24$; median = 16; mode = 12
b $M = 123$; median = 143; mode = 150
c $M = 6$; median = 6; mode = 6
d $M = 19$; median = 19; mode = 9 and 29
e $M = 56$; median = 66; mode = 70
f M = 48; median = 36; mode = 32

10 A school psychologist obtains the following sample of IQ scores from a local high school. What are (a) the mean and (b) the median? (c) Is there a mode?

98	111	101	100	99
99	123	100	134	101
96	102	102	101	105

11 Which measure has the most difficulty with extreme scores? Why?

12 Which measure is the best to use for ordinal data? Why?

13 Which measure is the best to use for nominal data? Why?

14 A population of scores includes 10 numbers ($N = 10$) and has a mean of 100. One of the scores is changed from an 80 to a 90. What is the value of the new mean?

15 A sample of $n = 6$ scores has a mean of 25. If one score with a value of 15 is removed, what is the value for the new mean?

16 Three friends sampled students at their university to measure daily time spent on social media. One asked 25 people and found an *M* of 45 minutes, another asked 50 people and found an *M* of 65 minutes, and a third asked 500 people with an *M* of 52 minutes. All three made efforts to sample randomly. What is our best guess of the actual population mean?

17 Below is a hypothetical sample of IQ scores from an incoming freshman class. Compute the mean, median, and mode. What does the relationship between these measures of centrality suggest about the shape of the distribution?

100	100	102	135	143	94	120	114	111	87	95
109	82	94	142	100	97	100	100	101	99	98
167	176	154	100	85	88	124	180	90	96	92
149	103	102	101	104	92	103	103	105	99	92

4

Measures of Variability

4.1 The Importance of Measures of Variability

Chapter 3 discussed measures of central tendency: the mean, median, and mode. Although the concept of "middle" is crucial to the description of a distribution, it is only part of the picture. Measures of central tendency do not tell us how scores are dispersed around a middle. Even distributions with the same mean can look very different. For example, although centered on the same value, Figure 4.1 depicts two very differently shaped distributions.

To complement measures of centrality, there is a need for statistical techniques which convey how spread out and dispersed scores are around a central point. Measures that reflect the variation within a distribution are called **measures of variability** (or **dispersion**). Several of these measures, along with their advantages and disadvantages, will be presented and discussed in this chapter.

4.2 The Range

The **range** reflects the overall span of the scores in a distribution, from the lowest value to the highest. It is easily calculated by subtracting the lowest score in a distribution from the highest.

Range

$$Range = X_H - X_L \quad (4.1)$$

where

X_H = highest score in the distribution
X_L = lowest score in the distribution

Essentials of Behavioral and Social Science Statistics, Third Edition.
K. Paul Nesselroade, Jr. and Laurence G. Grimm.
© 2025 John Wiley & Sons, Inc. Published 2025 by John Wiley & Sons, Inc.
Companion website: https://www.wiley.com/go/Essentials_of_Behavioral_and_Social_Science_Statistics3e

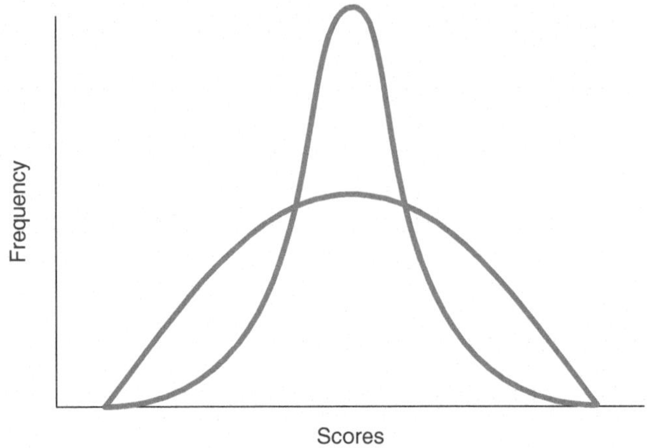

Figure 4.1 Two distributions with different variabilities yet having the same middle point.

■ **Question** *What is the range of this distribution?*

17, 44, 50, 23, 42

Solution 50 − 17 = 33 (organize the data so the lowest and highest scores can be identified) ■

The next worked problem illustrates one of the main issues that can arise when using the range as a measure of variability.

■ **Question** *What is the range of this distribution?*

2, 4, 5, 7, 9, 34

Solution 32

In this situation, the range gives a misleading impression of dispersion. Most of the scores are close together, but one is quite discrepant from the others (34). The extreme score generates a range that misleadingly suggests the distribution is rather spread out. If we are going to use a measure of dispersion that reflects the span of scores, then it would be helpful if we could use a measure that is less affected by extreme scores that lie at either end of the distribution. ■

4.2.1 The Interquartile Range and Semi-Interquartile Range

Distributions can be divided into different equal sections. One method uses quartiles. A **quartile** is one-fourth of a distribution of scores. The bottom 25% of the values in a distribution make up the first quartile. The second quartile includes the next 25% of scores up to the median. The third quartile runs from the median to the score, marking the bottom 75%. The upper 25% of scores comprise the fourth quartile.

A **percentile** identifies the percentage of scores that fall below a given value. In other words, the 20th percentile is the value at which 20% of the distribution's scores fall below. The first quartile ends at the 25th percentile, the second quartile ends at the 50th percentile, and so on.

The **interquartile range (IQR)** measures the width of the middle half of a distribution. As stated in terms of percentiles, the *IQR* is the span of scores between the 25th percentile (Q_1 marks the end of the first quartile) and the 75th percentile (Q_3 marks the end of the third quartile). This measure effectively lops off the upper and lower 25% of the distribution (see Figure 4.2). By removing the outermost quartiles, the *IQR* eliminates the problem of extreme scores by only focusing on the middle half of the distribution.

Interquartile range, IQR

$IQR = Q_3 - Q_1$ (4.2)

---

where

Q_3 = the third quartile (75th percentile)
Q_1 = the first quartile (25th percentile)

The **semi-interquartile range (SIQR)** is the *IQR* divided by 2.

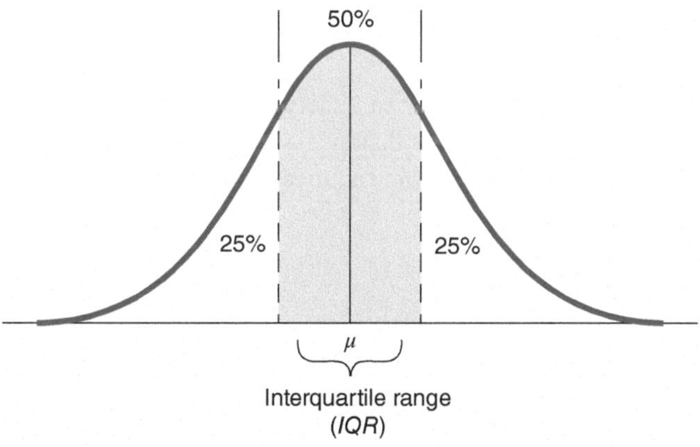

Figure 4.2 The interquartile range (*IQR*) spans the width of the middle 50% of the distribution.

Semi-interquartile range, SIQR

$$SIQR = \frac{Q_3 - Q_1}{2} \quad (4.3)$$

Like the *IQR*, the *SIQR* is more stable than the simple range because it is unaffected by an extreme score. Although neither the *IQR* nor the *SIQR* is affected by a single extreme score, they are influenced by distributions with numerous extreme scores, like skewed distributions.

■ **Question** *The hypothetical "Highbridge Aptitude Test" has a median of 100. The score at the 75th percentile is 130, and the score at the 25th percentile is 70. What are the* IQR *and* SIQR?

Solution

$$IQR = Q_3 - Q_1 = 130 - 70 = \mathbf{60}$$

$$SIQR = \frac{Q_3 - Q_1}{2} = \frac{60}{2} = \mathbf{30} \; ■$$

The range and related concepts are helpful measures of variability. For instance, the *SIQR* is commonly included when describing the distribution characteristics of standardized psychological or intellectual tests (e.g. the Wechsler Adult Intelligence Scale). (Recall that many psychological concepts, like intelligence, use scales that have both ordinal-like and interval-like properties.) The *IQR* and *SIQR* count every number but do not speculate on the quantitative space in between them. A more precise measure would consider the specific distance between the mean and every score.

4.3 The Mean Deviation

Each raw score in a distribution of interval or ratio scores sits at some distance from the mean. In Chapter 3, we labeled this distance a *deviation*, or *error*. The degree to which distribution scores deviate from the mean is a measure of variability. Consider these two distributions:

$$\text{Distribution } A\text{:} \quad 11, 12, 13, 14, 15, 16, 17 \qquad \mu = 14$$

$$\text{Distribution } B\text{:} \quad 5, 8, 11, 14, 17, 20, 23 \qquad \mu = 14$$

The mean of each distribution is 14. However, scores in *B* are more spread out than scores in *A*. In other words, Distribution *B* has an overall greater amount of deviation among the scores. But how can we measure overall deviation? We cannot simply sum the deviation scores of each distribution because Σx always $= 0$.

The negative deviation scores from values sitting below the mean always *balance* the positive deviation scores from those sitting above the mean. However, taking the *absolute value* of each deviation score will remove the negative signs and solve this problem. Taking the average of the absolute values of all deviation scores will give us a measure of variability. This arithmetic manipulation is called the **mean deviation**. Formulas 4.4 and 4.5 and the question that follows show how we can calculate this measure of dispersion.

Mean deviation for population, *MD*

$$MD = \frac{\Sigma\,|X-\mu\,|}{N} \quad (4.4)$$

Mean deviation for sample, *MD*

$$MD = \frac{\Sigma\,|X-M\,|}{n} \quad (4.5)$$

■ **Question** *Find the mean deviations for the two preceding distributions with the same mean?*

Distribution *A*				Distribution *B*			
Scores	μ	$(X-\mu)$	$\|X-\mu\|$	*Scores*	μ	$(X-\mu)$	$\|X-\mu\|$
11	14	−3	3	5	14	−9	9
12	14	−2	2	8	14	−6	6
13	14	−1	1	11	14	−3	3
14	14	0	0	14	14	0	0
15	14	1	1	17	14	3	3
16	14	2	2	20	14	6	6
17	14	3	3	23	14	9	9
$N=7$			$\Sigma\|X-\mu\| = 12$	$N=7$			$\Sigma\|X-\mu\| = 36$

Solution

$$MD_A = \frac{\Sigma\,|X-\mu\,|}{N} = \frac{12}{7} = 1.71 \quad MD_B = \frac{\Sigma\,|X-\mu\,|}{N} = \frac{36}{7} = 5.14$$ ■

As we can see in the examples above, smaller deviation values reflect tighter distributions, and larger deviation values reflect more dispersed distributions. Any formula that uses deviation scores as a measure of variability has the advantage of using the *magnitude* of the difference between each score and the mean in its calculations, unlike the range, *IQR*, and *SIQR*, which only use the relative position of scores. And yet, the mean deviation has some adverse properties that preclude it from being used in most statistical formulas.[1] What is needed is a formula that capitalizes on the conceptual basis of deviation scores but does not have the disadvantages of the mean deviation. Is there some other way to transform deviation scores to keep them from summing to zero?

4.4 The Variance

Another way to remove negative signs when summing deviation scores is to square them. This allows us to calculate the variance. The **variance** is defined as the average squared deviation score; it is symbolized as σ^2 (pronounced "sigma squared"; σ is the Greek lowercase of Σ) for population variances and s^2 for sample variances. Please note that the two formulas are slightly different.

Population variance, σ^2

$$\sigma^2 = \frac{\Sigma(X-\mu)^2}{N} \quad (4.6)$$

---

where

σ^2 = the symbol for the population variance
X = a raw score
μ = the population mean
N = the number of scores in the population

Sample variance

$$s^2 = \frac{\Sigma(X-M)^2}{n-1} \quad (4.7)$$

---

where

s^2 = the symbol for sample variance
M = the sample mean
n = the number of scores in the sample

---

1 The main problem with the sample mean deviation is using it to estimate the variability of a population. It does not bear a consistent relation to the mean deviation of the population from which it was drawn.

4.4.1 Sample Statistics as Unbiased Estimates of Population Parameters

Recall that a sample is a subset of scores drawn from a population. Researchers are interested in the characteristics of populations; samples are often used to make inferences about populations. Suppose we want to know the mean of a population, but we only have a sample. The best estimate of the population mean is the sample mean. Usually, the estimate will not be correct; rarely is the sample mean identical to the population mean. But the sample mean is just as likely to be larger than the population mean as it is to be smaller. This makes the sample mean an *unbiased* estimate.

In comparing the two variance formulas, note that the sample variance denominator is $n - 1$, instead of n (compare Formulas 4.6 and 4.7). This is a correction factor designed to adjust for a bias that occurs when sample variances are used to estimate population variances. Dividing by $n - 1$ produces a sample variance that is an *unbiased* estimate of a population variance just as likely to overestimate as to underestimate.

Finally, the error associated with sample statistics as estimates of population values decreases as the size of the sample increases. As has been mentioned before, larger samples yield more accurate estimates of population parameters.

■ **Question** *What is the variance of this sample of scores?*

3, 4, 6, 8, 9

Solution

z	*M*	*X* − *M*	(*X* − *M*)²
3	6	−3	9
4	6	−2	4
6	6	0	0
8	6	2	4
9	6	3	9
0	0	0	26

$$s^2 = \frac{\Sigma(X - M)^2}{n - 1} = \frac{26}{4} = 6.5 \ \blacksquare$$

4.4.2 Equivalent Formulas for the Variance

If we were to take a random sample of introductory-level statistics textbooks and turn to the chapters covering measures of variability, we would be surprised, and maybe confused, by the many different formulas that can be used to compute the variance of a distribution. The formulas yield the same answer (provided we note the distinction between a population and sample), but they serve different purposes. Some formulas are easier to use when performing hand calculations with raw data, and others are helpful in reminding us of the conceptual basis of variance. For example, Formulas 4.6 and 4.7 are the basic formulas for the variances of the population and sample. They are called *definitional formulas* since, in reading them, we can be reminded of the concept behind the measurement: the average squared deviation score. In Chapter 3, we learned that a single score minus the mean, $X-\mu$, could be expressed as x, ("little x") a deviation score. Therefore, Formulas 4.6 and 4.7 can be rewritten as Formulas 4.8 and 4.9, respectively. These are called the deviation score formulas.

The deviation score formulas

Population variance

$$\sigma^2 = \frac{\Sigma x^2}{N} \quad (4.8)$$

Sample variance

$$s^2 = \frac{\Sigma x^2}{n-1} \quad (4.9)$$

The numerators of both sets of variance formulas direct us to sum all the squared deviation scores. For this reason, the numerator of a variance formula is referred to as the *sum of squares* (or ***SS***). Hence, $SS = \Sigma(X-\mu)^2$ or $\Sigma(X-M)^2 = \Sigma x^2$. Substituting SS in the numerator of the population and sample formulas defines two additional ways variance formulas can be represented.

The sum of squares (*SS*) formulas

Population variance

$$\sigma^2 = \frac{SS}{N} \quad (4.10)$$

Sample variance

$$s^2 = \frac{SS}{n-1} \quad (4.11)$$

When working with raw scores, a *computational* (or *raw score*) *formula* eases the calculation task. Formulas 4.12 and 4.13 are used to *compute* the population and sample variances. (Yes, they look more involved, but they are easier to use when hand calculating, especially as the sample size grows.) When using a computational formula, pay close attention to the difference between ΣX^2 and $(\Sigma X)^2$. The ΣX^2 is found by first squaring each raw score and then summing up all the squared values. The quantity $(\Sigma X)^2$ requires that we first sum the raw scores and then square the sum. This algebraic distinction is a frequent component in hand calculations of statistical values.

If calculating by hand, it is recommended to create two columns, one containing the raw data (labeled X) and the other containing the square of each raw number (labeled X^2). Simply sum up both columns. The sum of the raw score column is (ΣX); by squaring this value we will get $(\Sigma X)^2$. The sum of the squared column is ΣX^2.

The computational formulas

Population variance

$$\sigma^2 = \frac{\Sigma X^2 - \left[(\Sigma X)^2/N\right]}{N} \quad (4.12)$$

Sample variance

$$s^2 = \frac{\Sigma X^2 - \left[(\Sigma X)^2/n\right]}{n-1} \quad (4.13)$$

Keep in mind that all sample formulas lead to the same answer, with any discrepancies accounted for by rounding errors. Of course, all population formulas also yield the same answer. Table 4.1 presents all of the formulas for the variance.

Table 4.1 Several equivalent expressions of the population and sample variances.

Variance formulas	
Population variance	**Sample variance**
Definitional formulas	
$\sigma^2 = \frac{\Sigma(X-\mu)^2}{N}$ (4.6)	$s^2 = \frac{\Sigma(X-M)^2}{n-1}$ (4.7)
Deviation score formulas	
$\sigma^2 = \frac{\Sigma x^2}{N}$ (4.8)	$s^2 = \frac{\Sigma x^2}{n-1}$ (4.9)
Sum of squares formulas	
$\sigma^2 = \frac{SS}{N}$ (4.10)	$s^2 = \frac{SS}{n-1}$ (4.11)
Computational formulas[a]	
$\sigma^2 = \frac{\Sigma X^2 - \left[(\Sigma X)^2/N\right]}{N}$ (4.12)	$s^2 = \frac{\Sigma X^2 - \left[(\Sigma X)^2/n\right]}{n-1}$ (4.13)

[a] Use these two formulas when working from raw data and calculating by hand.

■ **Question** *Use the computational formulas to determine the variance of this distribution, first treating it as a sample and then as a population.*

X	X^2
2	4
4	16
5	25
7	49
9	81
$\Sigma X = 27$	$\Sigma X^2 = 175$

Solution

Sample Computational Formula

$$s^2 = \frac{\Sigma X^2 - [(\Sigma X)^2/n]}{n-1} = \frac{175 - \left[(27)^2/5\right]}{5-1} = \mathbf{7.30}$$

As a sample, the variance is 7.30.

Population Computational Formula

$$\sigma^2 = \frac{\Sigma X^2 - [(\Sigma X)^2/N]}{N} = \frac{175 - \left[(27)^2/5\right]}{5} = \mathbf{5.84}$$

As a population, the variance is 5.84. ■

Sometimes an investigator will learn something important about a phenomenon when the dispersion of scores is examined. Box 4.1 presents a finding in which the variability of scores reflects an interesting aspect of aging.

Box 4.1 The Substantive Importance of the Variance

In experimental research, comparisons are typically made between the *means* of two conditions. However, sometimes between-group differences in *variability* are important as well. They reveal an important facet of the phenomenon under investigation. An example in which variability has substantive importance comes from the literature on aging. Chronological age is intrinsically a poor predictor of almost any measure of psychological functioning (Woods and Rusin 1988). However, as an investigator compares different age groups, they would find that the within-group variability, on several cognitive and physiological measures, increases with age (Krauss 1980). In other words, older people are more unlike each other than younger people; their distributions are more spread out on a whole host of psychological measures compared with the distributions of those younger. As a result, researchers investigating questions related to the aged need to pay more attention to individual differences as the age of their samples increases.

4.5 The Standard Deviation

The variance measure of dispersion is especially important because it is used in many statistical formulas. However, it is not the best measure when we want to communicate variability to others. This is because the variance is a squared value; it is not stated in the original units of the measured variable. The value seems inflated when compared with the raw scores or other descriptive statistics like the mean. For example, if someone told us that a sample mean of IQs was found to be 100 with a variance of 225, we might wonder if there was a small, medium, or large amount of variation among the scores. To resolve this, we can take the square root of the variance; this is called the **standard deviation**. The definitional, deviation score, SS, and computational formulas for the standard deviation are identical to the formulas for the variance, *with the exception that the formulas for the standard deviation are under square root signs.* These two important measures of variability are very similar. If we have one, we are one mathematical step away from the other. The symbols for the population and sample standard deviation are σ and s, respectively; this makes sense given that the symbols for the population and sample variance are σ^2 and s^2, respectively. Using the distribution in the preceding worked example, the sample and population standard deviations would be

$$s = \sqrt{7.30} = \mathbf{2.70} \quad and \quad \sigma = \sqrt{5.84} = \mathbf{2.42}$$

Table 4.2 offers the same formulas found in Table 4.1. Here, however, each formula is under a square root sign, thereby making them standard deviation formulas.

Table 4.2 Several equivalent expressions of the population and sample standard deviations.

Standard deviation formulas		
Population standard deviation		**Sample standard deviation**
	Definitional formulas	
$\sigma = \sqrt{\frac{\Sigma(X-\mu)^2}{N}}$		$s = \sqrt{\frac{\sum(X-M)^2}{n-1}}$
	Deviation score formulas	
$\sigma = \sqrt{\frac{\Sigma x^2}{N}}$		$s = \sqrt{\frac{\Sigma x^2}{n-1}}$
	Sum of squares formula	
$\sigma = \sqrt{\frac{SS}{N}}$		$s = \sqrt{\frac{SS}{n-1}}$
	Computational formulas[a]	
$\sigma = \sqrt{\frac{\Sigma X^2 - [(\Sigma X)^2/N]}{N}}$		$s = \sqrt{\frac{\Sigma X^2 - [(\Sigma X)^2/n]}{n-1}}$

[a] Use these two formulas when working from raw data and calculating by hand.

4.5.1 The Standard Deviation and the Normal Curve

In Chapter 5, much more will be said about the characteristics and importance of the normal distribution. For now, recall that a normal distribution is symmetrical and bell-shaped. When depicted on a graph, a normal distribution is called a normal curve. In a normal curve, approximately 68% of the scores will fall between one standard deviation below and one standard deviation above the mean. For instance, if a set of IQ scores has a mean of 100 and a standard deviation of 15, then approximately 68% of the scores in that distribution fall between 85 (100 − 15) and 115 (100 + 15). Furthermore, approximately 95% of the scores will fall between plus and minus two standard deviations from the mean. Finally, nearly all the scores in a distribution (approximately 99.7%) will fall within plus and minus three standard deviations of the mean. See Figure 4.3 for a visual depiction of what has been termed the **68-95-99.7 rule**. This figure displays how the area under a normal curve is allocated probabilistically. For instance, the percentage of a normal curve that is between the mean and a score that is two standard deviations below the mean is approximately (34% + 13.5%) 47.5%. Chapter 5 will further develop this idea.

Please note the standard deviation, or any other measure of variability, can never be a negative number. Variation is a measure of distance, and there is no such

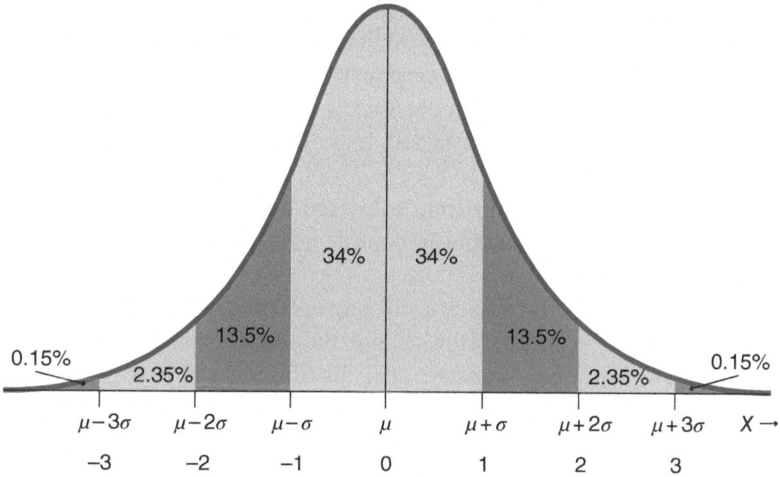

Figure 4.3 The relationship between standard deviations and probability for a normal curve.

Box 4.2 The Origins of the Standard Deviation

Karl Pearson (featured in Spotlight 14.1) proposed the standard deviation as a measure of dispersion in 1894. Before Pearson, statisticians used a closely related index of variability: the *probable error* (*pe*). The formula for the *pe* is:

$$pe = 0.6745\sqrt{\frac{\Sigma(X-\mu)^2}{N}}$$

The mean, ±1 *pe*, includes 50% of the scores of the distribution. (The *pe* concept is similar to the interquartile range.) Pearson believed that multiplying by .6475 was unnecessary. Moreover, as a measure of variability, there was no compelling advantage to the *pe*; both the standard deviation and *pe* reflect variability equally well. Pearson dropped the multiplier, named the new measure the *standard deviation*, and symbolized it as σ.

thing as negative distance. (See Box 4.2 for a bit more information about the origins of the standard deviation concept.)

4.6 Simple Transformations of the Mean and Variance

To make numbers easier to work with, researchers sometimes transform them by adding, subtracting, multiplying, or dividing all scores in a dataset by the same value. Performing simple arithmetic operations on a dataset has predictable effects on the mean and variance:

1) Adding, subtracting, multiplying, or dividing every score in a dataset by a constant will change the *mean* by the value and associated operation of that constant. For example, if the original mean is 100, adding 10 to every score will change it to 110, subtracting 10 from every score will change it to 90, multiplying every score by 2 will change it to 200, and dividing every score by 5 will change it to 20.
2) Adding or subtracting a constant from every score in a distribution *will not* influence the variance; these transformations do not alter the relative positions among the numbers.
3) Multiplying or dividing each score by a constant *will* change the variance exponentially. Multiplication and division change the relative spacing among the numbers. If multiplying by a constant, the distribution will spread out; the new variance will be the old variance multiplied by the constant squared. If dividing by a constant, the distribution will tighten up; the new variance will be the old variance divided by the constant squared. For example, if a distribution has a variance of 20, multiplying every number by 2 will give us a new variance of (20×2^2) 80, and dividing every number by 2 will give us a new variance of ($20/2^2$) 5. (See Table 4.3 for a worked example of these different transformations.)

Table 4.3 The effect of simple transformations on means and variances.

Original scores	+100	−100	×100	÷100
$\mu = 105.5$	205.50	5.50	10 550	1.055
$\sigma^2 = 9.67$	9.67	9.67	96 700	0.000 967

4.7 Deciding Which Measure of Variability to Use

Just as we looked at the strengths and weaknesses of the different measures of central tendency in Chapter 3, here we will look at the strengths and weaknesses of different measures of variability.

4.7.1 Extreme Scores

The presence of extreme scores in a distribution, depending on the degree of extremity and the percentage of scores considered extreme, can affect most of the measures of variability. Researchers often look at extreme scores with some degree of suspicion, wondering whether they are an accurate measurement.

If they are not accurate, then any measure of variability that is influenced by extreme scores will convey a misleading impression of the actual dispersion of scores in the distribution. The range is clearly the statistic that is most vulnerable to extreme scores. After all, only the highest and lowest scores of a distribution are used to compute it; one extreme erroneous score can lead to a very inaccurate view of dispersion.

The *IQR* and *SIQR*, however, are not much influenced by a *small* number of extreme scores, thereby offering a more reasonable statement of variability when extreme scores are found in a distribution. The variance and standard deviation are also affected by extreme scores. Since both measures use squared deviations, an extreme score that is a great distance from the mean will have a disproportionate effect on the variance, especially for small datasets. Caution is advised whenever measuring the variability of a dataset containing extreme scores.

Sometimes researchers, who suspect an extreme score is erroneous, consider discarding that score so as to generate a more accurate measure of variability. However, what if there are several extreme scores, as in a skewed distribution? There is no justification for discarding several scores. In these instances, the variability of a distribution is best described by the *IQR* or *SIQR* statistic.

4.7.2 An Arbitrary End Point to the Distribution

Recall the example mentioned in Chapter 3 about self-control and pain tolerance. If a participant can keep their hand submerged in ice-cold water for an exceptionally long time, a researcher might decide to terminate the task at an arbitrary end point (say, 300 seconds). Another example might involve a problem-solving task. What score should be assigned to someone who cannot figure out the answer? At some point, the researcher must stop them and assign a score. These situations present a problem for variability measures. When high or low scores in a dataset are arbitrary, any measure of variability based on those scores underrepresents the actual variability and thus misrepresents the concept. The *IQR* and *SIQR*, however, are relatively impervious to arbitrary cutoffs at the ends of distributions, provided there are only a few of these scores.

4.7.3 Common Practice

In common practice, it is rather rare for a researcher in the social or behavioral sciences to report the *IQR* or *SIQR*. Investigators in these disciplines simply do not have an "intuitive feel" for these measures. The range, despite all its vulnerabilities, is much more likely to be identified than the *IQR* or *SIQR*. However, although often presented, it is typically only used for descriptive purposes. It is simply noted. The variance, despite its essential role in statistical formulas, is also rarely reported. The squaring involved in the calculation makes it difficult for many to interpret.

If the data are normally distributed, by far, the most reported measure of variability is the standard deviation. Articles in scientific journals often include a table of means and standard deviations. We will rarely see a table of means and variances or other measures of variability.

Although the standard deviation is the most popular variability measure, we should not ignore the other measures. Indeed, researchers may err in relying too much on the standard deviation. There is no rule in the social or behavioral sciences requiring only one measure of variability. If it would be helpful, more than one measure of variability or central tendency can be reported. Methodological choices can also impact measures of variability. See Box 4.3 for more information.

Box 4.3 Is the Scientific Method Broken? Demand Characteristics

Throughout the text, a series of boxes ask whether the scientific method is broken in light of the replication problem currently plaguing the social, behavioral, and medical sciences. In this box, we will explore some of the problems that occur in the data-gathering process.

Sometimes researchers, even those with the best of intentions and who take precautions to remove their own biases, influence the responses of participants. In general, these are referred to as "demand characteristics" or "experimenter effects." Sometimes, the unintentional involvement of a researcher can lead to a response that would ordinarily be quite varied and shrink this variability down to almost nothing. Perhaps the most famous historical example of this concept is "Clever Hans," the horse that could do mathematics (Pfungst 1911). Hans was owned by Wilhelm von Osten, a German schoolteacher who claimed to have taught his horse to add, subtract, multiply, divide, and work with fractions. Hans answered questions by tapping with his hoof; his accuracy, though not perfect, was remarkable; so remarkable that the initial 1904 investigation by a team of academics concluded that it was not a trick. Three years later, however, a local psychologist, Oskar Pfungst, concluded that Hans was indeed clever, but was not doing mathematics. Instead, Hans had learned both a "start" and "stop" cue. The "start" cue was the act of being addressed by von Osten, who then looked down at Hans' hooves. As Hans approached the proper answer, von Osten, believing Hans was about to stop stomping, would make subtle straightening movements of the body and head. This served as the "stop" cue for the clever horse; his reward of food would be forthcoming. Hans usually landed on the right answer or very close to it. The normal variation of a horse stomping the ground was now being unintentionally managed by cues from the handler, such that the variance of stomps less than or greater than the "right" number shrank dramatically.

The same problem can take place when researchers are gathering data from other people. Numerous studies suggest that the hopes, expectations, and fears of data gatherers can subtly influence people's responses (e.g. Nichols and Maner 2008; Rosenthal and Fode 1963; Rubin et al. 2010), taking what might ordinarily be a diffuse set of responses and pulling them tightly together around the researchers desired or expected response.

Demand characteristics are potential problems for many of the topics that social and behavioral scientists investigate. Those of us in these fields of study need to take this human tendency into careful consideration when gathering and interpreting data.

Summary

Statistics that reflect the amount of variation in a distribution set are called measures of variability or dispersion. The range, calculated by subtracting the lowest score from the highest score in a distribution, defines the overall span of scores. Unfortunately, it is highly sensitive to extreme scores. The *IQR* avoids this problem by measuring the width of the middle half of the distribution. The *SIQR* is the IQR divided by 2.

Another family of measures utilizes the magnitude of the difference between each raw score and the mean, the deviation score. For instance, the *mean deviation* is the average of the absolute values of all deviation scores. The average of the squared deviation scores is called the *variance*. The formula for the sample variance differs slightly from the population formula. A correction factor is added to make the sample variance an unbiased estimate of the population variance.

The variance is particularly important because it is used in many statistical formulas. However, because it is a squared value, it is not the best descriptive measure of variability. The standard deviation is the square root of the variance. This is attractive as a descriptive measure since it converts the value back into the original units of the measured variable.

If the scores are normally distributed, the standard deviation can be used to analyze a dataset in terms of probability. The 68-95-99.7 rule states that the mean, plus and minus one standard deviation, encompasses roughly 68% of the total number of scores in a distribution; plus and minus two standard deviations include approximately 95% of the total number of scores; and plus and minus three standard deviations comprise virtually all scores (99.7%), given a normal distribution.

Transforming the scores of a distribution by the four basic arithmetic operations will have a predictable effect on the mean and variance. Adding, subtracting, multiplying, or dividing by a constant will alter the mean by the value and associated operation of that constant. Multiplication or division will also alter the variance by the constant squared or square rooted, respectively.

Deciding the most appropriate measure of variability to use depends on various features of the distribution. Factors such as extreme scores, arbitrary end points of a distribution, and common practices will influence a researcher's decision as to which measure of variability is most desirable.

Microsoft® Excel and SPSS®

Information regarding how to use Microsoft® Excel and SPSS® to assist in calculating various descriptive statistics can be found in Appendix C.

Key Formulas

Range

$$Range = X_H - X_L \quad (4.1)$$

Interquartile range, IQR

$$IQR = Q_3 - Q_1 \quad (4.2)$$

Semi-interquartile range, SIQR

$$SIQR = \frac{Q_3 - Q_1}{2} \quad (4.3)$$

Mean deviation formulas, MD

Population mean deviation, MD

$$MD = \frac{\Sigma\,|X - \mu|}{N} \quad (4.4)$$

Sample Mean Deviation, MD

$$MD = \frac{\Sigma\,|X - M|}{n} \quad (4.5)$$

The definitional formulas for the variance

Population variance, σ^2

$$\sigma^2 = \frac{\Sigma(X - \mu)^2}{N} \quad (4.6)$$

Sample variance, s^2

$$s^2 = \frac{\Sigma(X-M)^2}{n-1} \quad (4.7)$$

The computational formulas for the variance

Population variance

$$\sigma^2 = \frac{\Sigma X^2 - [(\Sigma X)^2/N]}{N} \quad (4.12)$$

Sample variance

$$s^2 = \frac{\Sigma X^2 - [(\Sigma X)^2/n]}{n-1} \quad (4.13)$$

Standard deviation formulas are the square root of the corresponding variance formula.

Key Terms

Measures of variability (or dispersion)
Range
Quartile
Percentile
Interquartile range
Semi-interquartile range
Mean deviation
Variance
Standard deviation
68-95-99.7 rule

Exercises

1 Given two samples, one with $n = 36$, the other with $n = 60$, which distribution will have the larger variance? Which variance is likely to better estimate the population variance?

2 Assume two samples: $M = 78$ and $M = 155$. Which sample would have the larger variance?

3 Why is the formula for a sample variance different from the formula for a population variance?

4 What does a distribution with a mean of 50 and standard deviation of zero look like?

5 What is the definition of the variance?

6 For each situation, specify whether we should use s or σ.

a A coach is interested in the variability of their basketball team's scores over the season.

b A clinician gathers a sample to evaluate a new therapy for anxiety.

c A teacher gives feedback to the class about their test performance.

d A manufacturer samples light bulbs to estimate life span variability.

7 A negatively skewed distribution has a mean of 500 and a standard deviation of 100. Can we determine what percentage of scores fall between 400 and 600? If so, what is it?

8 What is the main disadvantage when using the range as a measure of dispersion?

9 Descriptively, which measure of variability is better, the variance or standard deviation? Why?

10 Suppose a newcomer to American football decided to record the yardage gained or lost on each play of a game in terms of feet instead of the more typical measure of yards. What would the coach need to do with the data to compare the team's performance with previous games?

11 Three friends randomly sampled students at their university to measure the variability in student social media usage. One asked 25 people and found an s of 10 minutes; another asked 50 people and found an s of 15 minutes; and a third asked 500 people and found an s of 25 minutes. What is our best guess of the actual population standard deviation?

Work Problems

12 A researcher who uses heart rate as the dependent variable finds the 75th percentile to be a heart rate of 111 and the 25th percentile to be at 81.

a Compute the *IQR*.

b Compute the *SIQR*.

13 For the following populations of scores, find the mean deviations:

a 5, 7, 9, 9, 13, 14, 15, 16

b 23, 25, 31, 34, 36, 39, 44, 56, 63, 69

c 6, 8, 3, 9, 1, 4, 7, 4, 1, 1, 8, 2

14 A school psychologist informs a teacher about the mean and standard deviation of their students' IQ scores. The sample data are below. What are these values?

IQ scores: 98, 111, 102, 100, 101, 109

15 Calculate the range, variance, and standard deviation of this sample of scores.

2, 4, 7, 4, 8, 5, 1, 4, 4, 5

16 Calculate the variance and standard deviation for this population of scores.

22, 32, 21, 20, 19, 15, 23

17 Which distribution of sample scores has the larger variance?

Distribution *A*: 2, 4, 5, 1, 1, 2, 3, 9
Distribution *B*: 34, 39, 34, 35, 33, 32

18 If a normal distribution has an M of 50 and a s of 10, what scores encapsulate the middle 68% of the distribution? The middle 95%? The middle 99.7%?

19 Suppose a normal distribution has the same mean as Problem 18, but a standard deviation of 2. What scores encapsulate the middle 68, 95, and 99.7% of scores?

20 If a normal distribution has a mean of 80 and 68% of the scores are between 68 and 92, what is the variance of that distribution?

21 If a normal distribution has a variance of 100 and 95% of the scores are between the values of 120 and 160, what is the mean?

22 For a set of 10 000 scores that is normally distributed and has a μ of 100 and a σ of 15, about how many of the scores will be:

a Greater than 145?
b Greater than 130?
c Greater than $\pm 3\ \sigma$ away from 100?
d Greater than $\pm 2\ \sigma$ away from 100?

23 If a distribution has a $M = 4.5$ and $s^2 = 1.6$, what would be the M and s^2 if all the raw scores have 10 added to them?

24 Complete the following table. $\mu = 50$ and $\sigma = 5$. The constants specified are used to transform the scores of the distribution.

$X + 10$	$X - 10$	$X \cdot 10$	$X \div 10$
$\mu = ?$	$\mu = ?$	$\mu = ?$	$\mu = ?$
$\sigma^2 = ?$	$\sigma^2 = ?$	$\sigma^2 = ?$	$\sigma^2 = ?$

25 Find the range, *IQR*, and *SIQR* for the following sample distribution.

13	5	11	17	8	10	13	12	15
15	18	19	16	14	11	12	11	10
14	4	5	15	11	10	19	13	14
7	8	5	11	11	9	9	14	15
8	11	17	17	10	9	8	16	14
7	18	6	17	18	18	11	6	13

26 Determine the M, s^2, and s for the following sample distribution.

1020	1000	990	810	750	1130	1000
1060	1140	820	790	880	1110	1040
1000	1060	850	990	820	1010	1000

27 Multiply each value in the #26 dataset by three. Find the new M, s^2, and s.

28 Divide each value in the #26 dataset by three. Find the new M, s^2, and s.

5

The Normal Curve and Transformations

5.1 Percentile Rank

How did Austin do on his last exam? This simple question raises the issue of how best to convey a person's level of performance. Stating that Austin received a score of 35 may not be particularly helpful since it fails to provide context for the score. Stating that Austin's 35 was a *B* is better because it provides a rough indication of how he did with respect to some absolute standard. Stating the mean as well as the lowest and highest scores of the distribution would be additionally helpful in defining context; however, this information will not *specifically* locate his score in the distribution. Locating the score based on its *relation to the other scores* in the distribution would further assist in giving meaning to the score of 35.

One method used to specify the relative position of a score in a distribution involves transforming it into a percentile rank. The **percentile rank** states the percentage of scores that fall at or below a given score in a distribution. For instance, if 54% of the scores fall at or below the score of 35, the percentile rank for a score of 35 is 54%.

5.1.1 Computing the Percentile Rank of a Score

The formula for transforming a score to its percentile rank is very easy to use.

Formula for finding the percentile rank of *X*, *PR*

$$PR\ of\ X = \left(\frac{B + 1/2E}{N}\right) \cdot 100 \quad (5.1)$$

Essentials of Behavioral and Social Science Statistics, Third Edition.
K. Paul Nesselroade, Jr. and Laurence G. Grimm.
© 2025 John Wiley & Sons, Inc. Published 2025 by John Wiley & Sons, Inc.
Companion website: https://www.wiley.com/go/Essentials_of_Behavioral_and_Social_Science_Statistics3e

where

B = the number of scores *below* the given score X
E = the number of scores *equal* to X
N = the total number of scores in the distribution

■ **Question** *For the following distribution, what is the percentile rank of 16?*

12, 13, 13, 14, 16, 18, 22

Solution $N = 7$, $B = 4$, $E = 1$. Using Formula 5.1, we find that

$$PR\ of\ 16 = \left(\frac{4 + 1/2(1)}{7}\right) \cdot 100 = \mathbf{64\%}$$

This means we can say that 64% of the scores of this distribution fall at or below a score of 16. The score of 16 is at the 64th percentile. ■

For small samples like in the example given above, the percentile rank concept becomes a bit fuzzy. On the face of it, we see that the score of 16 is the fifth number up from the bottom. Since we are looking to find the percent of scores at or below 16 and there are seven numbers, we should be dividing 5 by 7 to get 71%, correct? Well, we need to remember that the score of 16, if on a continuous scale, is at the midpoint of the interval between 15.5 and 16.5. In a larger sample (with, say, 1000 values), we might have multiple 16s. It is reasonable to assume that about half of them, if measured more precisely, would have values between 15.5 and 16; therefore, at or below 16. The other half would have values between 16 and 16.5; therefore, above 16. This is why we use half of this frequency number when calculating percentile rank.

Percentile ranks are versatile; they can be used with any shaped distribution, skewed or otherwise. However, percentile ranks only communicate relative position. A different system, one that measures *how much* more or less one value is from another, will be needed to more precisely locate a given score within a distribution of scores.

5.2 Normal Distributions

Earlier in the text we were introduced to the notion of a normal distribution (curve). Now we will explore this concept in much greater detail. From this point in the text through Chapter 15, the normal curve will be heavily relied upon and extensively used, so much so that datasets will be assumed to be normally distributed unless there is specific information to the contrary.

5.2.1 The Importance of Normal Distributions

Normal distributions are of fundamental importance in the field of statistics for at least two reasons. First, they are common. Measurements of many naturally occurring phenomena, including psychological concepts like anxiety, mood, self-esteem, and so on, are normally distributed. Second, the means of repeatedly drawn samples from a population of scores of any distribution shape (normal or skewed) are normally distributed. These collections of sample means are called sampling distributions. This concept will be developed further in Chapter 7.

5.2.2 Characteristics of Normal Distributions

There is not a single normal distribution. Rather, there are an infinite number of possible normal distributions. However, each one shares the following five characteristics:

1) A normal distribution is unimodal, meaning it has one hump.
2) A normal distribution is symmetrical; the right half of the curve is a mirror image of the left. If we were to fold the curve at the midpoint, the two halves would coincide.
3) A normal distribution has the same value for the mean, median, and mode. This follows from the fact that a normal distribution is unimodal and symmetric.
4) A normal distribution is **asymptotic**; the tails of the distribution never quite touch the abscissa. This is only a theoretical point. As a practical matter, the chance of finding data points that fall, say, 7 or more standard deviations away from the mean is vanishingly small. Still, theoretically, the chance of this happening is not zero.
5) In a normal curve, roughly 68% of the scores in the distribution lie between $\mu \pm \sigma$, 95% of scores in the distribution lie between $\mu \pm 2\sigma$, and 99.7% of scores in the distribution lie between $\mu \pm 3\sigma$. This is the 68-95-99.7 rule mentioned in Chapter 4.

There may be other distributions having some of these characteristics, but only normal distributions will share all five. It is easy to forget that there are many normal distributions because statistics books often use similar drawings. Moreover, a normal curve is often referred to as *the* normal curve, as if there is just one. Keep in mind that any curve possessing all five foregoing characteristics is normal. Figure 5.1 illustrates three different normal curves.

The concept of the normal curve has broad applications. It is the basis for many statistical analyses used in the behavioral and social sciences. Spotlight 5.1 walks us through this story.

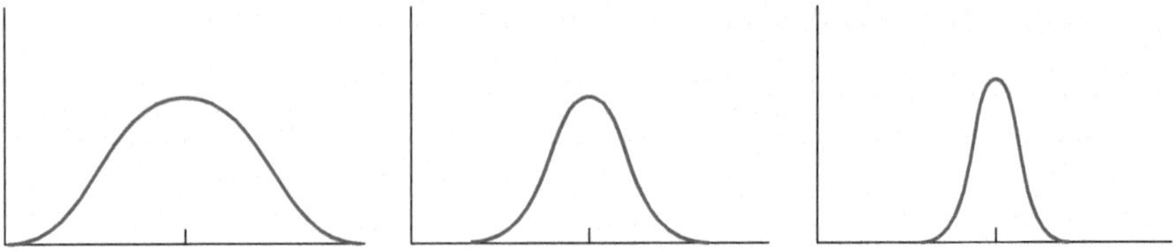

Figure 5.1 Three normal distributions that differ in their *M*'s and *s*'s.

Spotlight 5.1 Abraham de Moivre and the History of the Normal Curve

The discovery of the normal curve is usually attributed to Abraham de Moivre (1667–1754), being traced to a publication of his from 1733 (de Moivre 1738/1959; English Translation). He was held in high esteem by the intellectuals of his day, a friend of people like Edmond Halley (of Halley's Comet fame) and Sir Isaac Newton (Walker 1934, p. 322). De Moivre's discovery grew out of his interest in the probability of chance occurrences in which an event could take on one of two values, like the flipping of a coin. Given repeated tosses, if the number of heads observed is plotted on the horizontal axis and the probability of obtaining any number of heads is plotted on the vertical axis, as the number of tosses increases, the graph begins to take on the shape of a normal curve. How de Moivre arrived at this formula is unclear since it was the writing style of the day to publish results and conceal methods. Interestingly, he used two constants not associated with statistics: π, the ratio of the circumference of a circle to its diameter, and e, a constant used in calculating financial growth rates.

Although the context of de Moivre's discovery had to do with binomial events (i.e. two possible outcomes), the curve has applications for measuring error, a fact that made it useful for the work of observing the movements of planets and stars. In addition, the curve approximates all sorts of raw score distributions found in the natural world, like the height of adult males, IQ scores, the weight of wild rabbits, and even the widths of the foreheads of crabs (Pearson 1906). It was de Moivre who worked out the area under the curve for distances up to three standard deviations. As a result of this work, de Moivre is duly credited with the normal curve's discovery. However, other mathematicians deserve credit for popularizing it.

The first person to apply the normal curve to continuous measures was the English mathematician, Thomas Simpson (1755). Simpson realized that the slightly different judgments observers made when determining the position of a star could be thought of as a distribution of errors. Further, he realized that the problem of error could be minimized by using the mean of this distribution. Simpson understood that the normal curve could be used to counterbalance the positive and negative errors out of a final judgment.

The idea of the normal distribution was extended even further when the French mathematician Pierre Laplace (1749–1827) proved the Central Limit Theorem (a concept we will explore more in Chapter 7). This theorem is the single most important theorem in statistics. In essence, the Central Limit Theorem states that the means of many samples taken from a population will be normally distributed. This allows us to use the normal distribution to figure out the probability of obtaining a mean by chance.

The German mathematician Carl Gauss (1777–1855) also popularized the normal curve. In fact, the normal distribution is also called the Gaussian distribution. Gauss was one of the greatest mathematicians who ever lived, showing promise right from the start. For instance, at the age of 3, it is said he discovered an error in his father's calculations of employee wages! Later in life, when working at the University of Gottingen, Napoleon's armies were advancing on the city. Laplace is reported to have contacted Napoleon, asking him to spare Gottingen because "the foremost mathematician of his time lives there" (Dunnington 1955, p. 251). Gauss' most lasting contribution is the relocation of an asteroid. With the use of only a few observations, Gauss predicted exactly where the asteroid Ceres would reappear. He was using a technique he invented, the *method of least squares*, which found its way into modern statistics. (We will learn about it in Chapter 15.) Such is the importance of de Moivre's normal curve.

5.2.3 Area Under the Normal Curve

Recall from Chapter 2 that a simple frequency distribution can be depicted as a graph reflecting the frequency of every X value. Since this involves all scores in a distribution, we think of a normal curve using terms related to probability or likelihood. Chapter 6 will focus on probability theory, but for the time being, we can understand it to mean the likelihood that a given event (e.g. flipping a coin and getting a "head") will occur or not occur. This likelihood is quantified using the range 0–1. A probability of 0 means the event in question cannot occur, while a probability of 1 means it most certainly will. Probability theory is most useful when thinking about events that may or may not occur. For instance, an event that is just as likely to occur as it is to not occur (say, getting a "head" on a coin flip) has a probability of .50. Since the entire distribution is included, the probability that a given score will be found under the curve is 1. Furthermore, because a normal curve is symmetrical, a randomly selected score is just as likely to be above the mean as it is to be below. Since normal curves are mathematically well understood, we will learn how to determine the probabilities associated with any value found under a normal curve.

Throughout this chapter and those that follow, references are made to both *percentages* and *probabilities*. Please understand that these concepts are used interchangeably such that any point made about a percentage is true for probability. In this way, saying that 50% of scores fall above the mean is the same as saying the *probability* is .50 of randomly selecting a score above the mean.

5.3 Standard Scores (*z* Scores)

It is difficult to overstate the importance of standardizing scores in the practice of statistical analysis. The ability to standardize allows us to take data from any set of normally distributed scores, no matter the particular value of the mean or the variance, and think about that distribution in a similar way as any other normally distributed dataset. Some may argue it even allows us to compare apples with oranges (see Box 5.1).

Box 5.1 With *z* Scores, We Can Compare Apples and Oranges

Is he taller than he is heavy? This question, at first glance, seems to be nonsensical, like comparing apples with oranges. Height and weight, after all, are different variables and measured in different units. How can we say that 6 ft 2 in. is more or less than 155 lb? However, in the world of statistics, we *can* compare the relative position of scores in different distributions by using standardized scores. By transforming scores, we can convert original scores from different scales to a common unit. The common unit is the *z* score, which is the number of standard deviations a raw score is from the mean of a given distribution. Now if we were told that a man's height transforms to a *z* of +1.3 and his weight to a *z* of −.42, could we answer the question, "Is he taller than he is heavy?" If we understand that we are comparing two values *relative to the distribution from which they came*, then "yes," we could. When his height is transformed into a *z* score, the mean and standard deviation of a distribution of heights is used to make the transformation. His weight is transformed into a *z* score using the mean and standard deviation from a distribution of weights. In this way, to say someone is taller than they are heavy is to say that a transformed height value locates *higher* on the z distribution of heights than a transformed weight score locates on the z distribution of weights.

We can also use z scores to compare things like test performances on two different tests. Suppose a roommate is gloating a bit because they scored an 88 on a history exam while their roommate only scored an 82 on a psychology exam. However, we suspect that the history exam was much easier than the psychology one. If we knew the means and standard deviations of both exams (and if both sets of tests are normally distributed), we could see which of the roommates performed better in relation to the rest of their respective classmates.

This way of comparing scores from different scales of measurement is very useful in the social and behavioral sciences, as well as in the field of education. We can ask, for instance, if a person is more depressed than anxious, more paranoid than manic, or better at math than at reading. Although the scales of the tests are designed to tap different traits and abilities, and although each scale has its own mean and standard deviation, by standardizing the raw scores, an examiner can easily make cross-scale comparisons.

A **z score** is a measure of how many standard deviations a raw score is from the mean of the distribution. Given a normal curve, suppose the mean is 20 and the standard deviation is 4. A score of 24 is one standard deviation above the mean. Consequently, a score of 24 would be 1 z score above the mean ($z = +1$). What z score would be assigned to a score of 16? Since 16 is one standard deviation below the mean ($20 - 4$), it would transform into a z score of -1.

A z score is also called a standard unit or **standard score**. When all the raw scores from a normal distribution have been transformed into z scores, the resulting distribution is called the **standard normal distribution**. The standard normal distribution is a special normal distribution; it has a mean of 0 and a standard deviation of 1. This point is so important that it bears repeating; any normal distribution of raw scores, no matter the mean or the standard deviation, when converted into z scores, will take the shape of the *standard* normal distribution, having a mean of 0 and a standard deviation of 1. This makes the standard normal distribution very special.

Two z-score formulas are provided; one is used to transform the scores of a population, while the other is used to transform the scores of a sample.

Formulas for transforming an *X* score into a *z* score

Population

$$z = \frac{X - \mu}{\sigma} \quad (5.2)$$

Sample

$$z = \frac{X - M}{s} \quad (5.3)$$

where

X = the raw score to be transformed
μ = the mean of the population
σ = the standard deviation of the population
M = the mean of the sample
s = the sample standard deviation

■ **Question** *If $\mu = 4.80$ and $\sigma = 2.14$, what is the z score for a raw score of 6?*

Solution

$$z = \frac{X - \mu}{\sigma} = \frac{6 - 4.8}{2.14} = 0.56$$

For this distribution, a raw score of 6 is .56 standard deviations above the mean. ■

The following table lists several characteristics of z scores.

5.3.1 Important Facts About *z* Scores

1) A *z-score* distribution is established by transforming every raw score into a z score.
2) A z-score distribution always has a mean of 0 and a standard deviation of 1.
3) All raw scores that fall below the mean have a negative z value; all raw scores that fall above the mean have a positive z value.
4) A raw score that is one standard deviation from the mean has a z score of either ± 1, depending on whether it is above or below the mean.
5) A raw score that is the same as the mean has a z score of 0.
6) The z-score distribution will have the same shape as the raw score distribution.

5.3.2 Area Under the Curve and *z* Scores

All scores in a distribution are contained in the area under the curve. When the distribution is normal, half the scores are above the mean and half are below. In Chapter 4, we learned that about 68% of scores in a normal distribution fall between plus and minus one standard deviation of the mean. This is also true for the standard normal distribution (see Figure 5.2). Recall that statements of percentages can be translated into probability statements. For instance, the probability that a score selected at random will correspond to a positive z score is .50. The probability that a randomly selected raw score will correspond to a z score between ± 1 is approximately .68.

The features of the standard normal curve are so well understood that it is mathematically possible to specify the probability of any given score being drawn from a specified distribution. Thankfully, we have been mercifully spared the arduous task of calculating these probabilities. We can simply make use of the z table found in Appendix A (Table A.1). With the aid of this table, we can answer questions such as "What percentage of scores fall below a given X score?" or "What is the probability that a score taken at random will fall between two particular scores?" The z table, however, can only be used when working with data from the standard normal distribution.

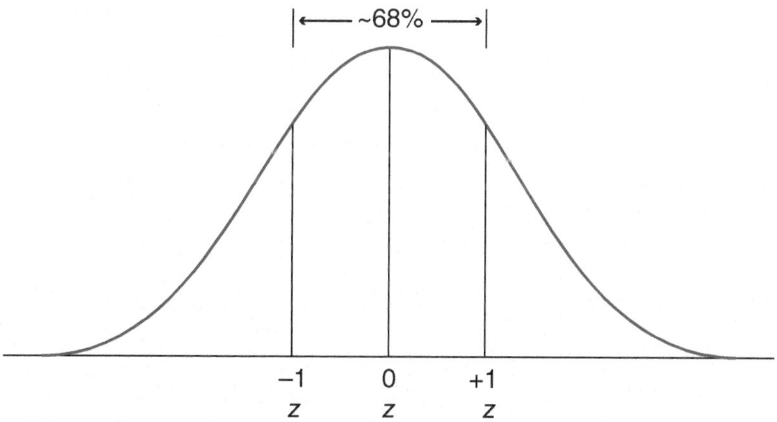

Figure 5.2 About 68% of the scores in a standard normal distribution have a z value between ±1.

5.3.3 Using the *z* Table

The following is a portion of the *z* table found in Appendix A (Table A.1).

(*A*) *z*	(*B*) Area between mean and *z*	(*C*) Area beyond *z*
⋮	⋮	⋮
1.00	.3413	.1587
1.01	.3438	.1562
1.02	.3461	.1539
1.03	.3485	.1515
1.04	.3508	.1492
⋮	⋮	⋮

Column *A* of the table lists *z* scores. Column *B* provides the probability that a single score will fall between the mean and the corresponding *z* value in column *A*. By moving the decimal point two places to the right, the numbers in column *B* reflect the percentage of scores falling between the mean and the given *z* score. Column *C* specifies the probability that a score will fall beyond a particular *z* score; that is, between that score and the end of the distribution. In this way, Columns *B* and *C* serve to cut one side of the distribution into two sections with the sum always equaling .50. Thus, for any *z* value, the corresponding probabilities found in column *B* plus column *C* will sum to .50. Finally, notice that the *z* table does not depict negative *z* scores. Recall that normal distributions are symmetrical; therefore, the probabilities associated with positive *z* scores match the probabilities associated with negative *z* scores. A table depicting probabilities for only one-half of a perfectly symmetrical distribution is sufficient.

Figure 5.3 shows a more precise measurement of the percentages of scores falling between various points of a standard normal distribution. Several worked examples will follow in order to familiarize us with the use of the *z* table.

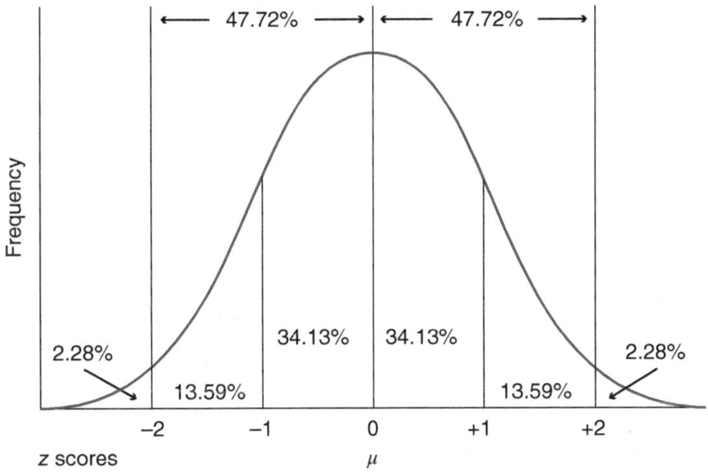

Figure 5.3 The percentages of scores that lie between whole number z values.

■ **Question** *How likely is a randomly selected score to fall between the mean and a z score of .39 (Figure 5.4)?*

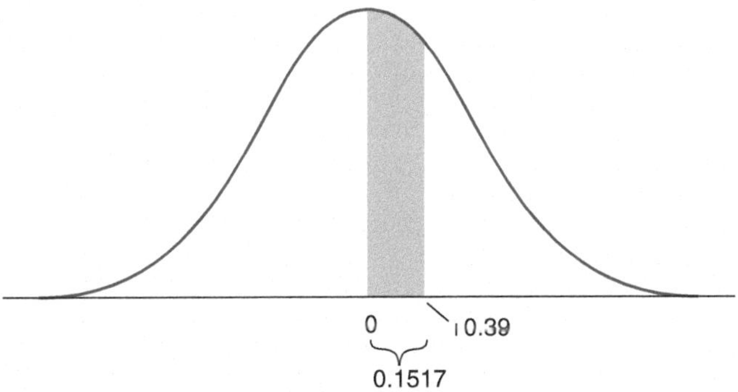

Figure 5.4 The probability of the shaded area is .1517 or 15.17%.

Solution .1517 ■

■ **Question** *How likely is a score to fall between a z score of +0.25 and +1.20 (Figure 5.5)?*

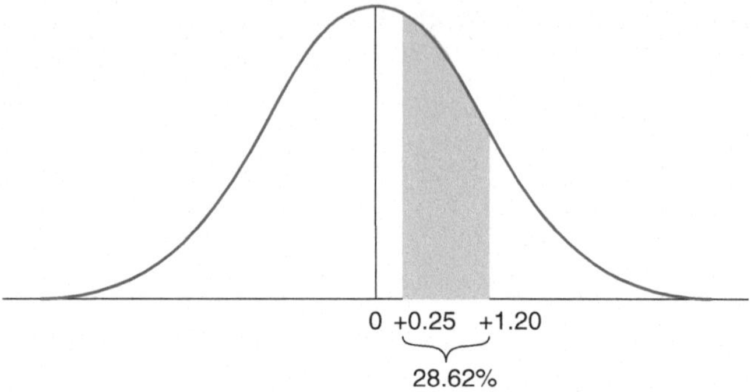

Figure 5.5 The shaded area contains 28.62% of the scores.

Solution The percentage of scores between the mean and a z of +1.20 is 38.49. This also includes the unwanted area between the mean and the z of +0.25. Subtracting the area between the mean and a z of +0.25 will isolate the proper area:

$$38.49-9.87\% = \mathbf{28.62\%}$$ ■

■ **Question** *How likely is a score to fall above a z score of +1.96 or below a z score of −1.96 (Figure 5.6)?*

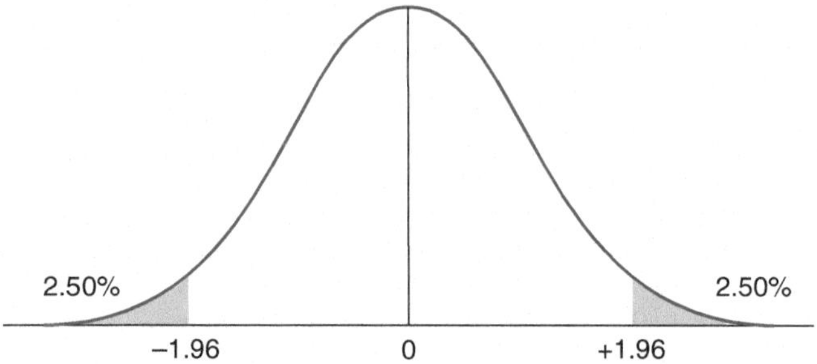

Figure 5.6 The shaded areas contain a total of 5% of the scores.

Solution Use the third column of the table when looking up 1.96. In total, 2.50% of scores are in each tail of the distribution beyond a z of 1.96. Therefore, the total percentage of scores in these two areas is 5%. Stated differently, the probability that a score drawn at random will fall *beyond* a z score ±1.96 is .05. Moreover, 95% of all scores fall within the boundaries of ±1.96 z scores. ■

■ **Question** *Find the z score cutoffs within which fall the middle 90% of scores (Figure 5.7).*

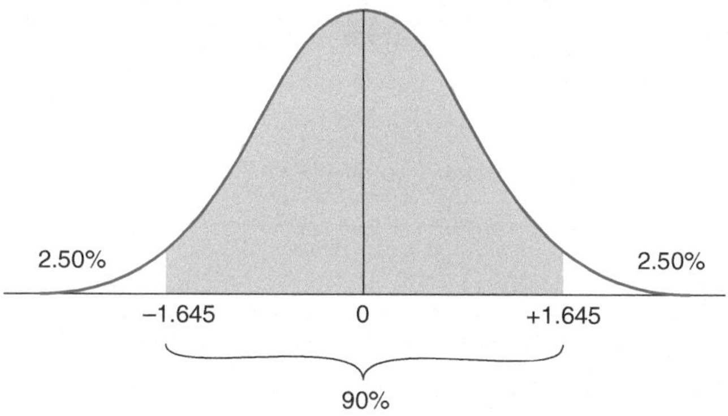

Figure 5.7 Approximately 90% of the scores fall between z scores of ±1.645.

Solution First, enter the second column of the table. Next, find the z score in column A that comes closest to the probability value of .4500 (i.e. half of the needed 90%). The z score we need is between 1.64 and 1.65. We could use either 1.64 or 1.65 and simply note that the area identified will be either slightly smaller or slightly larger than 90%. Alternatively, since the two values seem to be equally close to .4500 (in actuality, because we are talking about the area under a *curved* line, they are not exactly equally close), we could choose to take the midpoint between 1.64 and 1.65 and state that approximately 90% of the scores are within ±1.645. ■

5.3.4 Using the *z*-Score Formula

■ **Question** *Given that M = 25 and an s = 5, what percentage of scores fall between 25 and 32?*

Solution To use the z table, the raw scores of 25 and 32 must be transformed into z scores. Using Formula 5.3 (the symbols M and s indicate that the scores are from a sample), we find

$$z = \frac{X - M}{s} = \frac{25 - 25}{5} = \frac{0}{5} = 0$$

$$z = \frac{X - M}{s} = \frac{32 - 25}{5} = \frac{7}{5} = 1.40$$

Now that the raw scores have been converted into z scores, the question can be rephrased as, "What percentage of scores fall between a z score of 0 and 1.40?" The answer is 41.92%. ■

Up to now, we have been using z scores to find the percentage of scores falling within a given area of the normal curve. However, sometimes questions are asked that require z scores to be converted into raw scores. Formulas 5.4 and 5.5 enable us to accomplish this conversion.

Formulas for transforming a z score into an X score

Population

$X = \mu + z\sigma$ (5.4)

Sample

$X = M + zs$ (5.5)

■ **Question** *A teacher uses a placement test to assign students to one of three classrooms: an accelerated, remedial, or regular class. The regular class includes students whose obtained scores fall within the middle 60% of the distribution. Students scoring in the upper 20% are assigned to the accelerated class, while those scoring in the lower 20% are assigned to the remedial class. The class has a mean of 75 and a standard deviation of 7. What raw score cutoffs should be used to make the assignments (Figure 5.8)?*

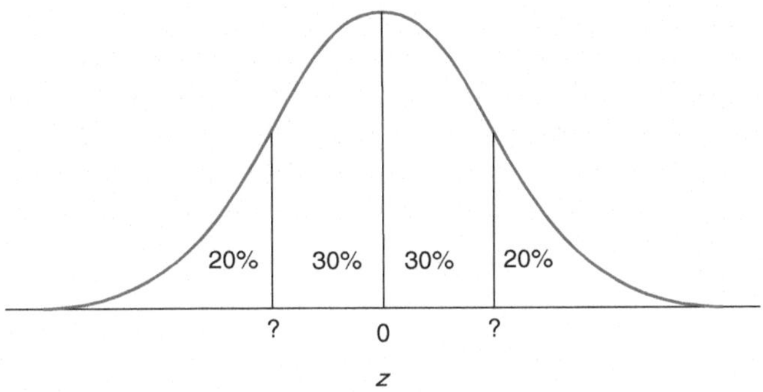

Figure 5.8 What are the cutoffs that bracket the middle 60% of scores in the distribution?

Solution This problem, at first glance, may seem overwhelming, but if we represent the problem visually, we can simplify matters. Our illustration should look like Figure 5.8. Since we want to find the values that cut off the top and bottom 20% of the distribution, we need to find the z scores that correspond to these points. It will be the positive and negative forms of the same value. Instead of using the first column of the z table, let us use Column C to find the percentage closest to 20% (.20). The percentage of 20.05 (.2005) is as close as this table allows. The z score that corresponds to .2005 is +0.84. This means that approximately 20% of the raw scores fall beyond a z of +0.84. Since the distribution is symmetrical, 20% of the scores also fall beyond a z score of −0.84. Stated differently, the middle 60% of the distribution falls between the z scores of ±0.84.

We are now able to convert the z scores to raw scores. Since we are looking for the raw score cutoffs that correspond to a positive *and* negative z score, two separate calculations are required. Whether we use Formula 5.4 or 5.5 depends on how we define the class: as a population or sample. Since the standard deviation was given, both formulas yield the same result:

$$Upper\ Cutoff = 75 + (+\,0.84)(7) = 80.88$$
$$Lower\ Cutoff = 75 + (-\,0.84)(7) = 69.12$$

Assuming the placement test yields whole number results, we can apply the findings to the distribution of test scores in the following manner: Students scoring 81 or higher are in the top 20% and should be assigned to the accelerated section, students scoring 69 or below are in the lower 20% and should be assigned to the remedial section, and students with scores from 70 to 80 should be assigned to the regular section. ■

5.3.5 Using *z* Scores to Calculate Percentile Rank

Recall that the percentage of scores falling below a given score is the percentile rank of that score. We have just learned that any score can be transformed into a *z* score. The *z* table can enable us to easily calculate percentile ranks via *z* scores. Bear in mind, however, that percentile ranks can only be computed with *z* scores *if the dataset is normally distributed.*

■ **Question** *In a normal distribution, what is the percentile rank of 15, when $M = 18$ and $s = 4$?*

Solution The *z score* of 15 is

$$z = \frac{15 - 18}{4} = \frac{-3}{4} = -0.75$$

Use the third column of the *z* table. This allows us to determine the percentage of scores *above* a *positive z* score or *below* a *negative z* score. Figure 5.9 uses shade to

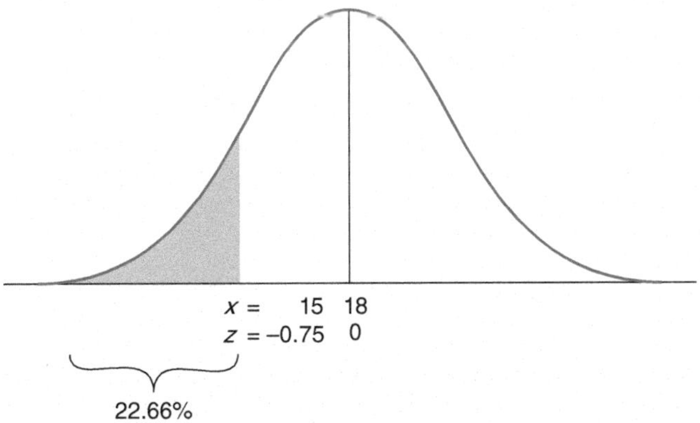

Figure 5.9 The shaded area is the percentage of scores that fall below a raw score of 15.

depict the area needed. The z score that corresponds to the raw score of 15 is −0.75. The percentage of scores that fall below a z of −0.75 is 22.66. Hence, the percentile rank of 15 is 22.66%. ■

■ **Question** *In a normal distribution, what is the percentile rank of 30, when $M = 27$ and $s = 2$?*

Solution The z score associated with a raw score of 30 is

$$z = \frac{30 - 27}{2} = \frac{3}{2} = 1.5$$

The percentage of scores between the mean and a z of 1.50 is 43.32. However, the percentile rank includes *all* the scores below X, so we need to add the lower half of the distribution to 43.32. Therefore, the percentile rank of 30 in this distribution is $43.32 + 50 = 93.32\%$. ■

5.3.6 A Couple Clarifications

A couple of issues should be addressed before we conclude the chapter. First, up to this point, we have been identifying the portion of a curve *above* or *below* given values in a dataset; however, it is unclear how to classify scores that correspond perfectly with a z score in question. In other words, if we are asked to find the percentage of scores in a dataset below a z of −1.96, column C in the z table would direct us to answer 2.5%; however, what about a raw score that corresponds perfectly with a z of −1.96? Is that raw score part of the lowest 2.5% or part of the upper 97.5%? This is not easy to answer. It depends, in part, on whether the raw scores are understood to be from a discrete or a continuous measure. The typical way to handle this situation is to suggest that data corresponding to the point itself should be included in the part of the curve being identified. In this way, it is appropriate to say that 2.5% of the scores in the distribution have a z score of −1.96 *or* below (not just below −1.96). Up until now, we have been using the language of *above* and *below*, but the phrases “at or above” and “at or below” are also appropriate.

Second, the z-score system is not the only method of standardizing scores. For instance, another similar standardizing method is the T-score system. It features a mean of 50, a standard deviation of 10, and has no negative values. We will not cover the T distribution since it has limited use. The z-score system is, by far, the most commonly used standardizing system.

Summary

Percentile rank and z scores are statistical transformations of original scores. They provide information about where a score stands in relation to other scores in a given distribution. The percentile rank of a score is expressed as the percentage of scores in the distribution that fall below that score. It is based on the rank order of scores and does not take the distance between scores into consideration.

The z score, however, uses distribution variability in the transformation formula. A z score is the number of standard deviations a raw score is from the mean of the distribution. All z scores above the mean are positive, while those below the mean are negative. In this way, z scores are mapped onto the standard normal distribution, a distribution with a mean of 0 and a standard deviation of 1. Formulas can be used to transform raw scores into z scores, and z scores into raw scores. Assuming the data are normally distributed, a z table can be used to find probabilities associated with selecting scores at random above- and below-given z scores.

Microsoft® Excel and SPSS®

Information regarding how to use Microsoft® Excel and SPSS® to assist in calculating z scores can be found in Appendix C.

Key Formulas

Formula for finding the percentile rank of *X*, *PR*

$$PR\ of\ X = \left(\frac{B + 1/2E}{N}\right) \cdot 100 \quad (5.1)$$

Formulas for transforming an *X* score into a *z* score

Population

$$z = \frac{X - \mu}{\sigma} \quad (5.2)$$

Sample

$$z = \frac{X - M}{s} \quad (5.3)$$

Formulas for transforming a *z* score into an *X* score

Population

$$X = \mu + z\sigma \quad (5.4)$$

Sample

$$X = M + zs \quad (5.5)$$

Key Terms

Percentile rank
Asymptotic
***z* Score**
Standard score
Standard normal distribution

Exercises

1 What is a percentile rank, conceptually?

2 What is a z score, conceptually?

3 Why are some z scores positive values, while others are negative?

4 Think of two examples of variables that are believed to be normally distributed across a population. Defend the answers.

5 Think of two examples of variables that are not believed to be normally distributed across a population. Defend the answers.

Work Problems

6 Using the following frequency distribution, what is the percentile rank of a score of:

a 4
b 6
c 3
d 2

X	f
7	3
6	4
5	7
4	12
3	10
2	7
1	6

Assume normality for all remaining questions in Chapter 5.

7 Transform these raw scores of a population distribution into z scores.

Raw scores: 4, 5, 7, 9, 10, 11

8 Given $M = 14$ and $s^2 = 16$, what is the z score of a raw score of 11?

9 In a distribution where $M = 25$ and $s = 3$, what raw score corresponds to a z score of .36?

10 If a distribution has a mean of 130 and a standard deviation of 13, what is the probability of randomly selecting a score above 140?

11 When $M = 34$ and $s = 3$, what percentage of scores are lower than 27?

12 What is the total percentage of scores that lie beyond z scores of ± 1.96?

13 What percentage of scores fall between the z scores ± 1.28?

14 What is the probability of randomly drawing a score between the z scores $+0.56$ and -1.2?

15 If a standardized anxiety questionnaire has a mean of 25 and a standard deviation of 5, what is the probability that an individual selected at random will score between 20 and 30?

16 A standardized test of reasoning ability has a $\mu = 70$ and a $\sigma = 7$. A principal would like to identify the students on the ends of the distribution. The principal determines the best students to be those in the top 10%, the worst being those in the bottom 10%. What raw score cutoffs should the principal use to identify the two groups?

17 Dr. Kilian gives an exam to abnormal psychology students and finds that $M = 56$ and $s = 5$.

a If the passing score is 38, what percentage of students will fail?

b If Dr. Kilian wants the "C" category to span the middle 30% of the distribution, what would be the cutoffs?

c If only the top 10% are to get an "A," what score serves as the cutoff?

18 A score that is 10 points above the mean corresponds to a z score of 2.50. What is s?

19 In a sample where $s = 15$, a raw score of 51 corresponds to a z of -1.00. What is the mean?

20 In a sample where $s = 5$, a raw score of 31 corresponds to a z of 2.00. What is the mean?

21 In a sample where $M = 75$, a raw score of 60 corresponds to a z of -2.00. What is s?

22 In a sample, a raw score of 35 corresponds to a z of -1.00, and a raw score of 40 corresponds to a z of -0.50. Find the mean and standard deviation for this sample.

23 In a population, a raw score of 77 corresponds to a z of 2.50, and a raw score of 41 corresponds to a z score of −5.00. Find the mean and standard deviation.

24 Suppose miles traveled per year by American drivers are normally distributed with a mean of 25 000 miles and a standard deviation of 6000 miles. If we wanted to find the miles traveled that would cut the distribution into five equally populated segments, what are the miles traveled that define the bottom 20%, the next 20%, and so on up to the top 20%?

25 Suppose the average American household generates 45 lb of garbage per week with a standard deviation of 11 lb. Suppose the local government wants to levy a tax on the worst offenders (top 15%) and offer a tax rebate as incentive for its most conscientious citizens (bottom 28%). What weekly garbage amounts correspond to these cutoffs?

26 For a sample where $M = 25$ and $s = 4$, what raw score corresponds to a z score of −1.75, and what percent of sample scores will be greater than that raw score?

27 For a population where $M = 1$ and $s = .15$, what percent of scores fall between .6 and .7?

28 Suppose Andrew and Lisa wanted to compare how well they each performed in their respective soccer games. They are, however, on different teams. Since both are defenders and neither score very much, they decided to compare the number of completed passes. Andrew completed 54 passes; his team average was 44 completed passes per player with a standard deviation of 6. Lisa completed 48 passes; her team average was 38 passes with a standard deviation of 7. Which one performed better relative to their teammates?

29 Suppose both Sarah and Justine think the other wastes too much time. Sarah feels Justine spends too much time on social media, while Justine feels Sarah spends too much time figuring out what to wear. Suppose further we know that people average 65 minutes per day on social media ($\sigma = 20$ minutes) and 15 minutes per day deciding what to wear ($\sigma = 4$ minutes). Justine spends 90 minutes a day on social media, and Sarah spends 20 minutes deciding what to wear. Who wastes more time compared with the rest of the population?

Part 2

Inferential Statistics

Theoretical Basis

6

Basic Concepts of Probability

6.1 Theoretical Support for Inferential Statistics

The concepts presented in this chapter and Chapter 7 build on what we have already learned regarding descriptive statistics and will, in conjunction, present the theoretical support for performing inferential statistics, the subject matter of the remainder of the text. **Inferential statistics** are a collection of mathematical techniques that use probability theory and hypothesis-testing logic to draw inferences about the characteristics of populations from the features of samples. The primary types of inferences concern whether a sample comes from a known population *or* whether two or more samples come from the same population. Operationally, inferential statistics are used to analyze the merits of different hypotheses.

This chapter will introduce us to the basics of probability theory – its terminology, formulaic principles, and conceptual limitations. Then, Chapter 7 will introduce us to hypothesis testing and sampling distributions. Together, these chapters help us understand the relationship between the features of randomly drawn samples and the features of the populations from which they come.

Probability theory helps us speculate about the nature of samples drawn from well-defined populations. For instance, imagine we are blindly selecting 10 jellybeans from a jar that contains an equal number of red and black jellybeans. We can imagine this jar to be a population and the selection of 10 to be a sample. Although we will not be able to predict the exact makeup of our sample, we might suggest that it should be close to about five of each color. We can make reasonable expectations of the sample based on our knowledge of the population from which it is drawn. Now, what if there was another jar of jellybeans from which we blindly selected 10 jellybeans, but this one contained 90% red and only 10% black? Just as before, we would not be able to predict how many red and black jellybeans would be in our sample, but because the population is well-defined, we could

Essentials of Behavioral and Social Science Statistics, Third Edition.
K. Paul Nesselroade, Jr. and Laurence G. Grimm.
© 2025 John Wiley & Sons, Inc. Published 2025 by John Wiley & Sons, Inc.
Companion website: https://www.wiley.com/go/Essentials_of_Behavioral_and_Social_Science_Statistics3e

use this knowledge to make some predictions. One, for instance, might be that we will almost certainly have more red than black jellybeans in our sample. By knowing the composition of the population, we can gauge how likely it is to get different kinds of samples. Probability theory helps us know what to expect when drawing samples from well-defined populations.

Inferential statistics reverse this logical flow of inference. Instead of inferring the features of a sample drawn from a well-defined population, inferential statistics starts with access to a well-defined sample and uses probability theory and various hypothesis-testing concepts to draw inferences about the features of the population from which the sample came. For instance, if we looked at our sample of 10 jellybeans and saw that they were all red, this information helps us think about from which of the 2 previously described jars the sample came. If, however, our sample contained seven red and three black jellybeans, we may not feel very certain at all. Additional ideas and terminology concerning hypothesis testing will need to be introduced to help us set up ways of making decisions.

This chapter and Chapter 7 contain numerous concepts that are theoretically important. Although full comprehension of these concepts is not a requirement to "run" inferential tests, a clear understanding is necessary to gain a deep appreciation of how these tests work, what the conclusions of the test mean (and do not mean), and why. The chapters following 6 and 7 can be roughly clustered into three different cohesive groups of analysis: (i) Part 3 – the z test, t tests, and power; (ii) Part 4 – analysis of variance tests; and (iii) Part 5 – correlation, regression analyses, and nonparametric tests. In terms of theoretical rationale, all the topics covered in Parts 3–5 are anchored in the twin pillars of probability theory and hypothesis testing. A firm grasp of the concepts in these two chapters will benefit us as we move through the rest of the text. Conversely, if we find ourselves struggling to understand the theoretical rationale of a given inferential procedure, a review of the material in this chapter and Chapter 7 is recommended.

6.2 The Taming of Chance

Historically, the process of decision-making in situations of uncertainty was often dealt with by techniques designed to let the fates or the gods control the outcomes; for instance, procedures like the casting or drawing of lots are often referenced in ancient writings. With the rise of Christianity in the West, this practice was largely replaced by the direct petitioning to the Christian God for guidance in the making of weighty decisions. However, starting in the seventeenth century, several natural philosophers and mathematicians started to speculate about how to deal with more mundane uncertainties associated with daily life (e.g. Reeves 2015). One such activity that spurred a good deal of interest concerned gambling. Many historians of science point to an exchange of letters between Blaise Pascal and Pierre de Fermat regarding how to fairly distribute the stakes in an interrupted gambling venture as a critical event in the development of probability theory (e.g. Katz 1993; Weisberg 2014). We can easily imagine a gambling situation where one contestant would be ahead and would be more likely to win if the contest was to continue, but this outcome would not be certain. Splitting the winnings in half would seem unfair to the one who is ahead. However, giving this person the full amount would seem unfair to the other contestant who, although sitting in a disadvantaged position, had not yet lost and had some hope of recovering to win. The discussion started by Pascal and Fermat generated a lot of further discussion over the next 300 years regarding how to quantify the uncertainty in situations such as these. (See also Box 2.1 presented earlier in this text.)

A breakthrough came when it was realized that "games of chance" were probably not that at all, but that the outcomes of rolls of the dice and the location of a card in a deck when shuffled were regulated and law-like, if only detailed information about preconditions and processes were available. However, given that such precise information could never be fully known, the key was to become *willfully ignorant* of the myriad of small effects that influence outcomes and choose to focus on the more general truths. If we choose to not focus on any particular roll of the dice and instead focus on what tends to happen across several events, we can construct a fairly accurate representation of the set of outcomes and their likelihood. For example, Figure 6.1 shows a representation of the set of outcomes associated with the rolling of two standard dice. Historians suggest this gained insight to be critical in the development of probability theory.

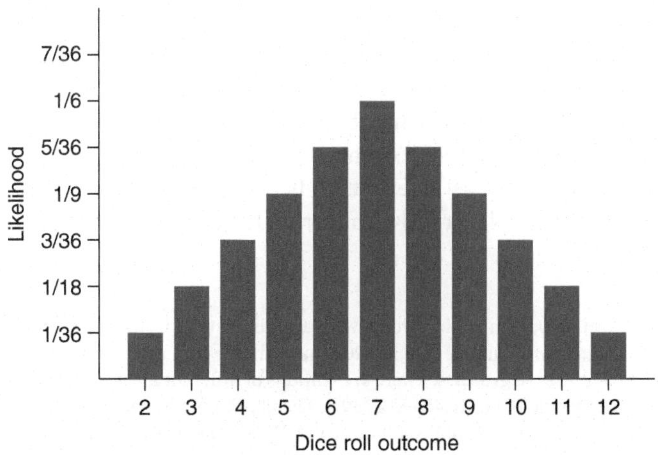

Figure 6.1 A probability distribution representing a single roll of a pair of six-sided dice.

6.3 What Is Probability?

Probability, as mathematicians use the term, can be defined as the likelihood of an event occurring. This likelihood can be represented by a number between 0 and 1. An event is anything that can either occur or not occur. For example, when we flip a coin, it either comes up as a head or it does not (i.e. it comes up as a tail). In coin flipping, "heads" is an event. Of course, "tails" is also an event. When we roll a die, it either lands on a 4 or it does not. Therefore, 4 is an event. Events that are single occurrences, like a "4" on a die roll or a "head" on a coin flip, are often called elementary events. An exhaustive set of elemental events is referred to as a "set."

When we assign a probability (i.e. a number between 0 and 1) to an event, we are stating how likely that event is to occur. If the probability is .5, it means that the event is as likely to occur as it is to not occur. A probability greater than .5 means that it is more likely to occur than not, and a probability of less than .5 means that the event is more likely to not occur than to occur. Events with a probability of 0 mean they cannot occur, while events with a probability of 1 mean that they most certainly will occur.[1]

A good place to start to understand how probabilities are determined is to look at situations where each elemental event is equally likely – for example, a roll of a die, a drawing of a card, or a flip of a coin. In each of these scenarios, it makes sense to think that each elementary event is equally likely to occur. In situations like this, relative frequency is used to determine likelihood. By relative frequency, we mean the comparison of the number of favorable events with the total number of possible events. Perhaps we can see that the likelihood of any elemental event is equal to the reciprocal of the total number of events in the set. For example, the likelihood of rolling a die and getting a 4 is 1/6 or .17 (actually, it is $.16\overline{6}$, but we will keep our answers to two decimals). There are six different events, and one of them is a 4. This is also the frequency of getting a 3, by the way. In fact, each of the six elemental events has the same likelihood of occurring. This is not always the case, of course. Sometimes, each elemental event is not equally likely. For instance, when we examine what will happen when a baseball player takes their turn batting, there are several different elemental events that could take place; for simplicity purposes, we can think of the set of events as being (i) an out, (ii) a walk, (iii) a single, (iv) a double, (iv) a triple, or (vi) a home run. Each of these six events,

1 Theoretically, this description is too simplistic. Events with a probability of 0 happen all the time. For instance, imagine a six-figure number (e.g. 587 202); now divide it by an imagined two-digit number with two decimal places (e.g. 33.91). There are trillions of different answers that can be found. The answer for this example is 17 316.484 812 739 6. The chance of anyone else picking these numbers is so close to zero as to be zero, and yet it occurred. We should distinguish between events with probability of zero and logically impossible events, like finding a square circle.

however, is not equally likely. It would be inappropriate for us to conclude that the likelihood of a batter getting a home run is 1 chance out of 6. Likewise, it would be inappropriate for us to decide that the chance of rain today is .5 because, after all, there are only two choices: it will either rain or not rain. In these cases, while we may have correctly identified the complete set of elemental events, we have failed to realize that each event is not equally likely.

When we examine finding the probability of an event if all elemental events are equally likely, we encounter our first significant probability formula. Let us introduce some notation information as well as a new term. In probability formulas, the uppercase letter *P* is used to represent "probability" and can be read as "The probability of..." Events are placed within parentheses. In this way, one can express the probability of flipping a coin and getting heads as *P*(*heads*). In our coin-flipping example, the *P*(*heads*) = .5. Often, mathematicians use an abstract symbol like a capital letter to represent an event. For the probability of event *A*, we could write *P*(*A*). Finally, note that the word "favorable" is used to identify the event or events that are being considered. So, assuming all elemental events are equally likely, the formula for determining the likelihood of a favorable event occurring is the ratio of favorable events compared to the set of all possible events.

Probability of favorable event (when all events are equally likely)

$$P = \frac{(number\ of\ favorable\ events)}{(total\ number\ of\ events)} \quad (6.1)$$

If, for instance, we wanted to know the likelihood of getting a spade with the draw of a single card from a deck of 52, since all elemental events are equally likely, the probability would be 13/52 (or .25 or 25%; each of these ways of representing this likelihood are acceptable). If the 2 Joker cards were also included in the deck, the ratio would be 13 favorable events over a total of 54 possible events (13/54, or .24).

Using the notion of relative frequency, we can say that a probability of .7 means that the event in question has a 7 out of 10 chance of occurring. Likewise, a probability of .25 would be a 1 out of 4 chance of occurring. We should be aware that the analogy breaks down if we try to apply this *literal* way of thinking to all things. For example, we can say that there is a .7 probability of measurable rain tomorrow. We should note, however, that no 10 elementary events of which 7 are favorable exist. Yes, it can be argued that on 7 out of 10 days with meteorological features like tomorrow, measurable rain will be recorded. But what does that really mean? Do we really have access to 10 days like tomorrow? No. However, we can still think of the .7 probability as a 7 out of 10 chance of rain. To make use of probabilistic information, it is not required for the user to identify the elemental events as real entities.

6.3.1 Sampling With and Without Replacement

Sampling with replacement is a method of sampling wherein a member of a population is randomly selected and then returned to the population before the next member is selected. Suppose we want to know the probability of selecting a red card from a deck of playing cards. Since half of the cards are red and half of the cards are black, the probability of selecting a red card at random is .50. If we return the selected card to the deck, the probability of the next selected card being red is still .50. However, suppose we do not return a selected red card to the deck and ask, "Now what is the probability of selecting a red card at random?" Now more black than red cards are in the deck. The probability of selecting a red card is *less than* .50. This is sampling *without* replacement.

The distinction between sampling with or without replacement is important for hypothesis testing. In particular, the sampling distribution concept, a centerpiece of Chapter 7, is derived from a sampling *with* replacement procedure. As a result, moving forward, we will be limiting ourselves to only sampling *with* replacement.

6.3.2 A Priori and A Posteriori Approaches to Probability

Up to this point, we have approached probability from an **a priori (or classical) approach**. This approach is based on a logical analysis of the probabilistic situation and the relative frequencies that are predicted. It is not based on the accumulation of data. Using logic alone, we can claim that we should get 100 "4's" if we rolled a single die 600 times.

The **a posteriori approach**, on the other hand, is an empirical approach to probability. It requires the collection of data. Using the preceding example, how would we determine the probability of getting a 4? We would need to collect some data. We could, for instance, roll a die 600 times and record how many times we got a 4. Suppose we did that and got a "4" 90 times. In this example, the a posteriori method yielded a probability that is close (.15) but not identical to the .17 probability determined by the a priori method. Theoretically, it is believed that if we roll a die an infinite number of times, this a posteriori approach will also result in a probability of .17 (or, stated more precisely, a ratio of 1/6).

Please note that when performing research, the a priori approach is often inappropriate. Without collecting data, we may not know the number of favorable events in a population. (See Box 6.1 for more about this topic.) For example, suppose we would like to know the probability that hospitalized patients have a diagnosis of depression. There is no way to use logic alone to determine the answer. We would have to take a random sample of hospitalized patients and observe what percentage received a diagnosis of depression. We could then make a statement about the likelihood of those hospitalized in the future having a diagnosis of depression.

Box 6.1 Is the Scientific Method Broken? Likelihood and Clarity

An aspect of probabilistic thinking that seems to have been lost in most modern discussions of probability and which may be partly responsible for the current reproducibility problem in science is the realization that uncertainty is not merely a question of likelihood but also one of clarity. Now "likelihood" (or "risk" as it is sometimes called) is usually understood as something that can be quantified numerically, like the odds of selecting a spade from a deck of cards. This predisposes that the conditions are well understood and the relative frequencies of favorable and unfavorable options are known. Every judgment call dealing only with likelihood has some known degree of risk. Modern probability theory, since the 1930s, has almost exclusively focused on only this aspect of uncertainty. However, uncertainty is not restricted to likelihood. Clarity is also a necessary component. By clarity, we mean the degree to which the situation one is speculating about is being properly conceptualized.

For instance, when we say that the likelihood of getting snake eyes when rolling two dice is 1/36, we are assuming that we have full clarity regarding the situation (e.g. what is known as "snake eyes," the features of the two dice, and so on). We realize that gambling on this outcome has a certain amount of risk associated with it (actually, a lot of risk associated with it!), but this risk is at least quantifiable. However, if we think about it for a few minutes, it should be easy to see that we may be overlooking several aspects of the probabilistic situation out of convenience. For instance, what if the side of each die is not equally likely? What if there are 10 sides to each die? What if the numbers on the die do not start with 1? Now, this is a simple example where there is most likely a lot of initial clarity regarding the situation, or, if not, issues of clarity could be easily resolved by simply looking at the dice to be rolled. However, in many research situations, gaining clarity is not as easy as inspecting the dice. Further compounding the situation, the clarity that is lacking may be best understood as a qualitative variable as opposed to a quantitative one, and as a result, not something that could even be factored into modern probability theory. Acknowledging this limitation, however, might leave the researcher without a clear path forward in their efforts to investigate. So, oftentimes researchers, perhaps unknowingly, look right past issues of clarity and focus only on the issue of likelihood when testing their hypotheses. By essentially ignoring issues of clarity and focusing exclusively on likelihood, modern probability theory has created a false sense of confidence in the stated outcomes of some statistical analyses. Some argue that our failure to address this fundamental issue has led to the challenging state in which many areas of scientific investigation currently find themselves (e.g. Byers 2011; Weisberg 2014).

To help illustrate this further, imagine a medical situation where a patient has arrived with a set of symptoms and an unknown diagnosis. There are two judgments that need to be made: First, what is the nature of the illness, and second, what form of treatment would be best? It might be helpful to consider the first of these two judgments as being representative of the issue of clarity. The doctor, relying on clinical expertise, personal experience, and reasoned judgment, increases clarity and offers a diagnosis. Some probabilistic information may have been used in this analysis, but this was not primarily an exercise in probability. The second question of *treatment*, however, may be more representative of the issues related to likelihood and probability. Based on the outcomes of previous studies, the doctor may choose to prescribe one form of treatment over another due to the likelihood of a favorable outcome derived from a quantification of relevant data.

Research situations oftentimes involve both forms of uncertainty: likelihood and clarity. However, when research replication efforts take place, issues of clarity may be overlooked because the original researchers had failed to consider them. This disconnect can lead to contradictory findings, thus fostering a distrust in the scientific method.

Since this chapter is merely a brief introduction to probability concepts, considering only the a priori approach will suffice. However, the a posteriori approach will be implicitly explored when hypothesis testing is introduced in Chapter 7.

6.4 The Addition Rule

The **addition rule** is used to determine the probability of occurrence of one of many possible events. It is typically applied when the question has the word "or" in it. For example, "What is the probability of rolling a die and obtaining a 4 *or* a 6?; What is the probability of drawing a club *or* a heart from a deck of cards?" We can represent this concept by using the term $P(A \text{ or } B)$. (In mathematics, this is usually written as $P(A \cup B)$, but we will just use the word "or" rather than introduce a new symbol.) Determining the proper formula to use when answering an "or" question depends upon another concept – mutual exclusivity. **Mutually exclusive events** occur when one event precludes the occurrence of another event. (The term "disjoint" may also be used to represent this concept.) For example, when we roll a die once, it is impossible to obtain a 4 and a 6; one precludes the other. When we select a card from the deck, it cannot have more than one suit. If the two events are mutually exclusive, the formula for the addition rule follows.

Addition rule formula for two mutually exclusive events

$$P(A \text{ or } B) = P(A) + P(B) \quad (6.2)$$

Formula 6.2 is read as "The probability of either event A or event B occurring equals the probability of event A occurring *plus* the probability of event B occurring."

■ **Question** *Assume that we roll one die. What is the probability of getting a 2 or a 5?*

Solution

Step 1. Determine $P(A)$. Let us call rolling a 2 event A. Is each event equally likely? Assuming the die is fair, yes. So, we will use Formula 6.1.

$$P(A) = \frac{(\textit{number of favorable events})}{(\textit{total number of events})} = \frac{1}{6} = .17$$

The probability of rolling a 2 is .17. (More accurately, the fraction is 1/6.)

Step 2. Determine $P(B)$. Let us call rolling a 5 event B. Each event is equally likely and using Formula 6.1 we get the same result, $P(B) = .17$

$$P(B) = \frac{(\textit{number of favorable events})}{(\textit{total number of events})} = \frac{1}{6} = .17$$

The probability of rolling a 5 is also .17. (More accurately, the fraction is 1/6.)

Step 3. Since events A and B are mutually exclusive, we can use Formula 6.2 to determine the probability of rolling a 2 or a 5.

$$P(A \text{ or } B) = P(A) + P(B)$$

$$P(A \text{ or } B) = 1/6 + 1/6 = 2/6 \textit{ or } .33$$

The probability of rolling a 2 or a 5 on a single toss of the die is .33. ■

The addition rule for mutually exclusive events is generalizable to situations where we want to determine the probability of occurrence of one of several events. The general equation for the addition rule with more than two mutually exclusive events follows.

Addition rule formula for more than two mutually exclusive events

$$P(A \text{ or } B \text{ or } C \text{ or} \ldots Z) = P(A) + P(B) + P(C) + \cdots + P(Z) \quad (6.3)$$

where

$P(Z)$ = the probability of occurrence of the last event

Formula 6.3 is merely an extension of the addition rule formula for two mutually exclusive events. Consequently, the computational steps follow the same format as outlined in the preceding worked example.

We must watch out, since not all cases of *P*(*A or B*) can be worked that easily. Sometimes events *A* and *B* cannot co-occur, but other times they can. For example, consider the likelihood of drawing a card that is either a heart or a queen. In this case, the probability of drawing a heart is 13/52 (or 1/4), and the probability of drawing a queen is 4/52 (or 1/13). But the probability of drawing a heart or a queen is 16/52; this is not the sum of 13/52 and 4/52 (which is 17/52). What is the difference? In this example, the two events (heart and queen) can co-occur; one can draw *both* a queen and a heart at the same time, namely, the queen of hearts. In mathematical language, that card represents a *co-occurrence*. An event being a "heart" is not mutually exclusive from an event being a "queen." If we leave the formula as is, the queen of hearts card will be counted twice, once when we tally up the queens and again when we tally up the hearts. We need a formula that keeps from double-counting events that qualify as being both events *A* and *B*. Here is the more general formula for finding the likelihood of events *A* or *B*, whether they co-occur or not.

Addition rule formula for two events

$$P(A\ or\ B) = P(A) + P(B) - P(A\ and\ B) \quad (6.4)$$

Notice the last expression in the formula, *P*(*A and B*). It removes (once) the events that would be double-counted by Formula 6.2. In the abovementioned example of selecting a queen or a heart, the probability of both a queen and a heart co-occurring is 1/52. This helps us understand why the correct answer is not 17/52 but rather 16/52. Perhaps it is helpful to point out that Formula 6.2 is just a special case of the more general Formula 6.4. Formula 6.4 always works, but Formula 6.2 works when *P*(*A* and *B*) = 0. Do we see that when *P*(*A and B*) = 0, the last expression in Formula 6.4 simply falls out, thus becoming Formula 6.2?

■ **Question** *Suppose we will win a game on the next roll if we get either an even number or a number greater than 4. What are our chances of winning?*

Solution

Step 1. First, determine the *P*(*A*). Let us call rolling an even number event *A*. Let us further assume that we know that the chance of rolling an even number is .5.

Step 2. Second, determine the *P*(*B*). Let us call rolling a number greater than 4 event *B*. Let us also assume that we know that the chance of rolling a number greater than 4 is .3333.

Step 3. Finally, determine *P*(*A and B*). This, for our problem, means the chance of rolling a 6, as that is the only value from the set that is both an even number and a number greater than 4. Let us assume we know that the chance of this occurring is .1667.

$$P(A \text{ or } B) = P(A) + P(B) - P(A \text{ and } B) = .5 + .3333 - .1667 = .6667 \text{ or } 2/3$$

The probability of rolling a die and getting either an even number or a number greater than 4 is .6667. We have a 2/3rd chance of winning on the next roll. ■

The addition rule for more than two events that are not mutually exclusive is complex. Since this chapter is an introduction to probability, this topic will not be covered.

6.5 The Multiplication Rule

The addition rule is used when we want to determine the probability of one of two or more events occurring. The **multiplication rule** is used to determine the probability of event *A and* event *B* occurring or $P(A \text{ and } B)$. (In mathematics, this is usually written as $P(A \cap B)$, but we will use the word "and" rather than introduce a new symbol.) The multiplication rule can be extended to situations where more than two events co-occur. However, since this is a basic textbook, we will only consider the co-occurrence of two events. We will, however, consider the multiplication rule under two different conditions: when events are independent of each other and when they are not.

Probabilistic independence between events is found when the occurrence of one event has no effect on the probability of a second event occurring. For instance, if we wanted to determine the likelihood of selecting a card from a deck of cards that is both a spade and a face card, the occurrence of a spade does not change the likelihood that the card will be a face card. If we did not know the selected card is a spade, the probability of a face card (including aces) is 16/52 or 4/13. If, however, we know the card selected is a spade, the probability that it is also a face card has not changed – it is still 4/13. In this case, we can say that event *A* (spade) and event *B* (face card) are *independent* of each other; the occurrence of one event did not change the likelihood of the other event occurring. (Even though we presented this relationship from the perspective of selecting a face card, this can also be shown from the perspective of selecting a spade. Namely, there is a 1 in 4 chance of selecting a spade knowing nothing else. If we learn that we have a face card, the ratio of having a spade stays the same: a 1 in 4 chance. Independence is a shared property.) The formula for the multiplication rule for independent events follows.

Multiplication rule for two independent events

$$P(A \text{ and } B) = P(A)P(B) \quad (6.5)$$

Formula 6.5 is read as "The probability of events A and B co-occurring is equal to the probability of event A occurring multiplied by the probability of event B occurring." Be sure to note that this formula is true only if the two events are independent. We will develop a more general rule for multiplication in the next section.

Let us look at an example. If we roll a die and flip a coin, certainly the outcome of each event is independent of the other. So, what is the probability of getting both a head on the coin flip *and* a 5 on the roll of the die? Well, the $P(head) = 1/2$ and $P(5) = 1/6$. So, using our rule, $P(head\ and\ 5) = P(head)P(5) = (1/2)(1/6) = 1/12$. If we think about this, we can see that it is correct. The elementary events in this example are the combined events of the two actions – the coin flip and die roll. There are 12 of them (e.g. *head and 1*, *head and 2*..., *tail and 1*, *tail and 2*...). Furthermore, all elementary events are equally likely, so the probability of any one of them (e.g. *head and* 5) is 1/12, just as we found using the formula.

Let us look at another example. Suppose we know the probability of Lisa earning an "*A*" in a collegiate course is .9, while the probability of Sam earning an "*A*" is .4. Suppose we also know that these students do not interact; in other words, their grades should be independent. So, the probability that both will get an *A* is (.9)(.4) = .36.

■ **Question** *What is the probability of randomly selecting a 4 and* then *an 8 on two successive draws from a deck of cards? (Assume sampling with replacement.)*

Solution

Step 1. First, determine the $P(A)$. Let us call drawing a 4 event A. Since all cards are equally likely to be selected, we can use Formula 6.1.

$$P(A) = \frac{(number\ of\ favorable\ events)}{(total\ number\ of\ events)} = \frac{4}{52} = .0769$$

The probability of randomly selecting a 4 is .0769 or about 7.7%.

Step 2. Determine $P(B)$. Let us call drawing an 8 event B. Note that $P(B)$ is unaffected by step 1 because the drawn card has been replaced.

$$P(B) = \frac{(number\ of\ favorable\ events)}{(total\ number\ of\ events)} = \frac{4}{52} = .0769$$

The probability of randomly selecting an 8 is also .0769 or about 7.7%.

Step 3. The multiplication rule is now applied.

$$P(A\ and\ B) = P(A)P(B) = (.0769)(.0769) = .0059$$

The probability of drawing a 4, replacing the card, and then drawing an 8 is .0059. Only 59 times out of 10 000 would these two events be expected to occur in this sequence. ■

Formula 6.5 can be extended to any number of independent events. For example, if we wanted to know the probability of drawing an ace, then a spade, and then the six of clubs on three successive draws (with replacement), we would merely multiply the three probabilities of occurrence.

Probabilistic dependence between events is found when knowledge of the occurrence of one event changes the likelihood of occurrence for the second event. For instance, if we wanted to determine the likelihood of selecting a person from a population that is both a biological female and under 2 meters tall, the occurrence of a biological female would change the likelihood that the person will also be under 2 meters tall. If we stipulate that about 8% of people are 2 meters tall or taller, we will have about a 92% chance of randomly selecting a person who is less than 2 meters tall. But if we stipulate that the person selected is a biological female, the likelihood has just jumped to about 99%. In this case, the occurrence of one event (female) changed the likelihood of the other event occurring. Yes, this is a bit confusing because both events are technically occurring at the same time (at the point of selecting the individual), but the way to calculate the likelihood of co-occurrence necessarily involves exploring the relationship between the two events. To determine the probability of two dependent events, we need to use the more general multiplication rule.

Multiplication rule for two events

$$P(A\ and\ B) = P(A \mid B)P(B) \quad (6.6)$$

The expression $P(A \mid B)$ is read as "The probability of event A occurring *given* the occurrence of event B." The symbol "|" is called a *pipe* and it means, "given the following has occurred." (It does not mean that $P(A)$ is to be divided by $P(B)$.) The pipe connotes *conditional probabilities*, a concept that will be further explored in the next section of this chapter. A worked problem clarifies the use of Formula 6.6.

■ **Question** *What is the probability of randomly selecting a person from the campus student population that is both a psychology major and a biological female?*

Given *Suppose we know that 10% of students are psychology majors, 80% of psychology majors are biological females, and 60% of the entire student body are biological females.*

Solution

Determine $P(A)$, $P(B)$, and $P(A \mid B)$. Let us call being a biological female $P(A)$. Let us call being a psychology major $P(B)$. This would mean that $P(A \mid B)$ is the probability of being a biological female given that one is a psychology major. (This is the only way to assign events for this problem – we do not have enough data to find $P(A \mid B)$ if events A and B are switched.) Since we are given all the needed values, there is no need for preliminary steps.

If being a psychology major and being a biological female were independent of each other (where the occurrence of one of these events did not change the rate of occurrence for the other), then we could use Formula 6.5 and determine that:

$$P(A\ and\ B) = P(A)P(B) = (.6)(.1) = .06$$

However, these events are not independent, and so we must use Formula 6.6. In this case,

$$P(A\ and\ B) = P(A \mid B)P(B) = (.8)(.1) = .08$$

The probability of randomly selecting a student on campus who is both a psychology major and a biological female is 8%. Since there is a dependency between events A and B, we should *not* use Formula 6.5; rather, we should use Formula 6.6. ■

Formula 6.6 allows us to incorporate into the calculations the realization that most psychology majors are biological females. And so, of the population of students, the likelihood of event A and event B co-occurring in the same person is increased a bit by the fact that events A and B already seem to co-occur a bit more than one might expect if the two events were independent of each other. Now, dependence does not imply increasing likelihood. In other situations, events A and B may be dependent in such a way that the occurrence of A *decreases* the likelihood of B occurring. The key term concept regarding independence is whether the likelihood of one event *changes* (increases *or* decreases) the occurrence of the other event.

6.6 Conditional Probabilities

As was stated earlier in the chapter, a **conditional probability** is an expression of likelihood given that another event has occurred. For instance, we can easily see that the probability of randomly selecting a person from a population who is pregnant will change if we know ahead of time if the person selected is a biological male or female. The $P(A \mid B)$ if event A equals "being pregnant" and B equals "being a biological female" is undoubtedly a low number, but $P(A \mid B)$ would be zero if event B was changed to "being a biological male." In a sense, we have been doing conditional probabilities all along. Oftentimes, the conditions surrounding a probability are simply assumed. For instance, what is the probability of rolling a die and having it land on a 4? The commonsense response is that the probability would be 1/6. But what if we were later told that the die was 10-sided? Well, we might be a bit upset to not have been told that up front, and we would instinctively know that we need to change our answer. It helps to think of conditional probabilities as "specifications of the situation." Sometimes the situation is so well known as to be assumed, yet other times it is either unclear or it changes based on other features of the problem. The probability of getting a 4 on a roll of a die changes based on the number of sides.

The formula for determining conditional probabilities is simply a reworking of the multiplication rule for two dependent events. The formula is as follows:

Conditional probability formula

$$P(A \mid B) = \frac{P(A \text{ and } B)}{P(B)} \quad (6.7)$$

By simply dividing both sides of Formula 6.6 by $P(B)$, we transform it into Formula 6.7.

Let us try to determine conditionals by looking at a form of a problem that is often used when teaching probability – the ball-in-basket scenario. Imagine a basket filled with 100 balls. These balls have different colors and have either an "*X*" or a "*Y*" on them. Let us further stipulate that each ball is equally likely to be drawn. Now suppose the basket has the following contents:

20	Red – *X*
20	Red – *Y*
10	Green – *X*
50	Green – *Y*

Based on the given information, we should be able to see the following:

$P(Red)$	= .4 (Formula 6.1)
$P(Green)$	= .6 (Formula 6.1)
$P(X)$	= .3 (Formula 6.1)
$P(Y)$	= .7 (Formula 6.1)
$P(Red\ or\ Green)$	= .6 + .4 = 1 (Formula 6.2)
$P(X\ or\ Y)$	= .3 + .7 = 1 (Formula 6.2)
$P(Red\ or\ X)$	= .4 + .3 − .2 = .5 (Formula 6.4)
$P(Green\ or\ X)$	= .6 + .3 − .1 = .8 (Formula 6.4)
$P(Red\ or\ Y)$	= .4 + .7 − .2 = .9 (Formula 6.4)
$P(Green\ or\ Y)$	= .6 + .7 − .5 = .8 (Formula 6.4)

Furthermore, we should be able to see that the following are true:

P(Red and X)	= .2 (given)
P(Red and Y)	= .2 (given)
P(Green and X)	= .1 (given)
P(Green and Y)	= .5 (given)

Let us try to find some conditional probabilities.

■ **Question** *Find these values? P(Red|X), P(Red|Y), P(Green|X), and P(Green|Y)*

Solution

For *P(Red|X)*, first assign "Red" and "*X*" to events *A* and *B*. Let us define "Red" as event *A* and "*X*" as event *B*. Formula 6.7 will be used for each of these solutions.

$$P(A \mid B) = \frac{P(A\,and\,B)}{P(B)}$$

$$P(Red \mid X) = \frac{P(Red\,and\,X)}{P(X)} = \frac{.2}{.3} = .67$$

The chance of selecting a red ball given an "*X*" ball has been drawn is .67. For *P(Red|Y)*, let us define "Red" as event *A* and "*Y*" as event *B*.

$$P(A \mid B) = \frac{P(A\,and\,B)}{P(B)}$$

$$P(Red \mid Y) = \frac{P(Red\,and\,Y)}{P(Y)} = \frac{.2}{.7} = .29$$

The chance of selecting a red ball given a "*Y*" ball has been drawn is .29. For *P(Green|X)*, let us define "Green" as event *A* and "*X*" as event *B*.

$$P(A \mid B) = \frac{P(A\,and\,B)}{P(B)}$$

$$P(Green \mid X) = \frac{P(Green\,and\,X)}{P(X)} = \frac{.1}{.3} = .33$$

The chance of selecting a green ball given an "*X*" ball has been drawn is .33. For *P(Green|Y)*, let us define "Green" as event *A* and "*Y*" as event *B*.

$$P(A \mid B) = \frac{P(A\,and\,B)}{P(B)}$$

$$P(Green \mid Y) = \frac{P(Green\,and\,Y)}{P(Y)} = \frac{.5}{.7} = .71$$

The chance of selecting a green ball given a "*Y*" ball has been drawn is .71. ■

Before we move on to the final section of this chapter, we need to go back to the topic of independence, but now from a more informed vantage point. When we introduced the concept of independence, we stated that two events are independent if the occurrence of one event does not change the likelihood of occurrence for the other event. Determining if two events are independent is often not a straightforward decision. With a little reflection, we may be able to see that selecting a spade and selecting a face card are independent events, but independence may be much more difficult to determine in other situations.

To help matters, it is important to realize that Formula 6.6 is the more generally true formula for determining multiplication problems in probability (the "and" problems). Formula 6.5 is merely a special case of Formula 6.6; when events A and B are independent. Look at the two formulas. Here we see that the only difference is that Formula 6.5 refers to $P(A)$, whereas Formula 6.6 refers to $P(A \mid B)$. However, if two events are independent, then $P(A)$ should equal $P(A \mid B)$. That is, the likelihood that event A will occur is the same whether or not event B has occurred. (Recall the card drawing question. The probability of selecting a spade is 1 out of 4 whether or not the card is a face card.) This means we can use the two formulas to determine if two events are independent of each other. We can simply run both formulas and see if they yield the same result. If they do, then we know that $P(A) = P(A \mid B)$, and events A and B must be independent. If, on the other hand, the formulas yield different answers, then we know that $P(A) \neq P(A \mid B)$; events A and B must be dependent upon one another. (We also know the answer provided by Formula 6.6 is the correct one.) In this way, we can use the two "and" formulas to figure out independence.

■ **Question** *Given the following situation involving balls in a basket, are events Green and X independent? What about Yellow and Y?*

15	Green – X
15	Green – Y
10	Red – X
20	Red – Y
25	Yellow – X
15	Yellow – Y

Solution

Step 1. To determine if events "Green" and "X" are independent, we need to first determine the values corresponding to Formulas 6.5 and 6.6. Let us assign "Green" to be event A and "X" to be event B. These are $P(Green)$, $P(X)$, $P(Green|X)$.

P(*Green*)	= .3 (Formula 6.1)
P(*X*)	= .5 (Formula 6.1)
P(*Green*|*X*)	= .15/.5 = .3 (Formula 6.7)

Step 2. Then we "run" both formulas.

$$P(A\ and\ B) = P(A)P(B)$$
$$P(Green\ and\ X) = P(Green)P(X) = .3(.5) = .15 \quad (6.5)$$

$$P(A\ and\ B) = P(A \mid B)P(B)$$
$$P(Green\ and\ X) = P(Green \mid X)P(X) = .3(.5) = .15 \quad (6.6)$$

Both formulas yield the same answer (.15). Events "Green" and "*X*" are independent. The chance of a "Green" ball occurring is not influenced by the occurrence of an "*X*" ball.

Now let us consider if "Yellow" and "*Y*" are independent.

Step 1. To determine if events "Yellow" and "*Y*" are independent, we need to first determine the values corresponding to Formulas 6.5 and 6.6. Let us assign "Yellow" to be event *A* and "*Y*" to be event *B*. These are *P*(*Yellow*), *P*(*Y*), *P*(*Yellow*|*Y*).

P(*Yellow*)	= .4 (Formula 6.1)
P(*Y*)	= .5 (Formula 6.1)
P(*Yellow*|*Y*)	= .15/.4 = .375 (Formula 6.7)

Step 2. Then we need to "run" both formulas.

$$P(A\ and\ B) = P(A)P(B)$$
$$P(Yellow\ and\ Y) = P(Yellow)P(Y) = .4(.5) = .2 \quad (6.5)$$

$$P(A\ and\ B) = P(A \mid B)P(B)$$
$$P(Yellow\ and\ Y) = P(Yellow \mid Y)P(Y) = .375(.5) = .19 \quad (6.6)$$

The formulas do not yield the same answer (.2, .19). The values are close, but technically, "Yellow" and "*Y*" are not independent. The chance of selecting a yellow ball is influenced by the occurrence of a "*Y*" ball (and vice versa). ■

A few more observations before we conclude. Please know that *P*(*A* or *B*) = P(*B* or *A*). Also, *P*(*A* and *B*) = *P*(*B* and *A*). These probability functions are commutative. However, $P(A|B)$ almost certainly does not equal $P(B|A)$. These are two very different statements. For instance, the probability that a New Yorker is Rude, *P(Rude|New Yorker)*, is not the same as the probability that a rude person is a New Yorker, *P(New Yorker|Rude)*.

Also, up to now, we have been asking questions about how likely something is to occur, but of course, we could be asking how likely it is for something to *not* occur. For instance, we could ask, "what is the probability that '*notA*' is going to happen?" To help clarify our thinking here, we need to recognize that the probability of event *A* and event *notA* combined is always 1. Together they make up what probability theorists refer to as an "exhaustive set." We can explore this a bit by asking ourselves questions like "What is the probability that we are wearing blue socks or not wearing blue socks?" Well, that probability is 1, correct? What is the probability that it is going to rain or not rain tomorrow? Again, the probability is 1. If the probability that a randomly selected person is of retirement age is .2, then the probability that a randomly selected person is not of retirement age must be .8. If we know the probability of event *A*, we can always deduce the probability of event *notA*.

Summary

This chapter and Chapter 7 serve as theoretical "bridges" connecting descriptive statistics to inferential statistics, the remaining material in the textbook. Inferential statistics, the ability to draw inferences about populations based on samples drawn from those populations, is dependent upon several concepts related to probability theory and hypothesis testing.

Modern probability theory started in the seventeenth century when several key thinkers decided it was best to approach questions of likelihood with a sense of willful ignorance regarding the many idiosyncratic issues associated with them, opting instead to focus on general tendencies. Probability is understood mathematically as a proportion ranging from 0 to 1: 0 being events that do not occur, 1 being events that are certain to occur. A distinction is made between sampling with and without replacement; "with replacement" requires a selected event to be returned to the pool of events prior to the next event's selection.

Various formulas are used to determine specific probabilities: the basic probability formula, the "or" formulas, the "and" formulas, and the conditional probability formula. To distinguish between the "or" formulas, the concept of "mutual exclusivity" is needed. To distinguish between the "and" formulas, the concept of "independence" is needed.

Key Formulas

Probability of favorable event (when all events are equally likely)

$$P = \frac{(number\ of\ favorable\ events)}{(total\ number\ of\ events)} \quad (6.1)$$

Addition rule formula for two mutually exclusive events

$$P(A\ or\ B) = P(A) + P(B) \quad (6.2)$$

Addition rule formula for more than two mutually exclusive events

$$P(A\ or\ B\ or\ C\ or ... Z) = P(A) + P(B) + P(C) + \cdots + P(Z) \quad (6.3)$$

Addition rule formula for two events

$$P(A\ or\ B) = P(A) + P(B) - P(A\ and\ B) \quad (6.4)$$

Multiplication rule formula for two independent events

$$P(A\ and\ B) = P(A)P(B) \quad (6.5)$$

Multiplication rule formula for two events

$$P(A\ and\ B) = P(A \mid B)P(B) \quad (6.6)$$

Conditional probability formula

$$P(A \mid B) = \frac{P(A\ and\ B)}{P(B)} \quad (6.7)$$

Key Terms

Probability
A priori (classical) approach
A posteriori approach
Addition rule
Mutually exclusive events
Multiplication rule
Probabilistic independence
Probabilistic dependence
Conditional probability

Exercises

1 Together, this chapter and Chapter 7 allow the researcher to: (select the best answer)
   a Understand how to run inferential statistics
   b Determine the nature of samples from populations and populations from samples

c Think philosophically about numbers
d Understand the relationship between numbers and people

2 Which of the following events are mutually exclusive?
a Being a resident of Country A; being a resident of Country B
b Making one toss of a die and obtaining either an even number or a 2
c Drawing a 5 from a deck of cards, replacing it, and then drawing a 2
d A person has black hair and blue eyes
e Obtaining a 1 and a 6 when rolling two dice at once
f Being a dog; being a cat
g Being pregnant; being not pregnant
h Being a Yankees fan; being a Red Sox fan
i Wearing an official Star Trek fan club T-shirt; getting a date

Work Problems

3 Assuming all events are equally likely, please determine the following probabilities:
a Selecting a spade from a deck of 52 cards
b Rolling a 6 on a 10-sided die
c Randomly selecting a pawn from a set of chess pieces
d Selecting North Carolina from a lottery draw involving all 50 states
e Missing a pop quiz in Psychology 101 due to skipping class

4 Suppose there is a bin with 40 red marbles and 60 white marbles. In each of the following problems, selections are made blindly, and the marbles are randomly distributed throughout the bin. Given this, what is the probability of:
a Drawing a red marble?
b Drawing a white marble?
c Drawing either a red or a white marble?
d Drawing two red marbles in a row, with the first marble replaced?
e Drawing two white marbles in a row, with the first marble replaced?

5 For each situation, specify whether the events are independent or dependent.
a The weekly state lottery drawing a successive selection of six balls
b The six lottery balls selected one week, and the six lottery balls selected the next
c Drawing two 5s from a deck of cards without replacing the first card
d Drawing a 3 from a deck of cards, replacing it, and then drawing a 6

6 Suppose we have a basket with the following set of balls:

10	Red – *X*
20	Red – *Y*
10	Green – *X*
30	Green – *Y*
15	Yellow – *X*
15	Yellow – *Y*

Find:

a *P*(*Red*)
b *P*(*Green*)
c *P*(*not Red*)
d *P*(*Red or Yellow*)
e *P*(*Red or X*)
f *P*(*Red and X*)
g *P*(*Red*|*X*)
h *P*(*X*|*Red*)
i Are events Red and *X* independent?
j Why or why not?

7 Suppose we have a basket with the following set of balls:

20	Green – *X*
15	Green – *Y*
10	Red – *X*
15	Red – *Y*
30	Yellow – *X*
10	Yellow – *Y*

Find:

a *P*(*Green*)
b *P*(*Y*)
c *P*(*Red or Yellow*)
d *P*(*Red or Yellow or Green*)
e *P*(*Red or X*)
f *P*(*not Green*)
g *P*(*X or Y*)
h *P*(*Yellow and X*)
i *P*(*Yellow and Red*)
j *P*(*Red*|*Y*)

k $P(Y|Red)$
l Are events X and *Green* independent? Why or why not?
m Are events *Yellow* and Y independent? Why or why not?

8 Suppose we know the following about a basket with red and green balls marked with an X or Y (i.e. four different kinds of balls: Red – X, Red – Y, Green – X, Green – Y):

$P(Red) = .2$	$P(Red\ and\ X) = .1$	$P(X) = .3$

Find:
a $P(Green)$
b $P(Red\ or\ X)$
c $P(Red|X)$
d $P(X|Red)$
e Are events *Red* and X independent? Why or why not?

9 Suppose we have a basket with the following set of balls:

15	Green – X
15	Green – Y
10	Red – X
20	Red – Y
25	Yellow – X
15	Yellow – Y

Find the following probabilities:
a $P(Green)$
b $P(Y)$
c $P(Red\ or\ Yellow)$
d $P(Red\ or\ X)$
e $P(not\ Green)$
f $P(X\ or\ Y)$
g $P(Yellow\ and\ X)$
h $P(Yellow\ and\ Red)$
i $P(Yellow|Y)$
j $P(Y|Yellow)$
k Are events *Green* and X independent? Why or why not?
l Are events *Yellow* and Y independent? Why or why not?

10 In the dice game *craps*, we win on the first roll if we get a 7 or an 11. (Two dice are rolled, and the numbers are added.) What is the probability of winning on the first roll? Hint: How many ways can you get a seven from the roll of two dice? What about 11?

7

Hypothesis Testing and Sampling Distributions

7.1 Inferential Statistics

As previously mentioned, probability theory and inferential statistics developed in the 1600s around questions related to gambling. The modern era of inferential statistics began in the late nineteenth century. One driving force was the interest of agricultural companies and breweries regarding the effectiveness of various treatments on crop yields. Another was the measurement needs of a growing number of social scientists interested in theorizing about a variety of economic, sociological, and psychological phenomena. In response to these needs, most of the basic statistical tools presented from Chapter 8 through the remainder of the text were developed within the span of just a few decades, from the 1880s to the 1920s.

7.1.1 Types of Inferential Estimation

To make a statistical inference is to use sample information to *estimate* a feature of a population. There are two kinds of estimations to be made: point estimation and interval estimation. Suppose we take a random sample and compute the mean. If someone were to ask us to estimate the mean of the population, we could use our sample mean to make a **point estimation.** Any sample statistic (e.g. mean, standard deviation) can be used to make a point estimation of a population parameter. The other kind of estimation is called a **confidence interval** (or interval estimation). In this procedure, a formula is used to generate two values that set the interval within which the population parameter can be found with a particular degree of confidence. Both forms of estimation will be explored in this textbook.

Essentials of Behavioral and Social Science Statistics, Third Edition.
K. Paul Nesselroade, Jr. and Laurence G. Grimm.
© 2025 John Wiley & Sons, Inc. Published 2025 by John Wiley & Sons, Inc.
Companion website: https://www.wiley.com/go/Essentials_of_Behavioral_and_Social_Science_Statistics3e

7.2 Hypothesis Testing

A **research** (or *scientific*) **hypothesis** is a formal statement or expectation about the outcome of a study. They are usually stated in terms of independent and dependent variables, and the relationship between them, or in terms of two variables and the degree of association between them. These statements are oftentimes derived from relevant preexisting theories and/or based on relevant previous research findings. For example, we might read something like the following, "Given that time pressures diminish the tendency to help (e.g. Darley and Batson 1973; Moore and Tenney 2012), participants in the low stress situation are predicted to be more helpful than participants in the high stress situation." The accumulation of knowledge in the behavioral and social sciences relies heavily on the process of formulating theories, stating hypotheses, using data to test hypotheses, and conducting further research. Although **hypothesis testing** occurs in many different contexts, most instances of it share two common characteristics: (i) the use of inferential statistics to extrapolate from samples to populations, and (ii) the effort to quantify the confidence that should be placed in these extrapolations.

7.2.1 The Use of Sample Data to Make Inferences About Populations

If an investigator is *only* interested in drawing conclusions about the participants in the study, then hypothesis testing is irrelevant. Confining one's interest in this way defines the participants *as* the population; thus, no inference is required. One goal of the researcher, however, is to acquire general knowledge about the world. The behavior of participants is interesting, only insofar as it allows us to make statements about the behavior of people beyond our study, a larger population. This requires representative samples.

To use samples to infer the characteristics of populations, mathematical tools are needed. Since the researcher is forced to work at the level of samples, anything concluded about populations always involves a degree of uncertainty. Inferences about population parameters are, therefore, probabilistic in nature. However, these probabilistic claims can be quantified.

7.2.2 Decision Errors: An Unpleasant Fact of Life in Hypothesis Testing

Whenever a conclusion is drawn about a population based on sample data, there is a chance the conclusion will be wrong. For example, suppose we wanted to use an upcoming timed test to compare two different anxiety reduction techniques for students. Suppose those given technique A were subsequently shown to be less anxious on the test than those given technique B. From this, we might conclude that technique A is more effective than technique B for the entire population of students. Although this is likely to be an accurate conclusion, there is always a chance it is not. Despite our best efforts, the samples may not represent the populations well, and our conclusion could be wrong. For this reason, there are concepts and principles presented later in the text to help us think about, reduce, and quantify the uncertainty associated with statistical conclusions. (One of these principles is replication; see Box 7.1.)

7.2.3 Research Hypotheses vs. Statistical Hypotheses

As noted previously, a *research hypothesis* is a statement based on relevant previous findings and/or theories regarding the expected outcome of a study. Each research hypothesis is accompanied by a statistical hypothesis. A **statistical hypothesis** is a numerical statement regarding the potential outcomes of a study. It serves as the vehicle for evaluating the research hypothesis. Some statistical hypotheses rely on known population parameters. These are sometimes referred to as *single-sample* research designs. For example, suppose an existing measure of stress has a known population mean of 50. A community psychologist, interested in the emotional effects of natural disasters, asks 100 randomly sampled people recently affected by an earthquake, to complete the stress measure. The research hypothesis may be stated as, "Natural disasters elevate stress levels among victims." (Note: Research hypotheses are typically based on predictions of differences; that is, that something will change.) The statistical hypothesis can be stated as, "The mean stress score of earthquake victims is greater than 50." The researcher then gathers data and decides if there is evidence to support the statistical hypothesis.

Other statistical hypotheses are based on similarities and differences between two groups; these hypotheses are sometimes referred to as *two-sample* research designs. An example of a statistical hypothesis involving two groups might read, "There is a difference in the mean number of arrhythmias between a 'beta-blocker' and a 'placebo' population." Notice that in these situations there are no known means; rather, the question at issue has to do with whether there is a *difference* between the two population means ("beta-blocker" and "placebo"). The job of the researcher is to gather data from both populations and decide if there is evidence that the population means are not the same.

Most current experimental research, however, involves more than two samples. For example, an experimental psychologist might be interested in the relationship between different types of reinforcement and the number of trials needed to learn a task. To explore this, several different kinds of incentives may be tested.

In the behavioral and social sciences, hypothesis testing does not always take place within an experimental context. The correlational approach is an alternative research method that differs from experimentation in that a presumed causal variable is not manipulated (recall the introduction of correlational designs in

Chapter 1). Because a variable is not manipulated, correlational research cannot identify a causal relationship. Instead, this approach measures the strength of a relationship. An example of a correlational research hypothesis is, "First-time parents of newborns who cry a lot have greater marital dissatisfaction than those with quiet babies." The investigator will search for a relationship between the amount of infant crying and parents' marital dissatisfaction. No attempt is made to manipulate variables.

Although the correlational approach does not manipulate variables, it does *not* alter the fundamental characteristics of hypothesis testing. The job of the researcher is to gather data and decide if there is evidence to conclude a relationship exists.

7.2.4 More on Statistical Hypotheses: The Null and Alternative Hypotheses

As previously noted, statistical hypotheses are numerical statements regarding the potential outcomes of an experiment. When conducting a study, statistical hypotheses always come in pairs: a null hypothesis, denoted H_0, and an alternative hypothesis, denoted H_1. In the context of an experiment, the **null hypothesis** states that there is *no effect* of the independent variable on the dependent variable. In a correlational context, the null hypothesis states that there is *no relationship* between two variables. It may help to realize that in hypothesis testing, the word "null" means either *no differences* between groups or *no relationships* between variables. The **alternative hypothesis** states that there *is* something going on. In the context of an experiment, it means that the independent variable has affected the dependent variable. In a correlational context, it means the two variables are related. The alternative hypothesis is the logical alternative and complement of the null hypothesis.

Box 7.1 Is the Scientific Method Broken? The Value of Replication

When researchers conclude the null cannot be rejected (also known as "failing to reject the null hypothesis"), the study's findings are deemed "nonsignificant." This term is a way of expressing the idea that differences between the sample means of the various conditions in a study are not substantial enough to warrant rejecting the null hypothesis. Most journals in the social and behavioral sciences are not interested in publishing research that has not found evidence of differences between conditions, so-called "nonsignificant findings." Each journal wants to include only articles that seem important and are likely to be read and referenced by others.

This policy, however, can create a problem. Imagine several researchers working separately, but each looking at a similar research question.

Furthermore, imagine that their research hypothesis is, in the end, not a good one. That is, perhaps the null hypothesis is true; the independent variable has no effect on the dependent variable. However, given the sampling error that naturally occurs within a sampling distribution, suppose one of the researchers gets an extreme sample mean that prompts them to reject the null. The researcher is in error; their sample mean was unusual and not representative of the population. However, they do not realize this and believe they are correct in rejecting the null hypothesis and claiming to have found an effect. If all the researchers who were looking at the same topic published their findings, readers might become suspicious of the one study showing an effect among the many others that do not. However, in current practice, readers will not be exposed to the nonsignificant findings. These other studies will not be published in journals. The only study that will be published is the one that found evidence to reject the null. Once we realize this, it is easy to see how the current publication practices in the behavioral and social sciences create the possibility that findings from some published studies may not be reliable.

A potential remedy for this is the replication of published studies. Unfortunately, not only are insignificant findings not valued by journals, neither are replication attempts. For example, according to Makel and others (2012), only about 1.6% of all published studies in the top 100 psychology journals from 1900 to 2012 were replication attempts. Furthermore, according to Pittelkow et al. (2023), funding for replication attempts is not keeping up with the exponential growth in published research. The backlog is growing, not diminishing.

Nonetheless, attempts in recent years to combat this problem are taking hold. A growing number of researchers are serving the scientific community by carefully and painstakingly replicating published research findings. A good example of this growing trend is the Center for Open Science (https://cos.io/). Hopefully, the value of replication efforts will continue to rise.

The null and alternative hypotheses are statistical hypotheses expressed numerically. For example, if an educational enrichment program is expected to increase IQ and we know that the average IQ in the population is 100, the null and alternative hypotheses are stated as

$$\mathrm{H}_0 : \mu = 100$$
$$\mathrm{H}_1 : \mu \neq 100$$

The null hypothesis can be determined as untenable whether the program increases *or* decreases IQ. To reject the null hypothesis in favor of the alternative hypothesis, we will need to find statistical evidence that the null is likely to be false.

Taken together, the null and alternative hypotheses are *mutually exclusive* and *collectively exhaustive*. Consider the hypothesis that a population mean is 100. Well, it either equals 100 or it does not. The hypotheses are mutually exclusive in that only one of them *can* be true and collectively exhaustive in that one of them *must* be true.

Notice also that statistical hypotheses are always made in reference to population parameters, not sample statistics. We will gain sample means or sample correlations from our data, but our inferential decision will always be in reference to a population parameter (usually a mean, a mean difference, or a correlation).

7.2.5 Truth in Hypothesis Testing

Once the data is analyzed, the researcher must make a decision about the null hypothesis. The two options available are to *reject* the null (in favor of the alternative) or to *not reject* the null – in other words, to *fail to reject the null*. Notice that there is no option to accept the null hypothesis. This is because it is not possible, on the strength of sample data alone, to demonstrate that it is true. To claim that the null is true is to make a strong statement. Analysis of samples, however, only provides *evidence* regarding population parameters; the evidence can suggest whether the null is false (in which case, we decide to reject it) or not (in which case we decide not to reject it). Please note: *Failure to reject the null does not mean the null has been determined to be true*. It could be, of course, but sample evidence alone cannot settle it. This is one of the limitations of using samples to make inferences about populations. For this reason, when we fail to reject a null, the conclusion should be that we simply do not know if the null is true or not; it is more like a scientific "shrug-of-the-shoulders" than a pronouncement that the null is correct. Unfortunately, this is a common error in statistical interpretation.

On the other hand, if the statistical analysis suggests that the null hypothesis should be rejected, this only means that we have *evidence* in favor of the alternative hypothesis. However, evidence is not proof; it could be misleading.

Of course, either the null hypothesis is true or it is not. And yet, we are not able to make this determination with certainty from sample data. The methods of hypothesis testing are probabilistic; probability, by definition, means that there is some level of uncertainty. This may seem unsettling, but it is unavoidable. However, even though decision errors can be made, we can, to some degree, quantify and understand the likelihood of making these errors. This topic will be more carefully addressed in the next chapter once we start using inferential statistics to make judgments regarding the null hypothesis.

Before leaving this topic, it is important to stress the posture difference between research hypotheses and statistical hypotheses. Research hypotheses are not stated numerically, are derived from theory and/or previous findings, and, most importantly, are typically predicting either a difference, an effect, or a relationship will be found. In short, research hypotheses are *hopeful* that something interesting will happen. Otherwise, why are we engaged in the research? Compare this with the posture of statistical hypotheses which start with the null as the default position. The null states that nothing is going on, and evidence to the contrary will have to be found before it is rejected. The initial posture is *skeptical* (as if to say, "Show me!"). While our heads and hearts may be predicting and hoping to find some effect or relationship, the mathematics of hypothesis testing starts with the presumption that nothing is going on.

One final point concerns the relationship between rejecting the null hypothesis and supporting the research hypothesis. Since the alternative hypothesis covers both sides of the null hypothesis, rejecting the null may or may not support the research hypothesis. It is possible to reject the null and yet fail to support the research hypothesis. For example, suppose we hypothesize Group 1 to outperform Group 2. Suppose further that evidence is found suggesting Group 2 outperformed Group 1. Even though the null hypothesis is rejected, this conclusion does not support our research hypothesis; the findings were opposite of the predicted direction. Rejecting the null does not necessarily mean one's inclinations were supported.

7.3 Sampling Distributions

Statistical hypothesis testing is the topic of interest for the rest of this textbook. Its conceptual foundation and the application of formulas to research data will be emphasized. These two topics are intertwined. A conceptual understanding is needed to understand why a formula is structured a certain way, when it should be used, and what the conclusion means.

The most important concept in inferential statistics is the *sampling distribution*. As we work through the rest of the chapter, be forewarned that the connection between sampling distributions and hypothesis testing may not be entirely obvious. Subsequent chapters, however, will develop this relationship.

7.3.1 Population and Sample Distributions

Every population of scores can be depicted as a probability distribution that reflects the frequency of occurrence of every score in the distribution. Furthermore, if the distribution is normal and we know the mean and standard deviation, we can make probability statements about the likelihood of selecting a score from a specified area of the distribution (remember all those z-score problems in Chapter 5?).

We already know the difference between a population and a sample, with the latter being a portion of the former. Furthermore, we know that samples have value only insomuch as they represent the population from which they come: the population we wish to study. Most importantly, when we make inferences from a sample to a population, we do not make inferences about a specific score. Instead, we make inferences about a population parameter from a sample statistic. For example, we may infer the mean or the variance of a population based on the mean or variance of a sample. It is meaningless to take a single score from a sample and try to estimate a single score of a population.

In the discussion of z scores in Chapter 5, questions were asked like, "What is the probability of selecting a *score* of less than 20, given that the mean of the population is 26 and the standard deviation is 4?" We were able to answer such questions because we used a standardized normal distribution, the z-score distribution: a distribution of transformed raw *scores*. However, in hypothesis testing, we are interested in population parameters – means (usually), never individual scores. If we want to make a probability statement about a sample *mean* coming from a given population, we need to create a *normal distribution of means* rather than of raw scores. This important topic is addressed in the following section.

7.3.2 A Sampling Distribution of Means

Generally stated, a **sampling distribution** is a theoretical frequency distribution of a statistic (usually a mean) based on a very large number of repeated samples of some specified size, n. With respect to means, a sampling distribution shows the relative frequency of all possible values of sample means, given a selected sample size. A researcher never actually goes through the process of constructing a sampling distribution. As was noted above, it is a *theoretical* mathematical concept; however, it is critically important to the hypothesis-testing process. The concept of a sampling distribution can be grasped by imaginatively walking through the steps that would be taken to construct it.

Step 1. Choose a population of scores. (The data must be interval or ratio.)
Step 2. Decide on a sample size, n. It can range from 2 to infinity.
Step 3. Take a random sample of size n, the sample size decided upon in step 2.
Step 4. Compute the sample mean and then replace the scores back. (This is called *random sampling with replacement*; see Section 6.3.)
Step 5. Repeat steps 3 and 4 until every possible combination of participants of size n has been drawn. (Now we see why this is an imaginative exercise!)
Step 6. Finally, plot the relative frequency distribution of all the means. This distribution is the sampling distribution of means *for that particular sample size*.

In this sampling distribution, the statistic of interest for us was the mean. The steps are the same no matter what statistic we select. If we need a sampling distribution of variances, use the same procedure, except compute and plot variances.

We could even take two samples, calculate the means, find the difference between them, and establish a sampling distribution of mean differences (techniques like this will be featured later in the text). There are, in fact, many occasions in hypothesis testing when a sampling distribution will not be a distribution of means.

7.3.3 Characteristics of Sampling Distributions

7.3.3.1 The Central Limit Theorem

This section begins with a theorem, the most important theorem in inferential statistics. It was formulated in 1810 by Pierre Laplace (1749–1827) and is called the **central limit theorem.**

There are several claims made by the central limit theorem that are critically important for understanding sampling distributions. First, if the population is normally distributed, the sampling distribution of means will be normally distributed. Second, even if the scores in the population are *not* normally distributed, assuming n is sufficiently large, the sampling distribution of means will be normally distributed. In other words, unless the raw population of scores is wildly non-normal *and* the selected sample size is rather small, the resulting sampling distribution will be normal. Third, the central limit theorem states that the mean of a sampling distribution, symbolized as μ_M (the population mean of all the sample means), *has the same value as the mean of the population of raw scores*. Therefore, $\mu = \mu_M$. This can be confusing. We are used to thinking of a mean as a feature of a set of *raw scores*. A sampling distribution is a set of *means*, and this set of means has its own mean, μ_M.

7.3.4 The Relationship Between σ and σ_M

One more claim from the central limit theorem is important to note: Whereas the mean of the sampling distribution is identical to the mean of the population, $\mu = \mu_M$, the standard deviation of the sampling distribution is not equal to the standard deviation of the population. The two measures are related, however, as Formula 7.1 shows. The standard deviation of the sampling distribution of means is called the **standard error of the mean** or simply the *standard error.*

Standard error of the mean

$$\sigma_M = \frac{\sigma}{\sqrt{n}} \quad (7.1)$$

Note the denominator of the standard error. The n refers to the sample size selected when taking repeated samples during the theoretical construction of the sampling distribution. Remember, when constructing a sampling distribution, every sample needs to be the same size. Therefore, the relationship between σ and

σ_M depends on the size of the samples being drawn. As a result, the variability of the sampling distribution is determined by the variability within the population of raw scores *and* the size of the samples used.

7.3.5 How and Why *n* Affects the Standard Error of the Mean

Suppose two different sampling distributions from the same population are established. In the first sampling distribution, the sample size is 4. In the second sampling distribution, the sample size is 25. The standard deviation of the population is 32. Using Formula 7.1,

$$\sigma_{M_1} = \frac{32}{\sqrt{4}} = 16 \quad \text{and} \quad \sigma_{M_2} = \frac{32}{\sqrt{25}} = 6.40$$

The standard error of the mean becomes smaller as the sample size increases. This is not the case with the mean of the sampling distribution; $\mu = \mu_M$, no matter the size of n. Step 2 in constructing a sampling distribution requires us to specify the size of the samples. By using Formula 7.1, we can see that n influences the variability of the sampling distribution; specifically, as n increases, the standard error decreases. This realization may prompt us to ask if there is a preferred sample size for sampling distributions. There is not. For a given set of scores, there are as many potential sampling distributions as there are potential sample sizes. Is there a preferred standard error value? Not really. However, it is true that the smaller the standard error, the more likely a null hypothesis can be rejected. For this reason, small standard errors are attractive.

Also, recall the second claim of the central limit theorem: As n increases, the resulting sampling distribution approaches normality, *even if* the population of raw scores is not normal. This means that as n increases, the sampling distribution from even a skewed raw data distribution becomes more normal. We will soon learn that most statistical tests assume a normal distribution, so a large n helps to meet that assumption. The size of n needed to achieve a normal sampling distribution depends on the degree of skewness or abnormality in the raw data. For most social and behavioral science contexts, an n of 30 is presumed to be sufficient, but oftentimes, much smaller sample sizes are uncontroversial.

One more point to be made: as the sample size increases, the sampling distribution will not only become more normal, but it will also look more and more like *the standard normal*, the same distribution described in detail in the z table (Table A.1).

Given two sampling distributions, why does the one with a smaller n have a larger standard error? As an example, let us contrast a sampling distribution in which $n = 2$ with one in which $n = 20$. Each one of the means of the sampling distributions is composed of scores. When $n = 2$, there are two scores that are averaged to arrive at the mean for that sample. When $n = 20$, there are 20 scores averaged to arrive at the mean. When thinking about unusually high or low sample means,

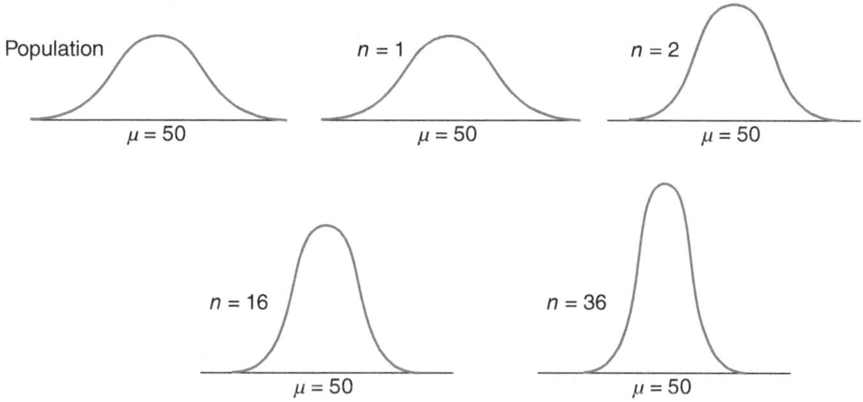

Figure 7.1 Several sampling distributions constructed with different sample sizes.

which is more *likely*, selecting 2 extreme scores to get an extreme sample mean or selecting 20 extreme scores to get an extreme sample mean? Although we do not usually get 2 extreme scores in a row, it is much more likely than getting 20 extreme scores within the same sample. In other words, with a small sample, there is less opportunity for other scores to compensate for extreme scores; in large samples, however, the role of extreme scores is diluted by the numerous common scores. Therefore, a sampling distribution based on $n = 2$ will contain many more extreme means than a sampling distribution where $n = 20$. (Try this thought experiment – imagine having a sample size that is almost the size of the population itself, say, $N - 1$. If we were to generate a sampling distribution by sampling the entire population −1, calculating the mean, replacing the scores, and doing this repeatedly, would not the resulting sampling distribution have an extremely small standard error? Every sample mean would be virtually the same score.) Figure 7.1 shows how the shape of a sampling distribution is affected by changes in the size of the samples.

The following list summarizes the main points of sampling distributions of means:

1) If a population of scores is normally distributed, the sampling distribution of means will be normally distributed.
2) Even if a population of scores is not normally distributed, the sampling distribution will approximate a normal curve as the sample size increases.
3) The sampling distribution mean equals the raw score population mean, $\mu = \mu_M$.
4) The standard deviation of the sampling distribution is called the standard error of the mean. The relationship between the standard error of the mean and the standard deviation of the population is $\sigma_M = \sigma/\sqrt{n}$.

Box 7.2 shows us how to use online software programs to create sampling distributions.

Box 7.2 Playing with the Numbers: Create Our Own Sampling Distribution

Programs found on the Internet allow us to see how changing the sample size, mean, and standard deviation of the population of raw scores changes the resulting sampling distribution. An online search for "sampling distribution applets" will produce several options. One that is quite flexible was created by Dr. Patrick Wessa (www.wessa.net/rwasp_samplingdistributionmean.wasp). In this program, we can input the number of replications we want to make (choose a number in the hundreds at least), the sample size (n), the mean of the population of raw scores, and the standard deviation of the population of raw scores. (It is recommended to leave the width and height of the chart as the given default values.) After inputting some values, click the "compute" button, wait for the program to run, and then scroll down to the second chart. Here we will see a graph reflecting the various sample means given our selected sample size, mean, and standard deviation.

By changing the values of the imputed numbers, we should be able to see the general principles of the sampling distribution concept as well as the claims of the central limit theorem on display. For instance, if we increase the sample size, we should see the standard deviation of the sampling distribution (or, the standard error) tighten (and vice versa if we decrease the sample size). (If it does not look like it has tightened, check the values on the X-axis to see if they were adjusted to better present the data.) If we change the value of the mean, the entire distribution should shift to be centered on the new mean. If we increase the standard deviation of the raw scores, the resulting sampling distribution should widen (and vice versa if we decrease the standard deviation). By exploring this program or other similar ones, we should be able to see that $\mu = \mu_M, \sigma_M = \sigma/\sqrt{n}$, and all sampling distributions approximate a normal distribution.

7.4 Estimating the Features of Sampling Distributions

Only in rare situations will a researcher know the value of population parameters. However, population parameters are needed to determine μ_M and σ_M. Thankfully, in most cases, population parameters can be estimated from sample statistics. If μ is known, then μ_M can be computed with certainty ($\mu = \mu_M$). If μ is not known, then a sample mean (M) can be used as an unbiased estimate of μ. ("Unbiased" means that the sample mean is just as likely to be larger than the population mean as it is to be smaller.) Of course, if σ is known, then σ_M can also be computed with certainty. However, when σ is unknown, σ must be estimated by using s (a sample standard deviation), an unbiased estimate of σ. Therefore, s_M becomes an estimate of σ_M. Formula 7.2 is the estimated standard error of the mean derived from the standard deviation of a single sample.

Estimated standard error of the mean

$$s_M = \frac{s}{\sqrt{n}} \quad (7.2)$$

---

where

s_M = the estimated standard error of the mean
s = the sample standard deviation
n = the sample size

Formula 7.2 allows us to estimate the amount of variability within a sampling distribution. As an estimate of σ_M, s_M becomes more reliable as the sample size increases.

Table 7.1 clarifies the various symbols and estimates that characterize sampling distributions.

Table 7.1 Important symbols of sampling distributions and population estimates.

1) M is an estimate of μ
2) s is an estimate of σ
3) μ_M is the same value as μ
4) σ_M is the standard deviation of a sampling distribution, the standard error of the mean
5) s_M is an estimate of σ_M

7.4.1 Sampling Distributions and Sampling Error

The notion of sampling error is fundamental in statistical inference. When discussing the mean in Chapter 3, an error was defined as the distance a score is from the mean of the distribution, $X - M$. The term "error" may be confusing as it usually implies a mistake; this is unfortunate. Within the context of a sampling distribution, "error" represents the amount by which a sample mean value is different from the mean value of the sampling distribution. Only those sample means that are equal to μ_M are perfect estimates of μ. Any sample mean that is different from the mean of the sampling distribution is in error in the sense that it is an inaccurate estimate of μ. A **sampling error** is the difference between a population parameter and the estimate of that parameter provided by a statistic. Thus, when estimating μ, a sampling error is $M - \mu$.

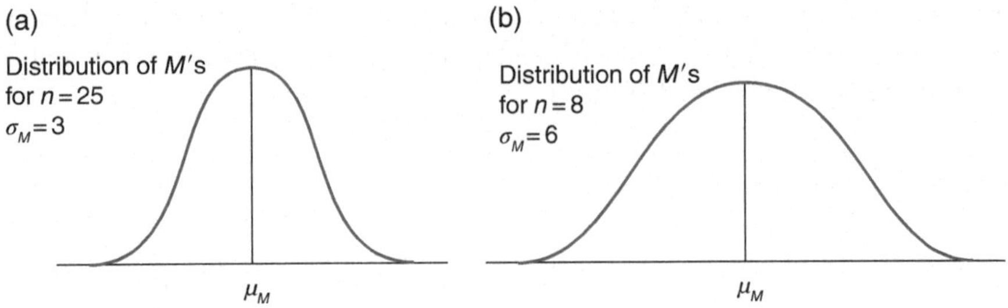

Figure 7.2 Sampling error increases from (a) to (b) as σ_M increases.

When working with a population distribution of raw scores, σ is the overall measure of error that is typically used. With a sampling distribution, the overall measure of error is σ_M. To demonstrate this, see that Figure 7.2 shows two sampling distributions with the same mean. Figure 7.2a has a smaller standard error, and as a result, the distribution is narrower than the sampling distribution in Figure 7.2b, where there are more sample means at a greater distance from μ_M. Suppose we select a mean at random from each distribution in Figure 7.2. With an error defined as $M - \mu$, from which distribution would we expect the larger error? Since the means in distribution 7.2b are more dispersed, we are more likely to have greater error with a mean selected from that distribution. In other words, a randomly selected mean from a broader distribution will likely be a poorer estimate of μ than one selected from a narrow distribution.

At this point, there is no way for us to fully appreciate the importance of sampling distributions. However, it is impossible to understand the statistical foundation of hypothesis testing if we do not first understand the characteristics and logic of sampling distributions. In large measure, the difference between being a statistical "number cruncher" and truly *understanding* inferential statistics depends on our ability to grasp the sampling distribution concept. The better we understand this chapter, the easier time we will have mastering much of the material presented in subsequent chapters. If there is confusion, please reread the necessary sections of the chapter to improve conceptual clarity.

Summary

Inferential statistics are used to make statements about the characteristics of a population from the characteristics of a random sample drawn from that population. Hypothesis testing is the primary form of inferential statistics to be explored in this text. Several characteristics of hypothesis testing were discussed:

1) A scientific research hypothesis is a formal statement or expectation about the outcome of a study. A research hypothesis precedes the collection of data and usually predicts a difference, effect, or relationship.
2) A statistical hypothesis is a numerical statement regarding the outcome of a study. Statistical hypotheses come in logically related pairs, the null and the alternative. They are mutually exclusive and collectively exhaustive.
3) Hypothesis testing uses samples to make inferences about populations. Statistical methods are used to make these inferences.
4) The null hypothesis states there *is no* effect or *no* relationship between the two variables. The alternative hypothesis states that there *is* an effect or relationship between variables.
5) Hypothesis testing yields a decision to either reject or fail to reject the null hypothesis. Since any decision is probabilistic, there is always a risk of committing a decision error.

Single-sample research uses one sample to test a hypothesis about the mean of a population. Two-sample research uses two samples to test a hypothesis about the difference between two population means. Many research designs are more complex, involving more than two samples.

A sampling distribution is a collection of some sample statistic, usually the mean. It is a theoretical distribution derived by computing the means of an almost infinite number of samples of size n. Sampling distributions are the conceptual and mathematical cornerstone of hypothesis testing. Sampling distributions are normal, unless the population of scores is not normal *and* the sample size selected is very small. The mean of the sampling distribution equals the mean of the population of raw scores ($\mu = \mu_M$), and the standard deviation of the sampling distribution (or standard error) is a function of the standard deviation of the raw scores divided by the square root of the selected sample size ($\sigma_M = \sigma/\sqrt{n}$.).

Key Formulas

Standard error of the mean

$$\sigma_M = \frac{\sigma}{\sqrt{n}} \quad (7.1)$$

Estimated standard error of the mean

$$s_M = \frac{s}{\sqrt{n}} \quad (7.2)$$

Key Terms

Point estimation
Confidence interval
Research hypothesis
Hypothesis testing
Statistical hypothesis
Null hypothesis
Alternative hypothesis
Sampling distribution
Central limit theorem
Standard error of the mean
Sampling error

Exercises

1 How is a theoretical sampling distribution of means constructed?

2 How is sampling distribution variability affected by the population standard deviation?

3 How is sampling distribution variability affected by sample size?

4 What is the relationship between a sample mean value and the standard error?

5 Distinguish between point estimation and a confidence interval.

6 How does hypothesis testing differ from estimation?

7 What is meant by single-sample vs. two-sample research designs?

8 What is the difference between research and statistical hypotheses?

9 Define null and alternative hypotheses.

10 Imagine it is known that American teenagers spend, on average, three hours a day on social media. Suppose a researcher wanted to test the effectiveness of one community's programs designed to curb social media usage by sampling the local teenager population and asking them about their social media involvement.

a What would the null and alternative hypotheses look like?
b What finding would result in failing to reject the null?
c What finding would support the goals of the community's programming?
d What finding would result in rejecting the null but would not support the objectives of the community program?

11 Suppose a research hypothesis predicts that Drug *A* will outperform Drug *B*.

a State the null and alternative statistical hypotheses.
b What finding would result in a failure to reject the null?
c What does a failure to reject the null mean for the research hypothesis?
d What finding would support the research hypothesis?
e What finding would result in rejecting the null but would not allow the researcher to support the research hypothesis?

Work Problems

12 A sampling distribution with an $n = 9$ is made from a population of scores with $\mu = 100$ and $\sigma = 10$. Describe the sampling distribution in terms of μ_M, σ_M, and shape.

13 Suppose we know an anxiety measure has a mean of 50 and a standard deviation of 10; $\mu = 50$; $\sigma = 10$. What is the standard error of a sampling distribution with $n = 20$?

14 Suppose we know university students average 6.5 hours of sleep during weeknights with a standard deviation of .5 hours. What is σ_M if a sampling distribution has an $n = 100$?

15 Suppose we know the average four-person family household generates 20 kg of garbage a week with a standard deviation of 4 kg. Suppose we wish to create a sampling distribution based on samples of 50 households. What is the mean of the sampling distribution?

16 Suppose we sample the local squirrel population and find that our sample generates a mean weight for squirrels of 17 ounces with a standard deviation of 2 ounces.

a What is our best estimate of the *population* weight and standard deviation?
b What is the estimate of s_M if we gathered 20 squirrels?

17 Suppose a local promoter, wanting to attract attention to the community, decides to try to create larger squirrels by spreading around town genetically modified nuts that have been supplemented with a growth hormone. Using data from Problem 16:

a What would be the research hypothesis?
b What would be the null and alternative hypotheses?
c What statistical finding would support the research hypothesis?
d What statistical finding would run counter to the research hypothesis?
e What would be a finding that would result in failing to reject the null?

Part 3

Inferential Statistics

z Test, *t* Tests, and Power

8

The Single-Sample z and t Tests

8.1 The Research Context

This chapter addresses the statistical analyses used in single-sample research projects. Recall that single-sample studies use one sample of participants to make an inference about whether the mean of the population from which the sample came is some specified value. Even though most studies in the social and behavioral sciences involve two or more samples, the relative simplicity of the single-sample method makes it ideal for learning how to run inferential tests. Once we gain familiarity with hypothesis testing in simple research designs, we will be more easily able to understand how to test hypotheses in complex research designs. Below are a couple of examples of single-sample research designs.

► **Example 8.1** Suppose a university administrator, who wants to know if incoming students have strong mathematical abilities, gathers the Scholastic Aptitude Test (SAT) Math scores from a sample of incoming students. Assume the national mean of the quantitative section of the SAT is 500 and the standard deviation is 100. ◄

► **Example 8.2** Suppose it is known that the average time it takes a graduate student to earn a PhD in psychology is 6.8 years, with a standard deviation of 1.2 years. The chairperson of a psychology graduate program gathers data from a sample of their recent graduates in order to compare their program to the national average. ◄

The statistical test that is used to decide if a sample does or does not come from a population, *when the standard deviation of the population is known,* is called a ***z* test**. Since only one sample of scores is taken, this is a single-sample or one-sample statistical test. Furthermore, since the standard deviation of the population is known, *we can determine the actual standard error*; we do not need to estimate

Essentials of Behavioral and Social Science Statistics, Third Edition.
K. Paul Nesselroade, Jr. and Laurence G. Grimm.
© 2025 John Wiley & Sons, Inc. Published 2025 by John Wiley & Sons, Inc.
Companion website: https://www.wiley.com/go/Essentials_of_Behavioral_and_Social_Science_Statistics3e

it using a sample standard deviation. If we did not know the population standard deviation, we would need to estimate it and then run the single-sample ***t* test** instead. This chapter discusses both tests, but we will look at z tests first.

8.2 The Sampling Distribution for the Single-Sample *z* Test

This section begins where Chapter 7 ends: the importance of sampling distributions. Recall that these distributions are theoretical and are derived by computing the means of an almost infinite number of samples of size n. The following four facts were presented:

1) If a population of scores is normally distributed, the sampling distribution of means will be normally distributed.
2) Even if a population of scores is not normally distributed, the sampling distribution will approximate a normal curve as the sample size increases.
3) The sampling distribution mean equals the raw score population mean, $\mu = \mu_M$.
4) The standard deviation of the sampling distribution is called the standard error of the mean. The relationship between the standard error of the mean and the standard deviation of the population is $\sigma_M = \sigma/\sqrt{n}$.

8.2.1 *z* Scores and the Sampling Distribution of Means

The z-score formula for transforming a raw score to a z score was given in Chapter 5 as:

$$z = \frac{X - \mu}{\sigma}$$

A z score specifies how many standard deviations the transformed raw score is from the mean of the distribution. If every score of a *normally distributed* population is transformed into a z score, the result is the standard normal curve, as described by Table A.1 in the Appendix A. The standard normal curve was used in Chapter 5 to solve z-score problems, and the focus of every problem was on *scores*. For example, a typical question was, "Given $\mu = 50$, and $\sigma = 5$, what is the probability of drawing a *score* at random above 60 or below 40?"

With the aid of the sampling distribution concept and after a slight adjustment to the z-score formula, we can create a statistical tool that will allow us to make probability statements about the likelihood of selecting a particular *sample mean* from a sampling distribution of sample means. To summarize, Chapter 5 showed us how to make likelihood statements about a given *score* within a *distribution of scores*, whereas this chapter will show us how to make likelihood statements about a given *sample mean* in a *distribution of sample means.*

8.2.2 The *z* Statistic

The formula for the z statistic transforms the sample means within a sampling distribution into z means. Formula 8.1 is used for this transformation.

---

z Statistic

$$z_{obt} = \frac{M - \mu}{\sigma_M} \quad (8.1)$$

---

where

M = the mean of the sample
μ = the hypothesized mean
σ_M = the standard error of the mean

The z_{obt} symbol is used to indicate that the z value is obtained from a mean rather than a raw score. The z statistic has the same basic form as the z-score formula. (Table 8.1 contrasts each value of the two formulas.) Just as with raw scores, when a sampling distribution of means is transformed into z values (using Formula 8.1), the standard normal distribution is produced. Table A.1 in Appendix A can now be used to make probability claims about *sample means.*

Table 8.1 A comparison of the *z*-score formula and the *z*-statistic formula.

***z*-Score formula (Formula 5.2)**	***z*-Statistic formula (Formula 8.1)**
$z = \frac{X - \mu}{\sigma}$	$z_{obt} = \frac{M - \mu}{\sigma_M}$
X = a single score	M = a single mean
μ = the population mean	$\mu = \mu_M$, the hypothesized population mean
σ = population standard deviation	σ_M = standard error of the sampling distribution

8.2.3 The Logic of the *z* Test

To walk through the logic of the z test, we can use a hypothetical study. Suppose an educational enrichment program is developed to teach math skills. The program entails the use of interactive web-based instruction to teach basic algebra. A *random sample* of 36 students is selected, given the interactive web-based instruction, and tested at the end of the school year. Suppose the national exam of math skills is known to have a mean of 100 and a standard deviation of 10. (That is, suppose we know both μ and σ.) A statistical test, based on a single sample, will be conducted to test the hypothesis that the population mean is 100. Therefore,

$$H_0 : \mu = 100$$
$$H_1 : \mu \neq 100$$

We are *not* testing the hypothesis that the mean of the national exam is 100. We have already been told that. The population we wish to make an inference about is a hypothetical population; the population of all students that *theoretically could* receive the educational enrichment program. We must assume that the sample of participants, once trained, is representative of the population of students who *could be* exposed to this program. The population of interest is *hypothetical*; it is *imaginary*. It will be helpful for us to gain comfort with the idea of getting real samples but understanding that they may represent imaginary populations.

The null and alternative hypotheses are statements that derive from the question, "Is the mean of the population from which the sample is taken 100?" In essence, we are asking if the program has any effect on math skills. If the program has an effect, we expect our sample mean to be sufficiently different from the hypothesized population mean of 100.

8.2.4 Sampling Error and Hypothesis Testing

Suppose the null hypothesis is true; that is, suppose the math enrichment program has no effect and the population mean of students who could experience the math enrichment program is still 100. When we gather the data from our sample and compute the mean, would we expect that sample mean to be *exactly* 100? Due to sampling error, even if the null hypothesis were true, it is extremely unlikely that the sample mean would be exactly 100. Suppose the sample mean drawn is 100.5. Are we willing to conclude that $\mu \neq 100$? Probably not. Suppose, however, the sample mean is 102. Or, what about a sample mean of 105, 115, or 130? As the distance from 100 increases, our confidence of a difference is likely growing, but how do we decide when a difference between a sample mean and the null population mean is big enough? A statistical test like the z test will help us with the decision.

When we conduct a z test, the sample mean is transformed into a z statistic, and this allows us to use the z table to judge the probability of obtaining that sample mean when the null hypothesis is true. If the z statistic is far away from 0, it suggests it is *very unlikely* that sampling error alone would produce that value. We would then infer that the population mean from which our sample is drawn is not what is specified by the null hypothesis, and the null would be rejected.

8.2.5 Sampling Distributions When the Null Hypothesis Is True or False

Figure 8.1 depicts two sampling distributions of a mathematics enrichment program. The sampling distribution on the left shows the mean of the population as 100; the one on the right is from a population with a mean of 104. As we continue reading, keep the following in mind: It is *not* the goal of hypothesis testing to infer the value of the mean for the hypothesized population. To make an inference of that sort, use a point or interval estimation (see Section 8.7 later in this chapter). The goal of hypothesis testing is to decide whether to reject the null hypothesis. In Figure 8.1, the sampling distribution on the left would be true if the null hypothesis is true ($\mu = 100$). In other words, if the enrichment program has no effect, we would expect the mean of the sampling distribution to be 100. If the null hypothesis is false, then the mean of the sampling distribution is some number other than 100. In this example, let us imagine that the true population mean for those who experience the enrichment program is 104. In reality, if the null hypothesis is false, the true mean could be anything other than 100.

Now suppose we take a sample, compute the mean, and find it to be 105. If the true situation is represented by the sampling distribution on the left, $\mu = 100$, then we must have unintentionally and randomly oversampled scores from the right tail of the distribution. However, if the true situation is represented by the sampling distribution on the right, the one with $\mu = 104$, then the sample mean of 105 is not at all unusual.

It is important to keep in mind when hypothesis testing that we do not know the mean of the population from which we are sampling, and therefore we do not know the mean of the sampling distribution of that population. Figure 8.1 gives us a behind-the-scenes glimpse into the nature of the decision we are making

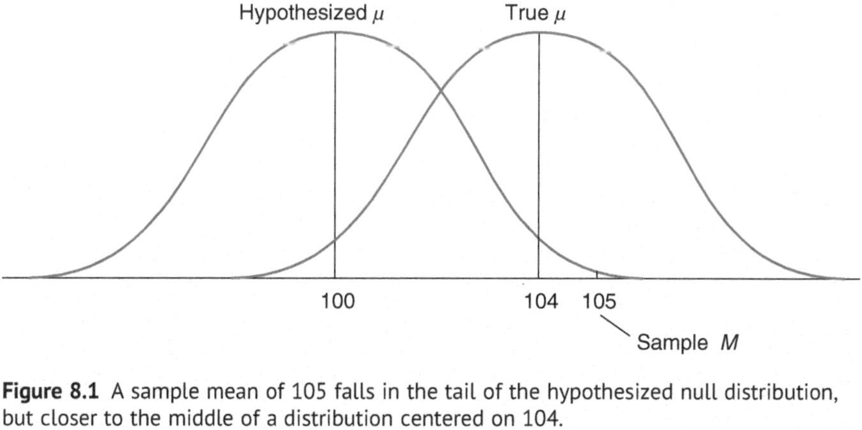

Figure 8.1 A sample mean of 105 falls in the tail of the hypothesized null distribution, but closer to the middle of a distribution centered on 104.

for explanatory purposes. In reality, we only have access to the null distribution (in this example, the distribution on the left).

Back to our example, we need to determine the likelihood of drawing a sample mean, by chance, from a distribution of sample means based on the null hypothesis. If the likelihood is low (determining what is meant by "low" will soon be discussed), then the null hypothesis should be rejected. In this example, a sample mean of 105 may lead us to conclude that the μ from which our sample was drawn is probably not 100, but this is not entirely clear. Others may feel differently. We cannot rely on gut feelings; we need to use an objective test. As mentioned earlier, thankfully, we have tools that can remove much of the guesswork. (Note: Other sampling distributions with a $\mu = 100$ may have a larger variance where a sample mean of 105 may fit quite well with a $\mu = 100$. Mean differences can be great or small depending on the standard error.)

8.2.6 Using the *z* Test in Deciding to Reject the Null Hypothesis

Let us continue to consider the effectiveness of that educational enrichment program. When testing the null hypothesis, we set up the test *as if* the null hypothesis were true. (Note: This is a very important point.) Since students are tested with a tool known to have a population mean of 100 and a standard deviation of 10, the null and alternative hypotheses are stated as:

$$\text{H}_0 : \mu = 100$$
$$\text{H}_1 : \mu \neq 100$$

A sample of 36 students is tested after completion of the program, and the sample mean is found to be 105. After the null and alternative hypotheses have been stated, the next step is to transform the sample mean into a z statistic using Formula 8.1:

$$z_{obt} = \frac{M - \mu}{\sigma_M}; \sigma_M = \frac{\sigma}{\sqrt{n}}$$

$$z_{obt} = \frac{105 - 100}{10/\sqrt{36}} = \mathbf{2.99}$$

How would we interpret a z_{obt} of 2.99? If we assume that the null hypothesis is true, $\mu = 100$, then our sample mean of 105 is 2.99 standard error units above the mean of the sampling distribution. What percentage of means is found at or above a z of 2.99? Refer to the third column in the z table (Table A.1). Only .14% of the means of the sampling distribution are found at or above a mean of 105 (a z of 2.99). Figure 8.2 visually represents this relationship.

Now there is a decision to make. We can assume that the mean of the sampling distribution (the mean of the population) is 100 and that we have experienced the highly unlikely event of oversampling from the extreme right tail of the

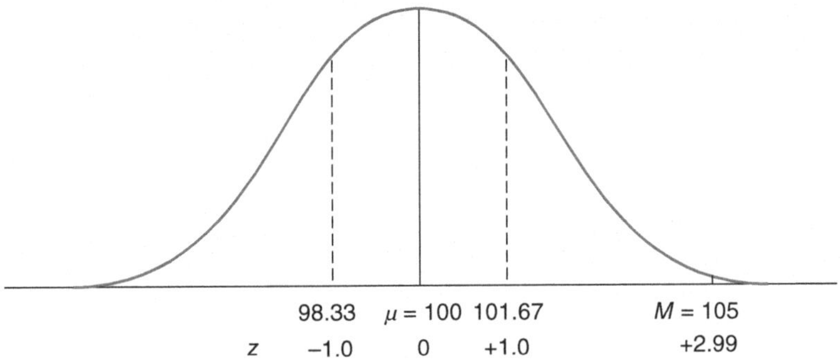

Figure 8.2 A sample mean of 105 is 2.99 standard error units above the mean.

distribution. On the other hand, we can reject the null hypothesis and consider it much more likely that our sample came from a different sampling distribution of means. In this situation, most would agree that the null hypothesis is probably false and should be rejected.

If it were left up to each individual researcher to decide what is and is not a rare statistical occurrence, hypothesis testing would lead to endless confusion. For this reason, social and behavioral scientists in most research situations follow a long-standing convention of using a probability value of .05 as the criterion for rejecting null hypotheses.

8.2.7 The Criterion for Statistical Significance

Over the years, statisticians have come to define the criterion for rejecting the null to be .05 (for a historical overview, see Cowles and Davis 1982). As applied to the single-sample z test, this means that if the probability of randomly drawing a given sample mean is less than .05, then we should reject the null hypothesis. In the z distribution, the z values of ± 1.96 mark the points that cutoff the most extreme 2.5% on each side of the curve (thus summing to 5% or .05). This means that if the z_{obt} is 1.96 standard error units away or more from the null population mean in either direction, then we should reject the null hypothesis.

8.2.8 Alpha Levels and Rejection Regions

When we use the cutoffs of ± 1.96, we are testing the null hypothesis at the 5% level of significance. The probability value of .05 is called the **alpha level,** symbolized as α. In hypothesis testing, the investigator chooses the alpha level. Other conventional alpha levels are .10 and .01. *Before* the data are analyzed, the researcher specifies the alpha level at which the null hypothesis is to be tested. In essence, when setting the alpha level, the researcher is defining the criterion for what will be considered a statistically rare event, given a true null hypothesis. An alpha value of .10

would be considered a permissive or liberal criterion, while an alpha value of .01 would be considered a conservative criterion.

As soon as an alpha level is selected, *rejection* and *fail to reject* regions are automatically determined. Figure 8.3 shows these regions for alpha levels of .10, .05, and .01. The rejection region is typically placed evenly in both tails of the sampling

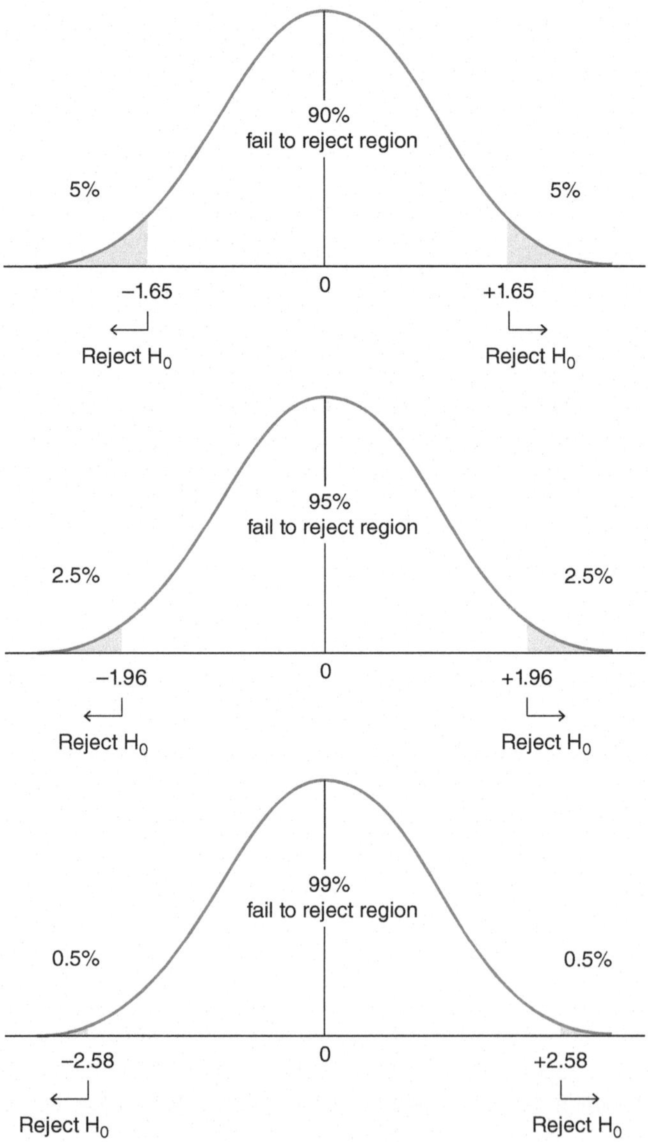

Figure 8.3 The acceptance and rejection regions for alpha levels of .10, .05, and .01.

distribution, that is, the farthest away from the mean. This is known as a *nondirectional* or *two-tailed* hypothesis test. The absolute values of the z scores that define the rejection regions are called the **critical values**. If the alpha level is set permissively at 10%, the z values of ± 1.65 mark the rejection regions. If the alpha level is set conservatively at 1%, the z values of ± 2.58 mark the rejection regions. The critical values of ± 1.65, ± 1.96, and ± 2.58 can be verified by using the z table (Table A.1).

If alpha is set at .05 and the null hypothesis is rejected, then we would state that $p < .05$, which means that the probability of obtaining a mean at or beyond the critical values *if the null hypothesis is true* is less than .05. In the educational enrichment program example, the z_{obt} was 2.99. Even if we had set alpha at .01 prior to collecting the data, we would still be rejecting the null hypothesis because 2.99 is larger than 2.58 ($p < .01$). Now that the null hypothesis has been rejected, what does this mean about the educational enrichment program?

8.2.9 The Implications of Rejecting the Null Hypothesis

What would we conclude if the null hypothesis were rejected? It means we have statistical evidence suggesting a difference has been found. In an experiment, if the research methodology is free from confounds, we would have reason to conclude that the independent variable influenced the dependent variable. In our example, we would have found evidence suggesting that the educational enrichment program improves the math abilities of those who are exposed to it. In other words, there is now statistical evidence that the program works.

What would we conclude if our sample mean was 102 and, as a result, we failed to reject the null hypothesis? As stated in Chapter 7, failing to reject the null hypothesis is not to accept the null hypothesis as correct. Even if the sample mean perfectly equals the hypothesized population mean, we cannot conclude that the null is correct. For one reason, the null may be false, but only by a little. Failing to reject the null simply means that the null *may* be correct; it cannot be rejected. Perhaps the strongest claim to be made when a null hypothesis is not rejected is something like the following: If the null is false, it does not look to be radically false. However, even a cautious statement like this is dependent upon other factors, like sample size. When a null hypothesis is not rejected, it is best merely to state that statistical evidence to reject the null was not found and to leave it at that.

8.3 Type I and Type II Errors

Since inferential statistics are based on probabilities, there is always the chance of making a decision error. There are two kinds of decision errors that can be made when hypothesis testing: rejecting the null when it should not be rejected (a false positive) and not rejecting the null when it should be rejected (a false negative).

These errors are analogous to the errors jurors can make in a courtroom. Imagine being a member of a jury that must decide the guilt or innocence of a defendant. Independent of what is decided, the defendant is either truly guilty or truly innocent. In this situation, there are four possible outcomes. First, if the defendant is truly guilty and we decide they are guilty, we are correct. Second, if the defendant is truly innocent and we decide they are innocent, we are correct. Third, if the person is truly innocent and we claim they are guilty, we have committed an error (a false positive). Finally, if the defendant is truly guilty and we conclude they are innocent, another type of error is committed (a false negative).

Sometimes one type of error is deemed to be more problematic than the other. For instance, if the case is a capital offense, a guilty verdict might mean the death penalty. In this case, we will want to be especially sure that we do not send an innocent person to their death. Also, note that there is a trade-off between the two types of errors. As we demand more evidence to make sure we do not find an innocent person guilty, we are simultaneously increasing the chance we will declare a guilty person to be innocent. The criteria we use to make our decision are influenced by the consequences we associate with committing each type of error.

In hypothesis testing, a false positive is called a **Type I error**. This occurs when a true null hypothesis is rejected. The other type of error, a false negative, is called a **Type II error**. This occurs when a false null hypothesis is not rejected. Table 8.2 specifies these outcomes. The problem is that we never know when we have made an error because we never know the real status of the null hypothesis.

In hypothesis testing, the investigator is in a better position than the juror because the investigator is able to directly control the probability of a Type I error. The alpha level chosen is the probability of committing a Type I error. Suppose alpha is set at .05. *Given a true null hypothesis,* there is a 5% chance the null will be mistakenly rejected. Recall that when the null is true, the z_{obt} will fall outside the critical values of ±1.96 only 5% of the time. If alpha is set at .01, the null will be mistakenly rejected only 1% of the time. We can make alpha as small as we would like to avoid making a Type I error.

Table 8.2 The decision grid depicting the four possible outcomes of hypothesis testing.

Our decision	The true situation	
	H_0 is true	H_0 is false
Fail to reject H_0	Correct	Type II error
Reject H_0	Type I error	Correct

At first blush, this might suggest that we should make alpha as stringent as possible. Why not set alpha at .0001? We would only commit a Type I error 1 out of 10 000 times! However, remember the trade-off for the juror: If we are too afraid to convict an innocent person, we will make it easier for a guilty defendant to be set free. As alpha is made more stringent, the probability of committing a Type II error increases.

In this way, adjustments to the Type I error rate reciprocally influence the likelihood of making a Type II error. The more permissive the Type I error rate, the less likely a Type II error will be made, and vice versa. However, the relationship between the error types is peculiar. Since the Type I error rate is established from a known null distribution, actions designed to increase or decrease the Type II error rate do not influence the Type I error rate. The likelihood of making a Type I error is simply whatever alpha value is decided upon, end of story. It is a strange world in which we live!

A final word of caution: Do not make the mistake of thinking that if the probability of a Type I error is .05, the probability of making a Type II error must be .95. The reciprocal relationship between the two types of errors does not mean they are complementary. Calculating the probability of making a Type II error is not straightforward. Actions the researcher can take to influence the Type II error rate will be discussed later in the text.

8.3.1 Deciding on Alpha

How do researchers decide on an alpha level? Just like jurors, they weigh the consequences. Suppose a researcher wants to investigate a new research area. There may be interesting effects to find there, but no one knows. Failing to reject a series of null hypotheses gets us nowhere. Researchers can become so disappointed after failing to reject that they may decide against performing further research in the area. For this reason, in new research areas, the investigator may choose to increase the alpha value to 10%. The smaller critical values ($z = \pm 1.65$) make it easier to reject the null. Once a finding is identified, subsequent studies can use the more standard alpha level of .05. A permissive alpha level reflects an investigator's belief that a Type II error is more problematic than a Type I error when exploring a new area of research.

Conversely, there are instances in which the consequences of making a Type I error are so serious that a researcher would want to set a very stringent alpha level. Consider a study testing for serious side effects when using a drug. If the null hypothesis is cast in such a way that rejecting it means that the drug is medically safe, the researcher might make alpha very conservative (for example, .001). In this situation, the researcher believes that it is much more serious to conclude that the drug is safe when it is not than to mistakenly conclude that the drug is not safe when, in fact, it is.

Another important issue is *when* to set alpha. To properly measure the probability of making a Type I error, the alpha level must be established *ahead of time*. It is highly inappropriate to first run a study and *then* decide where to set alpha so that the null hypothesis can be rejected. This is a form of "statistical cheating."

One final point about Type I errors: Suppose a researcher runs a study and fails to reject the null. Would the investigator be justified in performing the exact same study repeatedly until the null can be rejected? No, because eventually even a true null hypothesis will be mistakenly rejected (see Boxes 7.1 and 8.1). However, it is acceptable for researchers to reflect on failed studies, revise experimental procedures, and then run the study again. Conducting the same failed study repeatedly is both a waste of time and scientifically dubious.

Box 8.1 Is the Scientific Method Broken? Type I Errors and the Ioannidis Critique

If there is a "ground zero" for the current reproducibility crisis, it may be found in the personhood of John Ioannidis, Professor of Medicine, Epidemiology, and Population Health at the Stanford University School of Medicine. In 2005, he published an article in PLoS Medicine entitled "Why Most Published Research Findings are False" (Ioannidis 2005). As one might imagine, this article created a firestorm of controversy as well as an avalanche of articles reacting to this claim – some supporting (e.g. Freedman 2010) and some critiquing (e.g. Leek and Jager 2017). As a result, Ioannidis is currently one of the most cited scientists in the world.

Several of the points Ioannidis makes in the paper involve misunderstanding the perils of Type I errors. The Type I error rate is an accurate reflection of the risk involved in rejecting a singular null hypothesis. However, the testing of a null hypothesis does not take place in a vacuum, and other factors must be considered. These other factors include (i) how many questions are being asked in a given research project, (ii) how many other similar projects may be taking place elsewhere by other researchers, and, most importantly, (iii) what is the ratio of null findings to actual findings in each area of inquiry.

To help illuminate the argument, Ioannidis (Wilson 2016) asks readers to suppose that there are 101 stones in a given field. Only one of them, however, contains a diamond (i.e. a true finding). Gratefully, we have at our disposal a diamond-detecting machine that advertises a 99% accuracy of detection. That is, when the machine is placed overtop a stone without a diamond in it, 99% of the time, it will not light up. It will only give us a false positive (or Type I error) 1% of the time. Further, imagine that after checking several stones and getting no reaction, the machine finally starts to flash with activity. What is the probability that this stone, if cracked open, will contain a diamond? We might initially suggest that there is a 99% chance. However, recall that there are 100 dud stones in this field. The machine, if functioning at a 1% false positive rate, will likely register 1 false positive if all stones are checked. This means that of the 101 stones in the field, 2 are likely to register as positive for the diamond (one false positive and one real positive). Therefore, there is only a 50% chance of finding a diamond when this particular stone is cracked open. This is a little disappointing.

Now, imagine a field that has several thousand stones in it – still only one of them containing a diamond. Do we see how, in this situation, even a false positive rate of 1% may lead persistent researchers who are testing numerous hypotheses to draw faulty conclusions far too frequently? One key factor here, which is impossible to answer, is the ratio of stones containing diamonds. As this ratio increases, the ratio of true positives to false positives will also increase. However, how do researchers know ahead of time how "diamond-rich" is the field in which they are working? Herein lies a big problem: the unknown ratio of real to null findings in a given area of investigation. Knowing the detection equipment has a 1% false positive rate sounds impressive, but it does not solve this problem.

Further, do we see how the repeated testing of several stones changes the meaning of the 1% false positive rate? If we were to walk up to a field of stones and just test one, then the false positive rate of 1% makes sense. However, as we test stone after stone, the probability that at least one of the dud stones will register as significant grows as we work our way across the field. Herein lies a second problem: the additive nature of the Type I error rate.

One way to combat these problems, in addition to valuing replication (see Box 7.1), is to publicly report nonsignificant findings. Only once researchers get a sense of how many or how few "diamonds" there are in a field of inquiry can they begin to process what a supposed finding might mean. If the field of inquiry seems to be chock-full of effects and relationships, then a significant claim seems more likely to be an actual finding, but if the field has repeatedly been shown to be lacking meaningful findings, then a claim of significance should be interpreted with a great deal of suspicion. Unfortunately, despite the current reproducibility crisis, there seems to be little interest in creating publication opportunities for null findings. Until this happens, the Type I error problem is going to continue to bring a cloud of suspicion around claims of findings, especially those coming from new, previously unexplored fields of inquiry.

8.4 Is a Significant Finding "Significant?"

When the decision rule directs the researcher to reject the null hypothesis, oftentimes the word "significant" is used in the interpretation. For instance, "The effect of the drug was found to be statistically significant." Even though this phrasing is quite common in professional literature, it is easily misunderstood. In ordinary speech, "significant" means important. However, it is possible to achieve statistical "significance" even when the research finding is quite trivial. Consider the educational enrichment program as an example. Instead of using a sample size of 36, suppose we had used a sample of 1000 students. Further, suppose that the sample mean turned out to be 101 instead of 105. Using Formulas 8.1 and 7.1, we get the following z_{obt} value.

$$z_{obt} = \frac{M - \mu}{\sigma_M}; \ \sigma_M = \frac{\sigma}{\sqrt{n}}$$

$$z_{obt} = \frac{101 - 100}{10/\sqrt{1000}} = \mathbf{3.13}$$

Even if we set a conservative alpha level of .01, which has a critical value of ±2.58, the null hypothesis is to be rejected. However, it may not be clear to everyone looking at the finding that all of the time, effort, and expense used to implement the program would be worth a mere one-point increase on the national exam. In this case, although the phrase "significant" is accurate in the statistical sense, it may not be accurate in a practical sense.

The other side of the issue arises when the null hypothesis is not rejected; the statistical test is deemed *nonsignificant*. Note this is not the same as *not important*. A nonsignificant finding may be considered important. For example, new treatments are advertised all the time: treatments for arthritis, obesity, stress, or whatever ails us. If we contrast a no-treatment control group with one receiving an ineffective treatment, we will likely end up failing to reject the null, a *nonsignificant* finding. Since it is impossible to prove a null hypothesis, we cannot conclude that the treatment has been *proven* ineffective. Nevertheless, in the absence of evidence that a treatment *is* effective, there is no compelling reason to use it. Assuming the study was well designed, the treatment had an opportunity to show its effect and failed. This scenario exemplifies a statistically nonsignificant finding that may be considered rather important.

Before leaving this topic, it is important to mention a couple of more often-used phrases regarding statistical significance that can be problematic. Sometimes we may read that a finding is described as *marginally significant*. This phrase can be used when a finding just fails to fall inside the critical scores. The z_{obt} is close to the rejection region (for example, getting a z_{obt} of 1.93), and in the mind of the researcher, a Type II error is being made if the null is not rejected. The use of this term does highlight the arbitrary nature of the conventional 5% alpha level. However, it is important to realize that the burden for rejecting the null was not met. It may be true that a Type II error is occurring, and it is fine and acceptable for the researcher to bring this to the attention of others, but this possibility should not cause us to reclassify a "fail to reject" finding. The confusion around the interpretation of marginally significant findings constitutes another reason for valuing replication studies (see Box 7.1).

Finally, the phrase *highly significant* is sometimes found in professional literature. It is often used when the alpha value was initially set at .05, but the findings would have allowed for a rejection of the null even if the alpha value had been set at .01 or even .001. This phrasing implies that readers should be more impressed with the effect as the size of the p value decreases (Bakan 1966; Cohen 1990). This phrase can also be misleading. For one thing, the p value reflects the degree of *certainty* for an effect, not necessarily the *size* of an effect. Consider our last version of the enrichment program example, where a sample mean of 101 still allowed for the rejection of the null hypothesis and at $p < .01$. This finding could be described as "highly" significant. However, this description is misleading. It is true that the concepts of *certainty* and *size* are not completely independent of each other; after all, a very powerful effect is more likely to yield an extreme mean. However, factors other than effect size can influence our sense of certainty that an effect exists. (This topic will be discussed more later in the text.) It is also simply inappropriate to redraw the lines of meaning *after* the findings have been analyzed. It may be true that a shot arrow did not only hit the bull's-eye but also hit it right in the middle. However, it is inappropriate to shrink the bull's-eye around the arrow *after* it has been shot to make it look more impressive. After all, we would probably not be willing to count the shot as a miss if the smaller bull's-eye had been drawn ahead of time and then just missed by the arrow.

Playing fair in hypothesis testing involves setting the standard *ahead of time* and then making a binary decision based on the evidence: either the criteria were reached for rejecting the null or they were not. Recognizing that findings can be close but not quite in the rejection region *or* can register far into the rejection region can be theoretically important and can help direct future research, but they should not cause us to go back and adjust the rules of the hypothesis test.

For these reasons, this textbook will only use the terms "significant" and "significance" sparingly. Admittedly, these terms are simply unavoidable for a statistics book, but minimizing their use will help us think clearly about the proper meaning of statistical analyses. Furthermore, this text will avoid using the terms "marginally significant" and "highly significant." Inferential decisions will only be described as a rejection of the null or a failure to reject the null.

8.4.1 A Measure of Effect Size: Cohen's *d*

As just noted, one problem with hypothesis testing concerns the impreciseness of a statistic to reflect the size of the effect. To address this problem, it is recommended for researchers to provide a measure of effect size whenever a significant finding is reported. As different tests are presented in the textbook, a corresponding means of measuring effect size will often be included.

One of the simplest and most often-used measures of effect size is **Cohen's *d*.** Cohen (1988) suggested that the size of an effect can be standardized by using the standard deviation of the population to quantify the difference between two means.

Cohen's *d* for single-sample *z* test

$$d = \frac{\textit{mean difference}}{\textit{standard deviation}} = \frac{M - \mu}{\sigma} \quad (8.2)$$

The standard deviation is used to standardize the size of mean differences in much the same way that it is used by *z* scores to standardize raw scores. A mean difference of 10 units might suggest a huge effect if the two distributions are tightly packed around their respective means, but it may be much less impressive if the scores in the distributions are widely distributed.

In some texts, certain effect sizes are classified as either small, medium, or large. These distinctions are rather arbitrary and should not be seen as authoritative. The value of an effect size measure comes from the realization that larger values reflect greater effects, differences, or relationships – as the case may be. Additionally, since this statistic is concerned with the size of the effect and not the direction, it is typical to simply ignore the valence (plus or minus sign) of the obtained value and always report it as positive.

Versions of Cohen's *d* can be used to measure the effect size of many different types of *t* tests. As a result, we will see versions of this statistic in other places throughout Part 3.

8.5 The Sampling Distribution for the Single-Sample *t* Test

8.5.1 The *t* Distributions

Up to this point, we have used the *z* distribution to establish cutoffs for given alpha levels. However, when σ is *unknown*, the standard error cannot be calculated, and the means of a sampling distribution cannot be transformed by the *z* statistic. Instead, the sample standard deviation, *s*, must be used as an estimate of σ. This necessitates a move to the *t* statistic.

t Statistic

$$t_{obt} = \frac{M - \mu}{s_M} \quad (8.3)$$

where

$$s_M = \frac{s}{\sqrt{n}} \quad (7.2)$$

A ***t* distribution** is theoretically established by transforming every mean of a sampling distribution into a *t* statistic. However, because we do not know the standard error and are merely estimating it, the *t* formula will not produce a standardized normal curve identical to the *z* distribution. This means that we can no longer use critical values that were used for *z* tests. However, *t* distributions are normal when the population of raw scores being sampled is normal or the sample size used is not extremely small. A *t* distribution can be formed for any sampling distribution composed of means that are based on an *n* of 2 to infinity. However, as *n* decreases, the tails of the *t* distribution stretch out farther down each direction of the *X* axis (with the critical values marking the most extreme 5% also moving farther and farther away from 0; see Figure 8.4). Since a different *t* distribution can be established for each sampling distribution of size *n*, *t* distributions are collectively referred to as a *family of distributions*. Figure 8.4 illustrates four *t* distributions based on sampling distributions of decreasing *n*. Note: *t* distributions approximate the standard normal curve as *n* increases (with the critical scores of the *t* distribution eventually matching those of the *z* distribution, i.e. ±1.96). The sample size (*n*) is *roughly* reflected by the symbol *df* – the degrees of freedom. This concept is discussed in the following section. For the time being, know that the degrees of freedom for the single-sample *t* test are equal to $n - 1$.

As we examine Figure 8.4, remember that the *t* distribution is not a sampling distribution of means. The *t* distribution is a distribution of *t* values. Each *t* value is a sample mean transformed by the *t* formula. The mean of the *t* distribution is 0. However, unlike the *z* distribution, the standard deviation of the *t* distribution is not 1. It changes depending, in part, on the size of the samples used to establish the sampling distribution. Sample size influences the degree of accuracy of our estimate of σ; as it increases, our estimate of σ improves, and, correspondingly, the error in the resulting sampling distribution shrinks.

As *n* decreases and the *t* distribution elongates, the rejection region for the null moves farther away from 0. For example, if $\alpha = .05$, the critical value when $df = 2$ *is* ±4.30, but when $df = 25$, the critical value is ±2.06. The broader *t* distribution associated with $df = 2$ requires that we move 4.30 standard error units in each direction from the mean to bracket the middle 95% of the distribution. When the $df = 25$, we need to only move 2.06 standard error units from the mean to bracket the middle 95%.

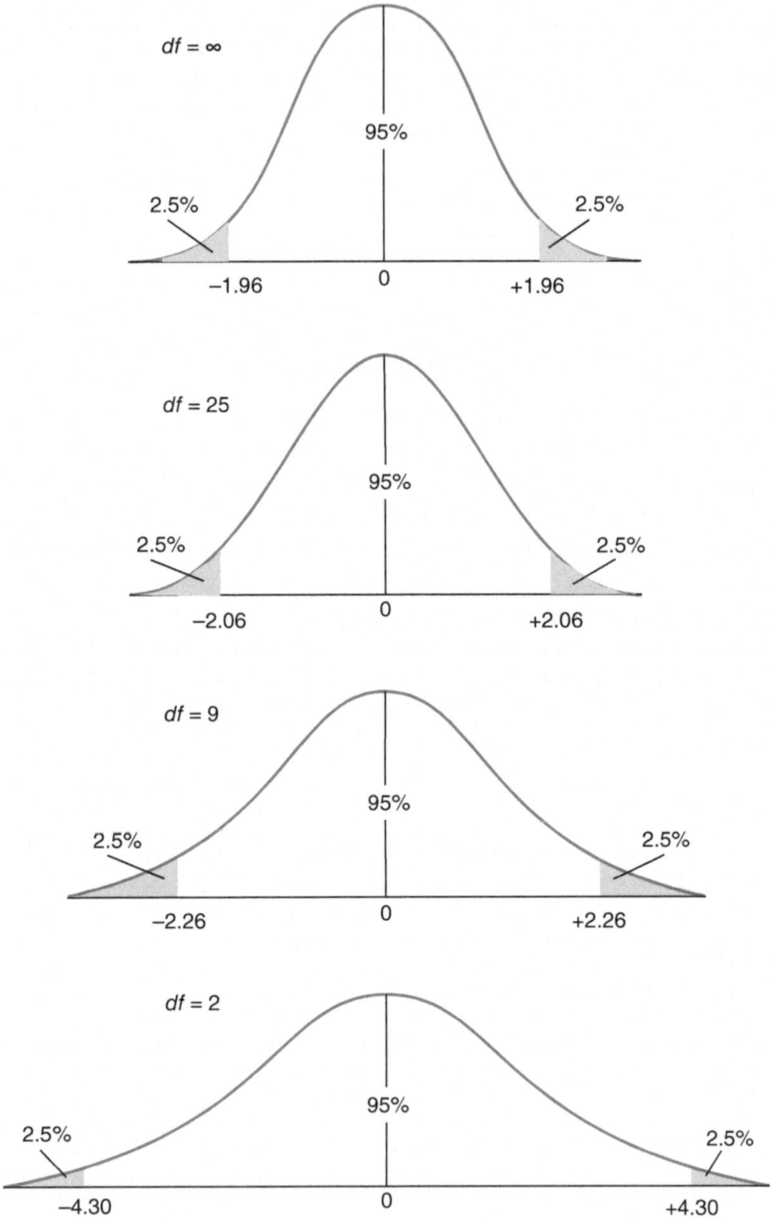

Figure 8.4 Examples of *t* distributions with various degrees of freedom.

8.5.2 Degrees of Freedom

Whenever a hypothesis test is conducted, an obtained value is computed and compared with a critical value. The critical value is found by referring to a table of critical values appropriate to the particular test statistic. We will get used to finding critical values for various kinds of inferential tests by using the **degrees of freedom** (***df***) associated with each test statistic. Moreover, degrees of freedom are also used to compute the obtained values. The reasons *why* mathematicians use degrees of freedom for inferential tests are complex and beyond the scope of this text (Walker 1940). Therefore, instead of discussing why *df* are used, only the concept will be explained.

Degrees of freedom are broadly defined as the number of data points in a set that are free to vary when estimating statistical values. For instance, suppose we are asked to pick 4 numbers that must sum to 10. We are free to pick any numbers for the first three, but the fourth number will be determined based on the value of the other three. The last number selected will be determined. Since three numbers are free to vary, we have 3 degrees of freedom. This is just a rough introduction to the concept of *degrees of freedom*. When using the *t* table to find critical values for the single-sample *t* test, the corresponding *df* value will be $n - 1$. Different statistical tests will have different ways of counting the degrees of freedom.

8.5.3 Using the *t* Table to Find Critical Values

Table A.2 is the *t* table, a portion of which is reproduced below.
α Values for two-tailed test

df	*.05*	*.01*
⋮	⋮	⋮
8	2.306	3.355
9	2.262	3.250
10	2.228	3.169
⋮	⋮	⋮
∞	1.960	2.576

The first column of the *t* table notes the degrees of freedom. Each *df* corresponds to a different *t* distribution. For each *df*, it would be possible to construct an entire table like the *z* table. However, to save space, only *t* values corresponding to typical rejection regions are presented.

The top two rows of the *t* table (Table A.2) state different alpha levels for one-tailed and two-tailed *t* tests. Only the two-tailed test is discussed in this chapter.[1]

1 The distinction between a one-tailed and two-tailed test is discussed in Chapter 9.

Suppose we set α at .05 and n at 10. What are the critical values (or t_{crit})? Find $df = 9$ (recall $df = n - 1$), and then find the .05 column for a two-tailed test. The critical value is 2.262 (which is understood as ± 2.262 since we are splitting the rejection region into the two tails of the distribution). What are the critical values when $\alpha = .01$ and the sample size is 9? The answer is ± 3.355 (recall $df = n - 1$).

Now, follow the *df* column to the bottom row, ∞. The corresponding critical value for $\alpha = .05$, two-tailed test is 1.96 – a number we should recognize. When n is very large, the t distribution assumes the shape of the standard normal curve. Therefore, the cutoffs for the 5% rejection region are the same as if we were using the z table. As we move from the bottom of the table to the top, the critical values increase. This reflects the fact that the t distribution becomes broader and the tails elongate as the sample size decreases. (Note: Table A.2 does not show every *df* value. If other t_{crit} values are needed, use an Internet search engine to find a more complete "t Table." Online tables can look different, but a careful reading should provide the proper value(s). Also, note that critical value differences between large *dfs* are negligible. Finally, note that the symbol "t_{obt}" refers to the t value obtained from Formula 8.3; "t_{crit}" refers to the critical value to which t_{obt} is compared. The subscripts clarify which t value is being referenced.)

8.5.4 A Worked Problem

Now we can use an example to put our t test knowledge to the test.

■ **Question** *A newspaper reports that the average high school student drinks six cups of coffee a week. The Northside High school board would like to know if their students drink that much coffee. Fifteen students were randomly selected and asked about their coffee habits. The mean number of coffee cups consumed per week is found to be 4.2 with a standard deviation of 1.5. Test the hypothesis that the newspaper report is accurate for the students at Northside High.*

Solution

Step 1. Identify the null and alternative hypotheses.

$$H_0 : \mu = 6$$
$$H_1 : \mu \neq 6$$

Step 2. Set alpha. Use an alpha of .05.

Step 3. Compute t_{obt} using Formulas 8.3 and 7.2.

$$t_{obt} = \frac{M - \mu}{s_M}; s_M = \frac{s}{\sqrt{n}}$$

$$t_{obt} = \frac{-1.8}{0.39} = \mathbf{-4.62}$$

Step 4. Using the t table, find the critical values for $df = 14$ and $\alpha = .05$. They are ± 2.145.

Step 5. Compare the t_{obt} of -4.62 with the critical values of ± 2.145. Since t_{obt} falls outside ± 2.145, the null hypothesis is rejected in favor of the alternative hypothesis ($H_1 : \mu \neq 6$).

Step 6. Interpret the findings. Statistical evidence suggests the amount of coffee consumed per week among this school's students is less than the amount stated in the newspaper. ■

8.5.5 Interpreting Inferential Findings

Before going any further, we need to make a few important comments about interpreting inferential tests. First, notice that the researcher is free to make a directional statement (i.e. the population mean in question is either less or more than the hypothesized mean value). Once a null is rejected, the researcher can look at the valence of the t_{obt} score and make a more specific interpretation. In this case, we conclude that students at this high school drink *less* coffee than high school students in general.

Second, notice the interpretation is made with cautious language. It starts with the phrase, "Statistical evidence suggests...." Although this specific wording is not required by any social science discipline, we will stick closely to it throughout the text, if, for no other reason than it creates good habits of thought. This phrase allows us to underscore three important points; it is *evidence* (not a hunch or opinion) that is *statistical* in nature (not logical, legal, or of some other form) and merely *suggestive* (probabilistic, not proof). All three of these are important concepts.

Finally, if an analysis results in a failure to reject the null, a good phrase to use is, "We do not have evidence to suggest...." We should *not* say, "We have evidence to suggest the null hypothesis is true." This would be to claim that we should accept the null, and we have already clarified that inferential tests do not allow for this strong of an inference. The proper way to communicate a failure to reject the null is to state that we *do not have evidence* suggesting a difference, effect, or relationship, whatever the case may be. Either we *have* evidence (when we reject the null) or we *do not have* evidence (when we fail to reject the null). Below is another worked example with a different type of finding to interpret.

■ **Question** *A bank president states that the average amount of money on deposit in savings accounts is $6500. To test this hypothesis, a random sample of nine deposits is examined. The sample mean is $7500, and the standard deviation is $1500. Can the claim be rejected?*

Solution

Step 1. Identify the null and alternative hypotheses.

$$H_0 : \mu = 6500$$
$$H_1 : \mu \neq 6500$$

Step 2. Set alpha. Use an alpha of .05.
Step 3. Compute t_{obt} using Formulas 8.3 and 7.2.

$$t_{obt} = \frac{M - \mu}{s_M}; \ s_M = \frac{s}{\sqrt{n}}$$

$$t_{obt} = \frac{1000}{500} = \mathbf{2.00}$$

Step 4. Find the critical values *for* $df = 8$ and $\alpha = .05$. The critical values are **±2.306.**
Step 5. Compare the t_{obt} of 2.00 to the critical values of ±2.306. Since t_{obt} does not fall outside ±2.306, the null hypothesis is not rejected.
Step 6. Interpret the finding. We do not have evidence to reject the null hypothesis. Keep in mind that this finding does not mean that we know $6500 *is* the average amount of money on deposit. Failing to reject the null hypothesis does not allow us to accept it. All we can say is that we have failed to find evidence rejecting the claim that $\mu = \$6500$. ■

8.5.6 Cohen's *d* for the Single-Sample *t* Test

As previously noted, one problem with hypothesis testing concerns the impreciseness of the resulting statistic to reflect the size of the effect when the null is rejected. Just as Cohen's *d* was used to determine the effect size for *z* test, it can also be used to *estimate* the effect size for a *t* test. For the single-sample *t* test, the formula is as follows:

8.5.7 Cohen's *d* for Single-Sample *t* Test

$$d = \frac{\text{mean difference}}{\text{sample standard deviation}} = \frac{M - \mu}{s} \quad (8.4)$$

8.6 Assumptions of the Single-Sample *z* and *t* Tests

Over the course of the remainder of the text, we will be exposed to several different inferential tests. The appropriate statistical test to use will depend on the research question of interest and the research method used. All statistical tests have assumptions. The ability to rely on the conclusions drawn from an inferential test is based on the degree to which the assumptions have been met. Some assumptions can be modestly violated without seriously compromising the interpretation of a statistical test. Other assumptions are critical. The assumptions for the single-sample t test, as well as the z test, are presented below.

1) **Representativeness.** It is assumed that the sample of participants is *representative* of the population in question. The goal of inferential statistics is to generalize from a sample to a population. If the sample is not representative of the population, it is possible that an untrue statement about the population will be made from the nonrepresentative sample. The best way to ensure representativeness is by *randomly sampling* the population. Obviously, not all studies can use random sampling; however, if other sampling methods are used, representativeness can become a concern. The representativeness assumption is *not* a mathematical assumption, but rather one of research methodology. If violated, the interpretive conclusions that follow from the t test may not be valid.
2) **Independent observations.** It is assumed that each score *within a sample* is *independent* of all other scores. It is possible to violate this assumption if the behavior of a participant in the study is influenced by the behavior of another participant. The scores of these two participants are no longer independent of each other.
3) **Interval or ratio scale of measurement.** The single-sample z and t tests utilize *means* and *standard deviations*. Both concepts require data to be from an interval or ratio scale.
4) **Normality.** It is assumed the sample is drawn from a normally distributed population. Recall that z and t tests require normal sampling distributions. If the population of raw scores is not normal, the sampling distribution may still be normal, *if n is sufficiently large*. For this reason, it is often claimed that z and t tests are *robust* to violations of normality. (A statistic is robust if it is resistant to mild violations of a given assumption.)

8.7 Interval Estimation of the Population Mean

In Chapter 7, we learned that there are two different kinds of inferential procedures: hypothesis testing and estimation. Up to this point, we have been focused on hypothesis testing. However, using a sample mean, the t distribution concept, and an estimate of the standard error, it is possible to generate an interval estimation of the population mean and quantify the confidence that it falls within that interval. Since each potential sample mean drawn from a population has a corresponding t value, we can use the t distribution and our obtained sample mean (which is an unbiased estimate of the population mean) to generate a probability function for the value of the actual population mean. Choosing t_{crit} values corresponding to different probabilities within the t distribution allows us to create intervals with differing degrees of certainty. The formula for a confidence interval is as follows:

The confidence interval for a population mean

$$LL = M - t_\alpha s_M \qquad (8.5)$$
$$UL = M + t_\alpha s_M$$

where

LL = the lower limit of the confidence interval
UL = the upper limit of the confidence interval
t_α = the t value marking the desired confidence, usually 95% ($\alpha = .05$)

Since we are generating an interval, two values are calculated, one at the lower end of the interval and the other at the upper end. As the interval widens and becomes less specific, the confidence grows that the actual mean falls within that window. A 95% confidence rate is typical, but the above formulas could easily be adjusted to find a 90% or 99% confidence interval simply by finding the corresponding t_{crit} values using the t table (Table A.2).

■ **Question** *Using the same data from the coffee-drinking habits of Northside High students, find the 95% confidence interval for the population mean ($M = 4.2$, $s = 1.5$, and $n = 15$).*

Solution

Step 1. Identify the null and alternative hypotheses.

$$H_0 : \mu = 6$$
$$H_1 : \mu \neq 6$$

Step 2. Set the confidence rate at 95%.

Step 3. Using the t table, find the cutoff values beyond which lie 2.5% in both the right and left tail of the distribution. The task is accomplished in the same way that we found t_{crit} for the t test. With $df = 14$ and $\alpha = .05$ (two-tailed test), the cutoff points are ± 2.145.

Step 4. Compute the confidence interval (recall Formula 7.2 is used to find s_M).

$$LL = M - t_{.05}s_M = 4.20 - 2.145\left(1.50/\sqrt{15}\right) = \mathbf{3.36}$$
$$UL = M + t_{.05}s_M = 4.20 + 2.145\left(1.50/\sqrt{15}\right) = \mathbf{5.04}$$

Step 5. Interpret the findings. Statistical evidence suggests 95% confidence that the mean number of coffee cups consumed per week by Northside students is between 3.36 and 5.04. ■

8.8 Formal Presentation of Findings

Proper reporting of inferential statistics can be challenging. The following are examples of how to report, in sentence form, a rejection of the null as well as a failure to reject the null. If rejecting the null, a sentence might read, "A single-sample *t* test found evidence that students from Northside High consume less coffee than high school students in general, $t(14) = -4.62, p < .05$." Notice the following, *t* and *p* are italicized, the degrees of freedom are placed within parentheses, a comma follows the actual statistical finding, the alpha value used is shown to have been eclipsed by the expression $p < .05$, t_{crit} is not reported, and a comma precedes the entire statistical expression that sits at the end of the sentence. Note also that a "0" is not typically placed to the left of the decimal in the probability statement. This accurately reflects what readers will find in most professional publications. If we also wanted to include a measure of effect size, the sentence could finish with, "... $t(14) = -4.62$, $p < .05$, $d = 1.2$."

If failing to reject the null, a sentence might read, "A single-sample *t* test did not find evidence that Northside High students consume a different amount of coffee from high school students in general, $t(14) = -1.62$, *n.s.*" First, notice the sentence wording. It did not *accept* the null by saying that evidence was found of no difference. Rather, it said that no evidence was found of a difference. Second, notice how the final part of the statistical expression used the letters *n.s.* (italicized). This stands for "not significant." Each inferential statement should end with either a description of the alpha value that was eclipsed, if the null is rejected (e.g. $p < .05$), or an expression that the null was not rejected (*n.s.*).

Summary

Single-sample statistical tests help to determine if a sample comes from a given population. The single-sample *z* test transforms a sample mean into a *z* value, which indicates the number of standard error units the sample mean is from the mean of the sampling distribution. If the z_{obt} equals or falls outside of the critical values, then statistical evidence exists to reject the null hypothesis; otherwise, the null should not be rejected.

The *t* statistic is used to test the null hypothesis when σ is unknown. In the *t* formula, *s* is used to estimate σ, and $s/\sqrt{n}$ is used to estimate $\sigma/\sqrt{n}$. The *t* statistic is used to transform a sampling distribution of means into a *t* distribution. The shape of the *t* distribution will approximate the standard normal curve as *n* increases. The critical value to which t_{obt} is compared is based on $n - 1$ degrees of freedom.

Hypothesis testing involves either rejecting or not rejecting the null hypothesis. The decision is probabilistic in nature, and the ultimate truth value of the null hypotheses can never be known. This decision risks two types of errors. A Type I error is committed when a true null hypothesis is rejected. The probability of making a Type I error is directly controlled by alpha, the criterion of significance. A Type II error is committed when a false null hypothesis is not rejected.

Statistical significance reflects the degree of certainty that the null hypothesis is false, but it does not necessarily reflect the size of the difference between the null mean and the sample evidence. To measure effect size, Cohen's *d* can be calculated.

The assumptions for the single-sample *z* and *t* tests are representativeness, independent observations, interval- or ratio-scaled data, and normally distributed populations. These tests are robust to violations of normality as *n* increases.

The standard error, sample mean, and *t* distribution can also be used to create a confidence interval for the actual value of an unknown population mean.

Microsoft® Excel and SPSS®

Information regarding how to use Microsoft® Excel and SPSS® to assist in performing single-sample *t* tests can be found in Appendix C.

Key Formulas

z Statistic

$$z_{obt} = \frac{M - \mu}{\sigma_M} \quad (8.1)$$

Cohen's *d* for single-sample *z* test

$$d = \frac{\textit{mean difference}}{\textit{standard deviation}} = \frac{M - \mu}{\sigma} \quad (8.2)$$

***t* Statistic**

$$t_{obt} = \frac{M - \mu}{s_M} \quad (8.3) \qquad (df = n - 1)$$

Cohen's *d* for single-sample *t* test

$$d = \frac{\text{mean difference}}{\text{sample std.deviation}} = \frac{M - \mu}{s} \quad (8.4)$$

The confidence interval for a population mean

$$\begin{aligned} LL &= M - t_\alpha s_M \\ UL &= M + t_\alpha s_M \end{aligned} \quad (8.5)$$

Key Terms

z Test
t Test
Alpha level
Critical values
Type I error
Type II error
Cohen's d
t distribution
Degrees of freedom

Exercises

1 How does a statistical test help a researcher make a decision about the null hypothesis? Can we know for sure that a decision is correct?

2 Provide two examples of single-sample research projects. In each instance, provide a research hypothesis as well as a corresponding pair of statistical hypotheses.

3 For each of the following situations, specify the null and alternative hypotheses:
   a The average respiration rate per minute is 8. Do smokers respire differently?
   b The average Beck Depression Inventory score is 12. Do mothers with young children differ from people in general?
   c The average miles per gallon (mpg) of cars used in the United States is 20. Is the observed mpg of cars in Japan different?

4 When conducting an inferential test, when should we use the t distribution?

5 If a researcher has access to both σ and s, which test should they use?

6 On what basis does a researcher decide on a given alpha level?

7 What type of error corresponds to a "false positive?" To a "false negative?"

8 Failing to reject the null when it is false makes what type of error?

9 Rejecting the null when it is true makes what type of error?

10 For which type of error can we set the risk rate (per test)?

Work Problems

11 If the difference between a mean from a treated sample and the known population mean is 5 and the standard deviation of the population is 10, what is Cohen's *d*?

12 If new shoes are claimed to elevate an athlete's jump by 3 in. and the standard deviation of heights jumped by athletes is 4 inches, what is the claimed effect size, in terms of Cohen's *d*?

13 If Cohen's *d* is .4, and the mean difference is 12, what is the standard deviation?

14 A publisher of a new statistics textbook claims an effect size of .2 (i.e. Cohen's *d* of .2). If we know the μ and σ of a nationally normed test are 100 and 20, respectively, what is the publisher's hypothesized mean for the users of their textbook?

15 Among trained typists, suppose the average typing speed on a standard keyboard is known to be 60 words per minute (wpm), with a standard deviation of 5 wpm. The manufacturer of an ergonomically designed keyboard claims their device improves typing speed. A random sample of 50 typists using the new keyboard yields a mean of 65 wpm. Does the new device affect typing speed? Set alpha at .05.

a Which test should be used? Why?
b State H_0 and H_1.
c What are the critical values?
d What is the obtained statistic?
e Reject the null hypothesis?
f What decision error might be made?
g Is the manufacturer's claim supported?
h If so, what is the effect size?

16 Suppose the mean weight of adult golden retrievers is known to be 90 pounds. A veterinarian claims a hormone in golden retriever puppies can increase their size. A sample of 41 pups is injected with the hormone, and their average weight at maturity is found to be 110 pounds, with a standard deviation of 30 pounds. Conduct an inferential test with $\alpha = .05$.

a Which test should be used? Why?
b State H_0 and H_1.
c What are the critical values?

d What is the obtained statistic?
e Reject the null hypothesis?
f If so, what is the effect size?
g What decision error might be made?
h Interpret the finding.

17 A researcher wants to determine if students at her university sleep more than most students. Suppose it is known that the university student sleep distribution is slightly skewed with a $\mu = 7.5$ hours/night and $\sigma = 2.4$. The researcher takes a sample of $n = 200$ students at her university and finds an average of 7.2 hours of sleep/night and a standard deviation of 1.8.
a What test should be used? Why?
b State H_0 and H_1.
c Does the skewness matter?
d If $\alpha = .05$, what are the critical values?
e What is the obtained statistic?
f Reject the null hypothesis?
g If so, what is the effect size?
h What decision error might be made?
i Interpret the finding.

18 An organizational psychologist believes that people who work at home experience greater job satisfaction. Imagine a job satisfaction rating scale that is claimed to have a mean of 50. The psychologist randomly sampled 20 people who work at home, finding $M = 63$ and $s = 17$.
a What test should be used? Why?
b State H_0 and H_1.
c If $\alpha = .05$, what are the critical values?
d What is the obtained statistic?
e Reject the null hypothesis?
f If so, what is the effect size?
g What decision error might be made?
h Interpret the finding.

19 Subjective life expectancy is a person's belief in how long they will live. Several years ago, Robbins (1988) found that young adult biological males tend to overestimate life expectancy. At the time of Robbins' research, the actual life expectancy for biological males was 72.4 years. The following hypothetical data are consistent with Robbins' findings.

Subjective life expectancy for males (years)		
77	72	78
74	82	75
80	76	79

- **a** What test should be used? Why?
- **b** State H_0 and H_1.
- **c** If $\alpha = .05$, what are the critical values?
- **d** What is the obtained statistic?
- **e** Reject the null hypothesis?
- **f** If so, what is the effect size?
- **g** What decision might be made?
- **h** Interpret the findings.

20 An anthropologist predicts that physical stress in childhood increases height (Landauer and Whiting 1964). The researchers located a tribe of people in which physical stress is a by-product of frequent tribal rituals (e.g. piercing and molding body parts, exposure to extreme temperatures, etc.). The mean height of people in the region who do *not* use physically stressful rituals is used as the population mean (males = 65 in.; females = 59 in.). The following hypothetical data are for adult biological males and females of the tribe in question. Conduct a t test for men and a t test for women. (Use an $\alpha = .05$.) Interpret both findings and present them in a professionally appropriate manner.

Men	Women
67	59
69	63
72	65
70	60
70	59
72	62
64	61
70	66

21 A director of mental health would like an idea of the mean number of patient assaults per month. A random sample of 18 inpatient units is taken. The average number of assaults is found to be 24.50 per unit, with a standard deviation of 2.61. Compute the 95% confidence interval for the population mean.

22 A drug manufacturer is researching the number of side effects generated by a new medication for high blood pressure. A random sample of 13 patients taking the drug is selected, and the mean number of side effects is found to be 4.2, with a standard deviation of .86. Compute the 95% confidence interval for the population mean.

23 A tire company is interested in knowing how many highway miles its tires can tolerate before the treads wear out. A random sample of 60 tires is selected and tested. The mean is found to be 56 000 miles, with a standard deviation of 4 300 miles. Compute the 95% confidence interval for the population mean.

24 An insurance company states that it takes them an average of 15 days to process an auto accident claim. A random sample of 40 claims is drawn from processed claims over the past six months. Based on the following data, is there any evidence that the mean number of days to pay claims is not 15? Set $\alpha = .05$.

Number of days to process a claim

22	11	7	9	9	8	7	23	45	9
23	21	8	8	5	16	9	22	17	6
12	29	6	5	9	23	7	33	24	5
15	14	9	7	3	17	8	19	15	8

25 Every day, a commuter recorded the amount of time the train was late. Over a period of two months, the mean number of minutes that the train was late is 24.5. The train authorities add extra cars during rush hour to address the problem. For the next 30 working days, the commuter records the amount of time that the train is late. Based on the following data, does there seem to be an improvement in service?

Number of minutes the train is late

22	11	25	19	9	8	7	23	45	20
23	21	8	1	5	16	3	22	17	24
12	16	22	32	9	10	7	11	10	2

9

The Independent- and Dependent-Samples *t* Tests

9.1 The Research Context for Between-Participants Designs

In Chapter 8, the *t* test was used to contrast one sample mean with a specified population value. A more common usage of the *t* test arises when means *from two different samples* are compared to infer whether there is a difference between the means of the two populations from which the samples came. If the two samples of scores are coming from two different sets of participants, the research context is known as a *between-participants design*, and the appropriate *t* test to use is called an **independent-samples *t* test**. If the two samples come from one set of participants measured under different circumstances, the research context is known as a *within-participants design* (or a *repeated-measures design*), and the appropriate *t* test is called a dependent-samples (or paired-samples) *t* test. Both tests will be explored in this chapter in a sequential fashion. First, we will explore the independent-samples *t* test. Below are two research examples in which an independent-samples *t* test can be used to analyze the data.

► **Example 9.1** Pham and others (2011) found a relationship between relaxation and money spent while shopping. They induced two states of relaxation in participants: The relaxed group watched a soothing video, and another watched a neutral video. Participants then assessed the monetary value of different items. The findings showed that relaxed participants bid significantly higher than the control participants for a whole range of auctioned items. ◄

► **Example 9.2** In 1972, Buffalo Creek, West Virginia, was the scene of a major flood due to the inappropriate removal of coal waste. The flood killed 125 people, and 5000 people lost their homes. Simpson-Housley and DeMan (1989) found that, 17 years later, the residents of Buffalo Creek scored higher on a measure of trait

Essentials of Behavioral and Social Science Statistics, Third Edition.
K. Paul Nesselroade, Jr. and Laurence G. Grimm.
© 2025 John Wiley & Sons, Inc. Published 2025 by John Wiley & Sons, Inc.
Companion website: https://www.wiley.com/go/Essentials_of_Behavioral_and_Social_Science_Statistics3e

anxiety in comparison with the residents of Kopperston, a nearby mining town that did not experience the flood. ◀

9.1.1 The Between-Participants Design

Between-participants designs (also called *between-groups, independent-samples, or independent-groups* designs) are defined by the fact that each group of participants comes from a different population. Moreover, no participant or group of participants are members of both populations. Oftentimes, this design is used for experimental purposes. We can imagine two groups of participants where the first is trained in one mnemonic technique (e.g. the peg-word system) and the second learns a different mnemonic technique (e.g. acronyms). Each is then given the same memory task, and the number of errors is recorded.

The independent-samples t test can be used when a researcher uses a between-participants design, whether an experimental manipulation is involved or not. In experimental settings, if participants are randomly assigned to two conditions and the internal validity is high, it is possible to make a causal statement about the relationship between the independent and dependent variables. However, the independent-samples t test can also be used in correlational designs. The Buffalo Creek study mentioned above is a case in point. Even though a t test can be used to compare means, no causal statement can be advanced about the relationship between variables. In general, the interpretation of any inferential test depends on the way the study is designed (see Chapter 1).

Spotlight 9.1 William Gosset

William Gosset (1876–1937) developed the t distribution as well as the independent- and dependent-samples t tests. After receiving a degree in chemistry and mathematics from Oxford, Gosset was hired by the Guinness brewery in Dublin in 1899. Around the turn of the century, many companies, especially in the agricultural industry, attempted to apply a scientific approach to product development. A typical research question would have been, "Which fertilizer will produce the largest corn yield?" Until Gosset's work, statisticians dealt with large numbers of observations, in the hundreds and thousands. Gosset wanted to develop statistical techniques that were nimbler and could be applied to even small sample sizes.

In his seminal 1908 article (Student 1908), "The Probable Error of a Mean," Gosset proposed a way to make inferences using smaller sample sizes where the standard error is not presumed, but rather estimated with different degrees of certainty based on the size of the sample. In this paper, Gosset also used the term "significance" to describe a reliable difference between means. Over the next 25 years, this terminology caught on, as did the conventional use of the 5% level of significance.

Gosset's 1908 article is one of the most important publications in the history of inferential statistics. "With one stroke, he: (i) discovered a new statistical distribution; (ii) invented a statistical test that became the prototype for a whole series of tests, including analysis of variance; and (iii) extended statistical analysis to small samples..." (Tankard 1984, p. 99). Although *t* tests are now one of the cornerstones of modern statistics, Gosset's work was not initially greeted with enthusiasm. Fisher, the originator of the analysis of variance (see Chapters 11–13), described the reaction of colleagues as "weighty apathy" (Fisher 1939, p. 5), and Cochran stated that "the *t* distribution did not spread like wildfire" (Cochran 1976, p. 13). Even Gosset underestimated the impact of his discoveries, as he wrote to Fisher, "I am sending you a copy of Student's Tables as you are the only man that's ever likely to use them!" (Gosset 1970; Letter 11).

Interestingly, Gosset used a pseudonym when publishing. Not wanting the competition to know about its scientific work, Guinness forbade its scientists from publishing. As a result, Gosset secretly published all his articles under the name of "Student." It is for this reason that the *t* test is also known as "Student's *t* test."

9.2 The Independent-Samples *t* Test

The independent-samples *t* test is used when two samples of participants provide scores on a measure. The *t* test compares the means of the two samples. The goal is not to determine whether the means of the two *samples* are different (simply comparing the sample means will accomplish this), but rather to make an inference about whether the *population* means from which the samples are taken are different.

Suppose we decide to conduct a study on the effects of stress. Previous research suggests that involvement in meditative exercises is beneficial. An experimental group practices meditation for 10 weeks, whereas a control group does not. After 10 weeks, both groups are asked to solve a series of arithmetic problems. Participants are told to work as quickly as possible and that a loud burst of sound will be played if the performance is inadequate. Stress is measured by the participants' heart rate during the task. The experimental hypothesis predicts that those in the meditation condition will have lower heart rates than those in the control condition.

The two samples represent two hypothetical populations. One sample represents the population of participants who theoretically could practice meditation. The control sample represents the hypothetical population of untreated participants.

9.2.1 The Null and Alternative Hypotheses

As with all inferential tests, when performing an independent-samples t test, the investigator specifies the null and alternative hypotheses beforehand. Recall that in single-sample research, the null hypothesis is stated as an equality, μ = some value known ahead of time, and the alternative hypothesis is stated as an inequality, $\mu \neq$ that same value. Similarly, in research involving two samples, the null hypothesis is stated as an equality, and the alternative hypothesis is expressed as an inequality. The null can be stated in two ways:

$$\text{H}_0 : \mu_1 = \mu_2 \text{ or } \text{H}_0 : \mu_1 - \mu_2 = 0$$

Note that these two statements are equivalent. The alternative hypothesis can also be stated in two ways, depending on how we choose to state the null hypothesis:

$$\text{H}_1 : \mu_1 \neq \mu_2 \text{ or } \text{H}_1 : \mu_1 - \mu_2 \neq 0$$

In the meditation and stress study, the research hypothesis is that meditation will reduce physiological reactions to a stressor. The statistical hypotheses (i.e. the null and alternative hypotheses) state that the samples are either taken from the same population (no treatment effect) or come from two different populations with different means (a treatment effect).

Note: Subscripts of 1 and 2 are typically used to represent the populations being compared. Researchers are free, however, to use other subscripts more tailored to fit the populations; for example, $\mu_{control}$ and μ_{exp}, or μ_{drug} and $\mu_{placebo}$, or even letters such as μ_A and μ_B.

9.2.2 The Sampling Distribution for an Independent-Samples *t* Test

9.2.2.1 Theoretically Constructing the Sampling Distribution

The sampling distribution for an independent-samples t test is derived from the *difference between independent sample means*. In Chapter 8, theoretical sampling distributions were constructed when one population was repeatedly sampled. In the present case, two populations will each be sampled, and the difference between the sample means will be used to create a *sampling distribution of sample mean differences*.

Returning to the meditation and stress study, the two hypothetical populations are "treated" and "untreated" participants. We sample these two populations in the sense that participants from one condition practice meditation and those from the other do not. Theoretically, to construct the sampling distribution, we would run the study, compute the mean heart rate under stress for each sample, and then *take the difference between the sample means*. This process would be repeated a near-infinite number of times. (Note: It is not required that $n_1 = n_2$, though the test assumes they are similar; however, n_1 must equal n_1 each time population 1 is sampled, and likewise for n_2.) For each iteration, the value $M_1 - M_2$ would be computed and used to create the frequency distribution. The result is a sampling distribution of sample mean differences.

9.2.2.2 Characteristics of the Sampling Distribution

Since statisticians have worked out the features of all sampling distributions, we have been spared the impossible task of constructing them. One of the characteristics of a sampling distribution of mean differences is that if the null hypothesis is true, that is $\mu_1 = \mu_2$, then the mean of the sampling distribution will be 0. (Can we see that for each pair of means, M_1 is just as likely to be larger than M_2 as it is to be smaller?) However, if the means of the populations differ, then the mean of the sampling distribution will differ accordingly.

Another characteristic of these sampling differences holds that if the two populations of raw scores are normally distributed, the sampling distribution will likewise be normal. However, even if the populations are not normal, if the sample sizes are sufficiently large, the resulting sampling distribution will be normal.

9.2.2.3 The Estimated Standard Error of the Difference

As with the single-sample *t* test, the independent-samples *t* test uses the estimate of the standard error as the denominator, symbolized as $s_{M_1 - M_2}$. However, because the sampling distribution is made up of mean differences, it is properly termed the *estimate of the standard error of the difference*, although it is often simply called the **estimated standard error**. (Since σ is rarely known for either population, the sample standard deviations, s_1 and s_2, are pooled together to get the best possible estimate.)

When calculating the estimate of the standard error by hand, use either Formula 9.1 (when the sample variances are given) or Formula 9.2 (when working from raw data).

Variance formula for the estimated standard error

$$s_{M_1 - M_2} = \sqrt{\frac{s_1^2(n_1 - 1) + s_2^2(n_2 - 1)}{n_1 + n_2 - 2}\left(\frac{1}{n_1} + \frac{1}{n_2}\right)} \quad (9.1)$$

Computational formula for the estimated standard error

$$s_{M_1 - M_2} = \sqrt{\frac{(\Sigma X_1^2 - (\Sigma X_1)^2/n_1) + (\sum X_2^2 - (\Sigma X_2)^2/n_2)}{n_1 + n_2 - 2}\left(\frac{1}{n_1} + \frac{1}{n_2}\right)} \quad (9.2)$$

9.2.3 Hypothesis Testing and the Sampling Distribution of Differences

Hypothesis testing operates statistically at the level of the sampling distribution. As the observed difference between the sample means increases, it becomes less and less likely that this difference is only due to sampling error, leaving us with the growing probability that there is a real difference between the populations.

Figure 9.1 shows three different scenarios regarding the relationship between sampling distributions and a corresponding sampling distribution of mean differences. In Figure 9.1a, the two sampling distributions are from identical populations. As a result, they fully overlap, and the resulting sampling distribution of mean differences, $(\mu_1 - \mu_2)$, centers on 0. In Figure 9.1b, the sampling distributions are from slightly different populations, and the sampling distribution of mean differences, $(\mu_1 - \mu_2)$, centers on 2. Finally, in Figure 9.1c, the two sampling distributions are from very different populations, and the sampling distribution of mean differences, $(\mu_1 - \mu_2)$, center on 10.

9.2.3.1 The Independent-Samples *t* Test

What the researcher would like to determine is if the difference between the sample means is so unusual as to suggest they came from different populations. The null sampling distribution concept (i.e. Figure 9.1a) is key to making this determination.

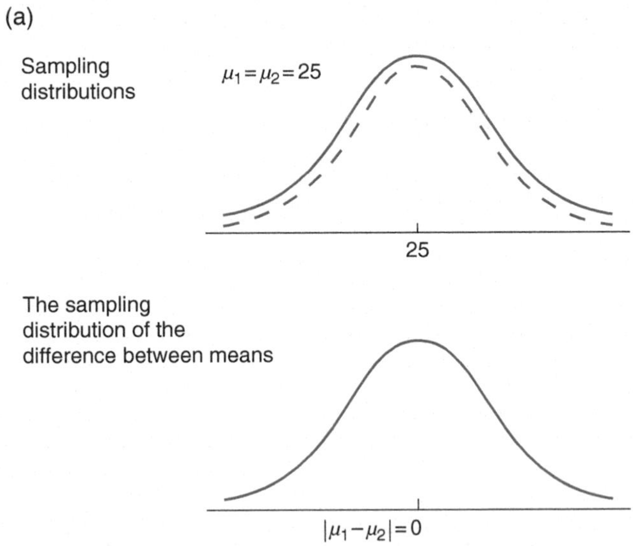

Figure 9.1 A sampling distribution of mean differences (i.e. the second curve in each pair, a, b, and c) has the same mean as the difference between the means of the two sampling distributions.

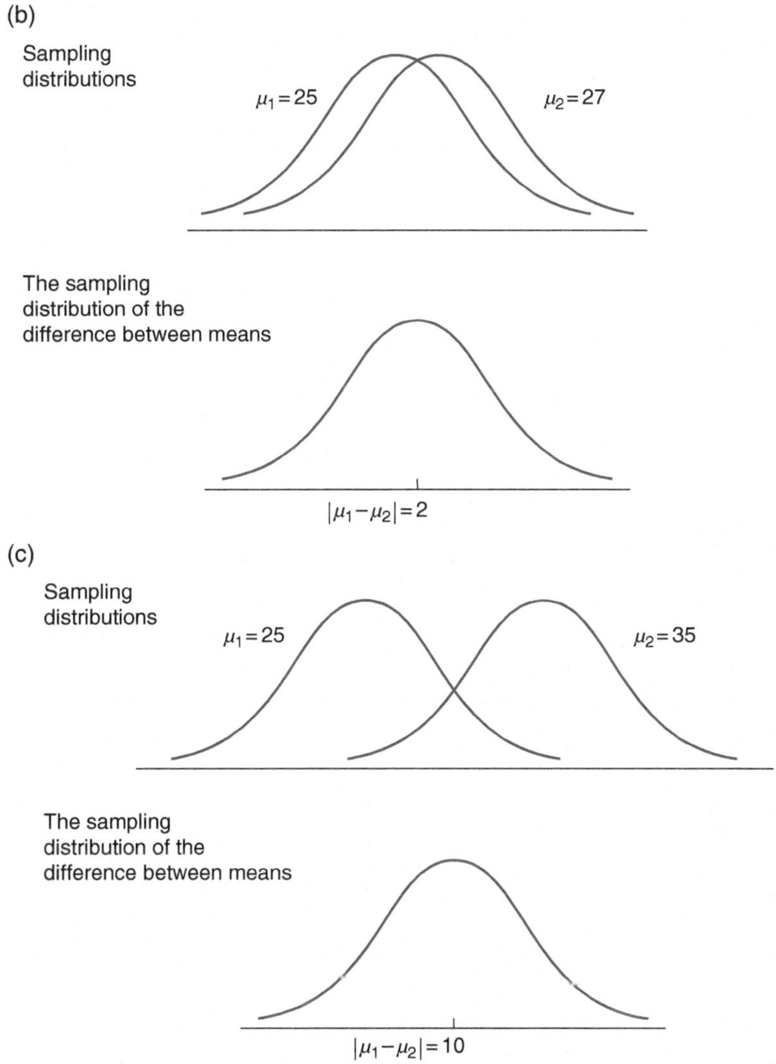

Figure 9.1 (Continued)

An estimate of its standard error, $s_{M_1-M_2}$, is found (see Formulas 9.1 and 9.2) and then used to interpret the sample mean difference as either expected, given sampling error, or not expected.

The logic of the independent-samples *t* test follows the logic of the tests explored in Chapter 8. The numerator measures the difference between two sample means, and an estimate of the standard deviation of the sampling distribution in the denominator interprets if the mean difference fits well within the null sampling distribution centered on 0 or not. If not, then evidence has been found that the sample means have come from different populations. Please note that the mean difference by itself is insufficient. A mean difference of, say, 10 may be seen as small or large. It depends on the size of the error term in the denominator. If there is a lot of variability in the scores, then a sample mean difference of 10 is small. However, if variability is small, then a difference of 10 is large.

Independent-samples *t* test

$$t_{obt} = \frac{M_1 - M_2}{s_{M_1 - M_2}} \quad (9.3)$$

To understand why Formula 9.3 looks as it does, it is helpful to recall the *t* ratio for a single-sample *t* test found in Chapter 8:

$$t_{obt} = \frac{M - \mu}{s_M}$$

We can see two differences. In the numerator, instead of comparing a sample mean to a known population mean, the independent-samples *t* test compares two sample means. Also, we have a new subscript for the variability measure of the sampling distribution. The subscripts remind us that this variability measure is from a sampling distribution featuring the difference between two sample means. Otherwise, the formulas are quite similar; they both generate a *t* value that will need to be interpreted.

9.2.4 The Independent-Samples *t* Test Distribution

Since the construction of the sampling distribution is conceptually similar to the single-sample *t* test, we can use the same *t* table in Appendix A.2 that was previously used. The degrees of freedom, however, are different. For an independent-samples *t* test, the degrees of freedom are $n - 1$ from both the first and second samples, or $n_1 + n_2 - 2$.

Table 9.1 summarizes the differences between the formulas for a *z* score, *z* statistic, single-sample *t* statistic, independent-samples *t* statistic, and the dependent-samples *t* statistic (to be covered later in this chapter). The purposes and the underlying distributions of the formulas are presented, as well as the degrees of freedom for the *z* test and the *t* tests.

9.2.5 A Worked Example

Armed with the *t* test logic and the requisite formulas, we are now ready to work through a problem and decide whether to reject the null hypothesis. Keep in mind that we are making an inference about whether the means of the two populations

Table 9.1 A summary comparison of transformation formulas.

z score (Chapter 5)

$$z = \frac{X - \mu}{\sigma}$$

Purpose. Transforms raw scores into z scores, placing them on the z distribution.

Distribution characteristics. The z distribution is the standard normal distribution. It has a mean of 0 and a standard deviation of 1. It is normally distributed if the population is normal.

z statistic (Chapter 8)

$$z_{obt} = \frac{M - \mu}{\sigma_M}$$

Purpose. Transforms a sampling distribution of means into a distribution of z values.

Distribution characteristics. This is also the standard normal distribution (see above).

Single-sample t statistic (Chapter 8)

$$t_{obt} = \frac{M - \mu}{s_M}$$

Purpose. Transforms a sampling distribution of means into a t distribution.

Distribution characteristics. All t distributions are symmetrical with $\mu = 0$ ($df = n - 1$).

Independent-samples t statistic (Chapter 9)

$$t_{obt} = \frac{M_1 - M_2}{s_{M_1 - M_2}}$$

Purpose. Transforms a sampling distribution of mean differences into a t distribution.

Distribution characteristics. All t distributions are symmetrical with $\mu = 0$ ($df = n_1 + n_2 - 2$).

Dependent-samples t statistic (Chapter 9)

$$t_{obt} = \frac{M_X - M_Y}{s_D}$$

Purpose. Transforms a sampling distribution of mean differences into a t distribution.

Distribution characteristics. All t distributions are symmetrical with $\mu = 0$ ($df = n_p - 1$).

are equal. In the context of an experiment, deciding the population means are *not* equal is to claim that the independent variable influenced the dependent variable, assuming no confounds. Consider once again the study about the effectiveness of meditation on participants' ability to tolerate stress.

Twenty participants are randomly assigned to either an experimental ($n_1 = 10$) or control condition ($n_2 = 10$). The experimental participants practice meditation 30 minutes a day, three times a week for 10 weeks. The control participants do not practice meditation. After 10 weeks, all participants are brought into the lab and asked to solve arithmetic problems under threat of a loud burst of sound for poor performance. The measure of stress is the participants' heart rate during the task. The experimental hypothesis is that the meditation group will have a lower heart rate than the control group. The null hypothesis is that the population means are the same. The data, along with the computation of t_{obt}, are presented in Table 9.2.

9.2.5.1 Testing the Null Hypothesis Using the Independent-Samples *t* Test

Step 1. Define the null and alternative hypotheses:

$$\mathrm{H}_0 : \mu_1 = \mu_2$$
$$\mathrm{H}_1 : \mu_1 \neq \mu_2$$

Step 2. Set alpha. Alpha is set at .05.

Step 3. Compute t_{obt} (see Table 9.2).

Step 4. Locate t_{crit} in Table A.2. (Recall $df = n_1 + n_2 - 2 = 10 + 10 - 2 = 18$.) Enter the left column of the *t* table and locate 18. Now, move to the column under alpha of .05 for a two-tailed test. The critical value is 2.10. (Since this is a two-tailed test, $t_{crit} = \pm 2.10$.)

Step 5. Compare the t_{obt} of −2.23 with a critical value of ±2.10. Since t_{obt} falls outside of ±2.10, the null hypothesis is rejected. Furthermore, the means are consistent with the experimental hypothesis (i.e. the meditation heart rate mean is slower than the control heart rate mean).

Step 6. Interpret the findings. Statistical evidence suggests that meditation leads to a reduction in the participants' heart rate when solving arithmetic problems under the threat of a loud burst of noise for poor performance, $t(18) = -2.23$, $p < .05$.

Table 9.2 Computing t_{obt} for an independent-samples *t* test.

Meditation scores	Control scores
84	88
78	97
67	74
87	80
80	87
78	90
78	90
79	86
82	84
81	78

Table 9.2 (Continued)

Calculated values	
$M_1 = 79.40$; $s_1 = 5.25$; $n_1 = 10$	$M_2 = 85.40$; $s_2 = 6.69$; $n_2 = 10$
$\Sigma X_1 = 794$; $\Sigma X_1^2 = 63\,292$	$\Sigma X_2 = 854$; $\Sigma X_2^2 = 73\,334$

$$t_{obt} = \frac{M_1 - M_2}{s_{M_1 - M_2}}$$

Insert Formula 9.2 for $s_{M_1 - M_2}$

$$t_{obt} = \frac{M_1 - M_2}{\sqrt{\frac{\left(\Sigma X_1^2 - \frac{(\Sigma X_1)^2}{n_1}\right) + \left(\Sigma X_2^2 - \frac{(\Sigma X_2)^2}{n_2}\right)}{n_1 + n_2 - 2}\left(\frac{1}{n_1} + \frac{1}{n_2}\right)}}$$

$$= \frac{79.4 - 85.4}{\sqrt{\frac{\left(63\,292 - \frac{(794)^2}{10}\right) + \left(73\,334 - \frac{(854)^2}{10}\right)}{10 + 10 - 2}\left(\frac{1}{10} + \frac{1}{10}\right)}}$$

$$= \mathbf{-2.23}$$

Raw scores are heart rates under stress. Since t_{obt} falls outside of $\pm t_{crit}$, the null hypothesis is rejected.

9.2.6 A Measure of Effect Size: Cohen's *d*

One secondary question that can be asked when a null is rejected is the size of the treatment effect. The t_{obt} value is not designed to measure effect size, but rather the likelihood that an effect exists; that is, the *certainty* of an effect, not the *size* of an effect. These are related concepts, but not identical. As noted in the previous chapter, a simple, direct, and often-used measure of effect size is Cohen's *d*. The formula is as follows.

Cohen's *d* for independent-samples *t* test

$$d = \frac{\textit{estimated mean difference}}{\textit{estimated standard deviation}} = \frac{M_1 - M_2}{\sqrt{\frac{s_1^2(n_1 - 1) + s_2^2(n_2 - 1)}{n_1 + n_2 - 2}}} \quad (9.4)$$

$M_1 - M_2$ is used as the best estimate of the mean difference between the two populations, and the denominator pools the two different sample standard deviations together to generate the best estimate of the population standard deviation. Notice, this only allows us to *estimate* the effect size. As with previous versions of the statistic, Cohen's *d* measures effect size in standard deviation terms, and larger *d* values reflect larger effect sizes.

9.3 Assumptions of the Independent-Samples *t* Test

There are five assumptions to be met when conducting independent-samples *t* tests. Since the first four are identical to the assumptions for the single-sample *t* test, they are simply listed. The fifth assumption is new. The first two are methodological and the final three are statistical:

1) **Representativeness.**
2) **Independent observations.**
3) **Interval or ratio scale of measurement.**
4) **The populations from which the samples are taken are normally distributed.**
5) **Homogeneity of variances.** The independent-samples *t* test assumes the variances of the two populations sampled are equal. A rule of thumb for judging homogeneity is to see if one sample variance is more than four times larger than the other. If so, the assumption is probably violated. The *t* test may be robust to violations of this assumption, particularly if $n_1 = n_2$. However, when sample sizes are unequal and variances are quite discrepant, this *t* test should not be run. Thankfully, other tests can be used (see Chapter 16).

9.4 Interval Estimation for Independent Samples

Recall that there are two kinds of inferential procedures: hypothesis testing and estimation. This chapter emphasizes hypothesis testing, but we can also use sample means, the *t* distribution, and an estimate of the standard error to generate an interval estimation of the population mean difference. Further, we can quantify the confidence we have that this difference falls within that interval. Since each potential sample mean difference drawn has a corresponding *t* value, we can use the *t* distribution and our obtained sample mean difference to generate a probability function for the value of the actual population mean difference. Choosing t_{crit} values corresponding to different probabilities within the *t* distribution allows us to create intervals with differing degrees of certainty. The formula for confidence intervals is as follows:

Confidence interval for a population mean difference for independent samples

$$\begin{aligned} LL &= (M_1 - M_2) - t_\alpha s_{M_1 - M_2} \\ UL &= (M_1 - M_2) + t_\alpha s_{M_1 - M_2} \end{aligned} \quad (9.5)$$

where

LL = the lower limit of the confidence interval

UL = the upper limit of the confidence interval
t_α = the t value marking the desired confidence, usually 95% ($\alpha = .05$)

Since we are generating an interval, two values are calculated, one value at the lower end of the interval and the other at the upper end. As the interval widens and becomes less specific, the confidence grows that the actual mean difference falls within that window. A 95% confidence rate is typical, but the abovementioned formulas could be used to find other confidence intervals by finding the corresponding t_{crit} values using the t table (Table A.2).

■ **Question** *Using the same data from the meditation and stress study explored earlier in the chapter, find the 95% confidence interval for the population mean difference in heart rates due to meditation* ($M_1 = 79.4; M_2 = 85.4; s_{M_1-M_2} = 2.69; n_1 = 10; n_2 = 10$).

Solution

Step 1. Identify the null and alternative hypotheses:

$$H_0 : \mu_1 = \mu_2$$
$$H_1 : \mu_1 \neq \mu_2$$

Step 2. Set the confidence rate at 95% (equivalent to $\alpha = .05$).

Step 3. Using the t table, find the cutoff values beyond which lie 2.5% of the distribution in each tail. With $df = 9$ and $\alpha = .05$ (two-tailed test), the cutoff points are ± 2.262.

Step 4. Compute the confidence interval. Having worked on the problem previously, we know:

$$M_1 - M_2 = -6; \text{and } s_{M_1-M_2} = 2.69$$
$$LL = (M_1 - M_2) - t_{.05}s_{M_1-M_2} = -6 - 2.262(2.69) = \mathbf{-0.08}$$
$$UL = (M_1 - M_2) + t_{.05}s_{M_1-M_2} = -6 + 2.262(2.69) = \mathbf{12.08}$$

Step 5. Interpret the findings. Statistical evidence suggests 95% confidence that meditation leads to a heart rate drop in this experimental situation of between –0.08 and 12.08 beats per min. ■

9.5 The Research Context for Within-Participants Designs

The independent-samples t test is used to contrast means computed from two *unrelated* samples of scores, that is, from *between-participants* designs. The **dependent-samples *t* test** (also known as a *paired-samples t test*, *related-samples t test*, or *correlated-samples t test*) is used to contrast means computed from two *related* samples of scores. A **within-participants design** (or *repeated-measures* *design*) evaluates one group of participants under two or more conditions. In this chapter, only the *two treatment* condition is addressed. Chapter 13 looks at analyses of within-participant designs with more than two conditions.

Most designs utilizing a dependent-samples *t* test are experimental in nature. For that reason, the language of experimentation will be used in the following sections. Two research examples from the psychological literature are presented below. In each instance, a within-participants design is used, and a dependent-samples *t* test can be used to analyze the data.

► **Example 9.3** Lehman and others (2013) tested several hypotheses regarding the features of music and the walking gait of participants. Eighteen individuals walked around while listening to different playlists. One hypothesis concerned tempo strength. The research team found that stronger-tempo songs influenced participants' walking gait more than softer ones. ◄

► **Example 9.4** Addison (1989) tested the hypothesis that people perceive bearded men differently than non-bearded men. Addison used one group of participants to rate pictures of both bearded and non-bearded men on several dimensions. The results indicated that males with beards are rated as more masculine, more aggressive, stronger, and more dominant than males without beards. However, they are not viewed as more intelligent. ◄

A within-participants design is efficient because a researcher can use fewer participants in comparison with a between-participants design. More importantly, a within-participants design also increases the *power* when testing the null hypothesis. The power of a statistical test is the probability of correctly rejecting a false null hypothesis, that is, the probability of avoiding a Type II error. (The concept of power is featured in Chapter 10.) All other things being equal, the power of a dependent-samples *t* test is greater than the power of the independent-samples *t* test. This is because dependent-samples tests reduce the variability associated with using different participants. (This concept will be explored in greater detail in Chapter 13.) As a result, the denominators in the dependent-samples tests tend to be smaller than those in independent-samples tests. Smaller denominators yield larger *t* values, and larger *t* values are more likely to fall into the rejection region of a distribution.

Another form of within-participants design occurs when participants supply scores both before and after a treatment. This type of design is called a pretest/posttest design.[1] It is useful when evaluating behavior change caused by introducing an independent variable. Studies that examine learning, performance, or therapy effects frequently employ pretest/posttest designs.

Dependent-samples *t* tests have also been used to analyze data from two independent but related samples of participants; this is called a **matched-samples** (or

1 Pretest/posttest designs are quasi-experimental. There are inherent weaknesses with these types of designs.

matched-participants) **design**. This design derives its name from the way participants are assigned to conditions. Participants are placed into pairs based on a prior measure (e.g. intelligence or extroversion), and then the pair is randomly assigned, one to each condition. Because of the prior matching on an important participant variable, it is argued that the two participants should be thought of as similar enough to justify the use of a dependent-samples *t* test. The statistical advantage of the dependent-samples *t* test has resulted in a variety of creative attempts to solve the methodological problems associated with within-participants designs.

Before leaving within-participants designs, it is important to note some of the methodological challenges that are introduced when participants are measured more than once. The primary issue concerns whether the second measurement of a participant is influenced by the first. If so, then the difference between the participant's two scores may be influenced by practice effects or fatigue effects. This explanation competes with the independent variable and creates a confound. There are potential methodological solutions for some of these situations, but they are not always applicable. For instance, in the walking gait study mentioned above, it would be wise for half of the participants to experience the soft tempo condition first followed by the strong tempo, while the other half of the participants experience these conditions in the reverse order. This is called **counterbalancing**. The statistical advantages of repeated-measures designs are attractive, but these are often offset by the corresponding methodological challenges that accompany them.

A full exploration of the reasons researchers use one design or the other is beyond the scope of this text. Please consult a research methodology resource for more information.

9.6 The Dependent-Samples *t* Test

The sampling distribution for the dependent-samples *t* test is based on differences between means of dependent samples. Consider a within-participants design in which one group of participants is exposed to two treatment conditions. From a participant's two scores, a difference score, D, can be computed. When averaged across the sample, a mean of the difference scores, symbolized as $\overline{D}$, is created. A theoretical sampling distribution is created by generating a vast collection of $\overline{D}$s.

The mean of the sampling distribution, $\mu_{\overline{D}}$, will be the same as the difference between the means of the two hypothetical populations. The null hypothesis states the mean of the sampling distribution is 0. This is the claim of no treatment effect. Under the null hypothesis, $\mu_X - \mu_Y = \mu_D = 0$. (The X and Y subscripts refer to the first and second sets of scores, respectively.)

The standard deviation of the sampling distribution is called the standard error of the difference, or just the standard error. The estimate of the population standard error is symbolized as $s_{\overline{D}}$. The formula divides the standard deviation of the difference scores, s_D, by the square root of the number of paired scores (see Formula 9.6).

The estimate of the standard error of the difference

$$s_{\overline{D}} = \frac{s_D}{\sqrt{n_p}} \quad (9.6)$$

where

s_D = the standard deviation of the difference scores
n_p = the number of *pairs* of scores

The computational formula for s_D is given in Formula 9.7.

Computational formula for the standard deviation of the difference scores

$$s_D = \sqrt{\frac{\Sigma D^2 - \left[(\Sigma D)^2/n_p\right]}{n_p - 1}} \quad (9.7)$$

where

$D = X - Y$, a participant's score in treatment 1 minus their score in treatment 2
n_p = the number of *pairs* of scores

A word of caution: Remember to calculate the standard error ($s_{\overline{D}}$) by dividing s_D by the *square root* of n_p. The standard deviation of difference scores, s_D, does not equal $s_{\overline{D}}$. Also, when finding D values, it is critical that we subtract a given participant's Y score from their X score. We must keep each participant's two scores aligned so the resulting D value reflects this particular participant's difference.

Before going further, let us use raw data to compute $s_{\overline{D}}$. Since $s_{\overline{D}}$ is the estimate of the standard error of the sampling distribution, it will be placed in the denominator of the t statistic.

■ **Question** *What is $s_{\overline{D}}$ for the following data set?*

Solution

	Scores			
Participant (*P*)	***X***	***Y***	***D***	***D*²**
P_1	5	3	2	4
P_2	6	4	2	4
P_3	7	7	0	0
P_4	6	7	−1	1
			$\Sigma D = 3$	$\Sigma D^2 = 9$

Using Formula 9.7 to compute s_D,

$$s_D = \sqrt{\frac{\Sigma D^2 - \left[(\Sigma D)^2/n_p\right]}{n_p - 1}} = \sqrt{\frac{9 - \left[(3)^2/4\right]}{4-1}} = \mathbf{1.50}$$

Now place the s_D value into the $s_{\overline{D}}$ formula, Formula 9.6.

$$s_{\overline{D}} = \frac{s_D}{\sqrt{n_p}} = \frac{1.50}{\sqrt{4}} = \mathbf{0.75}$$

The estimate of the standard error is .75. Remember, this estimate is the standard deviation for a sampling distribution of mean differences of a given sample size (n_p). ■

9.6.1 The *t* Distribution for Dependent Samples

To see if there is a difference between the means of two dependent samples, the sampling distribution must be transformed into a *t* distribution, and the sample mean difference must be positioned on it. Recall that the mean of the sampling distribution will equal the difference between the means of the populations. The null hypothesis typically assumes this difference to be 0, and the *t* test is set up accordingly. The difference between the obtained sample means is divided by the estimate of the standard error. This step registers how likely it is to get that particular sample mean difference *if* the null hypothesis is true. In this way, a sample mean difference is transformed and placed onto a *t* distribution. The dependent-samples *t* test formula is similar to all preceding hypothesis testing formulas (see Formula 9.8 below).

Dependent-samples *t* test

$$t_{obt} = \frac{M_X - M_Y}{s_{\overline{D}}} \quad (9.8)$$

As with the preceding hypothesis testing formulas, t_{obt} is the number of estimated standard error units the sample mean difference is from the mean of the *t* distribution. Therefore, if $t = 1.45$, the mean difference ($M_X - M_Y$) is 1.45 estimated standard error units above the mean of the *t* distribution.

The degrees of freedom for the dependent-samples *t* test are $n_p - 1$, the number of *pairs* of scores −1. There is a different *t* distribution for every sample size of paired scores. As with the previous *t* tests, the *t* table (Table A.2) is used to find t_{crit}.

9.6.2 The Hypotheses for the Dependent-Samples *t* Test

The null hypothesis states that the population means are the same. The alternative hypothesis states that the means are not the same.

$$\mathrm{H_0}: \mu_1 = \mu_2 \text{ or } \mu_D = 0$$
$$\mathrm{H_1}: \mu_1 \neq \mu_2 \text{ or } \mu_D \neq 0$$

The requisite formulas are now in place to conduct a dependent-samples *t* test.

9.6.2.1 A Worked Example

A cognitive psychologist believes caffeine has a deleterious effect on learning and recall. To test this assertion, a repeated-measures design is used where one group of participants is exposed to both conditions. In the experimental condition, participants consume two energy drinks within the space of an hour. Combined, these drinks well exceed the recommended daily limit. Immediately afterward, slides of geometric designs with a nonsense word printed below each design (e.g. "geostamp" or "gravoserv") are presented. After 20 minutes, the slides are presented again, in random order, without the nonsense word. Participants are to provide the word associated with each design. The dependent variable is the number of incorrect responses.

The control condition is administered one week later. Different designs and nonsense words are presented, but now the participants are given drinks that taste like energy drinks but contain no caffeine (a placebo). To avoid practice effects, one-half of the participants experience the conditions in the reverse order. Table 9.3

Table 9.3 A worked problem using the dependent-samples *t* test.

Participant (*P*)	Caffeine (*X*)	Placebo (*Y*)	*D*	D^2
P_1	8	4	4	16
P_2	12	8	4	16
P_3	6	2	4	16
P_4	4	6	−2	4
P_5	11	8	3	9
P_6	15	9	6	36
P_7	8	5	3	9
P_8	7	4	3	9
	$M_x = 8.88$	$M_y = 5.75$	$\Sigma D = 25$	$\Sigma D^2 = 115$

$$t_{obt} = \frac{M_X - M_Y}{s_{\overline{D}}};\ s_{\overline{D}} = \frac{s_D}{\sqrt{n_p}};\ s_D = \sqrt{\frac{\sum D^2 - \left[\left(\sum D\right)^2/n_p\right]}{n_p - 1}}$$

$$s_D = \sqrt{\frac{115 - \left[(25)^2/8\right]}{8-1}} = 2.30$$

$$s_{\overline{D}} = \frac{2.30}{\sqrt{8}} = .81$$

$$t_{obt} = \frac{8.88 - 5.75}{0.81} = \mathbf{3.86}\ df = n_p - 1 = 7;\ t_{crit} = \pm 2.365$$

Since t_{obt} falls outside the t_{crit} values, reject the null hypothesis that claims $\mu_X = \mu_Y$.

presents the raw data (number of incorrect responses) for each condition, as well as the computation of t_{obt}. Note: The X and Y score from each participant are aligned so as to find their particular difference score (D).

A dependent-samples t test is used to analyze the data. The procedural steps follow:

Step 1. Define the null and alternative hypotheses:

$$H_0 : \mu_X = \mu_Y$$
$$H_1 : \mu_X \neq \mu_Y$$

Step 2. Set alpha. Alpha is set at .05.

Step 3. Compute t_{obt} (see computations in Table 9.3).

Step 4. Locate the critical t value in Table A.2 in the Appendix A. Recall $df = n_p - 1 = 8 - 1 = 7$. Enter the left column of the t table and locate 7. Move to the column under alpha of .05 for a *two-tailed* test. The critical value is 2.365 (understood as $\pm$2.365 for a two-tailed test).

Step 5. Compare the t_{obt} of 3.86 with the critical value of $\pm$2.365. Since t_{obt} falls beyond the critical values, the null hypothesis is rejected. Furthermore, the means are consistent with the experimental hypothesis (i.e. fewer errors in the placebo condition).

Step 6. Interpret the findings. Statistical evidence suggests learning and/or recall is negatively affected by the ingestion of large amounts of caffeine, $t(7) = 3.86$, $p < .05$.

9.6.2.2 A Measure of Effect Size: Cohen's *d*

One secondary question that can be asked when a null hypothesis is rejected concerns treatment effect size. As noted in previous chapters, t_{obt} is not designed to measure effect size. A simple, direct, and often-used effect size measure is Cohen's *d*. Here is the formula:

Cohen's *d* for dependent-samples *t* test

$$d = \frac{\textit{sample mean difference}}{\textit{standard deviation of difference scores}} = \frac{M_X - M_Y}{s_D} \quad (9.9)$$

$M_X - M_Y$ is used as the best estimate of the mean difference between the two populations, and s_D is the best estimate of σ_D, the standard deviation of the population of difference scores. Notice, this only allows us to *estimate* the effect size. As noted earlier, Cohen's *d* measures effect size in standard deviation terms; larger *d* values reflect larger effect sizes. A word of caution, however, is needed: Cohen's *d* is likely to overestimate effect size in this test.

9.7 Assumptions of the Dependent-Samples *t* Test

The assumptions for the dependent-samples *t* test are nearly identical to the assumptions for the independent-samples *t* test (see Section 9.3). However, two observations coming from the same participant (assumption #2) are clearly not independent. For the dependent-samples *t* test, this assumption is restated as scores *within* a given treatment condition will be independent of each other. In addition, the normality assumption for the dependent-samples *t* test refers to the population of *difference* scores. Finally, just as in previous tests, as n_p increases, the test becomes robust to violations of the assumption of normality.

9.8 Interval Estimation for Dependent Samples

Just as in previous chapters, we can use statistical information from the dependent samples to generate an interval estimate for the mean difference between the populations. Since each potential sample mean difference drawn has a corresponding *t* value, we can use the *t* distribution and our obtained sample mean difference to generate a probability function for the value of the actual population mean difference. Choosing t_{crit} values corresponding to different probabilities within the *t* distribution allows us to create intervals with differing degrees of certainty. The formula for confidence intervals is as follows:

Confidence interval for a population mean difference for dependent samples

$$LL = (M_X - M_Y) - t_\alpha s_{\overline{D}}$$
$$UL = (M_X - M_Y) + t_\alpha s_{\overline{D}} \quad (9.10)$$

---

where

LL = the lower limit of the confidence interval
UL = the upper limit of the confidence interval
t_α = the critical value for a *t* distribution of a given sample size

Since we are generating an interval, two values are calculated, one value at the lower end of the interval and the other at the upper end. As the interval widens and becomes less specific, the confidence grows that the actual mean difference falls within that window. A 95% confidence rate is typical, but the above formulas could be used to find other confidence intervals by finding the corresponding t_{crit} values using the *t* table (Table A.2).

■ **Question** *Using the same data from the caffeine and learning study explored earlier in the chapter, find the 95% confidence interval for the population mean difference in performance between the two conditions.* ($M_X = 8.88$; $M_Y = 5.75$; $s_{\overline{D}} = 0.81$; $n_p = 8$)

Solution

Step 1. Identify the null and alternative hypotheses.

$$H_0 : \mu_X = \mu_Y$$
$$H_1 : \mu_X \neq \mu_Y$$

Step 2. Set the confidence rate at 95% (equivalent to $\alpha = .05$).

Step 3. Using the t table, find the cutoff values beyond which lie 2.5% of the distribution in each tail. With $df = 7$ and $\alpha = .05$ (two-tailed test), the cutoff points are ± 2.365.

Step 4. Compute the confidence interval. Having worked on the problem previously, we know:

$$M_X - M_Y = 3.13; \text{and } s_{\overline{D}} = 0.81$$
$$LL = (M_x - M_y) - t_{.05}s_{\overline{D}} = 3.13 - 2.365(0.81) = \mathbf{1.21}$$
$$UL = (M_x - M_y) + t_{.05}s_{\overline{D}} = 3.13 + 2.365(0.81) = \mathbf{5.05}$$

Step 5. Interpret the findings. Statistical evidence suggests 95% confidence that ingesting large amounts of caffeine negatively affects learning and/or recall in this experimental situation by increasing the error total between 1.21 and 5.05 items. ■

9.9 Comparing the Two Tests

As stated earlier, the probability of rejecting an incorrect null hypothesis is greater using the dependent-samples t test compared with the independent-samples t test, all other things being equal. A good way to demonstrate this point is to imagine that the caffeine and learning study was conducted using independent samples, that is, two unrelated groups of participants. Table 9.4 presents the summary statistics of the data in Table 9.3 and the computation of t_{obt} for an independent-samples t test. The obtained t value is 2.06, which just misses the critical value of ± 2.145. In this case, the difference in designs and type of t test makes the difference between rejecting and failing to reject the null. This is largely because a within-participants design reduced the variability of the data (standard error of .81 instead of 1.52), generating a larger t_{obt} value. The lesson to be learned? If

Table 9.4 An independent-samples t test on the raw data given in Table 9.3.

Caffeine	Placebo
$M_1 = 8.88$	$M_2 = 5.75$
$s_1^2 = 12.67$	$s_2^2 = 5.90$
$n_1 = 8$	$n_2 = 8$

Formula for independent-samples t test

$$t_{obt} = \frac{M_1 - M_2}{\sqrt{\frac{s_1^2(n_1-1) + s_2^2(n_2-1)}{n_1 + n_2 - 2}\left(\frac{1}{n_1} + \frac{1}{n_2}\right)}}$$

$$t_{obt} = \frac{8.88 - 5.75}{\sqrt{\frac{12.67(8-1) + 5.90(8-1)}{8+8-2}\left(\frac{1}{8} + \frac{1}{8}\right)}} = \mathbf{2.06}$$

$df = n_1 + n_2 - 2 = 8 + 8 - 2 = 14$; $\alpha = .05$

$t_{crit} = \pm 2.145$

Since t_{obt} does not fall beyond the t_{crit} values, fail to reject the null claiming $\mu_1 = \mu_2$.

we can manage the methodological problems of within-participants designs, it is statistically advantageous to use dependent-samples t tests for analysis.

9.10 The Appropriateness of Unidirectional Tests

Up to this point, hypothesis testing has been discussed from a nondirectional perspective, which means that our interest is in detecting differences between population means, irrespective of whether $\mu_1 < \mu_2$ or $\mu_1 > \mu_2$. The null hypothesis has always been stated as an equality (i.e. $\mu_1 = \mu_2$ or $\mu =$ some known value), and the alternative hypothesis has always been stated as an inequality (i.e. $\mu_1 \neq \mu_2$ or $\mu =$ some known value).

Research hypotheses are usually stated as predictions about the expected direction of an effect or relationship. For example, "Persuasion technique *A* will induce *greater* attitude change than persuasion technique *B*." Researchers, however, typically frame their *statistical hypotheses* in a nondirectional form. In other words, even though the research hypothesis makes a prediction about which of the two means will be larger, the null and alternative hypotheses allow the investigator to discover if a treatment effect or relationship between variables is opposite to the predicted effect. For instance, if a researcher thinks that an advertisement will *increase* the sales of a product, a nondirectional test allows for the discovery that the advertisement actually *decreased* sales. However, there are times when researchers choose to use a directional test. A directional test can detect a difference between means *only* in one direction. If a researcher uses a directional test to see if an advertisement increases sales, then it will be blind to the possibility that the advertisement decreases sales. As we will soon learn, the decision to adopt a directional test has challenging implications, and for that reason, it is controversial.

9.10.1 One-Tailed and Two-Tailed Tests

Typically, the rejection region for a null hypothesis is divided equally between both tails of the *t* distribution. This allows us to detect a difference when $\mu_1 > \mu_2$ *as well as* when $\mu_1 < \mu_2$. Tests constructed with this logic are called **two-tailed** (or **non-directional**) **tests**.

Two things change when conducting a **one-tailed** (or **directional**) **test**: the way the null and alternative hypotheses are stated and the placement of the rejection region. Consider the following research situation. Suppose a standard drug for treating heart disease exists, but this drug has a certain number of unfortunate side effects. A new drug has emerged which may produce fewer side effects. Since the new drug is more expensive, the old drug will remain the treatment of choice if they both have the same number of side effects *or* if the new drug has *more* side effects. The only finding that would be acted upon is if the new drug has *fewer* side effects.

Recall that null and alternative hypotheses must be mutually exclusive and collectively exhaustive. The null and alternative hypotheses for a directional test can be stated as:

$$\text{H}_0 : \mu_{now} \geq \mu_{old}$$
$$\text{H}_1 : \mu_{new} < \mu_{old}$$

These hypotheses are collectively exhaustive. Whether the new drug has fewer, the same, or more side effects, one of these two hypotheses will be correct. They are also mutually exclusive because they both cannot be true at the same time. (Note: $\mu_{new} \geq \mu_{old}$ is claimed as the *null hypothesis*. This is a bit of a misnomer since "null" implies no difference.)

Since a difference between means in only one direction can be detected, many researchers feel it is proper to place the entire rejection region in one tail of the distribution. Figure 9.2 uses the *t* distribution to illustrate rejection regions for two- and one-tailed tests when alpha is .05.

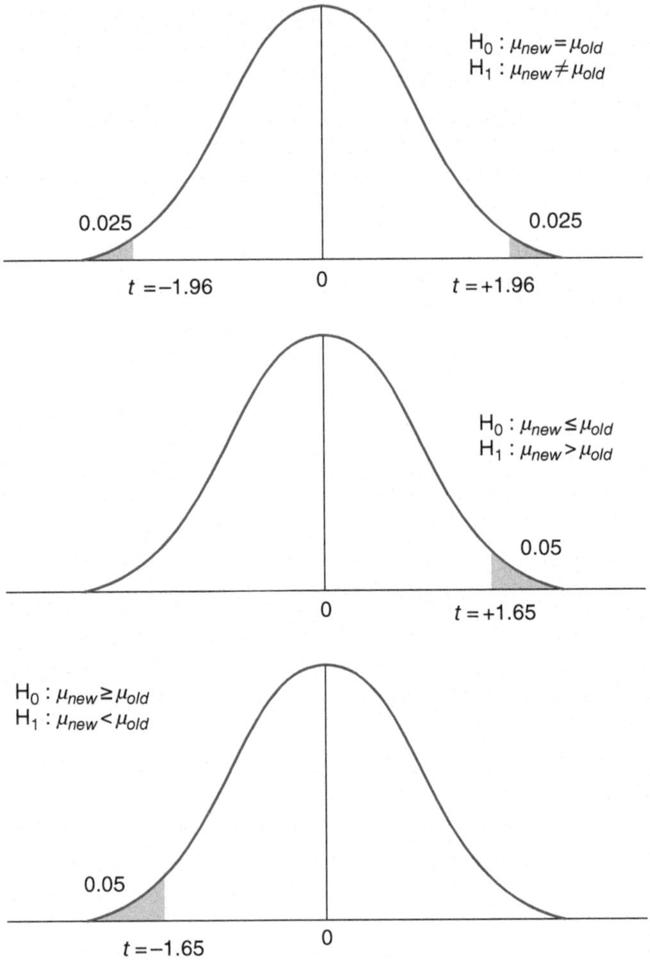

Figure 9.2 The rejection region of a *t* distribution for a one-tailed and two-tailed test, $\alpha = .05$.

9.10.2 The Sign of t_{obt} is Important in a Directional Test

Up to now, the decision to reject the null has been independent of the valence of t_{obt}. For this reason, we always state the t_{crit} value as $\pm$ for nondirectional tests. The situation is different with a directional test; whether t_{obt} is positive or negative matters because the rejection region is only on one side. For this reason, it is important to be careful when subtracting means in the numerator when running directional tests.

9.10.3 Using the *t* Table for a Directional Test

Below is a small section of the *t* table (Table A.2). Note that it has two sets of headings, one for two-tailed tests and another for one-tailed tests. Notice that for a given level of alpha and a given *df* value (say, $\alpha = .05$; $df = 7$), the critical value for a one-tailed test is closer to zero (1.895) than the value for a two-tailed test (2.365). This means that decisions to reject the null are more likely to occur for one-tailed tests if the mean difference is in the predicted direction.

α levels for two-tailed test		
df	*.10*	*.05*
α levels for one-tailed test		
	.05	.025
⋮	⋮	⋮
6	1.943	2.447
7	**1.895**	**2.365**
8	1.860	2.306
9	1.833	2.262
⋮	⋮	⋮

9.10.3.1 Use Caution in Deciding to Conduct a One-Tailed Test

Research hypotheses are almost always framed as a directional prediction. Therefore, why not always use a directional *t* test? For one reason, if we use a one-tailed test, we will be unable to detect differences between population means opposite of the predicted direction. Unexpected findings, however, may have theoretical implications, and, in any case, are important. Performing a one-tailed test implies that the researcher believes findings contrary to the research hypothesis are meaningless. However, this is shortsighted. After all, the history of science is littered with stories of unexpected yet truly meaningful findings being uncovered without prior intention. Unexpected findings stimulate future research. For this reason, a one-tailed should never be used simply because the research hypothesis is stated directionally.

The strongest justification for using a one-tailed test occurs when a new course of *action* is to be taken *only* if the result of the test is in one direction. We might think of this as a "theory versus application" issue. If the research is theoretically important, there is simply no debate; use a two-tailed test. Many argue, however, that if the research is about application *and* if that application will only be taken if one type of finding is uncovered, then a directional test is justified. However, even in these situations, the use of a one-tailed test is controversial (e.g. Lombardi and Hurlbert 2009). The purpose of a test is to measure the likelihood that *sampling error* can explain a difference between means. Moving half of the rejection region from one side over to the other reduces the ability for sampling error to account for a mean discrepancy in the selected direction. This *misrepresents what we are trying to measure* – the explanatory power of sampling error. Do not forget that hypothesis testing explores how likely it is that sampling error can account for differences between means. If it is unlikely, then we *infer* the presence of an effect. Contrary to what is often stated, statistical analyses do not directly *find* effects, but rather they direct us to *infer* them only when sampling error seems insufficient to explain an observed difference. This important point is lost on those who push for one-tailed tests.

Further, many argue that the use of directional tests results in a Type I error rate that is effectively greater than .05. If the critics are right, using a one-tailed test constitutes a form of statistical cheating, whether the purpose of the research is application or not. (See Box 9.1 for more information about the use of one-tailed tests.)

One final point concerns *when* a decision is made to use a one-tailed test. If a researcher believes a one-tailed test is warranted, the decision *must* be made and, of course, justified before the data are gathered and analyzed. Suppose we conduct a two-tailed test and find that the t_{obt} is close but does not fall in the rejection region. If we then move the entire rejection region over to one tail so that the t_{obt} value falls within it, we will be engaging in professional misconduct. Despite these serious concerns, one-tailed tests continue to be used. For this reason, the tables in Appendix A include references to one-tailed critical values.

Box 9.1 Is the Scientific Method Broken? The Disputed Use of One-Tailed *t* Tests

Lombardi and Hurlbert (2009) reviewed the use of one-tailed tests in published articles from two journals in the years 1989 and 2005 and found, "Of those articles with data sets susceptible to one-tailed tests, at least 24% in *Animal Behaviour* and at least 13% in *Oecologia* used one-tailed tests at least once." The authors note that, "Debate in the psychological literature of the 1950s established the logical criterion that one-tailed tests should be restricted to situations where there is interest only in results in one direction. 'Interest' should be defined; however, in terms of collective or societal interest and not by the individual investigator." According to this criterion, they contend, "all uses of one-tailed tests in the journals surveyed seem invalid." They go on to note that much of the confusion regarding the appropriateness of one-tailed tests stems from the apparent support given to them in many recent statistics textbooks. They conclude with this statement, "One-tailed tests rarely should be used for basic or applied research in ecology, animal behavior or any other science." The libertarian view taken by many researchers toward the use of one-tailed tests may be another reason for the current reproducibility crisis in the social, behavioral, and medical sciences.

9.11 Formal Presentation of Findings

Proper reporting of inferential statistics can be challenging. The following are examples of how to report, in sentence form, a rejection of the null as well as a failure to reject the null. If rejecting the null, a sentence might read, "An independent-samples *t* test found evidence suggesting meditation leads to a reduction in participants' physiological reaction to stress, $t(18) = -2.23$, $p < .05$." If we also wanted to include a measure of effect size, the sentence could finish with "...t $(18) = -2.23$, $p < .05$, $d = 1.05$." If failing to reject the null, a sentence might read, "A dependent-samples *t* test did not find evidence suggesting meditation leads to a change in participants' physiological reaction to stress, $t\ (18) = -1.62$, *n.s.*" For a more detailed analysis of the style, symbols, and punctuation used in these sentences, please see Section 8.8.

Summary

The independent-samples *t* test is used for between-participants designs when two separate samples of participants are measured. The purpose of the *t* test is to help us decide whether the two samples come from the same or different populations. The sampling distribution for an independent-samples *t* test is a distribution of the differences between independent sample means. The *t* test uses the weighted average of two sample variances to estimate the population variance, which is then used to estimate the standard error.

The independent-samples *t* test relies on the *t* distribution, a family of normal distributions that vary based on sample size. The *t* distribution, represented in the *t* table (Table A.2), is used to identify the critical values that t_{obt} must be equal to or greater than (in absolute value) to reject the null hypothesis. If t_{obt} is smaller, the proper decision is to fail to reject the null. In these cases, sampling error is a viable explanation for the differences between the sample means. If t_{obt} is larger, the proper decision is to reject the null. If a null hypothesis is rejected, Cohen's *d* can be calculated and used as a measure of effect size. The assumptions of the independent-samples *t* test are representativeness, independent observations, interval or ratio measures, normally distributed populations, and the homogeneity of population variances.

In *dependent sampling*, each score in one sample is related to a score in a second sample. Pairs of scores are formed either by matching or using a within-participants design, where each participant is exposed to every condition in the study. The sampling distribution for the dependent-samples *t* test is based on differences between means of matched or paired samples.

A dependent-samples *t* test functions similarly to an independent-samples *t* test, and yet, because there is less variability associated with paired scores, it is more powerful. This means the resulting t values tend to be larger and, as a result, are more likely to allow for the null to be rejected. However, with this statistical advantage come methodological problems, namely, the challenge of practice of fatigue effects associated with a second measurement of participants. If the null hypothesis can be rejected, a version of Cohen's d can be used to measure effect size.

The assumptions of the dependent-samples t test are similar to those for an independent-samples t test. However, dependently sampled scores are obviously not independent between conditions. For this test, the assumption pertains only to scores *within* a given condition.

For both tests introduced in this chapter, the standard error, sample means, and t distribution can also be used to create a confidence interval for the actual value of the difference between the means of the two populations.

Some researchers argue that there are times when it is appropriate to use a one-tailed or directional test of the research hypothesis, specifically when dealing with a question of application. However, even in these situations, the use of one-tailed test is problematic, controversial, and generally discouraged by statisticians. When conducting one-tailed (or directional) tests, the entire rejection region is placed in one tail. As a result, the critical value moves closer to zero, making a decision to reject the null more likely to occur. Many theorists believe this procedure is problematic; one reason being, that it inflates the alpha value beyond the stated level.

Microsoft® Excel and SPSS®

Information regarding how to use Microsoft® Excel and SPSS® to assist in performing independent- and dependent-sample t tests can be found in Appendix C.

Key Formulas for Independent-Samples *t* Test

Variance formula for the estimated standard error

$$s_{M_1-M_2} = \sqrt{\frac{s_1^2(n_1-1) + s_2^2(n_2-1)}{n_1+n_2-2}\left(\frac{1}{n_1}+\frac{1}{n_2}\right)} \quad (9.1)$$

Computational formula for the estimated standard error

$$s_{M_1-M_2} = \sqrt{\frac{\left(\Sigma X_1^2 - (\Sigma X_1)^2/n_1\right) + \left(\Sigma X_2^2 - (\Sigma X_2)^2/n_2\right)}{n_1+n_2-2}\left(\frac{1}{n_1}+\frac{1}{n_2}\right)} \quad (9.2)$$

Independent-samples *t* test

$$t_{obt} = \frac{M_1 - M_2}{s_{M_1-M_2}} \quad (9.3) \quad (df = n_1 + n_2 - 2)$$

Cohen's *d* for independent-samples *t* test

$$d = \frac{\textit{estimated mean difference}}{\textit{estimated standard deviation}} = \frac{M_1 - M_2}{\sqrt{\frac{s_1^2(n_1-1)+s_2^2(n_2-1)}{n_1+n_2-2}}} \quad (9.4)$$

Confidence interval for a population mean difference for independent samples

$$\begin{aligned} LL &= (M_1 - M_2) - t_\alpha s_{M_1 - M_2} \\ UL &= (M_1 - M_2) + t_\alpha s_{M_1 - M_2} \end{aligned} \quad (9.5)$$

Key Formulas for Dependent-Samples *t* Test

The estimate of the standard error of the difference

$$s_{\overline{D}} = \frac{s_D}{\sqrt{n_p}} \quad (9.6)$$

Computational formula for the standard deviation of the difference scores

$$s_D = \sqrt{\frac{\sum D^2 - \left[(\Sigma D)^2/n_p\right]}{n_p - 1}} \quad (9.7)$$

Dependent-samples *t* test

$$t_{obt} = \frac{(M_X - M_Y)}{s_{\overline{D}}} \quad (9.8) \qquad (df = n_p - 1)$$

Cohen's *d* for dependent-samples *t* test

$$d = \frac{\textit{sample mean difference}}{\textit{standard deviation of difference scores}} = \frac{M_X - M_Y}{s_D} \quad (9.9)$$

Confidence interval for a population mean difference for dependent samples

$$\begin{aligned} LL &= (M_X - M_Y) - t_\alpha s_{\overline{D}} \\ UL &= (M_X - M_Y) + t_\alpha s_{\overline{D}} \end{aligned} \quad (9.10)$$

Key Terms

Independent-samples *t* test
Between-participants designs
Estimated standard error
Homogeneity of variances
Dependent-samples *t* Test
Within-participants design
Matched-samples design
Counterbalancing
Two-tailed (or nondirectional) test
One-tailed (or directional) test

Exercises

1 Inferential statistics use _____ means to draw inferences about _____ means.

2 In which description should an independent-samples t test be used?
a Two samples gathered, each from a separate population.
b One sample, measured twice under different conditions.
c One sample, compared with a known population mean.
d None of the above.

3 In which description should a dependent-samples t test be used?
a Two samples gathered, each from a separate population.
b One sample, measured twice under different conditions.
c One sample, compared with a known population mean.
d None of the above.

4 A _____ _____ design measures each participant under each condition of the study, while a _____ _____ design premeasures participants, pairs them up based on this measure, and then randomly assigns them to conditions.
a Within-participants; matched-samples.
b Matched-samples; within-participants.
c Repeated-measures; within-participants.
d Within-participants; repeated-measures.

5 Why is data alignment of X and Y values for each participant important for dependent-samples t tests?

6 Which of the following is an accurate way to represent the null hypothesis for an independent-samples t test? (assume a two-tailed test)
a $M_1 - M_2 = 0$
b $M_1 - M_2 \neq 0$
c $\mu_1 - \mu_2 = 0$
d $\mu_1 - \mu_2 \neq 0$
e $\mu_1 = M_1$ and $\mu_2 = M_2$
f $\mu_1 = M_2$ and $\mu_2 = M_1$

7 Which of the following is an accurate way to represent the alternative hypothesis for an independent-samples t test? (assume a two-tailed test)
a $M_1 - M_2 = 0$
b $M_1 - M_2 \neq 0$
c $\mu_1 - \mu_2 = 0$
d $\mu_1 - \mu_2 \neq 0$

e $\mu_1 \neq M_1$ and $\mu_2 \neq M_2$
f $\mu_1 \neq M_2$ and $\mu_2 \neq M_1$

8 For a between-participants study comparing two conditions, if we knew one sample has an $n = 12$ and the other has an $n = 10$, what *df* would we use for the *t* test?
a 22
b 44
c 9
d 20

9 For a within-participants study comparing 2 conditions, with 12 scores in each condition, what is the *df* for the *t* statistic?
a 11
b 12
c 22
d 24

10 If the null hypothesis for an independent-samples *t* test is true, then any observed difference between the sample means is due to _____ _____.

11 Must researchers only use subscripts 1 and 2 when writing out null and alternative hypotheses? Why or why not?

12 Which of the following statistical expressions reflects the proper presentation of an independent-samples *t* test analysis?
a $t(24) = 2.21$, $d = .29$, $p > .05$
b $t(24) = 2.21$, $p < .05$, $d = .29$
c $t(24) = 2.21$, $d = .29$, $p < .05$
d $t(24) = 2.21$, $p > .05$, $d = .29$

13 Explain the difference between s_D and $s_{\overline{D}}$. Why is this difference important?

14 Which of the methodological assumptions of the independent-samples *t* test is effectively dealt with by randomly sampling from the populations in question?

15 The sampling distribution for a dependent-samples *t* test is a distribution of:
a Mean differences gathered from two independent samples.
b Mean differences gathered from two dependent samples.
c Means gathered from one population of raw scores.
d None of the above.

16 How does the sample size influence t_{crit} and why?

17 State the critical values and the proper decision regarding the null hypothesis for each of the following independent-samples t test results (all tests are two-tailed).

a $n_1 = 8$, $n_2 = 7$, $\alpha = .05$, $t_{obt} = 1.90$
b $n_1 = 30$, $n_2 = 30$, $\alpha = .01$, $t_{obt} = 7.82$
c $n_1 = 9$, $n_2 = 9$, $\alpha = .10$, $t_{obt} = -4.55$
d $n_1 = 14$, $n_2 = 12$, $\alpha = .05$, $t_{obt} = -.63$

18 A dependent-samples t result was as follows: $t(21) = 2.31$, $p < .05$. Which of the following analyses is consistent with this statement?

a The study used a total of 23 participants and the null was not rejected.
b The study used a total of 23 participants and the null was rejected.
c The study used a total of 22 participants and the null was not rejected.
d The study used a total of 22 participants and the null was rejected.

19 Which is an advantage of an independent-groups design over a repeated-measures design?

a It requires fewer participants to get more power.
b It eliminates methodological problems from measuring participants multiple times.
c It eliminates the need for a two-tailed test.
d There will be smaller error terms and larger t values.

Work Problems

(For all work problems, use a two-tailed test unless otherwise specified.)

20 A hypothetical study is conducted to evaluate the hypothesis that two-year-old children with no siblings will show more fear around unfamiliar children than two-year-olds with one or more siblings. Each child is put in a playroom with an unfamiliar child, and fear ratings are made by an evaluator who is blind to the hypothesis. Fear ratings can range from 1 (no fear) to 10 (a great deal of fear). For the following data:

a Specify the hypotheses.
b Perform the appropriate test.
c If $\alpha = .05$, what are the critical values?
d Reject the null hypothesis?

e If so, what is the effect size?
f What decision error might be made?
g Properly present the findings.

No siblings	Siblings
10	7
6	3
8	2
4	4
9	1
7	2

21 A social psychologist is interested in evaluating the hypothesis that anxiety increases interpersonal attraction. Participants are told they will hear several random bursts of noise in a learning experiment. Half are led to expect a loud noise (high anxiety condition), and the other half are led to expect a soft noise (low anxiety condition). While waiting for the experiment to begin, participants are placed in a room with a confederate. Later, participants are asked to rate the attractiveness of this person, from 1 (low) to 5 (high). (Assume the scale has been determined to be interval.) Hypothetical summary statistics are presented.
a Specify the hypotheses.
b Perform the appropriate test.
c If $\alpha = .05$, what are the critical values?
d Reject the null hypothesis?
e If so, what is the effect size?
f What decision error might be made?
g Properly present the findings.

High anxiety	Low anxiety
$M_1 = 4.2$	$M_2 = 2.2$
$s_1^2 = 0.5$	$s_2^2 = 0.7$
$n_1 = 10$	$n_2 = 10$

22 For Question 21, suppose the attraction rating variances are $s_1^2 = 5.2$ and $s_2^2 = 5.4$.
a Perform an independent-samples t test with $\alpha = .05$.
b Interpret the findings.
c What effect has the increased variability had on t_{obt}?

23 Using the means from Question 21 and the variances from Question 22, suppose the number of participants is increased to 30. Below are the summative statistics.

High anxiety	Low anxiety
$M_1 = 4.2$	$M_2 = 2.2$
$s_1^2 = 5.2$	$s_2^2 = 5.4$
$n_1 = 30$	$n_2 = 30$

a Perform an independent-samples t test with $\alpha = .05$.
b Interpret the findings.
c What effect has the increase in sample sizes had on t_{obt}?

24 Biaggio (1989) administered a Personal Incidents Record to biological male and female college students to assess the frequency of anger reactions. The author found statistical evidence that biological males reported more anger reactions in comparison with biological females. The following hypothetical data are consistent with Biaggio's findings.
a Specify the hypotheses.
b Perform the appropriate test.
c If $\alpha = .05$, what are the critical values?
d Reject the null hypothesis?
e If so, what is the effect size?
f What decision error might be made?
g Properly present the findings.

Frequency of anger reactions	
Males	**Females**
16	9
18	10
15	8
20	4
9	14

25 A psychologist working for a marketing firm wants to know which of two cheeses college students prefer, Gouda or Swiss. After participants taste both, ratings are obtained that can range from 1 (lousy) to 9 (fantastic). (Assume the scale has been determined to be interval.)

a State the hypotheses.
b Perform the appropriate test.
c If $\alpha = .05$, what are the critical values?
d Reject the null hypothesis?
e If so, what is the effect size?
f What decision error might be made?
g Properly present the findings.
h Are there methodological issues?

Participant	Gouda cheese	Swiss cheese
P_1	5	3
P_2	7	6
P_3	9	4
P_4	8	7
P_5	6	8

26 For the same data as in Question 25, find the 95% confidence intervals for the actual difference in students' ratings between the two types of cheeses.

27 When comparing firstborn college students with laterborns, Jourbert (1989) found that firstborns were more narcissistic. Hypothetical summary data are presented. Conduct an independent-samples t test between the narcissism scores of firstborns and laterborns for (a) biological males and for (b) biological females. Set alpha at .05.

a Biological males

Firstborns:	M_1:	23	s_1:	7.84	n_1:	10
Laterborns:	M_2:	16	s_2:	6.43	n_2:	15

b Biological females

Firstborns:	M_1:	17	s_1:	6.52	n_1:	19
Laterborns:	M_2:	12	s_2:	6.57	n_2:	28

28 For the same data as in Question 27, find the 95% confidence intervals for the actual narcissism difference between both firstborn and laterborn biological males and females.

29 Seery and others (2010) seem to have found statistical support for the old adage, "that which doesn't kill us makes us stronger." In their study, they found that individuals marked by moderate amounts of adversity in life reported better mental health and higher measures of subjective well-being than those with little or no history of adversity. The following are data consistent with their findings. A group of 15 individuals with few negative life experiences recorded a subjective well-being mean of 41 with a s of 4, while 14 individuals with several negative life events recorded a subjective well-being mean of 47.2 with a s of 5.

- **a** Specify the hypotheses.
- **b** Perform the appropriate test.
- **c** If $\alpha = .05$, what are the critical values?
- **d** Reject the null hypothesis?
- **e** If so, what is the effect size?
- **f** What decision error might be made?
- **g** Properly present the findings.

30 A child psychologist believes that children who do not spend much time with peers during recess have problems starting conversations. A group of 60 children are randomly assigned to 2 conditions (30 per condition). In the experimental condition, children learn how to begin a conversation. Those in the control condition are given talks about the importance of having friends. For 30 days after treatment, the children are observed during recess and their time spent with peers is recorded. The playtime means are: Exp = 17 minutes. Control = 13.5 minutes. The estimated standard error ($s_{M_1 - M_2}$) is 2.0.

- **a** State the hypotheses.
- **b** Run an independent-samples t test.
- **c** What is t_{crit}?
- **d** Reject the null hypothesis?
- **e** Properly present the findings.

31 Using the summary data from Question 30 and conduct a one-tailed test that will detect a mean difference only if the experimental condition generates greater peer interaction.

- **a** State the hypotheses.
- **b** Conduct the t test.
- **c** Interpret the findings.
- **d** Is a directional test justifiable?

32 Where students study may be important. Students who study in one setting may perform differently than students who have no regular study location. To test this, a random sample of 15 Introductory Sociology students is asked to study their class material for one hour every day in a special quiet room in the university library. A second sample of 15 students from the same class is also asked to study class material for one hour every day, but rotating among various settings (dorm room, cafeteria, and library). At the end of the semester, point totals for the class are obtained and compared, with the following summary statistics:

One setting	Various settings
$M_1 = 80;\ s_1^2 = 106.9;\ n_1 = 15$	$M_2 = 69; s_2^2 = 156.25; n_2 = 15$

a Specify the hypotheses.
b Perform the appropriate test.
c If $\alpha = .05$, what are the critical values?
d Reject the null hypothesis?
e If so, what is the effect size?
f What decision error might be made?
g Properly present the findings.
h Are there confounds in this study?

33 A psychologist tested a new drug for insomnia. The average amount of time it takes participants to fall asleep before treatment (Pretest), is compared with the average amount of time it takes participants to fall asleep after treatment (Posttest).
a State the hypotheses.
b Perform the appropriate test.
c If $\alpha = .05$, what are the critical values?
d Reject the null hypothesis?
e If so, what is the effect size?
f What decision error might be made?
g Properly present the findings.

Participant	Pretest	Posttest
P_1	120	30
P_2	60	40
P_3	90	30
P_4	100	80

34 A preschool teacher would like to ensure students rest during quiet time. The teacher wonders if children relax more quickly when a story is read, or soft music is played. For one week, a story is read every day at quiet time, and the average time it takes each child to fall asleep is recorded. This is repeated a second week, but with soft music being played. Assume all methodological challenges have been addressed. Data for each child is below.

a State the hypotheses.

b Perform the appropriate test.

c If $\alpha = .05$, what are the critical values?

d Reject the null hypothesis?

e If so, what is the effect size?

f What decision error might be made?

g Properly present the findings.

Participant	Story	Music
P_1	6	4
P_2	6	8
P_3	9	7
P_4	8	6
P_5	8	10
P_6	10	6
P_7	12	5

35 We observe that people seem to be happier when they are wearing a new article of clothing. To test this, we provide a small random sample of students with new t-shirts and instruct them to wear them all day. At the end of the day, we asked them to rate their happiness. A control group of students is also asked for this self-rating but without experimental manipulation. Ratings for each participant are reported below. Higher scores indicate greater happiness. Perform an independent-samples t test on the following data set ($\alpha = .05$) and interpret the findings. Are there any assumptions that should give the researcher cause for concern?

Happiness rating	
New t-shirt	**Control**
8	4
7	6
9	5

Happiness rating	
New t-shirt	**Control**
6	4
8	6
4	4
5	6
3	8
8	3
5	5
7	3
9	7

10

Power Analysis and Hypothesis Testing

10.1 Decision-Making While Hypothesis Testing

In an experimental context, the researcher attempts to reject the null hypothesis, which claims there is no effect due to treatment. (Although hypothesis testing is not limited to experiments, we will use the language of experimentation as we discuss statistical power.) The researcher uses hypothesis testing tools to make an inference about the null hypothesis, but with every decision, there is always the possibility of making a decision error. In this chapter, hypothesis testing is approached from another angle. Instead of asking questions about the probability of making an error, we will explore ways of improving our chances of arriving at a *correct decision*. More specifically, if there is a treatment effect, how can we give ourselves the best chance of detecting it? The **power** of a statistical test is the ability it has to reject a false null hypothesis. We can think of power as *test sensitivity*. How sensitive is our inferential test at detecting an effect? It is helpful to recollect the concept of power in visual enhancement systems like microscopes. The more powerful the microscope, the more likely small objects will be seen when inspected. The same is true for statistical power and small effects.

In terms of probability, there is *not* a complementary relationship between Type I and Type II errors; the probabilities associated with each do not sum to 1. These errors inhabit different worlds. In one world, the null is true, and in the other world, the null is false. Within each world, however, there is a trade-off, as indicated in Table 10.1. For example, in the world where the null is true, if the probability of a Type I error is .05, the probability of avoiding that error is .95.

This chapter will focus on the other world, where the null is false, (i.e. the bolded right-hand side of Figure 10.1). In this world, we want to demonstrate that the null is false. In other words, we want to avoid making a Type II error. The probability of committing a Type II error is symbolized as β (beta). As such, the probability of

Essentials of Behavioral and Social Science Statistics, Third Edition.
K. Paul Nesselroade, Jr. and Laurence G. Grimm.
© 2025 John Wiley & Sons, Inc. Published 2025 by John Wiley & Sons, Inc.
Companion website: https://www.wiley.com/go/Essentials_of_Behavioral_and_Social_Science_Statistics3e

Table 10.1 The four outcomes when rejecting or failing to reject a true or false null hypothesis.

		The true situation	
		H_0 is true	**H_0 is false**
Our decision	Fail to reject H_0	Correct $1-\alpha$ (95%)	**Type II error** $\boldsymbol{\beta}$ **(20%)**
	Reject H_0	Type I error α (5%)	**Correct** $\boldsymbol{1-\beta}$ **(Power)(80%)**
		100%	**100%**

The values given to the Type I and Type II error are just examples.

correctly rejecting a false null hypothesis is $1-\beta$. In other words, if a given test has a 20% chance of failing to find an existing effect, there must be an 80% chance of finding it.

10.2 Why Study Power?

Understanding the concept of power will increase our "research IQ." For example, now that we know how *t* tests work, we have probably adopted the attitude that large samples are preferred to small ones. Large samples give us better estimates of population parameters, thus smaller standard error terms, thereby making it easier to reject a false null. However, in some situations, small effects are not particularly meaningful. Consider a colleague studying a new depression medication. Running a time-consuming and costly study involving thousands of participants may effectively detect if a new drug has a slight effect on mood, but perhaps it would be better to run a more streamlined study with only a few participants to see if it produces a dramatic effect. We could more quickly and less expensively determine if this new medication will be a "game changer." Understanding the concept of power and the procedures of a power analysis can lead us to offer intelligent advice to the researcher regarding the number of participants needed for a given research situation. In our example, after a power analysis, we might be able to tell our colleague, "If you create a sample size of 40 patients, there is an 80% chance you will be able to detect an effect size that is one standard deviation or more. Reduce your sample size; save your money, time, and resources." That would be helpful information to know ahead of time, would it not? (See Box 10.1 for more information about the value of running power analyses.)

Here is another example. Suppose a colleague is frustrated over the failure of their study to reject the null. They wanted to see if cell phone use contributes to social anxiety and depression (which, by the way, is the case; for example, see Ting and Chen 2020). Our understanding of power allowed us to inform them that they had given themselves almost no chance of finding an effect. By encouraging them to redesign their study and explaining to them that even a modest effect is of theoretical and practical importance, we can help their research project get back on track. Power analyses are valuable for both planning a study ahead of time as well as evaluating one afterward, planning ahead being the preferable of the two.

10.3 The Five Factors that Influence Power

There are several factors influencing power, and some of them can be easily adjusted by the researcher. The factors are: (i) the *size of the mean difference* between the two populations being studied, (ii) the *size of the standard deviations* of the populations being studied, (iii) the *sample size* used, (iv) the *alpha value* selected, and (v) the *nature of the inferential test* (one-tailed or two-tailed). The last two factors are decisions made by the researcher; they are *decision driven.* The first three factors are *data-driven.* These involve the mathematical mechanics of an inferential test. Think of a generic *t* test; it is composed of a numerator comparing two means and a denominator composed of measures of variability and sample sizes. Below is the formula for the single-sample *t* test. The denominator s_M has been replaced with an equivalent expression $s/\sqrt{n}$, so that each of the data-driven factors that influence power can be identified. Let us symbolize $M - \mu$ as "1," s as "2," and $\sqrt{n}$ as "3."

$$t_{obt} = \frac{M - \mu}{\frac{s}{\sqrt{n}}} \quad \begin{matrix} \rightarrow (1 \\ \rightarrow (2 \\ \rightarrow (3 \end{matrix}$$

We will look at each of these three factors separately. As a working example to help us explore them, let us suppose we want to evaluate the effectiveness of an imagined study program we will call ACE, designed to increase scores on the Scholastic Achievement Test – Verbal (SAT-V). Suppose we know that the national mean for the SAT-V is 500. We want to know whether our sample of students, once having participated in the ACE program, will now show evidence that the program has helped improve their SAT-V performance. The null and alternative hypotheses can be stated as:

$$H_0 : \mu = 500$$
$$H_1 : \mu \neq 500$$

If large t values are key to rejecting null hypotheses, one way to increase the t value is to make the difference between the two means as big as possible by using the treatment strategically. For instance, if we believe that sleep deprivation will impact creativity, we do not want to deprive our participants of only one hour of sleep. We will want to keep them up all night. This will give us the best chance to test our claim. In the same way, if we believe exposure to the ACE study program will positively affect a student's SAT-V score, we will not want them to be exposed to it minimally, but rather maximally. If this program is effective, substantial exposure will give us the best chance of generating a big difference between the mean of the control group ($\mu = 500$) and the mean of the treated group. This rationale corresponds to point "1" ($M - \mu$) noted above.

The second data-driven factor (point "2") is the amount of variability in the distributions of raw data, which is reflected in the size of the standard deviation. If the two populations being sampled have similar means, high power can still be achieved if the variability of the scores in the populations is low. This concept is hard to represent in the SAT-V example because the standard deviations are known to be around 100. However, if we imagine the distribution of SAT-V scores to be much narrower (e.g. $\sigma = 10$), then an improvement of only 5 points would be easily noticed. We will not spend too much time with this factor since it is usually very hard, if not impossible, for the researcher to influence.

The third data-driven factor (point "3") is the sample size. As it increases, the estimate of the standard error decreases. If we are having trouble understanding this, imagine a situation where the sample size is so large that it is equal to $N - 1$ (i.e. one less than the total population). If a sampling distribution were created using a sample of that size, would not the standard error be virtually zero? Every sample mean found would be virtually identical. As n increases, the error term decreases, and t values get larger. Smaller error terms yield increased statistical power.

The first two data-driven factors, the difference between the means of the populations and the standard deviation(s) of the distribution(s), are often combined and referred to as the "treatment effect." Treatment effects are often described in standard deviation units. For instance, a drug may be hypothesized to increase mood by one standard deviation. We have already seen this concept when computing Cohen's d.

The second two data-driven factors, the standard deviation(s) of the distribution(s) and the sample sizes, can also be thought of as being combined; these two constitute the standard error. From this perspective, the researcher can think about a proposed study in terms of the difference between the means (point 1) and the standard error (points 2 and 3). There is not one proper way to conceptualize the data-driven factors that influence power, but it is important for researchers to develop some strategy for thinking about measuring and adjusting the power of a study prior to running it.

10.4 Decision Criteria that Influence Power

As noted earlier, in addition to the data-driven factors, power is also influenced by decisions made by the researcher; namely, the selection of an alpha level and the type of *t* test used (one-or two-tailed). For the decision factors, a relationship exists between the two types of errors. If the null hypothesis is false, every decision made to avoid a Type I error increases the likelihood of making a Type II error. For instance, smaller alpha values correspond to more extreme t_{crit} values, which, in turn, require larger *t* values to reject the null. In these situations, a false null is harder to detect. Similarly, the placement of the rejection regions influences the Type II error rate. These two decisions determine the critical value for what is needed to reject the null, and rejecting the null is what must be done to avoid a Type II error.

A hypothetical study based on our working example is presented to further clarify the concept of power and illustrate its calculation. The way to compute power varies, depending on the experimental design and type of statistical analysis used. The illustrative worked example below involves a one-tailed, single-sample *z* test.

10.4.1 Worked Example

As noted earlier, the ACE company would like to offer a college entrance preparation course that will help students increase their SAT-V (verbal) scores. Before marketing the program, the company asks us to evaluate its effectiveness. We already know that the national average of SAT-V scores is 500, with a standard deviation of 100. The company is not interested in finding evidence of small effects on performance. Rather, they want to show that their program will result in a big difference: one of 25 points or more. The company would like to do this by using a recent set of 36 students who completed the course.

As we work toward the solution, examine the problem from the perspective of the sampling distributions implied by the problem. Figure 10.1 depicts two sampling distributions. Figure 10.1a represents the case when H_0 is true; the distribution in Figure 10.1b represents the case when H_0 is false by 25 raw points. For simplicity purposes, we will use a one-tailed test. (Let us temporarily set aside our misgivings about the use of one-tailed tests.) For a one-tailed test, the critical value when $\alpha = .05$ is $+1.65$ (instead of ± 1.96). Figure 10.1a is the null distribution because it depicts the sampling distribution of means when H_0 is true. Figure 10.1b is the alternative distribution because it illustrates the sampling distribution of means when the null hypothesis is incorrect by 25 raw points. Do we see that

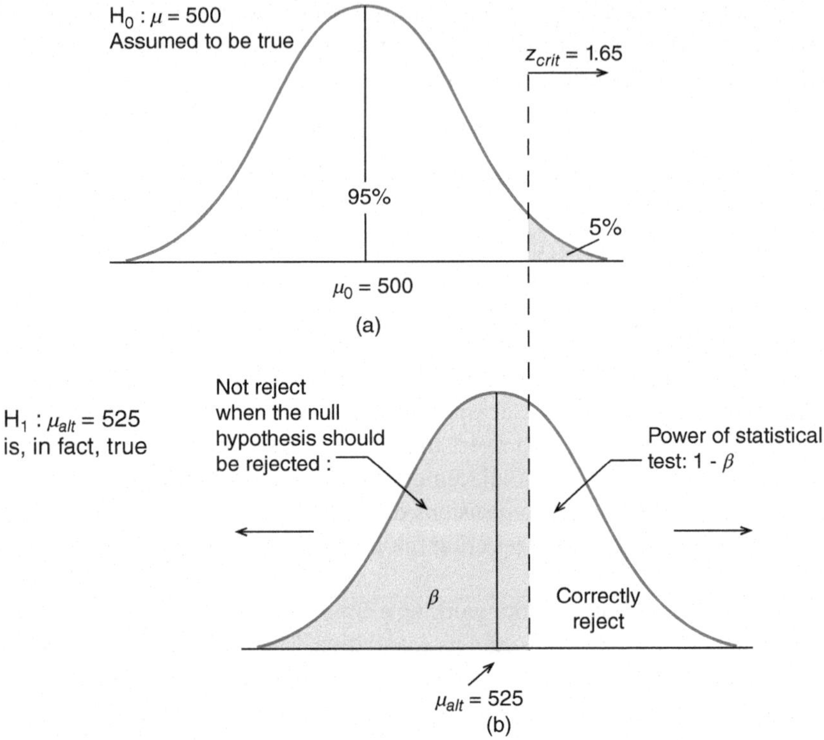

Figure 10.1 Sampling distributions of means. (a) Under H_0: $\mu = 500$. (b) Under $H_1: \mu = 525$.

the two decision criteria determine the critical score, which sets the location for the up and down dotted line? Do we also see that the three data-driven criteria determine the position of the alternative sampling distribution (i.e. the bottom figure)? Power will be determined by how much of the bottom figure falls into the rejection region (right side of the dotted line) of the top figure (i.e. the null distribution).

Power analyses can be difficult and involve many factors. For introductory purposes, we will explore a simple, straightforward method. A couple of simple calculations and a power table (Table A.3) are all that is needed. To use the table, we must introduce two new terms: δ (delta) and γ (gamma).

Formula for delta

$$\delta = \gamma\sqrt{n} \quad (10.1)$$

---

where

γ = the size of the treatment effect

Note that δ combines the size of the hypothesized treatment effect (γ) and the sample size (n) into a single measure. Delta will be used by the power table (Table A.3) to determine the proportion of the alternative sampling distribution that falls into the rejection region of the null sampling distribution (recall Figure 10.1).

Formula for gamma

$$\gamma = \frac{\mu_{alt} - \mu_0}{\sigma} \quad (10.2)$$

---

where

μ_{alt} = the mean of the alternative (treated) population
μ_0 = the population mean when the null hypothesis is true
σ = the population standard deviation; if σ is unavailable, use s

Note in Formula 10.2 that γ, the size of the treatment effect, determines the number of standard deviations between the null and alternative hypothesized means. If, by chance, the larger mean is subtracted from the smaller mean, gamma will be a negative value. However, it should always be understood as positive. There is no concept in statistics called negative power.

Keep in mind, this process does not *discover* the true population mean; rather, this mean is hypothetically specified ahead of time. A power analysis does not identify the difference between population means, and it does not test to see if there is a difference between population means. The power analysis merely calculates the probability of detecting a specified difference between population means, *assuming* that specified difference exists. Here is how the method works.

Step 1. Use Formula 10.2 to specify an effect size, γ (gamma).

$$\gamma = \frac{\mu_{alt} - \mu_0}{\sigma} = \frac{525 - 500}{100} = 0.25$$

The number .25 is the number of standard deviation units the treated population mean is from the mean of the population stated in the null hypothesis.

Step 2. Compute delta using Formula 10.1.

$$\delta = \gamma\sqrt{n} = .25\sqrt{36} = 1.50$$

Step 3. Determine the power by using the power table (Table A.3). In the worked example, use a one-tailed test, with $\alpha = .05$. With δ computed as 1.50, find δ in the first column of the table, and move to the column for $\alpha = .05$, one-tailed test. The power of the test is .44.

With the power analysis completed, we are now able to advise the company. Do we think the study should be conducted as proposed? There is only a 44% chance of getting a sample mean that would prompt a rejection of the null. In fact, there is a higher probability of making a Type II error (.56) than of rejecting the null (.44)! We should probably advise against it.

At this point, two questions can be raised. First, what can the company do to increase the power of the test? To answer this question, consider the three data-driven factors that influence power. Can any of them be altered? Perhaps the effectiveness of the study program can be improved, thereby increasing the difference between μ_{null} and μ_{alt}. More easily, however, the sample size of 36 can be increased.

Second, how much power is enough? There is no set or easy answer to this question, but most researchers prefer power to be in the range of .70–.90, with .80 as the conventional probability value. When a statistical test has a .80 probability of correctly rejecting the null hypothesis, there is a .20 probability of making a Type II error. We may note that a probability of .20 is four times greater than the conventional .05 probability of making a Type I error. Researchers are *usually* comfortable with that kind of trade-off. It reflects a conservative approach to hypothesis testing; concluding there is an effect when none exists is considered a more serious mistake than failing to identify an effect when one, in fact, exists. However, each study requires an investigator to consider what the acceptable probabilities are for different types of errors. The nature of the consequences for each type of error will guide the decision-making process.

As a review, let us use the power table to demonstrate the relationship between treatment effect size and power. (We will look at the relationship between sample size and power later.) Suppose the effect that we want to detect is 50 raw score units from the population mean of 500. The power question now becomes, "What is the power of my test to detect a population difference of 50 raw score units?" Use the data and sample size of the worked example.

$$\gamma = \frac{\mu_{alt} - \mu_0}{\sigma} = \frac{550 - 500}{100} = .50$$

$$\delta = \gamma\sqrt{n} = 0.50\sqrt{36} = \mathbf{3.00}$$

Now, with $\delta = 3.00$ and $\alpha = .05$, power jumps to .91, according to the table. If the population mean for the alternative hypothesis is 550, we have a 91% chance of detecting the effect.

Let us try one more. Recalculate δ when $\mu_{alt} = 505$. The power of the test drops to .12. There is only a paltry 12% chance of detecting a false null hypothesis if the effect of the ACE training course only moves the mean up to 505. These exercises demonstrate how treatment effect size influences power. As the effect size increases, so does statistical power.

These hypothesized effect sizes do not need to be drawn out of thin air. Many times, researchers establish them based on existing theory and/or previous research findings (see Section 10.5 for a broader discussion). A power analysis essentially says, "Given a particular effect size, sample size, alpha, and test type, here is the probability that a false null hypothesis will be rejected." This is valuable information when considering a research project.

10.5 Determining Effect Size: The Achilles Heel of Power Analyses

Researchers can directly control the alpha level, the type of test (one- or two-tailed), and, within reason, the sample size. The biggest challenge is predicting the size of the treatment effect. This difficulty often keeps researchers from conducting power analyses. However, we are not rendered helpless when having to state an effect size. There are two good sources of information for making a reasonable statement about effect size.

First, most studies have a heritage. Although we may be the first to conduct this particular study, it is likely that others have done work in this area or studied a similar phenomenon. By familiarizing ourselves with the effect sizes found by other researchers, we can better estimate an effect size for our experiment. Second, we may be able to conduct a pilot study: a trial run, so-to-speak. By first running the study with a small sample, we can both test out the experimental procedure and gain a feel for the strength of our independent variable and the amount of variability in the data.

Box 10.1 Is the Scientific Method Broken? The Need to Take Our Own Advice

Using the 1960 volume of the *Journal of Abnormal and Social Psychology*, Cohen (1962) conducted an interesting study. Although the authors of the articles did not use power analyses, Cohen computed the power for each study contained in that volume. Assuming researchers would want to detect a medium treatment effect size, Cohen found the average power of the tests to be just .46. This means that, on average, there was only a 46% chance of detecting a medium effect size! This realization prompts us to contemplate just how many other studies might be potentially included in the literature but are abandoned because not enough power is marshaled to detect a treatment effect. In response, Cohen admonished researchers to use power analyses (Cohen 1977), and yet, 24 years later, a study like Cohen's, using the 1984 volume of the *Journal of Abnormal Psychology*, found that the average statistical power for detecting a medium effect size had actually gone down to .37 (Sedlmeier and Gigerenzer 1989)! There is little reason to believe the situation is different today. Despite the urging of statisticians, power analyses have not become standard practice for researchers.

If a performed and reported power analysis were to become a standard step in the research process, it would both give studies that are investigating a genuine treatment effect a better chance of finding supporting evidence and give studies that end up failing to reject the null hypothesis more validity. For instance, a study possessing an appropriate amount of power to detect a small difference, that fails to reject the null could be theoretically and practically important to other researchers. For one thing, it might keep others from spending time and energy asking the same question, and second, it might stimulate the development of different ideas about how that part of the world works. Unfortunately, power analyses tend not to be performed, and null findings tend not to be published (see Box 8.1). Scientists can be a stubborn bunch, and the standards of the scientific process can be hard to change. However, with each new generation of scientists comes a new opportunity to do things differently. Will the next generation of researchers choose to use the tools of *statistical* power?

10.6 Determining Sample Size for a Single-Sample Test

Power analyses can also be used to determine the needed sample size for a study. To use power analyses in this way, we must first ask ourselves, "What size of a treatment effect do we *want* to detect?" Once this is determined, we can rework formulas and use them to calculate the number of participants needed to have a reasonably good chance of detecting an effect of that size.

To determine the needed sample size, we will need to state, beforehand, the alpha level (α), the treatment effect (γ), the desired power, and our preference for a one- or two-tailed test. Desired power can be set at any level, but as stated earlier in the chapter, the conventional value is .80.

Let us return once again to the SAT-V preparation course example. We found that if 36 participants are used, power is only .44. Since this is unacceptably low, we should provide the company with the number of participants needed to have an 80% chance of detecting the desired 25-point difference. Formula 10.3 reworks Formula 10.1 to solve for n.

Formula for determining sample size for a single-sample t test

$$n = \left(\frac{\delta}{\gamma}\right)^2 \quad (10.3)$$

Remember that γ is not the number of points between 500 and 525; rather, γ is the number of standard deviations 525 is from 500. In our example, γ was calculated as .25: $(525 - 500)/100 = .25$. Formula 10.3 also needs delta, δ. We can either use Table A.3, by locating the desired power and working backward to determine δ, or we can use Table A.4, input the desired power, and let the table determine δ. Let us use Table A.4, as it generates a more precise δ. Since our desired power is .80, look down the left-hand column and find .80. The column immediately to the right is for one-tailed tests when $\alpha = .05$. (Recall that we are using a one-tailed test for this example.) Here we find $\delta = 2.49$. We now have the values needed for Formula 10.3.

$$n = \left(\frac{\delta}{\gamma}\right)^2 = \left(\frac{2.49}{.25}\right)^2 = \mathbf{99}$$

For an 80% chance of detecting a 25-point difference, we will need 99 participants.

■ **Question** *Suppose we wanted to be almost certain of detecting a treatment effect size of .25. Set power at .99. How many participants would we now need?*

Solution Using Table A.4 to find δ for a power of .99, δ is found to equal 3.97.

$$n = \left(\frac{\delta}{\gamma}\right)^2 = \left(\frac{3.97}{0.25}\right)^2 = \mathbf{252}$$

Increasing the probability of detecting a 25-point difference to .99 requires 252 participants. If participants are easy to come by and the cost of gathering data is low, the investigator may want to use a level of power greater than the conventional 80%. ■

Readers should also know that a variety of easy-to-use power analysis software programs can be found online. Users input available information and solve for missing values.

Summary

The probability of rejecting a false null hypothesis is the power of a statistical test. Power increases as the magnitude of the treatment effect increases. (The treatment effect is the number of standard deviations between μ_{alt} and μ_0.) Increasing the sample size in a study will also increase the power of a statistical test. Furthermore, two features of the decision criteria for hypothesis testing influence power: the desired alpha value (Type I error rate) and the choice of a one- vs. two-tailed statistical test.

Conventionally, the desired power for a statistical test is .80, or 80%. A power analysis typically involves either solving for power or solving for the sample size needed to obtain a desired amount of power. A power analysis performed before a study begins helps a researcher set up a study to maximize the prospects of finding a desired effect without wasting time and energy.

Key Formulas

Formula for delta

$$\delta = \gamma\sqrt{n} \quad (10.1)$$

Formula for gamma

$$\gamma = \frac{\mu_{alt} - \mu_0}{\sigma} \quad (10.2)$$

Formula for determining sample size for a single-sample *t* test

$$n = \left(\frac{\delta}{\gamma}\right)^2 \quad (10.3)$$

Key Term

Power

Exercises

1 Power is the ability of a statistical test to:
- **a** Incorrectly fail to reject a null.
- **b** Incorrectly reject a null.
- **c** Correctly fail to reject a null.
- **d** Correctly reject a null.

2 How are the lenses of a microscope analogous to the concept of statistical power?

3 If β equals .10:
- **a** The Type I error rate = .90.
- **b** The Type II error rate = .10.
- **c** Alpha = .10.
- **d** Alpha = .90.

4 As the power of a statistical test increases, what happens to the Type I error rate?

5 As the power of a statistical test increases, what happens to the Type II error rate?

6 How is a hypothesized treatment effect calculated?

7 How does the size of the treatment effect influence power?

8 A researcher is interested in studying the relationship between sleep deprivation and cognitive performance. A power analysis of pilot data shows low power for detecting an effect when sleep loss is three hours (power = .24). If sample size cannot be increased, how else might the researcher increase power?

9 How does sample size influence the power of an inferential test?

10 Which of the following three factors influencing power does the researcher typically have the least ability to adjust? Why?
a The H_0 and H_{alt} difference.
b The sampling error.
c The sample size.

11 How does increasing or decreasing alpha influence power?

12 How does a choice of a one- vs. two-tailed statistical test influence power?

Work Problems

13 Compute treatment effect sizes for each of the following problems.
a $\mu_0 = 300, \mu_{alt} = 345, \sigma = 70$
b $\mu_0 = 300, \mu_{alt} = 345, \sigma = 20$
c $\mu_0 = 300, \mu_{alt} = 310, \sigma = 20$
d $\mu_0 = 300, \mu_{alt} = 310, \sigma = 50$

14 Using the numbers found in Problem 13, how many participants are needed to detect each effect in the population? (Assume $\alpha = .05$, two-tailed test, and desired power is .80.)

15 A researcher performs single-sample t tests on each set of summary data described below. For each case, compute the power of the statistical test. (Use s to estimate σ, an $\alpha = .05$, two-tailed tests, and round delta to the first decimal place.)

a $\mu_0 = 130, \mu_{alt} = 120, s = 15, n = 10$
b $\mu_0 = 130, \mu_{alt} = 120, s = 15, n = 40$
c $\mu_0 = 50, \mu_{alt} = 52, s = 10, n = 15$
d $\mu_0 = 50, \mu_{alt} = 52, s = 10, n = 100$

16 A researcher is interested in running a power analysis before the research is started. Use the means and standard deviations in Problem 15, parts a and c and determine what sample size is needed for each study to achieve power = .80, $\alpha = .05$, two-tailed test. (Round to the nearest whole number.)

Part 3 Review

The *z* Test, *t* Tests, and Power Analysis

Review of Concepts Presented in Part 3

The purpose of this brief review section is to revisit both the similar concepts from Chapters 8–10 as well as those that distinguish them one from another. First, let us look at the similarities. All four of the inferential tests presented in these chapters (single-sample z test, single-sample t test, independent-samples t test, and dependent-samples t test) are based on the same basic logic. That is, each one is designed to test a null hypothesis by taking a found difference between means and interpreting that difference in terms of the amount of sampling error we might expect to find if the null were true. In mathematical terms, a ratio is created where a difference between means is placed in the numerator and a measure of sampling error is placed in the denominator. The resulting value of this ratio is then compared to all the values one might expect to find if we were working under conditions where the null is true; that is, where no differences between the population means actually exist. If it is determined that the difference between the means is not likely to be explained by sampling error, a "treatment effect" is inferred. In these situations, a follow-up analysis measuring the size of the treatment effect can be run. Cohen's d measures the size of a mean difference in standard deviation units; various versions of this statistic exist which correspond to the various z and t tests.

Another common theme running through these tests is that each one is an inference; that is, an extrapolation from known sample data to unknown population data. As a result, there is always a likelihood that the sample data used for the test may misrepresent the population and render a misleading conclusion. Nonetheless, decision rules are employed which lead the researcher to cautiously

Essentials of Behavioral and Social Science Statistics, Third Edition.
K. Paul Nesselroade, Jr. and Laurence G. Grimm.
© 2025 John Wiley & Sons, Inc. Published 2025 by John Wiley & Sons, Inc.
Companion website: https://www.wiley.com/go/Essentials_of_Behavioral_and_Social_Science_Statistics3e

conclude that either the null is wrong or that the null may be true. This means that two types of decision errors can occur. A Type I error occurs when we conclude there is a difference when, in fact, there is not; a Type II error occurs when we do not conclude there is a difference when, in fact, there is one. To help avoid making Type II errors, the concept of a power analysis is introduced in Chapter 10. This is an a priori technique using several factors; two key ones being the expected sample size and the expected treatment effect size. The analytical tool is designed to help researchers predict ahead of time how likely a difference between means will be generated that will allow the null to be rejected. If the likelihood is low, the researcher may opt not to conduct the study (thereby saving time and energy) or they may decide, prior to running the study, to adjust the research design to improve the chance of gathering data that would lead to rejecting the null.

Each inferential test, however, is different and is used in different methodological situations. Chapter 8 tests compare a known population mean with a gathered sample mean. The null claims the sample mean came from the same population that produced the given population mean. For this test, the numerator is a sample mean subtracted from a known population mean. If the standard deviation of the population is known as well, then the standard error can be determined, and a single-sample z test can be run. The inferential decision is based on the values in the z table (Appendix A.1), with the typical critical values being ± 1.96. If the population standard deviation is unknown, then the sample standard deviation is used to estimate the standard error. This technique is called the single-sample t test, and the resulting t value is compared against critical values found in the t table (Appendix A.2). For situations when a sample mean is compared to a known population mean, the key diagnostic question concerns whether σ is known.

Chapter 9 tests compare two sample means when no population means are available. This is a more common research situation than those presented in Chapter 8. The independent-samples t test is employed when comparing two sample means coming from two independent samples. (Independent samples occur when two separate and unmatched groups of participants are used to create the sample means.) The error term in the denominator uses the standard deviation from both samples and the two sample sizes to help estimate the standard error. If the two means to be compared are composed of scores that are matched or come from the same set of participants, the test to use is the dependent-samples t test. Relative to independent-samples t tests, dependent-samples tests typically yield smaller standard error terms, thus generating larger t values. In other words, the dependent samples t test has more statistical power (i.e. ability to reject correctly false null hypotheses). Therefore, for situations when two samples are being compared, the key diagnostic question concerns whether the samples come from independent or dependent groups.

Since real-world research problems do not come with a label informing the researcher of which test to use for analysis, it is important for us to work on our diagnostic skills. Understandably, the exercises at the end of each chapter require only the use of test(s) found within that chapter to answer the questions. They are designed to get us familiar with solving a known type of problem. However, they are not designed to challenge our diagnostic skills (i.e. knowing which test to use for a given situation). The following review section, however, is designed to help us with this skill.

The exercises below will help us review the statistical differences between the various tests. The hypothesis testing exercises (numbers 3–11) will not identify which test is appropriate for the described scenario. We will need to use the available information to make that determination.

Part 3 Review Exercises

1 Which of the previously presented tests can be run prior to the gathering of any data?

2 Identify the critical values for the following situations (use online table if needed).
   a Dependent-samples t test, $\alpha = .05$, $df = 19$, two-tailed test
   b Independent-samples t test, $\alpha = .10$, $df = 16$, two-tailed test
   c Single-sample z test, $\alpha = .01$, $n = 40$, one-tailed test
   d Dependent-samples t test, $\alpha = .10$, $df = 7$, two-tailed test
   e Dependent-samples t test, $\alpha = .01$, $df = 100$, two-tailed test
   f Single-sample t test, $\alpha = .01$, $df = 8$, two-tailed test
   g Dependent-samples t test, $\alpha = .05$, $n_p = 5$, two-tailed test
   h Independent-samples t test, $\alpha = .05$, $n_1 = 6$, $n_2 = 4$, one-tailed test
   i Single-sample z test, $\alpha = .05$, $n = 22$, two-tailed test
   j Single-sample t test, $\alpha = .10$, $n = 30$, two-tailed test
   k Independent-samples t test, $\alpha = .01$, $n_1 = 14$, $n_2 = 15$, two-tailed test
   l Dependent-samples t test, $\alpha = .01$, $n_p = 3$, two-tailed test

3 A psychologist is concerned that children in the local grade school do not have enough social interaction with peers. Suppose previous research suggests grade-school children average 2.60 hours/day playing with friends. The psychologist samples 18 children at the school and records the time each spent playing with others. What is the appropriate inferential test and why?

4 A researcher believes people experience different numbers of dreams due to room temperature. Volunteers are asked to sleep for ten nights in an 80 °F. room and then for ten nights in a 65 °F. room. Total number of dreams is recorded. What is the appropriate inferential test and why?

5 A professor believes the current class of students is more intelligent than most previous classes. To test this, the psychologist administers an intelligence test to a random sample of 12 students. Previous research suggests the mean score for students is 110 with a standard deviation of 15. What is the appropriate inferential test, and why?

6 Oishi and Schimmack (2010) found that people who move homes frequently as children tend to have lower levels of subjective well-being (happiness) as adults. To explore this, 15 people are sampled who moved homes 4 or times before they were 12. They were given a subjective well-being questionnaire with a known $\mu = 50$. What is the appropriate inferential test, and why?

7 A physician compares the weight gain difference between an anabolic steroid and a natural growth stimulant. Sixteen patients are randomly assigned to two conditions. One group receives the steroid for 30 days and a second group receives the growth stimulant. Weight gain over 30 days is measured. What is the appropriate inferential test, and why?

8 Another researcher asks the same question in Exercise #7. But, in this study four patients are given the steroid for 30 days, followed by a 30-day period with the growth stimulant. The other four receive the treatments in reverse order. What is the appropriate inferential test and why?

9 A men's collegiate soccer program wants to see if their student-athletes are more fit than student-athletes in general. A nationally normed fitness test claims the mean mile time for biological male collegiate athletes is 5 minutes ($s = 30$ seconds). The coach randomly samples the team and measures mile times. What is the appropriate inferential test and why?

10 A researcher is interested in the effect of emotion on concentration. A two-sample study is designed in which anger is induced in one sample by a confederate provoking an argument in the lab waiting room. The control group does not undergo this mood induction. Both samples are then tested on a computer stunt driving game, and the number of times the participant runs the vehicle into an object (crashes) is counted. What is the appropriate inferential test and why?

11 A psychologist is comparing different exercise programs for cardiovascular fitness. The fitness measure is resting heart rate after program completion. Twelve college students are randomly assigned to three groups (four per group). Participants in the aerobic condition walk a treadmill for 30 minutes, three times a week. Those in the circuit condition perform exercises on weight machines for 30 minutes. Control condition participants are asked to simply maintain their usual amount of exercise. What is the appropriate inferential test and why?

Part 4

Inferential Statistics

Analyses of Variance

11

One-Way Analysis of Variance

11.1 The Research Context

In the preceding chapters, we have learned several tools that can be used to test null hypotheses between two means. However, frequently, researchers use more than two groups in a study. Many studies compare a control condition against two or more other conditions (e.g. a memory study comparing an imagery mnemonic, a repetition mnemonic, and a control group). Studies with more than two conditions may require an **analysis of variance (ANOVA)**. It is true, we could avoid using an ANOVA and simply conduct t tests on all possible two-group (also called *pairwise*) comparisons. For example, in a three-group study, we could generate a t value for groups 1 vs. 2, 1 vs. 3, and 2 vs. 3. However, running multiple t tests raises a serious statistical problem.

11.1.1 Multiple *t* Tests and the Type I Error

Recall that Type I errors are committed when a true null hypothesis is mistakenly rejected. The probability of making a Type I error is determined directly when setting the alpha level. This means that when the alpha level is set at .05, and a t test is conducted, the probability of mistakenly rejecting a true null hypothesis is .05 *for that one t test*. Over a *series* of t tests, however, the probability of making at least one Type I error becomes *inflated* (recall the lesson of Box 8.1). For example, if we performed three t tests, the probability of making at least one Type I error is closer to .14, or 14%. If 10 t tests were conducted, the probability of making at least 1 Type I error would be an astonishing .40. Over a series of t tests, the Type I error rate inflates by $1 - (1 - \alpha)^c$, where c is the number of independent comparisons. (To help envision this problem, imagine a 20-sided die, each side representing the statistical outcome when a null hypothesis is being tested. Nineteen of the sides would be white, representing a proper decision to fail to reject the null hypothesis.

Essentials of Behavioral and Social Science Statistics, Third Edition.
K. Paul Nesselroade, Jr. and Laurence G. Grimm.
© 2025 John Wiley & Sons, Inc. Published 2025 by John Wiley & Sons, Inc.
Companion website: https://www.wiley.com/go/Essentials_of_Behavioral_and_Social_Science_Statistics3e

However, one of the sides, representing 5%, is colored red and marked "Type I error." With each *t* test run where the null is true, the die must be rolled. The cumulative probability of the die, at least once, landing on the "Type I error" side increases in an almost additive fashion as the number of rolls accumulates.)

This alpha inflation is eliminated by using an ANOVA since only one test is performed: the *F* test. The *F* test, named after its originator, Sir Ronald Fisher (see Spotlight 11.1), provides a comparison of all population means in one test: just one roll of the 20-sided die. A large *F* value means there is statistical evidence suggesting at least two of the sample means come from different populations. The problem is that a large *F* value does not tell us which of the pairs of means are statistically different from one another. To make *pairwise* comparisons among all the means, special follow-up (or secondary) analyses are used, that control the Type I error rate. A couple of different versions of these analyses are presented near the end of this chapter.

A **One-way ANOVA** is used for research designs having independent samples and *one* independent variable (or factor) of three or more levels. Interestingly, one-way ANOVAs can be used on designs having two levels; in fact, they are equivalent in power to independent samples *t* tests. Conventionally, however, *t* tests are used for two-group designs. If our design has *two* independent variables (or factors), a two-way ANOVA is used; this technique is covered in Chapter 12. If our design uses the same participants for each level of the factor, a repeated-measures ANOVA is used, which is the subject matter for Chapter 13.

The test's name, an analysis of variance, may seem to suggest that it is a test of variances. However, means are still being compared, but the comparison is based on the sources of variation within the data, including the variation between-group means. Just as with the *t* tests, the ANOVA will help us decide if the sample group means are coming from different populations. How this is accomplished will become clear as the chapter progresses.

Spotlight 11.1 Sir Ronald Fisher

Ronald Fisher (1890–1962) was born in England and was considered a child prodigy (Box 1978). His mathematical ability, evident from childhood, flourished throughout his life and led to the invention of the most popular inferential test in experimentation: the ANOVA.

Fisher received his training in mathematics at Cambridge and subsequently taught math in public schools. Apparently, he was a lousy teacher and did not enjoy the profession. In 1917, he married Ruth Guinness, the cousin of the well-known Irish brewery operators. (Recall that Gossett, developer of the *t* test, was employed by Guinness.) When he quit teaching, Fisher went to work for the

Rothamsted Experimental Station, the oldest agricultural research center in the world. It was during his tenure here that Fisher made many of his most brilliant statistical and research design contributions.

One of the problems in agricultural experimentation at that time was how to determine the effects of a multitude of variables on plant yields; factors for consideration were soil, fertilizer, weeds, seeds, and weather. Furthermore, each of these variables had multiple expressions. The desire to make multiple comparisons at that same time led Fisher to develop the ANOVA. To appreciate Fisher's context, it is interesting to note that the article that presented the ANOVA is titled "Studies in Crop Variation. II. The Manurial Response of Different Potato Varieties" (Fisher and MacKenzie 1923).

Fisher's contributions were broad. He is responsible for coining many of the common terms in statistics, including *variance*, *randomization*, and even the term *statistic* itself. Fisher also worked out many of the sampling distributions required for hypothesis testing and was an influential figure in establishing the 5% alpha level for rejecting null hypotheses. Unfortunately, like many other statisticians of that time who were carving out the mathematical tools for the new social and behavioral sciences (e.g. Galton, Pearson, and Spearman), Fisher was a strident eugenicist, holding, among other distasteful beliefs, the strong opinion that there were intellectual differences between various races of humans. This conviction was so strong that it resulted in him registering a dissenting opinion to the United Nations Educational, Scientific, and Cultural Organization (UNESCO) project on racism (UNESCO 1952).

Although the personal beliefs of so many of the pioneers of inferential statistics are offensive, the power of the statistical tools they created should not be diminished by these associations. These tools, though originally created from motives to show inherent differences between-groups and classes of people, are thankfully not limited to those sorts of questions.

11.2 Hypotheses

The ANOVA directly tests the null and alternative hypotheses. These always reference populations and are typically stated as:

$H_0: \mu_1 = \mu_2 = \mu_3 = \mu_k$, (where "k" means "to the last condition")

H_1: at least two of the means are different.

Note that H_1 is *not* $\mu_1 \neq \mu_2 \neq \mu_3 \neq \mu_k$. This erroneous statement of the alternative hypothesis would indicate that all the means must be substantially different from one another. The null hypothesis can be rejected when just two of the population means are different.

Although this chapter does not give much attention to Type I and Type II errors, these concepts are still active for ANOVA tests. Any rejection of a null hypothesis carries with it a chance of making a Type I error, and any failure to reject a null runs the risk of making a Type II error.

11.3 The Conceptual Basis: Sources of Variation

11.3.1 Between-Group Variation

If we conduct a study with two or more conditions, the resulting sample means across the different conditions will almost never be identical. This would be true even if the null hypothesis were true. Our task is to decide if the differences between the sample means can be explained by sampling error or if we should also infer the action of the independent variable (or "factor," if the context is not an experiment). Our decision will be based on the outcome of an inferential test. The differences between the means across the conditions are termed **between-group variation.** (The conditions are typically termed *groups*; we will often use this term.) When considering what can account for between-group variation, three possible explanations emerge:

1) **Treatment** (or **primary) variance**. This is the effect due to treatment; that is, the degree to which the mean differences are caused by the features of the condition. This source of variation is what researchers attempt to maximize and then detect with statistical analysis.
2) **Individual differences**. Each research participant comes to the experiment with a unique constellation of personality traits called *participant variables* (see Chapter 1). As a result, each participant responds somewhat uniquely to the features of the experiment. The random assignment of participants usually does a good job of balancing these individual differences across the conditions. However, even when random assignment is used, it is possible for group means to differ due to these individual differences. For example, consider a study comparing three teaching techniques. If, *by chance*, a few more of the sharper students are randomly assigned to one condition than the other two, this may create unwanted variance between the group means. The larger the group of randomly assigned participants in the study, the less likely this will be a problem.
3) **Experimental error**. This variance reflects the difference between a measurement and the true value. There are three potential sources of experimental error: the unreliability of measuring instrumentation (e.g. intelligence tests do not generate the same value each time they are administered to a person), inconsistent interactions between an experimenter and participants (e.g. slight differences in how an experimenter states the instructions to different

participants), and random forms of environmental disturbances during the experiment (e.g. the distracting effects of the sound of a lawn mower outside the laboratory walls). In short, any uncontrolled aspect of the experiment could account for the variation among group means. These uncontrolled aspects of the experiment are assumed to be somewhat equally spread across the breadth of the research design and do not intrude *systematically*. (Do not confuse experimental error with confounding error. Confounds *systematically* vary among the groups and compete with the independent variable to explain mean differences. Because experimental error influences data in an *unsystematic* manner, confounding variance is not introduced.)

Individual differences (2) and experimental error (3) are considered random factors. As a result, they are often combined and termed **error variance** (or **secondary variance**), as opposed to treatment, or primary, variance. They are considered random because they are not intentionally or systematically manipulated by the experimenter, yet they may intrude and produce differences between-group means. In summary, any combination of the three noted reasons above can influence the variation *between*-group means.

11.3.2 Within-Group Variation

Between-group variation considers variability between means. **Within-group variation** refers to variability among scores *within* a group or condition. Would we expect every person in a given group to produce the exact same score? Of course not, for the following two reasons:

1) **Individual differences.** Individual differences will account for much of the variation of *scores within any group*. Once again, let us use the teaching technique example. Within any condition, there will be individuals with varying levels of adeptness regarding the material being covered, not to mention differing motivation levels, and a host of other factors. Please note: Individual differences with a group will be present even in conditions where a treatment effect is occurring. Students may be performing better in one condition compared to others, but within a given group they will not all be performing identically.
2) **Experimental error.** The variation among scores within a group could also be due to experimental error. These errors are the same as those that occur in between-group variation: incidental, unsystematic, and unintended.

It is important to realize that the treatment administered to each group *does not* contribute to *within-group* variability since each participant in a group is exposed to the same treatment. Also, since individual differences and experimental error are random factors, the only sources of within-group variation are random. The sources of variability are depicted in Figure 11.1.

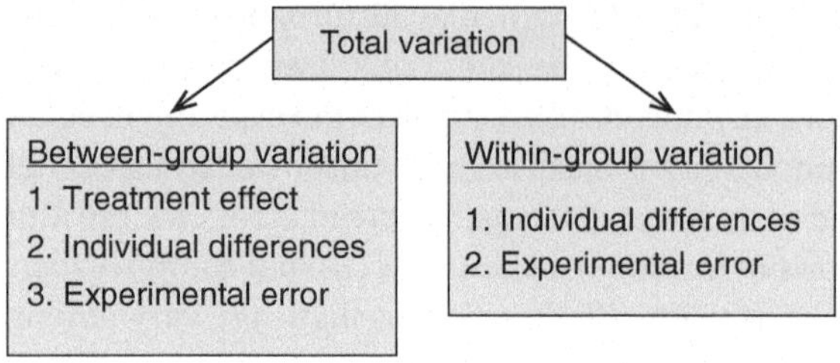

Figure 11.1 The ANOVA partitions the variation into between-group and within-group sources.

11.3.3 Sources of Variation and the Null Hypothesis

The F test is the ratio between between-group variance and within-group variance.

$$F = \frac{between - group\ variance}{within - group\ variance}$$

Stated differently,

$$F = \frac{treatment\ variance + error\ variance}{error\ variance}$$

Suppose there is no treatment variance, only error variance. In this instance, any difference between the means is completely due to random factors. In the absence of a treatment effect, the between-group variance will be nothing but error variance. This means that when there is no treatment effect, the F ratio will be close to 1 (one measure of error variance divided by another measure of error variance). As treatment variance appears, the numerator increases, and the F ratio becomes larger. The larger the value of F, the more likely treatment variance exists. Figure 11.2 summarizes the sources of variance when the null hypothesis is correct and incorrect.

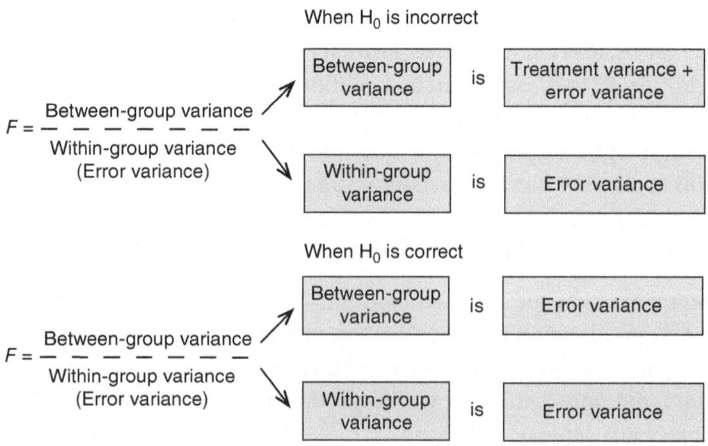

Figure 11.2 Sources of variance when the null hypothesis is correct and incorrect.

11.4 The Assumptions

As with the z and t tests, ANOVA's have assumptions that must be met. The assumptions for the one-way ANOVA are the same as those for the independent samples *t* test. As a result, they will simply be listed.

1) **Representativeness.**
2) **Independent observations.**
3) **Interval or ratio scale of measurement.**
4) **The populations from which the samples are taken are normally distributed.**
5) **Homogeneity of variances.**

It is important to note that the *F* test is *robust* to violations of the last two assumptions, particularly if each sample contains a sufficiently large and equal number of observations. However, gross violations of these assumptions will adversely affect the validity of the *F* test. Strong violations would require the use of analysis tools that do not make assumptions about the shape of the population distributions (see Chapter 16).

11.5 Computing the *F* Ratio

To find the *F* statistic that is used to test the null hypothesis, we will introduce some intermediate concepts and formulas. We will also work through an example. Please note, even if we are using a statistical software package to run the one-way ANOVA, the following material will help us understand what the program does and how the output should be interpreted.

Working from the *F* statistic backward, we need to find a measure of both within-group variance and between-group variance. Recall that if there is no treatment effect, these two measures will be similar since both measure only random factors at work in the data. If that is the case, the resulting *F* ratio will be somewhat close to 1. Figure 11.3 reflects this situation. Here, we see four means, all coming from the same distribution (or distributions with the same mean). As you can see, the four means (M_A, M_B, M_C, and M_D) are not that far apart from each other. This is even more evident when we look at Figure 11.4. Here we have one of our four group means (M_D) that clearly comes from a population with a different mean. This represents a study where one of the conditions has introduced treatment variance. Although the four means in Figure 11.3 are not identical, their variability can be explained by random factors. But random factors alone are insufficient to explain the variability we see in Figure 11.4. The *F* value for the study reflected in Figure 11.3 is probably a number not too far from 1. However, the *F* value for the study reflected in Figure 11.4 is probably much greater than one.

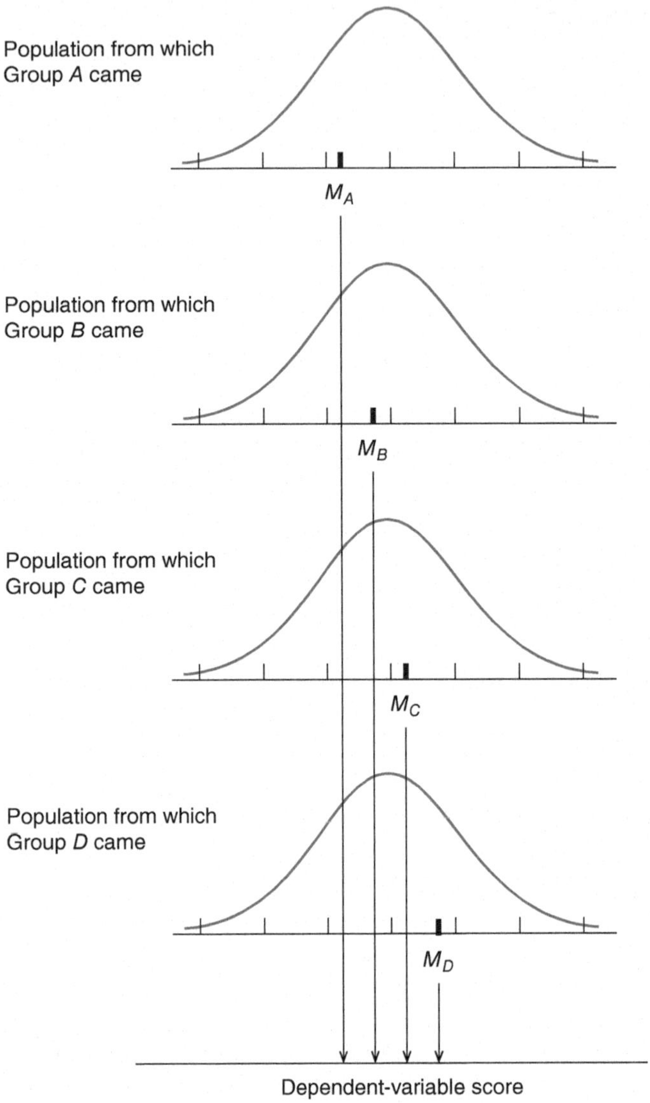

Figure 11.3 When H_0 is true, the sample means are drawn from identical populations, and little variation is expected between them.

The measure of the between-groups variance is termed the **mean square between** (or mean square between-groups) and is symbolized as $\boldsymbol{MS_{BG}}$. The measure of within-groups variance is termed the **mean square within** and is symbolized as ($\boldsymbol{MS_W}$). They are appropriately termed *mean square* because they reflect

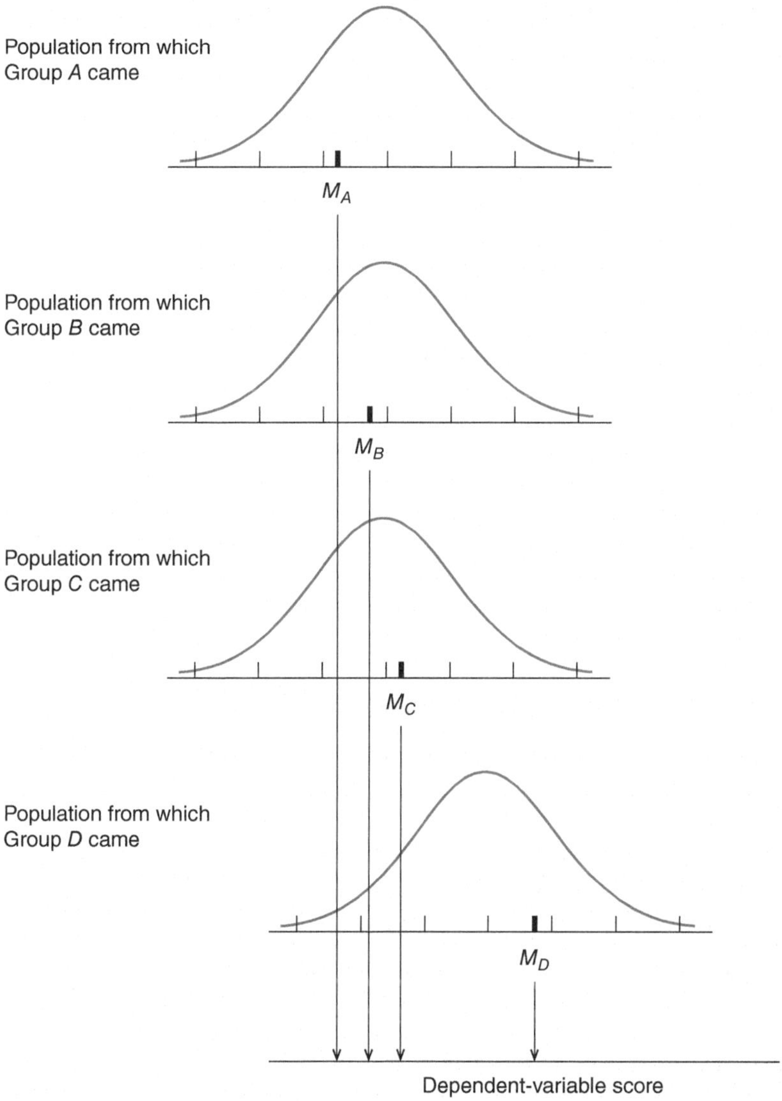

Figure 11.4 When H_0 is false, at least one sample is drawn from a population different from the rest.

the average (mean) amount of variation. $\boldsymbol{MS_{BG}}$ is the average variation of the group means around the grand mean (the grand mean being the mean of all the scores in the experiment, irrespective of individual groups). $\boldsymbol{MS_W}$ is the average amount of variation of the individual scores with respect to the group from which the scores are taken.

11.5.1 MS_{BG} and the Sum of Squares Between-Groups (SS_{BG})

The MS_{BG} is a ratio of the sum of squares between the groups (SS_{BG}) divided by the degrees of freedom for SS_{BG}. (Degrees of freedom will be discussed later.) Formula 11.1 is the computational formula for calculating SS_{BG}. Use this formula when calculating SS_{BG} by hand.

Computational formula for SS_{BG}

$$SS_{BG} = \frac{(\Sigma X_1)^2}{n_1} + \frac{(\Sigma X_2)^2}{n_2} + \cdots + \frac{(\Sigma X_k)^2}{n_k} - \left[\frac{(\Sigma X)^2}{N}\right] \quad (11.1)$$

where

ΣX_1 = the sum of scores in group 1
ΣX_2 = the sum of scores in the second group and so on
ΣX_k = the sum of scores in the last group; the kth group means the last group
ΣX = sum of all the scores in the study
n_1 = the number of participants in group 1
n_k = the number of participants in the last group
N = total number of participants in the study

11.5.2 MS_W and the Sum of Squares Within-Groups (SS_W)

The MS_W is a ratio of the sum of squares within-groups (SS_W) divided by the degrees of freedom associated with SS_W. The computational formula is given in Formula 11.2. Use this formula when calculating SS_W by hand.

Computational formula for SS_W

$$SS_W = \Sigma X^2 - \left[\frac{(\Sigma X_1)^2}{n_1} + \frac{(\Sigma X_2)^2}{n_2} + \cdots + \frac{(\Sigma X_k)^2}{n_k}\right] \quad (11.2)$$

where

ΣX^2 = sum of *all* squared scores
$(\Sigma X_1)^2$ = the sum of the scores in group 1, quantity squared
$(\Sigma X_2)^2$ = the sum of the scores in group 2, quantity squared
$(\Sigma X_k)^2$ = the sum of the scores in the last group, quantity squared
n_1 = number of participants in group 1
n_k = number of participants in the last group

11.5.3 The Total Sum of Squares (SS_T)

The total sum of squares (SS_T) is not used when computing the F ratio. Nonetheless, we need to calculate it for use in secondary analyses. The total sum of squares equals the sum of squares between-groups plus the sum of the squares within-groups:

$$SS_T = SS_{BG} + SS_W$$

Independently computing SS_T also allows us to make sure the SS_{BG} and SS_W calculations are accurate. Formula 11.3 is the computational formula for SS_T. Use this formula when computing SS_T by hand.

Computational formula for SS_T

$$SS_T = \Sigma X^2 - \frac{(\Sigma X)^2}{N} \quad (11.3)$$

---

where

ΣX^2 = the sum of *all* squared scores
$(\Sigma X)^2$ = the sum of *all* scores, quantity squared
N = the total number of participants

Note the similarities between the three *SS* formulas; each of the subcomponents is used twice. This realization will speed up hand calculations.

11.5.4 Degrees of Freedom

To arrive at the *F* ratio, SS_{BG} and SS_W must be divided by their appropriate degrees of freedom (*df*). This step generates MS_{BG} and MS_W. Dividing by *df* turns a sum of squares into an average sum of squares, which is what "mean squares" means. The *df* for between-groups (df_{BG}) is $df_{BG} = k - 1$, where k equals the number of groups or conditions. The degrees of freedom for the within-groups (df_W) is $df_W = (n_1 - 1) + (n_2 - 1) + \cdots + (n_k - 1) = N - k$. The *df* for SS_T is $N - 1$. It is not needed for any later calculation. However, $df_T = df_{BG} + df_W$.

Computing MS_{BG} and MS_W is accomplished in the following manner.

Calculating MS_{BG}, MS_W, and F

$$MS_{BG} = \frac{SS_{BG}}{df_{BG}} \quad (11.4)$$

$$MW_W = \frac{SS_W}{df_W} \quad (11.5)$$

$$F = \frac{MS_{BG}}{MS_W} \quad (11.6)$$

11.5.5 Illustrating the Computational Steps with Raw Data

Table 11.1 presents the raw scores and summary statistics for a hypothetical study with three groups. This data will be used to illustrate the computational steps

Table 11.1 Hypothetical raw data and summary statistics for a study with three groups.

Group 1	Group 2	Group 3
P_1 4	P_5 8	P_9 3
P_2 5	P_6 8	P_{10} 2
P_3 4	P_7 9	P_{11} 1
P_4 6	P_8 10	P_{12} 3
$M_1 = 4.75$	$M_2 = 8.75$	$M_3 = 2.25$
$\Sigma X_1 = 19$	$\Sigma X_2 = 35$	$\Sigma X_3 = 9$
$\Sigma X_1^2 = 93$	$\Sigma X_2^2 = 309$	$\Sigma X_3^2 = 23$
$n_1 = 4$	$n_2 = 4$	$n_3 = 4$

required to arrive at the F ratio. The by-hand calculations can be tedious, but we have had extensive practice in performing the arithmetic operations needed to compute F.

Step 1. Compute SS_{BG} using Formula 11.1.

Computational formula for SS_{BG}

$$SS_{BG} = \frac{(\Sigma X_1)^2}{n_1} + \frac{(\Sigma X_2)^2}{n_2} + \cdots + \frac{(\Sigma X_k)^2}{n_k} - \left[\frac{(\Sigma X)^2}{N}\right]$$

$$SS_{BG} = \frac{(19)^2}{4} + \frac{(35)^2}{4} + \frac{(9)^2}{4} - \left[\frac{(19 + 35 + 9)^2}{12}\right] = \mathbf{86}$$

Step 2. Compute df_{BG}: $df_{BG} = k - 1 = 3 - 1 = 2$.

Step 3. Find MS_{BG} using Formula 11.4.

$$MS_{BG} = \frac{SS_{BG}}{df_{BG}} = \frac{86}{2} = \mathbf{43}$$

Step 4. Compute SS_W using Formula 11.2.

Computational formula for SS_W

$$SS_W = \Sigma X^2 - \left[\frac{(\Sigma X_1)^2}{n_1} + \frac{(\Sigma X_2)^2}{n_2} + \cdots + \frac{(\Sigma X_k)^2}{n_k}\right]$$

$$SS_W = (4)^2 + (5)^2 + (4)^2 + \cdots + (2)^2 + (1)^2 + (3)^2 - \left[\frac{19^2}{4} + \frac{35^2}{4} + \frac{9^2}{4}\right] = \mathbf{8.25}$$

Step 5. Compute df_W: $df_W = N - k = 12 - 3 = 9$.

Step 6. Find MS_W using Formula 11.5

$$MS_W = \frac{SS_W}{df_W} = \frac{8.25}{9} = \mathbf{0.92}$$

Step 7. (Optional) Compute SS_T using Formula 11.6.

Computational formula for SS_T

$$SS_T = \Sigma X^2 - \frac{(\Sigma X)^2}{N} = 4^2 + 5^2 + 4^2 + \cdots + 2^2 + 1^2 + 3^2 - \left[\frac{63^2}{12}\right] = \mathbf{94.25}$$

If the calculations are correct, then SS_T should equal $SS_{BG} + SS_W$:

$$SS_T = SS_{BG} + SS_W$$

$$94.25 = 86 + 8.25$$

Even though SS_T is not used to calculate F, finding SS_T can serve as a computational check. Additionally, we may need this value for secondary analysis purposes.

Step 8. Compute the F ratio using Formula 11.6.

$$F = \frac{MS_{BG}}{MS_W} = \frac{43}{0.92} = \mathbf{46.74}$$

11.6 Testing Null Hypotheses

11.6.1 The *F* Sampling Distributions

Like the t distribution, the shape of the F sampling distribution is affected by the sample size. Unlike the t distribution, the F sampling distribution is not normally distributed; rather, it is positively skewed. Like the t tests, the F test utilizes the degrees of freedom concept to determine the critical value marking the most extreme part of the ***F* distribution**. Unlike the t test, the F test requires two different measures of degrees of freedom but only generates one critical score.

As is the case with all sampling distributions, an F distribution is theoretical; we do not need to go through the arduous task of creating it. Since the F statistic is a ratio of variances, it should make sense to us that the sampling distributions of F are based on ratios of variances.

Recall that the F ratio is:

$$F = \frac{MS_{BG}}{MS_W} = \frac{\textit{treatment variance} + \textit{error variance}}{\textit{error variance}}$$

When the null hypothesis is true, there is no treatment effect. Therefore, when H_0 is true:

$$F = \frac{MS_{BG}}{MS_W} = \frac{error\ variance}{error\ variance}$$

The sampling distributions used to test hypotheses are developed under the assumption that the null is true. For the one-way ANOVA, this means $\mu_1 = \mu_2 = \mu_3 = \mu_k$. The F sampling distribution for a one-way ANOVA is theoretically created in a similar fashion to the t sampling distribution. It involves establishing a sample size for each of the various populations to be sampled, then a sampling of those populations, followed by the generation of an F value (via the calculation of a MS_{BG} and MS_W), a replacement of the sampled scores, and then a repeating of this process a near infinite number of times. The resulting distribution of F values have the following features:

1) Since all measures of variability are positive, there are no negative F values. In fact, the value of 0 is a boundary which serves to create a positively skewed distribution of F values.
2) Since null F distributions only measure random factors, the two measures of variability (MS_{BG}, MS_W) are likely to be similar, thus creating a distribution around the value of 1.

When testing a null hypothesis, the critical value for a sampling distribution with a given number of groups and given number of participants can be found. All that is needed are the degrees of freedom. Figure 11.5 depicts two F

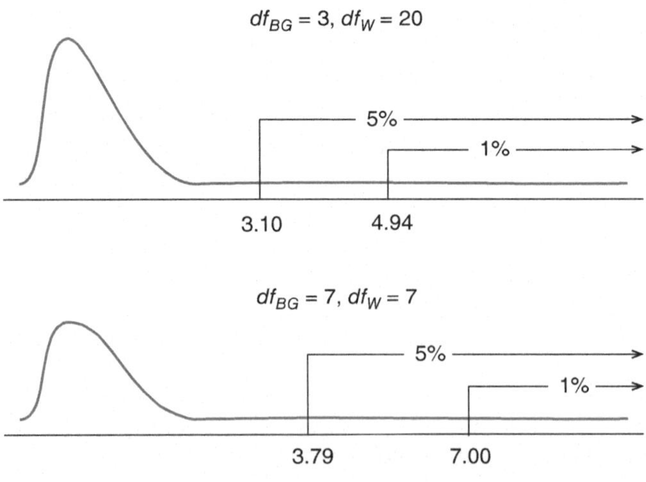

Figure 11.5 Two F distributions. One with $df_{BG} = 3$, $df_W = 20$, another with $df_{BG} = 7$, $df_W = 7$.

distributions based on different sets of degrees of freedom. Note that as the degrees of freedom change, the shape of the *F* distribution changes, and the critical values marking the ends of the distribution change.

As we examine Figure 11.5, keep in mind that these are *F* distributions obtained when the H_0 is true. Where the *F* value obtained from our calculations falls on the distribution marks how well it fits with the null set of *F* values. If the obtained *F* falls in the right tail of the distribution, say, beyond the point marking the outermost 5%, that *F* does not fit well with the set of expected *F*'s if the null is true. Either the null is true, and sampling error has given rise to an unlikely large *F* value (a Type I error has occurred), or the null is false and MS_{BG} is being influenced by primary variance (i.e. a treatment effect). In this situation, we are to reject the null. We have found statistical evidence that it is false.

11.6.2 Using the *F* Table

The *F* table (Table A.5) provides critical values to determine if an *F* ratio falls in the tail on the right-hand side, the rejection region. The following is a portion of the table. The boldface values in the body of the table are used when $\alpha = 1\%$. The lightface values are used when $\alpha = 5\%$.

.05 (lightface) and .01 (boldface) α levels for the F distribution

Degrees of freedom (denominator)	Degrees of freedom (numerator) ··· 3	4	5 ···
⋮			
19	3.13 **5.01**	2.90 **4.50**	2.74 **4.17**
20	3.10 **4.94**	2.87 **4.43**	2.71 **4.10**
21	3.07 **4.87**	2.84 **4.37**	2.68 **4.04**
⋮	⋮	⋮	⋮

To find the relevant critical value, we must use both the degrees of freedom from the numerator, df_{BG}, and the denominator of the *F* ratio, df_W. The corresponding df_{BG} value is found in the row at the top of the table, and the corresponding df_W is found down the lefthand column of the table. When testing the null hypothesis with an *F* ratio where $df_{BG} = 4$ and $df_W = 20$, the critical values for an alpha of 5% and 1% are 2.87 and 4.43, respectively.

Let us test a null hypothesis with the *F* ratio found in the worked example using an alpha of .05. The obtained *F* was 46.74. The $df_{BG} = 2$ and $df_W = 9$. Referring to the F table (Table A.5), the critical value is stated as 4.26. Since the obtained F value is larger than the critical value of 4.26, we reject the null hypothesis that $\mu_1 = \mu_2 = \mu_3$.

11.7 The ANOVA Summary Table

Table 11.2 is one customary way to summarize the results of an ANOVA. The values in the table are taken from the worked example. Most statistical software programs generate tables that look similar. The ANOVA summary table presents the most relevant information from the F value calculations. It will also be a resource for values needed for secondary analyses. A close inspection of the table reveals that the MS values result from dividing an SS value by its corresponding df. Furthermore, the F value is the MS_{BG} divided by the MS_W. Finally, the p value represents the likelihood of the obtained F value occurring if the null hypothesis is true.

It is important to note that the one-way ANOVA does not require each level of the independent variable to contain the same size n. The formulas **still** work, even if the ns are not all equal. However, to maximize power, it is best to have sample sizes that are roughly equal.

Table 11.2 An ANOVA summary table.

Source of variation	*SS*	*df*	*MS*	*F*	*p*
Between-groups	86	2	43	46.74	<.05
Within-groups (error)	8.25	9	.92		
Total	94.25	11			

11.8 Measuring Effect Size

The F test provides information about evidence of a difference between means. A large F ratio indicates that the differences between sample means are unlikely to have occurred by chance. However, the size of the F ratio concerns the *certainty* of an effect, not necessarily the size of the effect. Studies with high power, for instance, may generate very large Fs even when the treatment effect is quite modest.

As with t tests in previous chapters, we need to use a different statistic to measure the size of an effect. Over the years, statisticians have developed several measures of effect size. We will look at two of them. The first statistic offered here is a measure of the estimated magnitude of the treatment effect in the population,

omega-squared (ω^2). This statistic is easy to calculate, requiring only values from the ANOVA summary table.

Formula for omega-squared, ω^2

$$\varpi^2 = \frac{SS_{BG} - df_{BG}(MS_W)}{SS_T + MS_W} \quad (11.7)$$

Using the values from Table 11.3, what is ω^2?

$$\varpi^2 = \frac{86 - 2(0.92)}{94.25 + 0.92} = \mathbf{0.88}$$

The interpretation of $\omega^2 = .88$ is that 88% of the variation among the scores is accounted for by the levels of the independent variable. We can think of it as the ratio of primary variance (or treatment variance) to the total amount of variation in the study (primary variance + secondary variance). Is 88% a large treatment effect? Yes, very large! Some researchers (e.g. Cohen 1977) have created guidelines for interpreting the size of ω^2; however, these are rather arbitrary designations. We simply need to understand that as ω^2 increases, so does the effect size.

A more commonly used measure of effect size is **eta-squared (η^2)**. Even though ω^2 generates a more accurate measure, η^2 is currently preferred by researchers. As the worked example demonstrates, η^2 tends to overestimate the effect size when compared to ω^2.

Formula for eta-squared, η^2

$$\eta^2 = \frac{SS_{BG}}{SS_T} \quad (11.8)$$

Using the values from Table 11.3, what is η^2?

$$\eta^2 = \frac{86}{94.25} = \mathbf{.91}$$

In an age of big data, where statistical tools have tremendous power to find even small effects, it is important to record effect size as well as an *F* value. The certainty of an effect (*F* value) is not the same thing as the size of the effect (ω^2 or η^2). With massive numbers of participants, we can have high certainty that an effect exists and yet wonder whether it is meaningful.

11.9 Locating the Source(s) of Significance

After rejecting a null hypothesis, the researcher has statistical evidence suggesting that at least two of the means have come from different populations. That, however, is all that is known. It is hard to imagine a situation in which a researcher would not want to discover *which* groups appear to be different from one another. *After* a null is rejected, follow-up comparisons of sample means can be made. Sometimes, these comparisons are referred to as "protected t tests," because the initial F result has protected them from inflating alpha. The topic of making **multiple comparisons** is very complicated. Some of the issues involved are *how many* comparisons are to be made, *when* the decision is made to make comparisons (either beforehand, termed **a priori tests** or **planned comparisons,** or after the data has been gathered, so-called **post hoc** or **a posteriori tests**), and *how conservative* to make the rejection region. Unfortunately, even cursory coverage of this topic is beyond the scope of this book; the interested reader is referred to any number of advanced statistics books or websites, Cohen (2013) being one example. As a result, we will only be introduced to two of the more common and general purpose secondary analyses, **Tukey's *honestly significant difference (HSD)* Test** and **Fisher's *least significant difference (LSD)* test.**

One of the most used follow-up tests is Tukey's *HSD*. (Spotlight 11.2 provides information about John Tukey.) This test allows us to compute a single value, called the *honestly significant difference*, or *HSD*, to use as the minimal difference between any two-group means (provided all groups have an equal n). Any difference between two means that exceeds Tukey's *HSD*, is statistical evidence the means came from different populations. Here is the formula:

Formula for Tukey's *HSD*

$$HSD = q\sqrt{\frac{MS_W}{n}} \quad (11.9)$$

where

q = the studentized range statistic (Table A.6)
n = the number of scores in each group (must be the same)

Spotlight 11.2 John Wilder Tukey

John Tukey (1915–2000) was born in New Bedford, Massachusetts. He was home-schooled by educator parents who responded to his childhood questions not with answers but with clues and follow-up questions designed to help him solve his own problems (McCullagh 2003). This philosophy produced a remarkable student, resulting in two degrees from Brown University in chemistry and a PhD in mathematics from Princeton in 1939, where he stayed on as a professor upon graduation.

During World War II, he decided to serve the war effort in the Fire Control Research Office (think "artillery fire"), where he tackled many mathematical issues related to ballistics, gun and artillery control, and range firing (e.g. Sande 2001). After the war, he stayed involved in several government projects, including the enrichment of uranium and the development of the U-2 spy plane.

Concurrent with his academic career, he worked for AT&T Bell Laboratories where he, among other contributions, created several neologisms, including "bit" for binary digit, and "software" as a contrast to hardware (e.g. Leonhardt 2000). Over the course of his career, Tukey was awarded entrance into the prestigious National Academy of Sciences, the National Medal of Science from President Nixon, and the Medal of Honor from the Institute of Electrical and Electronic Engineers.

As an academic, Tukey's contributions were numerous and significant (pun intended!), publishing in the fields of chemistry, mathematics, environmental research, philosophy, and statistics (forming and chairing Princeton's Department of Statistics in 1966) (McCullagh 2003). He also served as a scientific adviser to several presidents and was tapped to be a critical reviewer of influential notable publications such as the *Kinsey Report* (1953) and *Silent Spring* (Carson 1962). His contributions to the field of statistics are numerous. Some of his most notable statistical creations include the box-and-whisker plot, the stem-and-leaf diagram, and the Tukey range test (now known as Tukey's *HSD* test).

To determine the proper q value, we must know the number of groups in the study (k), the degrees of freedom for S_W (df_W), and the desired alpha level.

We can use the data from Table 11.2 to determine the *HSD*. Given that $k = 3$, $df_W = 9$, and $\alpha = .05$, $q = 3.95$ (see Table A.6),

$$HSD = q\sqrt{\frac{MS_W}{n}} = 3.95\sqrt{\frac{0.92}{4}} = \mathbf{1.89}$$

The group means from Table 11.2 are $M_1 = 4.75$, $M_2 = 8.75$, and $M_3 = 2.25$. Applying Tukey's *HSD* as our post hoc analysis tool directs us to conclude that statistical evidence for population differences exist between all three groups: groups 1 and 2 ($4.75 - 8.75 = -4$), groups 1 and 3 ($4.74 - 2.25 = 2.5$), and groups 2 and 3 ($8.75 - 2.25 = 6.5$). (Interpret mean differences in absolute value terms when comparing to the *HSD*.)

The second posttest we will explore is Fisher's *LSD*. The formula for this test is somewhat similar to previous t tests. However, even though only two means are contrasted in Formula 11.10, the error term used, MS_W, is derived from *all* the groups in the study. The result is a more stable estimate of the population variance.

This t test, consequently, has more power than an independent samples t test when multiple comparisons are performed.

Formula for Fisher's *LSD* test

$$t = \frac{M_i - M_j}{\sqrt{MS_W\left(\frac{1}{n_i} + \frac{1}{n_j}\right)}} \quad (11.10)$$

---

where

M_i, M_j = the means for the two groups being compared
n_i, n_j = the number of scores in each of the two groups being compared

We can use this test, if we choose, to make all possible comparisons. Notice that, although the means change with each comparison, if the *n*s are the same, the denominator will not change. However, if we have different n values, unlike Tukey's *HSD*, Fisher's *LSD* can still be run. The t_{crit} value is based on $N - k$ degrees of freedom and is found in the t table, *not* the F table. Using the data from Table 11.1, Table 11.3 provides an example of how the *LSD* t tests are conducted. Each comparison uses an alpha level of .05.

The results of the protected t tests show that statistical evidence has been found suggesting differences for all three comparisons, $t(9) = -6.56$, $p < .05$; $t(9) = 4.09$, $p < .05$; and $t(9) = 10.66$, $p < .05$, respectively. The three t test comparisons have allowed us to locate the source(s) of significance detected by the F test.

The t tests are based on the data found in Table 11.1.

Table 11.3 Protected t tests following a significant ANOVA.

Group 1	Group 2	Group 3
$M_1 = 4.75$	$M_2 = 8.75$	$M_3 = 2.25$
$n_1 = 4$	$n_2 = 4$	$n_3 = 4$

$MS_w = .92$ (as determined in the worked problem)

$$M_1 \ vs.\ M_2 = \frac{4.75 - 8.75}{\sqrt{.92\,(1/5 + 1/5)}} = \frac{-4}{\sqrt{.37}} = \frac{-4}{.61} = -\mathbf{6.56}\ (p < .05)$$

$$M_1 \ vs.\ M_3 = \frac{4.75 - 2.25}{\sqrt{.37}} = \frac{2.5}{.61} = \mathbf{4.09}\ (p < .05)$$

$$M_2 \ vs.\ M_3 = \frac{8.75 - 2.25}{\sqrt{.37}} = \frac{6.5}{.61} = \mathbf{10.66}\ (p < .05)$$

df for t's $= df_W = N - k = 12 - 3 = 9$; $\alpha = .05$; $t_{crit} = \pm 2.26$

11.10 Formal Presentation of Findings

Every journal has a standard format for reporting results in an article. Most social and behavioral science journals rely on the format offered by the American Psychological Association (2020). When formally reporting the rejection of a null hypothesis, we must include the df_{BG}, df_W, F value, and alpha level used to make our decision. For instance, "Statistical evidence suggests the type of drug administered influenced its effectiveness, $F(3, 69) = 9.59$, $p < .05$. (Follow-up t tests would be presented in the same manner as described in Chapters 8 and 9.) A failure to reject might read, "There was no statistical evidence to suggest one type of drug was more effective than the others, $F(3, 69) = 1.59$, *n.s.*" Measures of effect size can be added at the end of the sentence when a null hypothesis has been rejected.

Many other principles common to the proper reporting of all types of statistical findings were first laid out in Section 8.8. For instance, notice once again that the critical value for the test is usually not presented.

Summary

A one-way ANOVA can be used when a study has two or more levels of one independent variable and participants are placed into independent groups (not repeatedly measured). When conducting an ANOVA, it is assumed that the samples are representative of the populations from which they come; these populations are normally distributed and have roughly equivalent variances. Minor violations of the statistical assumptions (normality and homogeneity of variance) are not serious since the F test is considered to be robust to these assumptions. However, it is essential that the sample data be representative of the population, that each participant's score on the dependent variable be independent of every other participant's score, and that the data be measured on an interval or ratio scale.

The variation among scores within each group is due to random factors: individual differences and experimental error. The influence of random factors creates error variance (also called secondary variance). Mathematically, it is referred to as *within-group variance. Between-group variance* is caused by error variance as well as the effect of the treatment variable (also called primary variance). To determine if the null hypothesis can be rejected, the between-group variance (MS_{BG}, primary and secondary variance) is compared in ratio form with the within-group variance (MS_W, secondary variance only). This ratio is called an F statistic.

If there is no primary variance, the resulting ratio is close to 1. However, as primary variance exists, it only inflates the numerator (MS_{BG}), making the resulting ratio increasingly larger than 1. If the resulting ratio falls into the outermost 5% of the null F sampling distribution, evidence of a treatment effect has been found.

The sampling distribution that is used to test the significance of the F ratio is determined by the df_{BG} $(k-1)$ and df_W $(N-k)$. Although the shape of the F distribution will vary depending on sample sizes and the number of conditions, the F distribution is always positively skewed.

The size of an F ratio does not indicate the magnitude of a treatment effect. To assess effect size, omega-squared (ω^2) or eta-squared (η^2) can be used. To locate the source(s) of a statistically significant F test, Tukey's *HSD*, Fisher's *LSD*, or any number of other tests can be used to make comparisons between pairs of group means.

Microsoft® Excel and SPSS®

Information regarding how to use Microsoft® Excel and SPSS® to assist in performing one-way ANOVAs can be found in Appendix C.

Key Formulas

Computational formula for SS_{BG}

$$SS_{BG} = \frac{(\Sigma X_1)^2}{n_1} + \frac{(\Sigma X_2)^2}{n_2} + \cdots + \frac{(\Sigma X_k)^2}{n_k} - \left[\frac{(\Sigma X)^2}{N}\right] \quad (11.1)$$

Computational formula for SS_W

$$SS_W = \Sigma X^2 - \left[\frac{(\Sigma X_1)^2}{n_1} + \frac{(\Sigma X_2)^2}{n_2} + \ldots + \frac{(\Sigma X_k)^2}{n_k}\right] \quad (11.2)$$

Computational formula for SS_T

$$SS_T = \Sigma X^2 - \frac{(\Sigma X)^2}{N} \quad (11.3)$$

Calculating MS_{BG}, MS_W, and the F ratio

$$MS_{BG} = \frac{SS_{BG}}{df_{BG}} \quad (11.4) \qquad MS_W = \frac{SS_W}{df_W} \quad (11.5) \qquad F = \frac{MS_{BG}}{MS_W} \quad (11.6)$$

Formula for omega-squared, ω^2

$$\varpi^2 = \frac{SS_{BG} - df_{BG}(MS_W)}{SS_T + MS_W} \quad (11.7)$$

Formula for eta-squared, η^2

$$\eta^2 = \frac{SS_{BG}}{SS_T} \quad (11.8)$$

Formula for Tukey's *HSD*

$$HSD = q\sqrt{\frac{MS_W}{n}} \quad (11.9)$$

Formula for Fisher's *LSD* test

$$t = \frac{M_i - M_j}{\sqrt{MS_W\left(\frac{1}{n_i} + \frac{1}{n_j}\right)}} \quad (11.10) \qquad (t_{crit} \text{ uses } df_w = N - k)$$

Key Terms

Analysis of variance (ANOVA)
Between-group variation
Individual differences
Error (or secondary) variance
Mean square between
***F* Distribution**
Eta-squared
A priori tests (or planned comparisons)
Tukey's *HSD*
One-way ANOVA
Treatment (or primary) variance
Experimental error
Within-group variation
Mean square within
Omega-squared
Multiple comparisons
Post hoc (or a posteriori) tests
Fisher's *LSD*

Exercises

1 What is the acronym for "analysis of variance?"

2 Use one-way ANOVAs for designs with ____ or more groups that are ____ of each other.

3 For designs with three or more conditions, why are ANOVAs preferred to *t* tests?

4 What term describes the variance from participants being in different conditions?

5 What are the three sources of variation that can account for group mean differences?

6 What term describes the variance found within-groups or conditions?

7 What are the two sources of variation that can account for within-group variability?

8 The symbol for between-group variance is _____; the symbol for within-groups is _____.

9 How does the MS_{BG}/MS_W ratio detect the presence of primary variance?

10 Describe, in brief terms, the shapes of the null F distributions.

11 The one-way ANOVA assumptions are the same as the _______ test.

12 The one-way ANOVA is robust to violations of which assumptions?

13 The table used to organize the ANOVA analysis is called the _______.

14 What variance is found in an F ratio numerator when H_0 is correct? When H_0 is incorrect?

15 When can a multiple comparison tool like Fisher's *LSD* be used?

16 *LSD* stands for: _____________.

17 *HSD* stands for: _____________.

18 Select the right answer: In Tukey's test, if the difference between two-group means meets or exceeds the *HSD* value, there *is/is not* evidence of a significant difference.

Work Problems

19 Complete the following ANOVA summary table. Test to see if the null can be rejected.

Source	*SS*	*df*	*MS*	*F*	*p*
Between-groups	280.3	3	______	______	______
Within-groups	______	______	7.28		
Total	527.98	______			

20 Complete the following ANOVA summary table. Test to see if the null can be rejected.

Source	*SS*	*df*	*MS*	*F*	*p*
Between-groups	______	4	1.47	______	______
Within-groups	6.30	______	______		
Total	______	15			

21 A psychologist is comparing different exercise programs for cardiovascular fitness. The fitness measure is resting heart rate after program completion. Twelve college students are randomly assigned to three groups (four per group). Participants in the aerobic condition walk a treadmill for 30 minutes, three times a week. Those in the circuit condition perform exercises on weight machines for 30 minutes. Control condition participants are asked to simply maintain their usual amount of exercise. The resting heart rates of all participants are taken after 10 weeks.

Aerobic	Circuit	Control
65	74	74
62	65	78
56	62	86
60	72	75

a State the null and alternative hypotheses.
Calculate the following values.
b SS_{BG}
c SS_W
d df_{BG}
e df_W
f MS_{BG}
g MS_W
h SS_T
i df_T
j F ratio
k What is F_{crit} when $\alpha = .05$?
l Should we reject or fail to reject the H_0?
m Make an ANOVA summary table.
n Calculate omega-squared.
o Conduct Fisher's *LSD* *t* tests if appropriate and properly report the findings.
p Does this F test allow for a causal interpretation?

22 A political scientist hypothesizes that people from the US Midwest are more politically conservative than those from the two coasts. Five participants between the ages of 21 and 60 are randomly selected from the US West, Midwest, and East ($N = 15$). A questionnaire measuring conservatism is administered, with higher scores reflecting greater conservatism.

West	Midwest	East
3	9	4
6	15	9
2	9	1
4	4	2
3	6	3

a State the null and alternative hypotheses.
Calculate the following values.
b SS_{BG}
c SS_W
d df_{BG}
e df_W
f MS_{BG}
g MS_W

h SS_T
i df_T
j F ratio
k What is F_{crit} when $\alpha = .05$?
l Should we reject or fail to reject the H_0?
m Make an ANOVA summary table.
n Calculate eta-squared.
o Conduct Tukey's *HSD* and interpret.
p Does this F test allow for a causal interpretation?

23 A clinical psychologist is interested in evaluating treatments for panic attacks. The number of reported attacks during a six-month program of treatment is used as the dependent variable. Fifteen clients suffering from a panic disorder are randomly assigned to one of three treatment conditions, breathing, medication, and control:

Breathing	Medication	Control
16	12	9
22	15	12
15	13	16
9	18	18
13	12	10

a State the null and alternative hypotheses.
Calculate the following values.
b SS_{BG}
c SS_W
d df_{BG}
e df_W
f MS_{BG}
g MS_W
h SS_T
i df_T
j F ratio
k What is F_{crit} when $\alpha = .05$?
l Should we reject or fail to reject the H_0?

m Make an ANOVA summary table.
n Provide an interpretation of the results.
o Calculate omega-squared and interpret.
p Conduct protected t tests if appropriate and interpret.
q Does this F test allow for a causal interpretation?

24 We observe that people seem to be happier when they are wearing a new article of clothing. We would like to test whether happiness depends on the particular type of new clothing worn. To test this, we provide a random sample of five classmates with new t-shirts and five others with new shoes and instruct them to wear the articles of clothing all day. At the end of the day, we ask these participants to rate their happiness. A control group of five classmates is also asked for this self-rating, but without the experimental manipulation. (Assume ratings are measured on an interval scale.) Higher scores indicate greater happiness. Conduct a one-way ANOVA on these data.

New T-shirt	New shoes	Control
6	8	4
6	7	6
7	9	5
5	7	3
8	10	5

a State the null and alternative hypotheses.
Calculate the following values.
b SS_{BG}
c SS_W
d df_{BG}
e df_W
f MS_{BG}
g MS_W
h SS_T
i df_T
j F ratio
k What is F_{crit} when $\alpha = .05$?
l Should we reject or fail to reject the H_0?

m Make an ANOVA summary table.
n Calculate eta-squared.
o Conduct Tukey's *HSD* and interpret.
p Provide a proper reporting of the findings.
q Does this F test allow for a causal interpretation?

25 A psychologist is interested in comparing antidepressant medication and biofeedback. Forty-five headache sufferers are randomly assigned to one of three conditions (a control group is added). During the treatment, the number of headaches is recorded. Conduct an F test ($\alpha = .05$) and use Fisher's *LSD* to make post hoc comparisons, if needed.

Medication	Biofeedback	Control
2	4	5
1	2	7
2	3	8
6	5	10
7	4	8
8	2	2
6	7	8
3	4	8
2	0	2
0	3	5
1	0	1
2	5	6
0	1	2
4	2	1
5	3	8

26 Sam believes there is a relationship between a person's wardrobe and the type of major pursued at university. The theory states that science majors attract people who are not particularly interested in self-presentation. Those students who are interested in the arts, however, have a greater interest in self-presentation. Humanities majors fall in between. Sam hypothesizes that the number of shoes brought to school might be a way to start to investigate the theory. Below are the data: a "hard" science, "social" science, humanity, and art. To try to control for any gender effects, Sam has only recruited biological males. Use α = .05 to test the null hypothesis. If warranted, use Tukey's HSD when making post hoc comparisons.

Chemistry	Psychology	History	Theater
2	5	5	7
3	2	7	4
2	5	8	9
6	5	10	8
4	4	8	9
5	8	4	5
6	7	8	11
3	4	9	4
2	9	2	6
4	5	5	11
2	3	6	7
4	5	6	9
3	1	3	5
4	2	9	10
7	6	8	9
3	5	5	8
4	4	7	5

12

Two-Way Analysis of Variance

12.1 The Research Context

12.1.1 Factorial Designs

In previous chapters, research designs were presented with only one factor, or in experimental terms, one independent variable.[1] However, we do not live in a "one-variable world." Our behavior is constantly affected by the combined influence of two or more variables. For this reason, many research designs allow researchers to evaluate the interplay between several factors and measure their combined influence on behavior. **Factorial (or complex) designs** are a blend of two or more single-variable designs. This allows one to learn about more than one independent variable, thereby saving the time, expense, and effort required to conduct separate, single-variable experiments. More importantly, complex designs allow information to be gained about combined effects on a dependent variable. When factorial designs are used, a prediction is typically made regarding a combined effect. The researcher believes the effect of one independent variable will be altered depending on the value of the second independent variable. An example serves to illustrate this point.

If we saw a person in trouble, would we come to their aid? As we think about this question, we may answer, "it depends." It may depend on how many people are present, how dangerous the situation is, or even what the person in trouble looks like (Latané and Darley 1970). "It depends" qualifies our answer, which, in effect, says our action is determined by the joint presence of certain conditions. For instance, we may only help if it is not too dangerous *and* we are the only person around to offer assistance. In this example, our behavior depends on the joint occurrence of two or more variables. It might also be true that we would help if

1 As in previous chapters, concepts introduced here will be presented from an experimental perspective, even though two-way ANOVAs can also be used to analyze data gathered in nonexperimental designs.

Essentials of Behavioral and Social Science Statistics, Third Edition.
K. Paul Nesselroade, Jr. and Laurence G. Grimm.
© 2025 John Wiley & Sons, Inc. Published 2025 by John Wiley & Sons, Inc.
Companion website: https://www.wiley.com/go/Essentials_of_Behavioral_and_Social_Science_Statistics3e

there were no one else around *or* if the person in need was dressed like us, but not if *both* conditions were true, thinking it might be a trap. It is one thing to learn how one variable (e.g. the presence of others) influences our behavior, but it is quite another to look at the effect of one variable among others. The opportunity to look at combined effects is the great attraction of factorial designs.

12.1.2 Examples of Factorial Designs and Cell Notations

A system of terminology and notation is employed to identify the features of a factorial design (see Examples 12.1 and 12.2). Each design is broken up into cells, each **cell** corresponding to a unique combination of treatment conditions and each containing its own group of participants. Each independent variable in a factorial design is referred to as a **factor** (thus the term "factorial design"); this term is more generally appropriate since many designs involve nonexperimental variables. When there are two independent variables, one variable is often designated Factor A and the other is Factor B. A cell designation of A_1B_3 refers to the group of participants who are in the first level of Factor A and the third level of Factor B. In Example 12.1, A_1B_3 refers to the participants who have an anxiety disorder with panic attacks (first level of Factor A) and receive Imipramine (third level of Factor B). By convention, when the design is depicted with two rows and three columns, it is referred to as a 2 × 3 (pronounced "2 by 3") factorial design (rows × columns). If one of the independent variables has more levels than the other independent variable, the variable with more levels is usually placed at the top of the figure, thereby creating more columns than rows. There is no special logic for designating the letters A and B to the two factors.

► **Example 12.1** A psychiatrist is interested in comparing the effectiveness of three different psychopharmacological treatments for anxiety with two types of patients, those with and without panic attacks. Factor A is *Patient Type*, Factor B is *Treatment*, and the dependent variable is the participants' self-reports of anxiety. This is a 2 × 3 factorial design.

		Factor B (treatment)		
		Valium	Alprazolam	Imipramine
Factor A (patient type)	Anxiety with panic	A_1B_1	A_1B_2	A_1B_3
	Anxiety without panic	A_2B_1	A_2B_2	A_2B_3

◄

► **Example 12.2** A social psychologist hypothesizes that two advertising techniques will be differentially effective, depending on the product. Factor A is *advertising technique*, Factor B is *product*, and the dependent variable *is attitude toward the product*. This is a 2 × 2 factorial design.

		Factor B (product)	
		Autos	Cell phones
Factor A (advertising technique)	Image appeal	A_1B_1	A_1B_2
	Technical information	A_2B_1	A_2B_2

◄

12.1.3 Main Effects and Interactions

To illustrate how two single-independent variable experiments can be combined into one factorial design, consider two hypothetical studies on memory.

In Experiment 1, a psychologist hypothesizes that memory will be better if a person is in a good mood when attempting to *recall* a list of words. On the first day of the study, all participants are given a list of words to memorize. A test for recall is conducted on the second day, with half of the participants tested immediately after reading a mood-elevating passage. The other half is tested immediately after reading a mood-depressing passage. If the psychologists' hypothesis is correct, participants who just finished the mood-elevating passage should remember more words than participants who just finished the depressing passage (see Experiment 1, Table 12.1).

In Experiment 2, a researcher hypothesizes that memory will be best if an individual is in a good mood while *memorizing* a list of words. This means the mood manipulation (reading a happy or depressing passage) is accomplished on the first

Table 12.1 Two separate between-group designs, each with one independent variable.

Day 1	**Day 2**
Experiment 1: Mood state during recall	
Both groups memorize list	Group 1: Recall while happy
	Group 2: Recall while sad
Experiment 2: Mood state during memorization	
Group 1: Memorize while happy	Both groups recall list
Group 2: Memorize while sad	

day, just prior to having the participants memorize the word list. On the second day, all participants are asked to recall as much of the list as they can. This experiment is also depicted in Table 12.1.

Now consider a state-dependent theory of memory in which it is predicted that memory will be facilitated when a person recalls a list of words while in the same emotional state that existed when the list was memorized. Neither of the experiments in Table 12.1 can test this hypothesis since the prediction requires a manipulated mood state during memorization *and* during recall. However, by combining the two studies, we can obtain all the information gathered from each single-variable experiment, as well as the interactive effects of the two independent variables.

An effect found among the conditions of one independent variable, independent of the influence of another independent variable, is called a **main effect**. An effect produced by the combination of independent variables is termed an **interaction**. The **two-way ANOVA** is an analytical procedure that allows us to investigate both *main effects* and *interactions*.

The two independent variables are Mood During Memorization and Mood During Recall. Each independent variable has two levels: Happy and Sad. The formal name of this design is a *Completely Randomized 2 × 2 Factorial Design*. It is completely randomized since participants are randomly assigned to the treatment conditions. It is factorial because there is more than one factor. It is 2 × 2 since there are two levels of each factor.[2] Table 12.2 illustrates this design.

Table 12.2 A 2 × 2 factorial design with significant interaction and no significant main effects.

		Factor *B*: Mood during recall		
		Happy	*Sad*	
Factor *A*: Mood during memorization	*Happy*	(1) 20	(2) 10	$M_{A_1} = 15$
	Sad	(3) 10	(4) 20	$M_{A_2} = 15$
		$M_{B_1} = 15$	$M_{B_2} = 15$	

The numbers in the cells represent group means, the numbers in parentheses are cell numbers.

2 Some descriptions may use the term "full factorial." This term means that every possible combination of the levels of the factors is realized. Since this is almost always the case, the simpler term "factorial" is typically used.

12.1.4 Using Diagrams and Graphs to Examine Main Effects and Interactions

Viewing diagrams and graphs of data from factorial designs allows us to speculate about the presence of main effects and interactions. Of course, we *must* conduct statistical analyses to discover if there really is statistical evidence for these effects. However, it can be helpful to graph group means before performing a statistical analysis, and researchers often display graphs in their publications as a visual complement to a verbal summary of study results.

Throughout this section, we will stay with this 2 × 2 memory experiment. Cell means will be altered in each successive example so that we can learn what the diagrams and graphs look like as the experimental results change. A word of caution: If we read the text without referring to the relevant diagrams and graphs, we can easily get lost. We need to bounce back and forth between the illustrations and the text to understand the discussion.

In Table 12.2, the numbers within each cell are group means, representing the average number of words recalled by the participants in that group. The means in the margins (called marginal means) represent the average number of words recalled for participants across that condition (identified by row or column). Let us first consider each independent variable separately. When addressing only Factor *B*, the design can be simplified by collapsing the conditions of Factor *A*, leaving us with just two cells (Factor *B* Happy; Factor *B* Sad) in the design, and the question is, "What is the effect of mood state during recall on the number of words remembered?" The average of collapsed cells 1 and 3 vs. 2 and 4 is found in the marginal means and reveals no difference ($M_{B1} = 15$ and $M_{B2} = 15$). Comparing the levels within one independent variable allows us to determine if there is a main effect for that factor. Since there is no evidence of a main effect for Factor *B*, we might be tempted to conclude that the mood state present *during recall* has no effect on the number of words remembered.

The same logic can be applied when considering Factor *A*. Now the design is simplified by collapsing the conditions of Factor *B*, leaving us with just two cells in the design (Factor *A* Happy; Factor *A* Sad), and the question is, "What is the effect of mood state during memorization on later recall?" The average of collapsed cells 1 and 2 vs. 3 and 4 is found in the marginal means and reveals no difference ($M_{A1} = 15$ and $M_{A2} = 15$). There is no evidence for a main effect for Factor *A*. Once again, we might be tempted to conclude that the mood state present *during memorization* has no effect on the number of words remembered. Bear in mind that these data are hypothetical. The marginal means in Table 12.2 have purposely been presented as identical to simplify the example. With real data, the marginal means are rarely identical, even when there is no main effect. Nevertheless, if each independent variable were examined in separate, single-variable studies, each study would be a washout. However, by using a factorial design, a third question is possible: "Does the mood state during memorization interact with the mood state during recall?" Although there are no main effects in this study, it would appear there is an interaction. By examining the pattern of means within the cells, we may conclude that recall is facilitated when there is congruence between Mood During Memorization and Mood During Recall. It does not seem to matter what the moods are as long as they are similar. These results lend support to the proposed state-dependent theory of memory.

When speculating about an interaction, a graph of cell means can be helpful. Figure 12.1 is a graph of the means presented in Table 12.2. When graphing means from a factorial design, the levels of one independent variable are indicated on the horizontal axis. (Which variable to select is arbitrary.) In Figure 12.1, Mood During Recall has been placed on the horizontal axis. The dependent variable is represented as means on the vertical axis. The second independent variable is placed *within* the graph. In the example, the second independent variable is Mood During Memorization. Since there are two levels of this variable, there will be two lines on the graph: one for each level of the factor that is embedded into the graph. Do not make the mistake of thinking that each line depicts each independent variable.

In Figure 12.1, the line drawn from the upper left to the lower right of the graph represents the *Happy* Mood During Memorization level of Factor *A*. The line connects the means of two groups: Happy During Memorization/Happy During Recall (Group 1) and Happy During Memorization/Sad During Recall (Group 2).

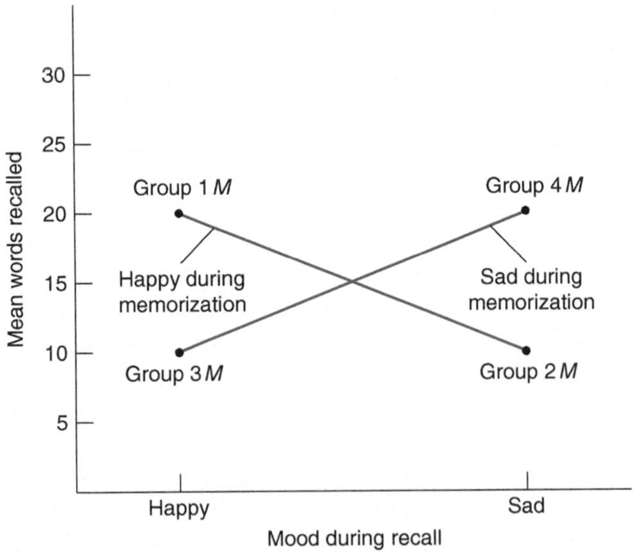

Figure 12.1 A graph of the group means in Table 12.2 suggests an interaction but no main effects.

Carefully examine the means in Table 12.2 and the graph in Figure 12.1. Match up the cell means from the table with the points on the graph. See how the graph lines follow the cell means from the Mood During Memorization perspective.

The data in Table 12.2 suggest no main effects but a potential interaction. On a graph, *an interaction is revealed when the lines are not parallel.* Figure 12.1 reveals a potential interaction; the lines are not only nonparallel, but they also cross. The lines, however, do not have to cross. How nonparallel the lines need to be to register an interaction depends on the statistical power in each particular study. In general, however, the more unparallel they are, the more likely it is that evidence for an interaction will be found when the statistical analysis is performed.

Up to this point, data have been used to show an interaction but no main effects. We will now use the same experiment, adjust the cell means, and illustrate various combinations of main effects and interactions.

The data in Table 12.3 show a main effect for Factor *B* but no main effect for Factor *A* or interaction. Although the marginal means for Factor *A* are not identical, they are rather close; most likely, there is not a main effect. Factor *B* is another matter. The marginal means of 29 and 11 are rather different, indicating that a main effect would likely be found upon analysis.

The interpretation of the presumed experimental results depicted in Table 12.3 would suggest memory is greatest when participants are happy during recall (main effect for Factor *B*). The Mood During Memorization does not seem to matter (no main effect for Factor *A*). Further, a congruency of mood states during memorization and recall does not seem to matter (no interaction effect).

Figure 12.2 graphically displays the means in Table 12.3. Notice that the lines are parallel, reflecting the absence of an interaction. When examining a graph for a main effect, look to see if one of the lines is higher than the other. If so, then there may be a main effect.

In Figure 12.2, Mood During Memorization is placed on the horizontal axis. Why the switch from Figure 12.1 to 12.2? By changing the variable on the horizontal axis, the main effect is shown more clearly. Figure 12.3 also illustrates the

Table 12.3 A 2 × 2 factorial design with one main effect (Factor *B*) and no interaction.

		Factor *B*: Mood during recall		
		Happy	*Sad*	
Factor *A*: Mood during memorization	*Happy*	(1) 30	(2) 12	$M_{A_1} = 21$
	Sad	(3) 28	(4) 10	$M_{A_2} = 19$
		$M_{B_1} = 29$	$M_{B_2} = 11$	

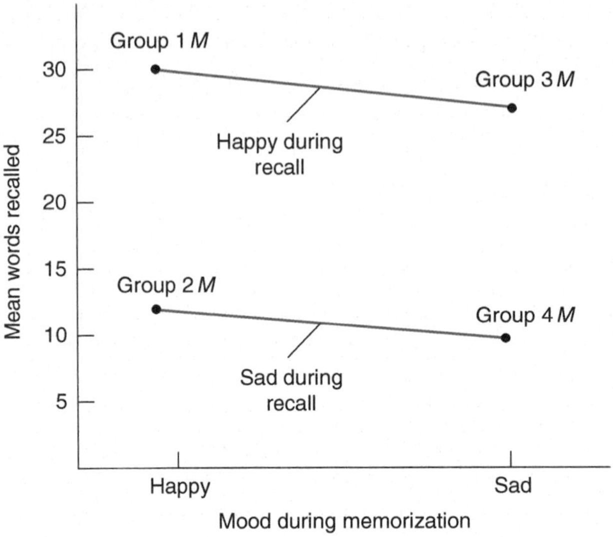

Figure 12.2 A graph of means from Table 12.3. There is one main effect but no interaction.

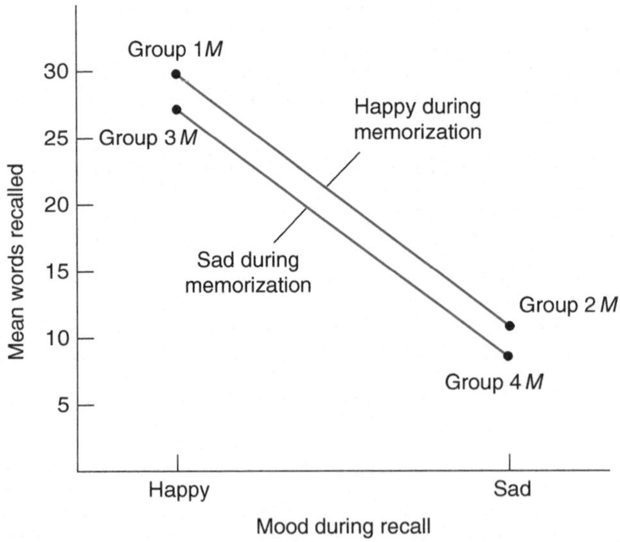

Figure 12.3 Another graph of means from Table 12.3, however, Mood During Recall is on the *X* axis. Both figures reveal evidence of a main effect and no interaction.

means from Table 12.3, but places Mood During Recall on the X axis. The lines are still parallel, indicating the absence of an interaction. The Happy During Memorization line is only slightly higher than the line drawn for Sad During Memorization. However, both means for "Happy" during recall are much higher than the means for "Sad" during recall. Either way, the evidence for a main effect can be seen.

At this point, we will adjust the means to illustrate a main effect for Factor *A*. Table 12.4 is one way to illustrate a main effect for Mood During Memorization.

The pattern of means in Table 12.4 shows that memory is facilitated when participants are happy while memorizing the list of words (main effect for Factor *A*). Memory is not affected by the mood state present during recall (no main effect for Factor *B*), and the combination of mood states during memorization and recall does not seem to matter (no interaction).

In Figure 12.4, the lines are nearly parallel: probably no interaction. The main effect for Mood During Memorization is revealed by the different heights of the lines. The Happy During Memorization line is highest, suggesting a main effect for Factor *A*.

Table 12.5 presents yet another analytical scenario. Here, we find a main effect for Factor *A*, as well as an interaction, but no main effect for Factor *B*.

Table 12.5 shows that mood state during recall does not influence memory (no main effect for Factor *B*). However, the main effect revealed for Factor *A* indicates that memory is enhanced when words are memorized while in a happy mood. Now, examine Figure 12.5. One line is higher than the other, revealing this main effect. Because the lines are not parallel, an interaction is also indicated. When more than one *type* of effect is found, keep this rule in mind: *Always interpret the* ***higher-order effect*** *first, and then interpret lower effects cautiously in light of the initial analysis.* The term *higher-order effect* means the effect involving the greatest number of factors. In a two-way ANOVA, it means we need to interpret an interaction before main effects. One interpretation of this interaction is to

Table 12.4 A 2 × 2 factorial design with one main effect (Factor *A*) and no interaction.

		Factor *B*: Mood during recall		
		Happy	*Sad*	
Factor *A*: Mood during memorization	*Happy*	(1) 30	(2) 28	$M_{A_1} = 29$
	Sad	(3) 20	(4) 20	$M_{A_2} = 20$
		$M_{B_1} = 25$	$M_{B_2} = 24$	

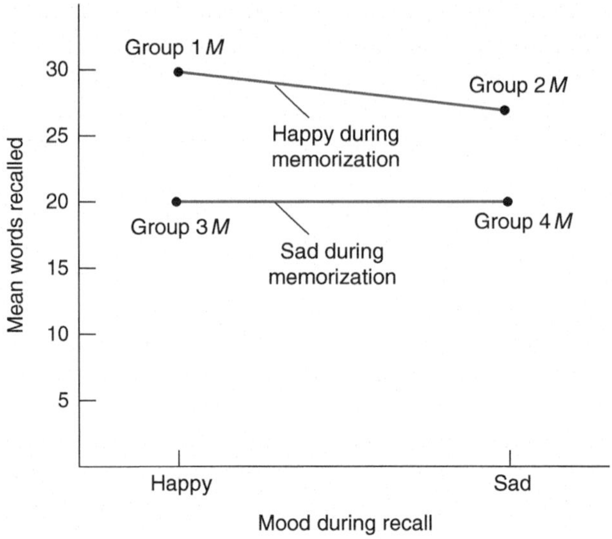

Figure 12.4 A graph of means from Table 12.4. A main effect for Factor *A* but no interaction.

Table 12.5 A 2 × 2 factorial design with one main effect (Factor *A*) as well as an interaction.

		Factor *B*: Mood during recall		
		Happy	*Sad*	
Factor *A*: Mood during memorization	*Happy*	(1) 40	(2) 30	$M_{A_1} = 35$
	Sad	(3) 20	(4) 26	$M_{A_2} = 23$
		$M_{B_1} = 30$	$M_{B_2} = 28$	

suggest that mood congruence between memorization and recall increases memory, but even more so for those in a positive mood.

Refer to Table 12.5. Notice the marginal means for Mood During Memorization (35 vs. 23). That is the main effect. However, the mean in cell 1 (40) is pulling up the marginal mean of 35. The interaction may be *creating* the main effect. The Happy/Happy condition shows the greatest influence on recall. The congruence between the two Sad conditions (cell 4) does not show a similar influence on recall ($M_4 = 26$). Therefore, not only does the Happy/Happy condition contribute to the interaction, but it may also account for the main effect for Factor *A*. In situations like this, we must be very cautious about drawing conclusions regarding main

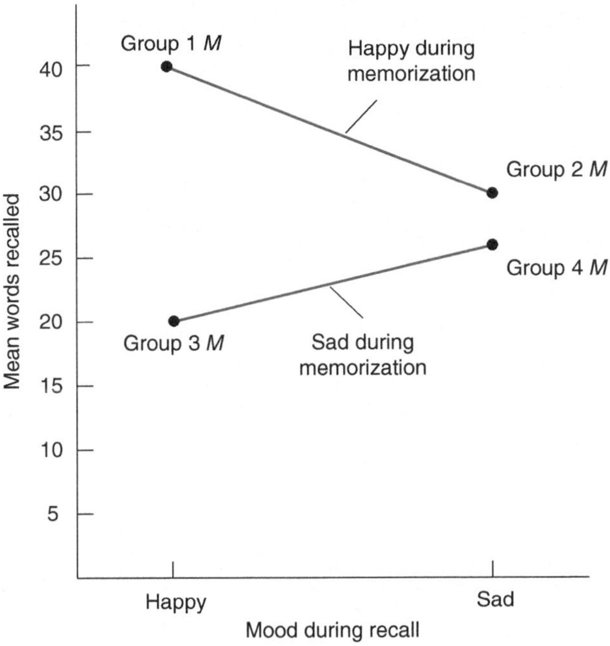

Figure 12.5 A graph of means from Table 12.5. A main effect for Factor *A* and an interaction.

effects. This main effect might be an *artifact* of the higher-order interaction. Main effects that can be explained by an interaction are referred to as *illusory* main effects.

Some methodologists take the extreme position that main effects should not be interpreted when an interaction is found. A more moderate position is to interpret the interaction first and then cautiously interpret any lower-order effects (main effects) within the context of the interaction analysis and the theory used to justify the study.

Our illustrations of some of the outcomes of a 2 × 2 factorial design have not exhausted all possible outcomes. The entire set of possible outcomes is:

1) Factor *A* main effect, no Factor *B* main effect, and no interaction.
2) Factor *B* main effect, no Factor *A* main effect, and no interaction.
3) Main effects for Factors *A* and *B*, no interaction.
4) Main effect for Factor *A*, no main effect for Factor *B*, but an interaction.
5) Main effect for Factor *B*, no main effect for Factor *A*, but an interaction.
6) Main effects for both factors and an interaction.
7) No main effects for either factor but an interaction.
8) No main effects and no interaction.

To recap, a main effect addresses the differences among levels of an independent variable. The number of independent variables is the number of potential main effects. An interaction is the combined influence of two or more independent variables. A significant interaction means that the influence of an independent variable changes based on the level of a second independent variable. Interactions may generate illusory main effects; interpret main effects cautiously.

12.1.5 Factorial Designs with More Than Two Independent Variables

When three independent variables are combined in a factorial design, the number of possible outcomes is increased. Not only are there potential main effects for each independent variable, but more than one interaction can occur. With three independent variables, the investigator can test for four interactions: $A \times B$, $A \times C$, $B \times C$, and $A \times B \times C$. The last interaction is called a three-way interaction; it is a higher-order effect than the two-way interactions. It is possible to design experiments with four- and five-way interactions. However, they are often very difficult to interpret. This text will only examine designs with no more than two factors.

12.2 The Logic of the Two-Way ANOVA

12.2.1 The Null and Alternative Hypotheses

There are three separate sets of hypotheses when conducting a two-way ANOVA, each requiring a separate F ratio to test them.

12.2.2 Main Effect for Factor *A*

The independent variable designated as Factor A has two or more levels. The null hypothesis states that there are no differences between the population means of these *levels*. Symbolically, this is noted as

$$H_0 : \mu_{A_1} = \mu_{A_2} = \mu_{A_k}$$

The subscript k refers to the last level of the factor. The alternative hypothesis states at least one difference between the population means. This can be expressed as

$$H_1 : \text{At least two population means from Factor } A \text{ are not equal}$$

12.2.3 Main Effect for Factor *B*

The hypotheses for Factor B follow the same logic as those for Factor A. They are:

$$H_0 : \mu_{B_1} = \mu_{B_2} = \mu_{B_k}$$

$$H_1 : \text{At least two population means from Factor } B \text{ are not equal}$$

12.2.4 The *A* × *B* Interaction

The null hypothesis for the interaction is that there is no interaction. That is, the effect of Factor *A* is independent of the effect of Factor *B*. This can be represented as

$$H_0 : \text{There is no interaction between population means}$$

The alternative hypothesis states that an effect, not reducible to a main effect, will be found. This can be expressed as

$$H_1 : \text{There is an interaction between population means}$$

Although this chapter does not give much attention to Type I and Type II errors, these concepts are still active for ANOVA tests. Any rejection of a null hypothesis carries with it a chance of making a Type I error, and any failure to reject a null runs the risk of making a Type II error.

12.2.5 Partitioning Variability

The structure of the variance in two-way ANOVAs depicted in Figure 12.6 is an extension of the one-way ANOVA. In fact, the first stage of the structure is identical: Total variance is due to between-group variance plus within-group variance. The second stage of the model is a further partitioning of the between-group variance into the variance due to Factor *A*, the variance due to Factor *B*, and the variance due to the interaction. The last stage of the model shows the constituent parts of the between-groups variance; that is, treatment (or primary) variance and error (or secondary) variance. An F test consists of the variation due to a factor or interaction being divided by the within-group variation noted near the top of the model. An explanation of each source of variability is provided in the following sections.

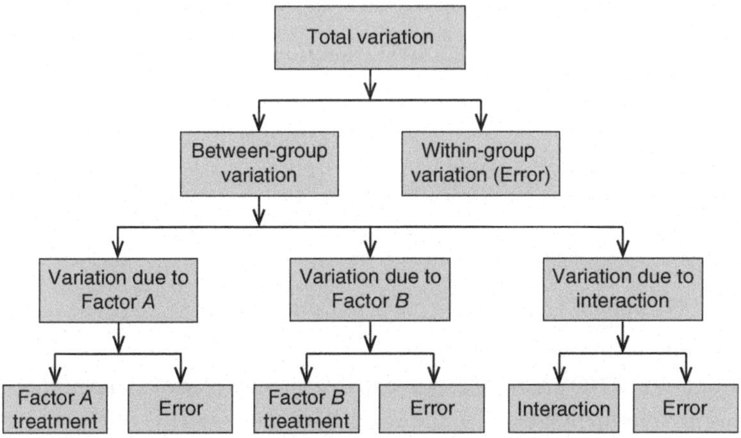

Figure 12.6 Partitioning the total variation in the two-way ANOVA.

12.2.6 Between-Group Variability

Between-group variability refers to variability among *all* the means in the study. In a two-way ANOVA, treatment variance is not allocated to only one factor but rather is allocated to either Factor *A*, Factor *B*, or a combination of Factor *A* and Factor *B* (the interaction).

12.2.7 Factor *A* Variability

Factor *A* variability refers to the difference among the means of the levels of Factor *A*. These mean differences are due to Factor *A* treatment variance plus error variance.

12.2.8 Factor *B* Variability

Factor *B* variability refers to the difference among the means of the levels of Factor *B*. These mean differences are due to Factor *B* treatment variance plus error variance.

12.2.9 Interaction Variability

Interaction variability is due to the combined influence of Factors *A* and *B*, plus error variance.

12.2.10 Within-Group Variability

Within-group variance is the average of the variances *within* each group. Within-group variance is also called *error variance* or simply *error*. The variability within each group is due to two sets of factors: individual differences and experimental error.[3]

12.2.11 The Conceptual Form of the Three *F* Ratios

The two-way ANOVA yields three different *F* ratios: one each for Factor *A*, Factor *B*, and an interaction. The conceptual basis of these *F* ratios should have a familiar look.

$$F_A = \frac{\textit{Treatment A effect} + \textit{error variance}}{\textit{error variance}}$$

$$F_B = \frac{\textit{Treatment B effect} + \textit{error variance}}{\textit{error variance}}$$

$$F_{A \times B} = \frac{\textit{Treatment } A \times B \textit{ effect} + \textit{error variance}}{\textit{error variance}}$$

As with the one-way ANOVA, the effect due to treatment is placed in the numerator. If there is no treatment effect (i.e. the H_0 is true), the *F* ratio is reduced to a

3 Refer to Chapter 11 to review the sources of variability within ANOVA analyses.

measure of error variance divided by another measure of error variance. Therefore, when the null is correct, the *F* ratio should be close to 1. As the effect due to treatment increases, the numerator grows, and the *F* ratio will become increasingly greater than 1. At some point, the *F* ratio will exceed the critical *F* value associated with the test, and our decision rule will direct us to reject the null hypothesis.

12.3 Definitional and Computational Formulas

We now turn to the computational steps and formulas used to calculate the three *F* ratios.

12.3.1 Summary of the Computational Steps

The two-way ANOVA involves the computation of the following values.

1. SS_A	7. MS_A
2. SS_B	8. MS_B
3. SS_{AXB}	9. $MS_{A \times B}$
4. SS_W	10. F_A
5. SS_T	11. F_B
6. *df* for each *SS*	12. $FA_{A \times B}$

The time demands associated with hand-calculating two-way ANOVAs are significant. The use of software such as Microsoft® Excel or SPSS® is recommended. (See Appendix C for more information on how to use these programs.) The information below is designed to help us interpret the output of these programs.

12.3.2 Sums of Squares

Just as with the one-way ANOVA, various sums of squares are calculated. In effect, the SS_{BG} calculated in a one-way is broken up into three constituent parts reflecting the three possible effects, SS_A, SS_B, and SS_{AXB}. These, when added to SS_W will sum to SS_T.

12.3.3 Partitioning Degrees of Freedom

As the abovementioned list indicates, each *SS* in the analysis of variance has corresponding degrees of freedom. Partitioning the degrees of freedom follows the same logic as partitioning the variability. Table 12.6 lists the various degrees of freedom and their computation. Note that

$$df_T = df_A + df_B + df_{A \times B} + df_W$$

Table 12.6 The degrees of freedom and their computation.[a]

Source	Degrees of freedom	Symbol
Total	$N - 1$	df_T
Within groups	$N - (k_A \times k_B)$	df_W
Factor A	$k_A - 1$	df_A
Factor B	$k_B - 1$	df_B
Interaction	$df_A \times df_B$	$df_{A \times B}$

[a] Note: N = total number of participants in the study; k = the number of conditions per factor

12.3.4 Calculating Mean Squares and *F* Ratios

The values listed above work together to generate the needed mean square values. The expressions below describe how these constituent parts are assembled to produce the three F values. All calculations follow the same form as the one-way ANOVA.

$$MS_A = \frac{SS_A}{df_A} \quad (12.1)$$

$$MS_B = \frac{SS_B}{df_B} \quad (12.2)$$

$$MS_{A \times B} = \frac{SS_{A \times B}}{df_{A \times B}} \quad (12.3)$$

$$MS_W = \frac{SS_W}{df_W} \quad (12.4)$$

When calculating the F ratios, the denominator of each F is MS_W.

$$F_A = \frac{MS_A}{MS_W} \quad (12.5)$$

$$F_B = \frac{MS_B}{MS_W} \quad (12.6)$$

$$F_{A \times B} = \frac{MS_{A \times B}}{MS_W} \quad (12.7)$$

12.3.5 An Example Problem

Our context for a two-way ANOVA is the following hypothetical experiment. Table 12.7 presents the design. A psychologist is interested in the relative effectiveness of two forms of therapy: cognitive-behavioral therapy (CBT) and

Table 12.7 The 2 × 2 factorial design of the worked problem, including cell notations.

		Therapy type		
		CBT	PDT	
Clinical problem type	Anxiety	A_1B_1	A_1B_2	A_1
	Depression	A_2B_1	A_2B_2	A_2
		B_1	B_2	

psychodynamic therapy (PDT). There is reason to believe that CBT is more effective for anxiety problems and PDT is more effective for depression. In other words, an interaction is predicted. Factor A, "Clinical Problem Type," has two levels: Anxiety and Depression. Since both factors have two levels, this is a 2 × 2 factorial design. The dependent variable is improvement ratings made by an independent observer. Table 12.8 contains the raw data. Table 12.9 shows the cell and marginal means. (Only the raw data are needed to run the analysis in Microsoft® Excel or SPSS®.) Figure 12.7 is a graph of the means.

Table 12.8 The raw data.

		Therapy type	
		CBT	PDT
Clinical problem type	Anxiety	8 6 6 9 8	3 4 1 6 2
	Depression	4 7 4 5 5	8 9 7 7 8

Table 12.9 Cell and marginal means for the raw data in Table 12.8.

		Therapy type		
		CBT	PDT	
Clinical problem type	Anxiety	$M_{A_1B_1} = 7.40$	$M_{A_1B_2} = 3.20$	$M_{A_1} = 5.30$
	Depression	$M_{A_2B_1} = 5.00$	$M_{A_2B_2} = 7.80$	$M_{A_2} = 6.40$
		$M_{B_1} = 6.20$	$M_{B_2} = 5.50$	$M_G = 5.85$

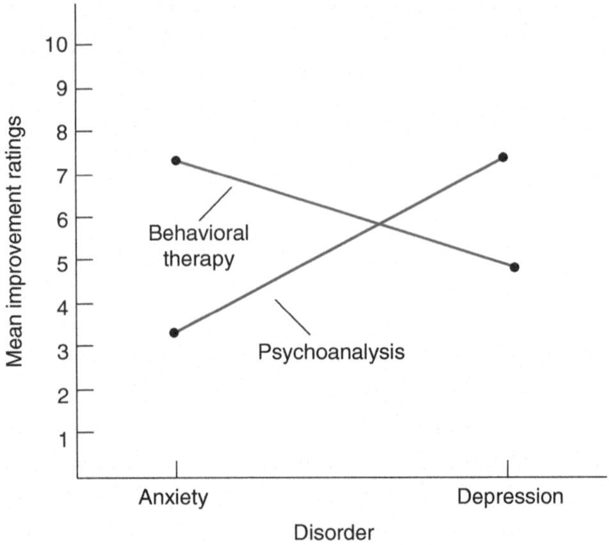

Figure 12.7 A graph of the cell and marginal means from Table 12.8.

12.4 The ANOVA Summary Table

The results of the foregoing analysis are summarized in Table 12.10. The structure of the two-way ANOVA summary table is similar to the one-way table. However, the "between-groups" row is replaced by rows for Factor A, Factor B, and the $A \times B$ interaction. Most software programs generate a similar table or have multiple tables from which to pull the information. In other words, this table can be constructed from the outputs of statistical software programs. As we examine the values in the table, verify that SS_T equals the sum of all the sum of squares, that df_T equals

Table 12.10 The ANOVA summary table based on the analysis of the data in Table 12.8.

Source	*SS*	*df*	*MS*	*F*	*p*
Factor A (disorder)	6.05	1	6.05	3.13	*n.s.*
Factor B (treatment)	2.45	1	2.45	1.27	*n.s.*
$A \times B$	61.25	1	61.25	31.74	<.05
Within groups	30.80	16	1.93		
Total	100.55	19			

the sum of all the *dfs*, that *MS* values are found by dividing *SS* values with corresponding *df* values, and that *F* values are found by dividing each *MS* factor variance by the within-groups *MS*. The *p* values reflect the likelihood of finding a given *F* value if the null is true.

12.4.1 Unequal Numbers of Participants

Researchers strive to include the same number of participants in each condition. While this is not a requirement, it does help *F* tests be more robust to minor violations of the statistical assumptions (e.g. normality). For a discussion of the conceptual issues related to factorial designs with unequal sample sizes, refer to Keppel and Wickens (2004).

12.5 Using the *F* Ratios to Test Null Hypotheses

To determine if *F* ratios are sufficiently large to reject null hypotheses, an obtained *F* must be compared with an appropriate F_{crit} value found in Table A.5. This is the same table used to test the *F* ratios of the one-way ANOVA. Each of the three *F* ratios has degrees of freedom for the numerator and the denominator (refer to Table 12.6).

The degrees of freedom for F_A are df_A, df_W; for F_B they are df_B, df_W; and for $F_{A \times B}$ they are $df_{A \times B}$, df_W. For a 2 × 2 design, the pair of *dfs* for each *F* ratio is the same. Other designs will generate different pairs of *dfs*. Since our example is a 2 × 2 design, each set of *dfs* is 1 and 16. As a result, only one F_{crit} value is needed. According to the *F* table, F_{crit} is 4.49 when $\alpha = .05$ and 8.53 when $\alpha = .01$. We will use the critical value for $\alpha = .05$.

Since F_A (3.13) and F_B (1.27) are smaller than F_{crit} (4.49), we will fail to reject the null hypothesis for Factor *A* and Factor *B*. However, $F_{A \times B} = 31.74$; this value is larger than F_{crit}. As a result, we will reject the null hypothesis of no interaction.

12.5.1 Interpreting the Findings

What do these findings mean for our example problem? Our only rejected null hypothesis concerns the interaction. Interpreting *this* interaction is straightforward. Examining the group means indicates that PDT seems to be a superior treatment for people suffering from depression, whereas the treatment of choice for people troubled by anxiety seems to be CBT. Discovering a significant interaction always leads to a qualified interpretation of a main effect. If one or both would have been rejected, we would need to interpret them carefully. They may be legitimate, or they may be illusory. An inspection of the cell means (perhaps in graphical form) can be helpful.

12.6 The Assumptions

The assumptions for the two-way ANOVA are the same as those for the one-way ANOVA. Stated succinctly, they are (i) the samples are representative of the populations from which they come, (ii) observations are independent of one another, (iii) gathered data come from an interval or ratio scale, (iv) the populations from which the data come are normally distributed, and (v) the variances of each population distribution are the same.

As before, F tests can tolerate situations when the last two assumptions are not *strictly* met. However, gross violations of these assumptions require the use of different statistical tests.

12.7 Measuring Effect Sizes

The F tests in the two-way ANOVA allow us to test the null hypothesis for each independent variable as well as the interaction. However, the F values do not provide information about the *size* of each effect. Chapter 11 used omega-squared (ω^2) and eta-squared (η^2) to determine the amount of variability in the scores due to the independent variable. With a two-way ANOVA, we can use either of these same two measures to compute a size for the effect of Factor A, Factor B, and the interaction. Of course, the issue of effect size only arises when a null hypothesis has been rejected and, if being a lower-order effect, it has been interpreted to be more than just an artifact of a higher-order effect. The formulas for calculating ω^2 are slightly different for each potential effect.

Formulas for omega-squared

$$\varpi_A^2 = \frac{SS_A - (df_A)MS_W}{SS_T + MS_W} \quad (12.8)$$

$$\varpi_B^2 = \frac{SS_B - (df_B)MS_W}{SS_T + MS_W} \quad (12.9)$$

$$\varpi_{A\times B}^2 = \frac{SS_{A\times B} - (df_{A\times B})MS_W}{SS_T + MS_W} \quad (12.10)$$

In our example, only the interaction was significant. Using values from Table 12.10, we find

$$\varpi_{A\times B}^2 = \frac{61.25 - (1)1.93}{100.55 + 1.93} = \mathbf{0.58}$$

This value reflects the ratio between the amount of primary variance associated with the interaction effect and the total variance in the study. Larger values reflect larger effect sizes.

Recall from Chapter 11 that although ω^2 is a more refined measure, the most commonly reported effect size statistic is eta-squared (η^2).

Formulas for eta-squared

$$\eta_A^2 = \frac{SS_A}{SS_A + SS_W} \quad (12.11)$$

$$\eta_B^2 = \frac{SS_B}{SS_B + SS_W} \quad (12.12)$$

$$\eta_{A \times B}^2 = \frac{SS_{A \times B}}{SS_{A \times B} + SS_W} \quad (12.13)$$

In our example, only the interaction was significant. Using values from Table 12.10 we find

$$\eta_{A \times B}^2 = \frac{61.25}{61.25 + 30.80} = \mathbf{0.67}$$

Comparing both measures, we see that η^2 tends to generate larger estimates of effect size.

12.8 Multiple Comparisons

Multiple comparisons are conducted to help interpret why a null hypothesis is being rejected. If we were to conduct a study with only two groups, perform a *t* test, and find evidence to reject the null, would we need to follow with other comparisons? No; with only two groups, a *t* test is all that is needed to locate the difference. Likewise, in a 2 × 2 factorial design, a significant main effect requires no further analysis. However, a significant interaction means that we have evidence of differences between the means of at least two cells that cannot be explained by main effects. Follow-up tests allow us to make multiple comparisons between pairs of group means that share one factor. A pattern of pairwise comparisons is interpreted by the researcher to gain a clearer understanding of the interaction.

The multiple comparison procedures used in this chapter are the same ones presented in Chapter 11: Tukey's *HSD* and Fisher's *LSD* (or protected *t* test). First, we will look at Tukey's *HSD* procedure. If the null hypothesis for an interaction is rejected, then an *HSD* for the interaction effect can be used to see which pair(s) of cell means is driving the interaction. If a null hypothesis for either Factor *A* or Factor *B* is rejected (and if that factor has more than two conditions), then an *HSD* can also be used to determine which pairs of means are driving the main effect. (Recall, a major drawback to the Tukey *HSD* test is the requirement of equal *ns* for all cells.) Below are the formulas for the various two-way ANOVA Tukey's *HSD* tests.

Formulas for Tukey's *HSD* values

$$HSD_A = q_A\sqrt{\frac{MS_W}{n_A}} \quad (12.14)$$

$$HSD_B = q_B\sqrt{\frac{MS_W}{n_B}} \quad (12.15)$$

$$HSD_{A \times B} = q_{A \times B}\sqrt{\frac{MS_W}{n_{A \times B}}} \quad (12.16)$$

---

where

q_A, q_B, $q_{A \times B}$ = studentized range statistic (Table A.6); for $q_{A \times B}$ use adjusted k as indicated below.
n_A = the number of scores in each level of Factor A (must be the same)
n_B = the number of scores in each level of Factor B (must be the same)
$n_{A \times B}$ = the number of scores in each cell (equal n in all cells required)

Design of study	Number of cell means	Adjusted value of k
2 × 2	4	3
2 × 3	6	5
2 × 4	8	6
3 × 3	9	7
3 × 4	12	8
4 × 4	16	10

To demonstrate how to make multiple comparisons in a study that is more complex than a 2 × 2, we will add a control group, making it a 2 × 3 factorial design, and introduce new data. We will also forego the computational steps of this study and proceed directly to the results of the analysis. Table 12.11 presents the cell and marginal means, as well as the sample sizes. Table 12.12 presents the ANOVA summary table.

The results of the ANOVA show a significant main effect for Factor *B*, Therapy Type. The Factor *A* main effect, Clinical Problem Type, and the interaction are not statistically significant. Therefore, the issue of locating the source(s) of significance only arises for Factor *B*.

Table 12.11 The sample sizes, cells, and marginal means for a hypothetical therapy study.

		Therapy type			
		CBT	PDT	Control	
Clinical problem type	Anxiety	$M_{A_1B_1} = 5.80$ $n_{A_1B_1} = 10$	$M_{A_1B_2} = 7.70$ $n_{A_1B_2} = 10$	$M_{A_1B_3} = 4.60$ $n_{A_1B_3} = 10$	$M_{A_1} = 6.03$
	Depression	$M_{A_2B_1} = 5.50$ $n_{A_2B_1} = 10$	$M_{A_2B_2} = 7.80$ $n_{A_2B_2} = 10$	$M_{A_2B_3} = 2.80$ $n_{A_2B_3} = 10$	$M_{A_2} = 5.37$
		$M_{B_1} = 5.65$	$M_{B_2} = 7.75$	$M_{B_3} = 3.70$	

Table 12.12 The ANOVA summary table based on the data in Table 12.11.

Source	*SS*	*df*	*MS*	*F*	*p*
Factor *A* (disorder)	6.67	1	6.67	1.98	n.s.
Factor *B* (treatment)	164.10	2	82.05	24.35	$p < .01$
$A \times B$	10.03	2	5.01	1.49	n.s.
Within groups	181.80	54	3.37		
Total	362.60	59			

Tukey's *HSD* for Factor *B*

$$HSD_B = q_B\sqrt{\frac{MS_W}{n_B}} = 3.41\sqrt{\frac{3.37}{20}} = \mathbf{1.40}$$

The q value was found using a table on the Internet (Table A.6 in Appendix A is incomplete). Once an *HSD* value is determined, the difference between each pair of means in Factor *B* can be compared (CBT vs. PDT, –2.1; CBT vs. Control, 1.95; and PDT vs. Control, 4.05). The findings suggest evidence exists for all three tested differences. Looking at the means, we can say statistical evidence suggests PDT worked better than both CBT and the control, while CBT worked better than the control.

Now we will perform multiple comparisons using another procedure, Fisher's *LSD* test. The formula for Fisher's *LSD* is:

Formula for Fisher's *LSD* test, two-way ANOVA

$$t = \frac{M_i - M_j}{\sqrt{MS_W\left(\frac{1}{n_i} + \frac{1}{n_j}\right)}} \quad (12.17)$$

where

M_i, M_j = the means for the two levels or cells being compared
n_i, n_j = the number of participants for each level or cell being compared

If we are testing the difference between two levels of a factor, M_i and M_j refer to the means of the two levels. Similarly, n_i and n_j refer to the number of participants in each level. If we are making a comparison between two cells, M_i and M_j refer to the two cell means and n_i and n_j are the number of participants in each cell.

Fisher's *LSD* test relies on the *t* distribution. The null hypothesis for each comparison is that M_i and M_j come from the same population. The critical value is the same for each pairwise comparison and is found in the *t* table (Table A.2). The critical value is found by entering the column for a two-tailed test at the desired alpha level and entering the row corresponding to the degrees of freedom. We will set alpha at .05. The degrees of freedom is df_W or $N - k$. From the ANOVA summary table, df_W is 54. The critical value is approximately ± 2.01 (Table A.2 in Appendix A is incomplete). Each protected *t* value is compared to this critical value of ± 2.01.

CBT vs. PDT

$$t = \frac{5.65 - 7.75}{\sqrt{3.37\left(\frac{1}{20} + \frac{1}{20}\right)}} = \frac{-2.10}{0.58} = \mathbf{-3.62}$$

CBT vs. control

$$t = \frac{5.65 - 3.70}{\sqrt{3.37\left(\frac{1}{20} + \frac{1}{20}\right)}} = \frac{1.95}{0.58} = \mathbf{3.36}$$

PDT vs. control

$$t = \frac{7.75 - 3.70}{\sqrt{3.37\left(\frac{1}{20} + \frac{1}{20}\right)}} = \frac{4.05}{0.58} = \mathbf{6.98}$$

Since all obtained *t* values fall outside of ± 2.01, we have statistical evidence for all three differences. This is the same result found using Tukey's *HSD*. As was noted in Chapter 11, there are numerous multiple comparison tools, each designed for use under different research situations. Which one to use in a given situation is a complex topic and beyond the scope of this text. However, two different and frequently used follow-up tests have been presented here for our use. Please consult advanced statistics resources for further information about follow-up tests. (Box 12.1 informs us about other statistical procedures related to ANOVAs.)

Box 12.1 Next Steps with ANOVA

This text allocates three chapters to ANOVAs, Chapters 11–13. (Chapter 13 will introduce us to the repeated-measures ANOVA, similar to the one-way ANOVA, but the participants are repeatedly measured across all conditions.) However, there is much more that could be said about this extremely important family of analyses. This box is going to introduce us briefly to other procedures in the ANOVA family that researchers can use to analyze more complicated research designs.

A *mixed-design ANOVA* is similar to a two-way ANOVA. It can be used to analyze data from a research design where one factor is between groups (just like the one- and two-way ANOVAs presented in Chapters 11 and 12), but the other factor is repeated measures (see Chapter 13). Imagine a situation in which depressed and nondepressed individuals are measured for their ability to use two different mnemonic techniques. "Depression state" would be a between-groups factor with participants either in one or the other condition, but "mnemonic device" could be used as a repeated-measures factor, where each participant is measured under both conditions.

A *multivariate analysis of variance (MANOVA)* analyzes more than one dependent variable. Imagine, in our example, if we measured not only "memory success" but also "time needed for recall." A MANOVA allows us to analyze both dependent variables in the same analysis.

Analysis of covariance (ANCOVA) allows the researcher to look for effects after a covariate (a variable that is suspected to covary with a factor) has been removed. Think of this technique as a way to remove the possible effects of a confounding variable. For instance, what if we wanted to remove variance associated with energy level (perhaps we fear that nondepressed people perform better simply because they are more energetic)? If "energy level" can be measured, an ANCOVA provides a mechanism to remove this variance so we can focus on the different therapy styles.

We can also combine the features of the MANOVA and ANCOVA and run a *multivariate analysis of covariance* (MANCOVA). This procedure analyzes multiple dependent variables, and it also factors out a concerning covariate. MANOVAs, ANCOVAs, and MANCOVAs can be used with between groups, repeated measures, and mixed designs. As this box shows, there are many ANOVA possibilities available to the researcher.

12.9 Interpreting the Factors in a Two-Way ANOVA

Factorial designs frequently use a participant variable as one of the factors. Personality type, biological sex, age, and psychodiagnosis are examples of participant variables that might be used as a factor in a two-way ANOVA design. It is important to remember that when a participant variable is used as a factor, cause–effect statements cannot be made regarding the relationship between this factor and the dependent variable. This component of the design is correlational because participants are selected based on their standing on that variable. In other words, the experimenter does not manipulate a participant variable.

For example, in the foregoing study, one of the factors used was "disorder type," anxiety or depression. These psychological states were not created by the experimenter. Participants were selected for the study because they were already either anxious or depressed. If anxiety or depression had been experimentally induced, then "disorder type" would not be a participant variable, rather it would be an independent variable.

Suppose one factor is a participant variable and the other is manipulated by the experimenter. Only a main effect for the second factor can be interpreted with cause-and-effect language. Now suppose that there is an interaction between a participant variable and an experimental variable. Can we make a cause-and-effect statement, or is the interaction correlational? Here, we must be very careful. For example, suppose we find that Type A people are more conforming than Type B people, but only when there is a clear payoff for conforming. When there is no identifiable payoff, Type As are less conforming than Type B people. (Imagine a graph with crossing lines forming an "×.") The payoff vs. no payoff is experimentally manipulated, but Types A and B are intrinsic to the participants. In this situation, we could suggest the payoff situation *causes* differential behavior effects among Type A and Type B participants. The dependent variable change is explained by the action of the independent variable for two different populations (Type A and Type B personalities). However, we should not conclude that personality type *causes* differential behavior as the payoff situation changes. There are any number of other participant variables correlated with personality type (e.g. competitiveness, hostility, time urgency, dominance) that could explain the relationship. In these situations, there is simply no way to nail down a causal connection between a participant variable and the dependent variable.

There are also two-way designs in which both factors are participant variables. In these instances, the entire study is correlational. Remember, the causal interpretation of research results is determined by the *design* of the study, not the statistical analysis used to analyze it. A researcher must always determine the methodological status of each factor in the design.

12.10 Formal Presentation of Findings

The proper reporting of two-way ANOVA findings is similar to what is presented in Section 11.10 regarding one-way ANOVA findings. However, remember to give priority to significant interactions prior to presenting information about main effects. (For this reason, the reporting may involve more than one sentence.) The reader needs to be made aware of both the presence of a rejected $F_{A \times B}$ null hypothesis as well as any follow-up tests that help to interpret this finding *prior* to being informed of main effect findings.

Many other principles common to the proper reporting of all types of statistical findings were first laid out in Section 8.8. Please consult this portion of the text for more general information about the proper reporting of statistical findings.

Summary

A factorial design combines at least two factors. This arrangement allows an investigator to examine the effect of each factor separately (called main effects) as well as the joint effect of the factors (called interactions). Each factor has at least two levels. A two-way design with two levels of each factor is termed a 2 × 2 design. If one of the variables has three levels, it is termed a 2 × 3 design, and so on. When visualized as a grid, each combination of levels of factors is termed a cell, for which a mean can be found. A marginal mean reflects all scores from one level of a factor.

When conducting a two-way ANOVA, there are three separate null hypotheses and three different F ratios to test them. The null hypotheses for Factors A and B state there are no differences in the population means for the levels of Factors A and B, respectively. The null hypothesis for the interaction is that there is no interaction.

The structure and logic of the two-way ANOVA are extensions of the one-way ANOVA. Total variance is due to between-group variance plus within-group variance. However, a second stage is introduced. The between-group variance is partitioned into the variances due to Factor A, Factor B, and the interaction; these correspond to three F tests (F_A, F_B, $F_{A \times B}$). The assumptions of the two-way ANOVA are identical to the one-way ANOVA.

An F value is used to test a null hypothesis; however, it does not provide information about the *size* of each effect. Omega-squared and eta-squared are statistical tools that compare primary variance with the overall variance in the study.

If the design has a factor with more than two levels, a significant main effect for that factor does not tell us which levels are driving the effect. In addition, the two-way ANOVA does not analyze differences between specific groups. Various follow-up comparisons can be used to locate the sources of significance between conditions and between cells. This chapter presents two such tests, Tukey's *HSD* and Fisher's *LSD*.

Finally, when interpreting a two-way ANOVA, consider whether a participant variable is used as a factor. Participant variables are correlational and cannot be used for a causal interpretation.

Microsoft® Excel and SPSS®

Information regarding how to use Microsoft® Excel and SPSS® to assist in performing two-way ANOVAs can be found in Appendix C.

Key Formulas

Formulas for MS_A, MS_B, MS_{AxB}, MS_W

$$MS_A = \frac{SS_A}{df_A} \quad (12.1) \qquad MS_B = \frac{SS_B}{df_B} \quad (12.2)$$

$$MS_{A\times B} = \frac{SS_{A\times B}}{df_{A\times B}} \quad (12.3) \qquad MS_W = \frac{SS_W}{df_W} \quad (12.4)$$

Formulas for F_A, F_B, and $F_{A\times B}$

$$F_A = \frac{MS_A}{MS_W} \quad (12.5) \qquad F_B = \frac{MS_B}{MS_W} \quad (12.6) \qquad F_{A\times B} = \frac{MS_{A\times B}}{MS_W} \quad (12.7)$$

Formulas for omega-squared

$$\varpi_A^2 = \frac{SS_A - (df_A)MS_W}{SS_T + MS_W} \quad (12.8)$$

$$\varpi_B^2 = \frac{SS_B - (df_B)MS_W}{SS_T + MS_W} \quad (12.9)$$

$$\varpi_{A\times B}^2 = \frac{SS_{A\times B} - (df_{A\times B})MS_W}{SS_T + MS_W} \quad (12.10)$$

Formulas for eta-squared

$$\eta_A^2 = \frac{SS_A}{SS_A + SS_W} \quad (12.11)$$

$$\eta_B^2 = \frac{SS_B}{SS_B + SS_W} \quad (12.12)$$

$$\eta_{A\times B}^2 = \frac{SS_{A\times B}}{SS_{A\times B} + SS_W} \quad (12.13)$$

Formulas for Tukey's *HSD* values

$$HSD_A = q_A\sqrt{\frac{MS_W}{n_A}} \quad (12.14)$$

$$HSD_B = q_B\sqrt{\frac{MS_W}{n_B}} \quad (12.15)$$

$$HSD_{A\times B} = q_{A\times B}\sqrt{\frac{MS_W}{n_{A\times B}}} \quad (12.16)$$

Formula for Fisher's *LSD* test, two-way ANOVA

$$t = \frac{M_i - M_j}{\sqrt{MS_W\left(\frac{1}{n_i} + \frac{1}{n_j}\right)}} \quad (12.17) \qquad (t_{crit} \text{ uses } df_W \text{ or } N - k)$$

Key Terms

Factorial (or complex) designs
Cell
Factor
Main effect
Interaction
Two-way ANOVA
Higher-order effect

Exercises

1 Describe the necessary design conditions needed to run a two-way ANOVA.

2 Specify the first and second partitioned stages of the two-way ANOVA.

3 Differentiate between a main effect and an interaction.

4 Use cells to draw the following research designs. Create meaningful labels for the factors. Make some of them independent variables and identify them as such.

   a 2×2
   b 3×2
   c 4×3
   d 3×6

5 What is the advantage of placing two factors in the same design?

6 How many types of *F*s does a two-way ANOVA generate, and how many of each type?

7 What are the null and alternative hypotheses for:
   a Factor *A*
   b Factor *B*
   c Interaction

8 Complete these ANOVA summary tables. Test the null hypotheses with $\alpha = .05$.

a

Source	*SS*	*df*	*MS*	*F*	*p*
Factor *A*	147.00	2	73.50		
Factor *B*	27.44	2	13.72		
$A \times B$	12.22	4	3.06		
Within groups	95.33	45	2.12		
Total	281.99	53			

b

Source	*SS*	*df*	*MS*	*F*	*p*
Factor *A*	3.36		3.36		
Factor *B*	66.67	2			
$A \times B$	56.89	2			
Within groups					
Total	238.75	35			

c

Source	*SS*	*df*	*MS*	*F*	*p*
Factor *A*		1	.45		
Factor *B*	6.05				
$A \times B$	84.05	1			
Within groups		16	1.28		
Total					

9 Draw a graph for each of the following datasets. Place Factor *B* on the *X* axis and the means on the *Y* axis. The two lines on the graphs correspond to the levels of Factor *A*. For each graph, indicate if there is visual evidence of an $A \times B$ interaction.

a

Factor *A*	B_1	B_2	
	35	5	A_1
	55	25	A_2

(Columns: Factor *B*)

b

Factor *A*	B_1	B_2	
	30	30	A_1
	10	50	A_2

(Columns: Factor *B*)

c

Factor *A*	B_1	B_2	
	25	10	A_1
	10	30	A_2

(Columns: Factor *B*)

10 Why does the size of the *F* not necessarily reflect effect size?

11 How do the values of effect size measures (such as ω^2 and η^2) reflect effect size?

12 Why are multiple comparisons not needed to explore main effects in a 2 × 2 design?

13 Which comparison tool (Tukey *HSD*, Fisher's *LSD*) is more flexible on sample sizes?

14 Why should we use caution when interpreting a main effect when there is an interaction?

15 Suppose a constant (e.g. 5) was added to each score in a 2 × 2 factorial design. What effect would this have on the main effects and interaction? How would MS_W be affected?

Work Problems

16 An experimental psychologist is interested in the effect of reinforcement and food deprivation on performance. Performance is measured by the time, in seconds, it takes a rat to run down an alley to a food box. Twenty rats are randomly assigned to four treatment conditions featuring deprivation (high/low) and incentive (high/low). The raw data are presented in the following 2 × 2 matrix. Set alpha at .05 and perform a two-way ANOVA.

a Identify cell and marginal means and summarize the results in an ANOVA table.

b Provide a graph with incentive on the X axis.

c Calculate ω^2 for any rejected null hypotheses.

		Factor *B*: Deprivation (Body weight)	
		85% (high)	95% (low)
Factor *A*: Incentive	45 mg (low)	7 8 6 7 7	10 7 6 8 6
	260 mg (high)	5 4 4 5 6	9 9 6 7 7

17 An educational psychologist is interested in the effect of delayed feedback on learning and whether delayed feedback operates differently as a function of educational level. All participants, composed of freshmen and seniors, are administered a 15-question test. After answering the questions, the participants are given the correct answers at various intervals, depending on the experimental condition. All participants are given the same test four days later. The dependent variable is the number of correctly answered questions. Set alpha at .05.

a Identify cell and marginal means and summarize the results in an ANOVA table.

b Graph the results with delayed feedback on the X axis.

c Calculate η^2 for any rejected null hypothesis.

d Conduct multiple comparisons if appropriate.

e Interpret the findings.

		Factor *B*: Feedback delay		
		No delay	2-H delay	1-D delay
Factor *A*: Education level	Freshmen	15 12 13 10 11	7 9 5 8 8	4 6 7 7 7
	Seniors	13 15 13 10 10	6 5 6 9 6	8 5 5 6 7

18 A psychologist is interested in whether Black defendants draw stiffer sentences than White defendants, whether White judges give stiffer sentences than Black judges, and whether there is an interaction between these factors. A hypothetical dataset was constructed. The first table shows group means and the second table is a partial ANOVA table. (Higher means reflect stiffer sentences.) Complete the ANOVA summary table and interpret the findings.

		Factor *B:* Defendant	
		Black	*White*
Factor *A*: Judge	*Black*	27.33	20.00
	White	29.50	23.67

Source	***SS***	***df***	***MS***	***F***	***p***
Factor *A* (Judge)	51.04	1			
Factor *B* (Defendant)	260.04	1			
A × *B*	3.38	1			
Within groups					
Total	1396.63	23			

19 A psychologist is interested in the following questions: (i) Can cognitive strategies increase the delay of gratification among children? (ii) Is there a difference between the biological sexes in the ability to delay gratification? (iii) Is there an interaction between biological sex and cognitive strategies?

The experimental task required the child participant to sit in front of a bowl of marshmallows placed on a table. Each child was told, "You can eat as many of the marshmallows as you want, but I would like you to try and wait until I return. If you can't wait, that's OK, but please try. To help you not eat any marshmallows, I am going to give you something to think about." In the cognitive transformation condition, the children were taught to imagine the marshmallows as white, fluffy clouds. In the self-talk condition, the children were told to repeat to themselves, "Don't eat the marshmallows." In the control condition, the children were not provided with any cognitive techniques. The experimenter left the room and observed the participant through a one-way mirror, recording the number of seconds elapsed before the child ate a marshmallow. Provide an ANOVA summary table, calculate η^2 for any found effects, and interpret the findings.

		Factor *B*: Cognitive strategy								
		Transformation			Self-talk			Control		
Factor *A*: Biological sex	Males	15	12	13	14	10	15	30	12	17
		15	15	30	16	17	28	19	22	25
		10	19	32	11	22	32	14	43	32
		20	25	29	19	18	29	10	18	39
		30	35	19	29	47	25	25	45	16
		75	60	25	74	55	18	70	50	20
		40	50		45	60		40	62	
	Females	65	89	61	72	90	65	85	80	60
		45	22	49	35	18	49	25	30	59
		50	78	35	60	75	39	60	70	25
		53	74	74	43	74	72	70	90	75
		75	99	77	75	89	60	67	80	80
		64	77	82	68	85	80	63	82	75
		55	43		50	48		50	42	

20 An experimental psychologist hypothesizes that a High-drive state will increase errors on a mental arithmetic task compared to a Low-drive state. Drive state is manipulated by telling half the participants that task performance is related to intelligence (High-drive state) while the other half are told the data will be used to establish norms. Participants in the Difficult condition receive more complicated problems than those in the Easy condition. The researcher is predicting a main effect for drive state and an interaction between task difficulty and drive state. The dependent variable is the number of errors made. Perform a two-way ANOVA on the following data, with alpha set at .05. Generate an ANOVA summary table, perform any useful follow-up tests, and interpret the findings.

		Task difficulty: Factor *B*					
		Easy			Difficult		
Driver state: Factor *A*	High drive	18	12	15	28	20	19
		10	16	18	30	15	15
		19	15	20	35	30	27
		15	12	17	37	37	29
		20	22	17	25	29	30
	Low drive	16	14	15	15	17	18
		12	29	20	20	25	16
		10	27	20	10	16	25
		22	30	25	18	13	11
		20	16	19	19	12	16

21 Kirschner and Karpinski (2010) found evidence that college students who are on social media while studying performed more poorly on academic assessments than students who are not. But might this effect be dependent on *what* material is being covered? Seventy-two university students are randomly sampled (36 self-professed "users" of social media and 36 who claimed not to use social media when studying), and each one is assigned to one of three topics of study. For consistency reasons, the registrar's office was asked to classify student academic performance by using a 7-point Likert scale (higher numbers reflecting better performance). Below are some hypothetical data. Generate an ANOVA summary table, perform any useful follow-up tests, and interpret the findings. Additionally, comment on any assumptions we should be concerned about for our analysis.

		Factor *B*: Type of academic material								
		Sciences			Classics			Arts		
Factor *A*: Social media	User	3	7	4	4	6	2	4	5	3
		5	3	5	3	2	5	4	2	3
		6	6	2	5	2	2	5	3	1
		4	4	1	2	3	3	6	5	4
	Nonuser	6	5	2	6	6	4	4	5	5
		4	7	5	2	6	5	5	4	5
		5	3	4	4	4	7	3	5	7
		3	6	6	5	5	6	4	2	6

22 Another form of distraction is listening to music. Might the type of material being studied be important? Students are selected from either a Shakespeare or German class. Study time is controlled. After six weeks, students take a 50-point test over assigned material. Independent judges determined the tests to be equally difficult. Hypothetical data are given below. Generate an ANOVA summary table, perform any useful follow-up tests, and interpret the findings.

		Factor *B*: Music	
		Music	No Music
Factor *A*: Class	Shakespeare	19 37 14 24	46 43 38 30
	German	25 14 14 39	33 46 35 48

13

Repeated-Measures Analysis of Variance

13.1 The Research Context

13.1.1 Repeated-Measures Designs

In a between-groups design, each participant receives one and only one treatment. This is true regardless of the number of factors. In a repeated-measures design (also called a *within-participants design*), every participant is exposed to *each* of the treatment conditions.[1] Using the same group of participants for each condition allows us to eliminate some of the errors due to random factors. This makes repeated-measures designs more statistically efficient than between-group designs, requiring fewer participants to achieve the same statistical power. In Chapter 9, we learned about the dependent-samples t test for repeated-measures designs with two conditions. In this chapter, the number of conditions can be greater than two. The appropriate statistical analysis is called a **repeated-measures ANOVA**. (The relationship between the independent-samples t test and the one-way ANOVA is paralleled by the relationship between the dependent-samples t test and the repeated-measures ANOVA.)

The main advantage of repeated-measures designs is greater control over participant variables. Chapter 1 explained that participant variables are the fixed attributes of a person at the time of an experiment. Intelligence, biological sex, and ethnicity are examples of participant variables. When using between-groups designs, participant variables can be problematic. For instance, suppose we want to evaluate a teaching technique designed to enhance learning. If, by chance, the intelligence differences of participants are not evenly distributed across the conditions, the variance due to "intelligence" will add to the random factors variance and make it harder to find evidence of primary variance. The problem virtually

1 As with previous ANOVA chapters, the repeated-measures ANOVA will be discussed experimentally, even though it can be used to analyze correlational data. Most repeated-measures designs, however, are experimental.

Essentials of Behavioral and Social Science Statistics, Third Edition.
K. Paul Nesselroade, Jr. and Laurence G. Grimm.
© 2025 John Wiley & Sons, Inc. Published 2025 by John Wiley & Sons, Inc.
Companion website: https://www.wiley.com/go/Essentials_of_Behavioral_and_Social_Science_Statistics3e

Table 13.1 A within-participants design: 20 participants are exposed to each treatment condition.

Treatment I	Treatment II	Treatment III
Participant 1	Participant 1	Participant 1
Participant 2	Participant 2	Participant 2
Participant 3	Participant 3	Participant 3
⋮	⋮	⋮
Participant 20	Participant 20	Participant 20

disappears, however, when we use the same participants. As a result, the variance due to random factors is reduced.

Suppose we are interested in testing the effectiveness of three different studying strategies on exam performance. In a between-groups design, we might randomly assign 60 participants to the three training programs (20 participants per group). In a repeated-measures design, we would take 20 participants and run them through each of the three programs (see Table 13.1). In this way, a repeated-measures design reduces random factor variance by eliminating individual differences between conditions.

This statistical advantage, however, comes with methodological challenges. Namely, the effects of previous measurements on future measurements, so-called *carryover* effects. How sure are we that the measurement of a participant's subsequent performances is only due to the changing conditions, not the fact that they have already been measured? For instance, some participants might show improvement on follow-up measures of a task because they are now more familiar with it; this is termed a *practice* effect. Other experimental situations might work the other way around. Participants might perform more poorly on subsequent measures due to weariness; this is termed a *fatigue* effect. Both situations introduce confounding variance that can interfere with an interpretation of the findings.

These problems may be addressed by carefully managing the order participants experience the conditions. If everyone experiences the same order, there is no way to determine if group differences are due to treatment effects or the order of progression through the study. To avoid **order effects**, researchers often *counterbalance* the conditions, a technique first introduced in Chapter 9. Here, in designs with more than 2 conditions, Treatment I may be presented first for some participants, second for others, and third for yet others. The same would hold true for Treatments II and III. Confounding variance due to order is controlled to the degree this technique equally disperses order effects across all conditions. If effective, the unwanted variance becomes error variance. There is much to say about the methodological challenges of repeated-measures designs. However, this is not a research methodology text.[2] For more information, please consult a research methodology resource.

2 Whenever a repeated-measures design is used in this chapter, assume there are no methodological issues.

Not all research questions lend themselves to a within-participants design. For instance, it is very unusual to find a psychotherapy study that uses this design. However, drug treatments are often investigated this way. Since drugs tend not to leave a lasting effect on behavior and learning, different drugs can be administered to the same participants at different times. Carryover effects are controlled by allowing the first drug to clear the patient's system before beginning the second drug trial. However, any treatment that produces a relatively permanent change in the measured variable is best evaluated using an independent-samples design.

13.1.2 Examples of Repeated-Measures Designs

► **Example 13.1** A marketing psychologist wants to see if people can tell the difference in the smoothness of the ride among three cars. A sample of participants is blindfolded and given a ride in a Lexus, a Cadillac, and a Rolls Royce. Ratings of smoothness are obtained after each ride. ◄

► **Example 13.2** A cognitive psychologist is interested in the effects of caffeine on memory. The same sample of participants is asked to memorize and recall nonsense syllables under three levels of caffeine intake. ◄

The type of repeated-measures design discussed in this chapter is limited to the case in which only *one* independent variable is used. However, factorial designs with a repeated-measures factor are common in the behavioral sciences; these are called *mixed designs* (see Box 12.1). For instance, any study that has a (between-group) treatment factor and a pretest/posttest second factor qualifies as a mixed design (see Box 13.1). Most advanced statistics textbooks cover the analysis of mixed designs.

Box 13.1 Next Steps for Repeated-Measures ANOVAs

We have already learned that the term "mixed design" refers to studies with multiple factors where at least one factor is between participants and at least one is repeated measures (within participants). Sometimes we will find the longer term "mixed between-within-participants design" used, or when speaking of the analysis of this design, the term "split-plot ANOVA" (or SPANOVA). What has not been talked about previously is the term "quasi-experimentation." A quasi-experiment is a study design that takes on some of the features of an experiment, but not all of them. There are many versions of quasi-experiments, and discussions of them can be found in most research methodology resources (e.g. Cook and Campbell 1979 and Shadish et al. 2002 are widely regarded as excellent resources).

Of particular interest are quasi-experiments where order effects are not counterbalanced; in fact, the order is of particular importance. The best example of this would be multiple-group pretest/posttest designs. In these designs, participants are assigned to different groups where a treatment is administered after a premeasure but before a post-measure. The question to be answered is, "which treatment type creates the most change over time?"

For example, suppose a claim is made that a certain type of treatment works initially but fades in effectiveness as time passes, while a different treatment has a less dramatic initial effect, but creates change that is more lasting. A design could be set up to test this claim where participants are assigned to one of two treatment conditions (between-groups factor) and measured at three different times: premeasure, post-measure 1, and post-measure 2. This would be quasi-experimental because threats to internal validity may come with the passage of time. Changes in participants' scores between the measures may be due to treatment but may also be caused by other factors which may vary somewhat systematically with the treatment conditions. (For example, imagine that one group of participants must travel to point A to get treatment and road construction has made the commute much longer; participants in the other group who are traveling to point B for treatment incur no such added frustration.) Quasi-experimentation designs are susceptible to these types of problems. Nonetheless, they are popular and very helpful for tracking change across time.

The 2 × 3 quasi-experimental design described above will be analyzed using a hybrid of the ANOVA presented in Chapter 12 (two-way ANOVA) and the ANOVA presented in this chapter (repeated-measures ANOVA). These types of designs are well represented in the behavioral and social science literature and statistical software programs such as SPSS can be used to analyze them.

13.2 The Logic of the Repeated-Measures ANOVA

13.2.1 The Hypotheses

A repeated-measures ANOVA generates one *F* ratio; it tests for population mean differences among the levels of the independent variable. Therefore, there is only one pair of hypotheses:

$$H_0 : \mu_1 = \mu_2 = \mu_3 = \mu_k$$

H_1 : at least two of the population means are different.

Although this chapter does not give much attention to Type I and Type II errors, these concepts are still active for ANOVA tests. Any rejection of a null hypothesis carries with it a chance of making a Type I error, and any failure to reject a null runs the risk of making a Type II error.

13.2.2 Partitioning Variability

The basic logic of the repeated-measures ANOVA is the same as with other ANOVAs. A ratio is created where the numerator contains variance due to a treatment effect (also called "primary variance") as well as random factors, and the denominator contains variance due only to random factors. In the context of a repeated-measures design, there is only one sample of participants; they form different groups only in the sense that they are exposed to different treatment conditions at different times. Figure 13.1 illustrates the sources of variability in the repeated-measures ANOVA.

13.2.3 Between-Group Variability

In a one-way ANOVA, between-group variability is due to three factors: treatment, individual differences, and experimental error. The last two are often termed *error* or *error variance*. The critical difference for repeated-measures designs is that variance due to individual differences is absent from between-group variation. Differences among treatment means cannot be due to individual differences, since each condition is composed of the same participants. Therefore,

between-group variance = *treatment effect* + *experimental error*

13.2.4 Within-Group Variability

In a one-way ANOVA, within-group variability is due to two factors: individual differences and experimental error. This is also true in repeated-measures designs. However, the repeated-measures ANOVA allows for the partitioning of the within-group variability into variance due to individual differences and variation due to experimental error. This is possible because individuals are being measured multiple times. Since these two sources of variance can be separated, the variability due to individual differences can be removed. This means we can create a denominator for the F ratio that only includes experimental error:

$$F = \frac{\textit{treatment effect} + \textit{experimental error}}{\textit{experimental error}}$$

Removing the variance due to individual differences from the denominator decreases the size of the denominator. As denominators decrease in size, F ratios increase. As F ratios increase, the likelihood of rejecting a null also increases. Consequently, the repeated-measures ANOVA has greater power than the between-groups ANOVA. Recall from Chapter 10 that power is influenced by several factors, one of them being the amount of variation in the data. Repeated-measures designs remove the variance due to individual differences and correspondingly increase the power.

Refer to Figure 13.1. Note that in the last stage of the model, within-group variability has been partitioned. A new term has been introduced: *between-participants variability*. (Do not confuse this with *between-group variability*.) This is variation due to individual differences. The denominator in a repeated-measures ANOVA refers *only* to experimental error.

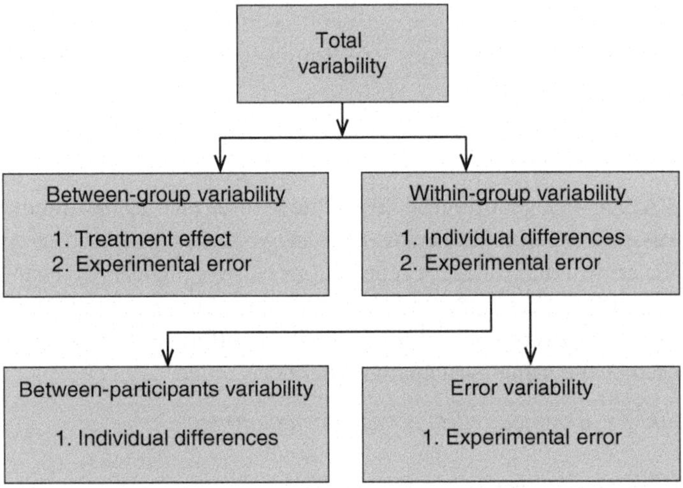

Figure 13.1 Partitioning the total variation in the repeated-measures ANOVA.

13.2.5 Does Removing Individual Differences Variability Really Matter?

In a one-way ANOVA, variance due to individual differences shows up in the numerator and the denominator. If, in a repeated-measures ANOVA, this variance is removed from both the numerator and the denominator, what is gained? A simple illustration answers this question. Suppose two experiments are contrasted, one using an independent-groups design and another using a repeated-measures design. Further, let us assign some units of variation to each component of the F ratio:

$$\begin{aligned} \textit{treatment effect} &= 200\ \textit{units} \\ \textit{individual differences} &= 300\ \textit{units} \\ \textit{experimental error} &= 50\ \textit{units} \end{aligned}$$

The F ratio for independent-groups design (one-way ANOVA) would be as follows:

$$F = \frac{\textit{treatment} + \textit{individual differences} + \textit{experimental error}}{\textit{individual differences} + \textit{experimantal error}}$$

$$F = \frac{200 + 300 + 50}{300 + 50} = \frac{550}{350} = \mathbf{1.57}$$

The F ratio for a repeated-measures design (repeated-measures ANOVA would be as follows: (Notice, the treatment variance and experimental error values are not altered.)

$$F = \frac{treatment + experimental\ error}{experimental\ error}$$

$$F = \frac{200 + 50}{50} = \frac{250}{50} = \mathbf{5.00}$$

The result in this example is dramatic. Removing individual differences variability is not always this substantial. The impact of using a repeated-measures design is determined by the relative size of the variance due to individual differences.

Since larger F ratios are more likely to lead to rejecting null hypotheses, repeated-measures ANOVAs have more *power* than ANOVAs from between-groups designs, all other things being equal. As a result, if methodological challenges can be addressed, researchers typically opt for repeated-measures designs over between-participants designs.

13.3 The Formulas

The repeated-measures ANOVA involves the computation of the following values.

1) SS_{BG}
2) SS_W
3) SS_{BP}
4) SS_{error}
5) SS_{total}
6) df for each SS
7) MS_{BG}
8) MS_{error}
9) F

The time demands associated with hand-calculating repeated-measures ANOVAs are significant. The use of software such as Microsoft® Excel or SPSS® is recommended. (See Appendix C for more information on how to use these programs.) The information below is designed to help us interpret the outputs of these programs.

13.3.1 Partitioning the Degrees of Freedom

Partitioning the total degrees of freedom follows the same form as partitioning the total variation of scores. Figure 13.2 illustrates this. Table 13.2 lists the various degrees of freedom and their computation.

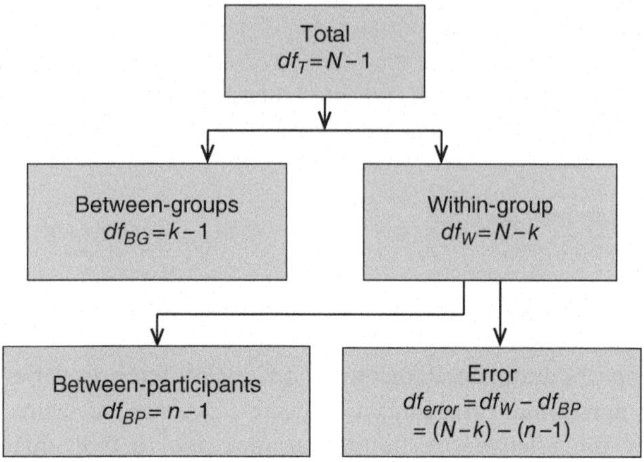

Figure 13.2 Partitioning the degrees of freedom in the repeated-measures ANOVA.

Table 13.2 The degrees of freedom for repeated-measures ANOVAs.[a]

Source	Computation	Degrees of freedom
Total	$N - 1$	df_T
Between groups	$k - 1$	df_{BG}
Within groups	$N - k$	df_W
Between participants	$n - 1$	df_{BP}
Error	$(N - k) - (n - 1)$	$df_{Error} = df_w - df_{BP}$

[a] For repeated-measures ANOVAs, $N =$ the total number of observations, while $n =$ the total number of participants.

13.3.2 The Mean Squares and *F* Ratio

The last step in the repeated-measures ANOVA is to calculate the various mean squares used in the *F* ratio. Since these designs have only one independent variable, there is only one *F* ratio:

$$F = \frac{MS_{BG}}{MS_{error}} \quad (13.1)$$

where

$$MS_{BG} = \frac{SS_{BG}}{df_{BG}} \quad (13.2)$$

$$MS_{error} = \frac{SS_{error}}{df_{error}} \quad (13.3)$$

Remember that the denominator of the F ratio in the repeated-measures ANOVA is MS_{error} rather than MS_W. With the effect of individual differences removed, using MS_{error} preserves the basic logic of the F ratio:

$$F = \frac{treatment + experimental\ error}{experimental\ error}$$

13.3.3 An Example Problem

Psychologists have noted that anxious people often have difficulty concentrating. Suppose a researcher is interested in the effects of different doses of a new psychotropic drug on cognitive performance. Ten participants are selected and exposed to four treatment conditions. In three treatment conditions, the participants are administered 2.5, 5, and 10 mg of the drug. A fourth condition, where participants receive a placebo, is added as a control. Participants are asked to solve a series of mental arithmetic problems within a given period. The dependent variable is the number of errors committed. The treatment conditions are spaced apart to ensure a drug has cleared the participants' system before a new dosage is assessed. In addition, treatment order is counterbalanced among the participants to control for practice effects. Table 13.3 presents the design of this experiment. In the table, each X symbolizes a score. The first subscript identifies the participant; the second identifies the condition. For example, X_{23} refers to the score for participant number 2 under condition number 3. Note: the total number of participants is 10 but the total

Table 13.3 The repeated-measures design for the hypothetical worked problem.

	Treatment			
	Placebo	**2.5 mg**	**5 mg**	**10 mg**
Participants	*(1)*	*(2)*	*(3)*	*(4)*
P_1	X_{11}	X_{12}	X_{13}	X_{14}
P_2	X_{21}	X_{22}	X_{23}	X_{24}
P_3	X_{31}	X_{32}	X_{33}	X_{34}
P_4	X_{41}	X_{42}	X_{43}	X_{44}
P_5	X_{51}	X_{52}	X_{53}	X_{54}
P_6	X_{61}	X_{62}	X_{63}	X_{64}
P_7	X_{71}	X_{72}	X_{73}	X_{74}
P_8	X_{81}	X_{82}	X_{83}	X_{84}
P_9	X_{91}	X_{92}	X_{93}	X_{94}
P_{10}	X_{101}	X_{102}	X_{103}	X_{104}

Table 13.4 Raw data and means for the worked problem.

	Treatment				
Participants	**Placebo**	**2.5 mg**	**5 mg**	**10 mg**	
P_1	14	5	9	15	$\Sigma P_1 = 43$
P_2	12	3	6	11	$\Sigma P_2 = 32$
P_3	10	2	7	9	$\Sigma P_3 = 28$
P_4	7	4	5	7	$\Sigma P_4 = 23$
P_5	9	4	6	8	$\Sigma P_5 = 27$
P_6	9	1	3	7	$\Sigma P_6 = 20$
P_7	10	3	4	9	$\Sigma P_7 = 26$
P_8	5	0	0	13	$\Sigma P_8 = 18$
P_9	6	4	4	6	$\Sigma P_9 = 20$
P_{10}	8	6	6	7	$\Sigma P_{10} = 27$
	$\Sigma_1 = 90$	$\Sigma X_2 = 32$	$\Sigma X_3 = 50$	$\Sigma X_4 = 92$	$\Sigma X = 264$
	$M_1 = 9.0$	$M_2 = 3.2$	$M_3 = 5.0$	$M_4 = 9.2$	

number of scores is 40. This observation will become important when we analyze degrees of freedom. The raw data and means are presented in Table 13.4. (Only the raw data is needed to run the analysis in Microsoft® Excel or SPSS®.)

13.4 The ANOVA Summary Table

The general form of the summary table for repeated-measures ANOVAs is shown in Table 13.5. The presentation of the findings from the example problem is shown in Table 13.6. This is not the only form a summary table for repeated-measures ANOVAs can take; however, it is the format we will be using in this text.

Table 13.5 General form of the summary table for a repeated-measures ANOVA.

Source	*SS*	*df*	*MS*	*F*	*p*
Between groups	SS_{BG}	df_{BG}	MS_{BG}	F	
Within groups	SS_W	df_W			
Between participants	SS_{BP}	df_{BP}			
Error	SS_{error}	df_{error}	MS_{error}		
Total	SS_T	df_T			

Table 13.6 The ANOVA summary table for the hypothetical study.

Source	*SS*	*df*	*MS*	*F*	*p*
Between groups (dose level)	266.40	3	88.80	22.09	$p < .01$
Within groups	227.20	36			
Between participants	118.60	9			
Error	108.60	27	4.02		
Total	493.60	39			

13.5 Using the *F* Ratio to Test the Null Hypothesis

The null hypothesis for the example problem is $H_0: \mu_1 = \mu_2 = \mu_3 = \mu_4$. The alternative hypothesis is that H_0 is false. If the null is rejected, we would conclude that at least two of the sample means come from different populations. To reject the null, the obtained *F* ratio must be larger than the critical value of *F* found in the *F* table (Table A.5). The degrees of freedom for *F* are the degrees of freedom associated with the numerator and denominator of the *F* ratio, df_{BG} and df_{Error}, respectively. The degrees of freedom for this study are $df_{BG} = 3$ and $df_{Error} = 27$. Entering the *F* table, the critical value when $\alpha = .05$ is 2.96. We have statistical evidence to reject the null.

13.6 Interpreting the Findings

What does a rejected null hypothesis mean for the study? The independent variable has four levels: placebo, 2.5, 5, and 10 mg of the new drug. A value of *F* that exceeds F_{crit} means we have statistical evidence suggesting that at least two group means come from different populations. In other words, the number of errors committed during the timed mental arithmetic test seems to be affected by dosage level. At this point, we have the same dilemma as when we reject an *F* ratio in a one-way ANOVA. There is no way to know which groups are the ones that are differing from one another. To locate the differences, pairwise comparisons among the means are required. This topic will be discussed shortly.

13.7 The Assumptions

The assumptions for the repeated-measures ANOVA are the same as those for the one-way and two-way ANOVA, apart from the second and last assumption; namely, (i) representativeness, (ii) observations *within* each condition are independent of each other, (iii) gathered data comes from an interval or ratio scale, (iv) normally distributed population data, and (v) the variances of the population of the difference scores are homogeneous.

The difference in the second assumption reflects the fact that the same participants are used for each condition. (This qualification to the assumption of independence was also made for the dependent-samples t test.) The fifth assumption is a bit more complex. We may find it described elsewhere as the *assumption of sphericity*. In brief, it is assumed that the variance of the difference scores between any two sets of sample scores is roughly equal to the variance of the difference scores between their corresponding populations. It is possible to test this assumption and make corrections if it is violated (see Keppel and Wickens 2004).

Just as with the previous ANOVAs, as n_p increases the F test becomes robust to minor violations of the last two assumptions and can be performed even when they are not strictly met. However, gross violations require the use of statistical tests (nonparametric tests) which do not make these assumptions.

13.8 Measuring Effect Size

In previous chapters, it was emphasized that the size of the observed statistic does not necessarily indicate the size of the effect. Effect size can be determined with the omega-squared, ω^2, or eta-squared, η^2, statistic.

Repeated-measures omega-squared

$$\varpi^2 = \frac{SS_{BG} - df_{BG}(MS_{error})}{SS_T + MS_{error}} \qquad (13.4)$$

Using the data from the hypothetical worked problem,

$$\varpi^2 = \frac{266.40 - 3(4.02)}{493.60 + 4.02} = \mathbf{0.51}$$

This statistic estimates that 51% of the dependent variable variation is due to dosage level.

Now we will look at how eta-squared measures the effect size.

Repeated-measures eta-squared

$$\eta^2 = \frac{SS_{BG}}{SS_T} \qquad (13.5)$$

Using the data from the hypothetical worked problem,

$$\eta^2 = \frac{266.40}{493.60} = \mathbf{.54}$$

Here again, we see that η^2 seems to overestimate the effect size relative to ω^2.

13.9 Locating the Source(s) of Statistical Evidence

A rejected null requires follow-up comparisons to locate the source(s) of the statistical evidence. As with the preceding two chapters, Tukey's *HSD* and Fisher's *LSD* or protected *t* test are presented. The formula for Tukey's *HSD* is similar to previous versions. Be sure to see that MS_{Error} is now used as the error term and to know that df_{Error} is needed to find q. Here is the formula.

Formula for Tukey's *HSD*, repeated-measures

$$HSD = q\sqrt{\frac{MS_{error}}{n}} \quad (13.6)$$

---

where

q = the Studentized range statistic (Table A.6)
n = the number of scores in each group

In the worked problem, there are four levels to the independent variable, but only one *HSD* value is needed to make all comparisons[3]:

$$HSD = q\sqrt{\frac{MS_{error}}{n}} = 3.87\sqrt{\frac{4.02}{10}} = \mathbf{2.45}$$

Applying this value to all six comparisons, we find evidence suggesting participants performed better under the 2.5 and 5 mg conditions than under the 10 mg and placebo conditions.

Fisher's *LSD* test is also very similar to the versions presented in previous chapters. However, MS_{error} instead of MS_W serves as the error term.

Formula for Fisher's *LSD*, repeated-measures

$$t = \frac{M_i - M_j}{\sqrt{MS_{error}\left(\frac{1}{n_i} + \frac{1}{n_j}\right)}} \quad (13.7)$$

---

3 To find q, an online table was used. The one found in Table A.6 is incomplete.

where

$M_i, M_j =$ the means for the two groups being compared
$n_i, n_j =$ the number of scores in each of the two groups being compared

The critical value is found in the t table (Table A.2). The df used to find the critical value is taken from the MS_{error} term ($df_{error} = dfw - df_{BS}$).

In the worked problem, there are four levels to the independent variable. To make all possible pairwise comparisons requires six t tests. Only one comparison is made for illustration.[4]

Placebo vs. 5 mg

$$t = \frac{M_i - M_j}{\sqrt{MS_{\text{error}}\left(\frac{1}{n_i} + \frac{1}{n_j}\right)}} = \frac{9 - 5}{\sqrt{4.02\left(\frac{1}{10} + \frac{1}{10}\right)}} = \mathbf{4.44}$$

The critical value in the t table that corresponds to a df of 27 is ± 2.05. Since the obtained value of 4.44 falls outside of the critical values, we should reject the null hypothesis that these two means come from the same population. We would interpret this statistically significant difference by saying that statistical evidence has been found suggesting the mental concentration of anxious participants is improved when they are administered 5 mg of the new drug compared with a pill that they *think* will help their concentration (placebo).

13.10 Formal Presentation of Findings

The proper reporting of repeated-measures ANOVA findings is similar to what is presented in Section 11.10, regarding the reporting of one-way ANOVA findings. When reporting a significant F, we must include the df_{BG} and df_{error}, the F value, and the alpha level used to make our decision. For instance, "Statistical evidence suggests the type of therapy administered influenced recovery, $F(2, 14) = 13.18$, $p < .05$. Further analysis found evidence suggesting Therapy C performed better than Therapy A, $t(14) = 4.71$, $p < .05$; and Therapy B, $t(14) = 3.15$, $p < .05$. No difference was found between Therapy A and Therapy B, $t(14) = 2.21$, *n.s.*" A failure to reject might read, "There was no statistical evidence to suggest the type of therapy used influenced recovery, $F(2, 14) = 1.95$, *n.s.*" Measures of effect size can be added at the end of the sentence when appropriate.

4 The protected t test is presented here to maintain consistency with Chapters 11 and 12. However, there is evidence that using Fisher's *LSD* for more than three pairwise comparisons increases the probability of a Type I error.

Many other principles common to the proper reporting of all types of statistical findings were first presented in Section 8.8. Please consult this portion of the text for more general information about the proper reporting of statistical findings.

Summary

In a repeated-measures (within-participants) design, every participant is exposed to each of the treatment conditions. Since this design uses the same group of participants, it requires fewer participants than a between-groups design and is also more statistically efficient. The efficiency occurs because a repeated-measures design eliminates individual differences as a potential explanation for the results of the study. By using the same participants in all conditions, this variance is logically eliminated from the numerator of the F ratio and can be mathematically partitioned out of the denominator. However, not all research questions lend themselves to within-participants designs and many methodological problems may occur.

The total variation of scores can be partitioned into between-group variance and within-group variance. Between-group variance is due to treatment effect and experimental error. Within-group variance is composed of individual differences variability (which is dropped from the analysis) and experimental error variability. The repeated-measures F ratio has the form:

$$F = \frac{\textit{treatment effect} + \textit{experimental error}}{\textit{experimental error}}$$

The assumptions of repeated-measures ANOVAs are nearly identical to those of other ANOVA tests. However, scores are only independent of each other *within* each condition, and the variances which are homogeneous are the various pairs of sample difference scores compared to their corresponding populations.

An F that leads to rejecting the null hypothesis does not provide direct information about the size of the effect. Omega-squared and eta-squared are statistics that estimate effect size. Furthermore, a rejected null hypothesis requires follow-up tests where pairwise comparisons can be made among the means of different levels to locate the source(s) of statistical evidence. Tukey's *HSD* and Fisher's *LSD* are two of the procedures that can accomplish that task.

Microsoft® Excel and SPSS®

Information regarding how to use Microsoft® Excel and SPSS® to assist in performing repeated-measures ANOVAs can be found in Appendix C.

Key Formulas

Formula for *F* ratio

$$F = \frac{MS_{BG}}{MS_{error}} \quad (13.1)$$

Formulas for MS_{BG}, MS_{error}

$$MS_{BG} = \frac{SS_{BG}}{df_{BG}} \quad (13.2) \qquad MS_{error} = \frac{SS_{error}}{df_{error}} \quad (13.3)$$

Repeated-measures omega-squared, ω^2

$$\varpi^2 = \frac{SS_{BG} - df_{BG}(MS_{error})}{SS_T + MS_{error}} \quad (13.4)$$

Repeated-measures eta-squared, η^2

$$\eta^2 = \frac{SS_{BG}}{SS_T} \quad (13.5)$$

Formula for Tukey's *HSD*, repeated-measures

$$HSD = q\sqrt{\frac{MS_{error}}{n}} \quad (13.6)$$

Repeated-measures Fisher's *LSD*

$$t = \frac{M_i - M_j}{\sqrt{MS_{error}\left(\frac{1}{n_i} + \frac{1}{n_j}\right)}} \quad (13.7) \qquad (t_{crit} \text{ uses } df_{error} = dfw - df_{BS})$$

Key Terms

Repeated-measures ANOVA **Order effects**

Exercises

1 How is a repeated-measures ANOVA similar to a dependent-samples *t* test?

2 Methodologically distinguish between repeated-measures and between-groups designs?

3 What are "order effects" and why are they a concern?

4 What is "counterbalancing" and how can it be helpful?

5 When should repeated-measures designs not be used?

6 In a repeated-measures design, what accounts for between-group variation, and what accounts for within-group variation?

7 How does the denominator of a one-way and repeated-measures ANOVA differ?

8 What values comprise the *F* ratio for a repeated-measures ANOVA?

9 How is variance due to individual differences removed from a repeated-measures design?

10 Suppose we have access to 15 participants, and we need to measure performance across three conditions. Compare the *df* values if we ran a one-way ANOVA with the *df* values if we ran a repeated-measures ANOVA. Compare F_{crit} values ($\alpha = .05$) as well.

11 If we add together df_{BP}, df_{Error}, and df_{BG} – what does this equal?

12 What are the *df* values for the *F* ratio of a repeated-measures ANOVA with 5 conditions and 20 participants?

13 Suppose repeated-measures ANOVA results are reported as $F(3, 24) = 4.25$, $p < .05$. How many participants were involved in the study?

Work Problems

14 A study examines sales performance and various incentive programs. For one month each, three incentive programs are used at a local car dealership. The number of cars sold by each salesperson under each incentive program is recorded. Conduct a repeated-measures ANOVA on the following data.

- **a** Find the program means and summarize the results in an ANOVA summary table.
- **b** Calculate a measure of effect size (either one), if appropriate.
- **c** If necessary, run Tukey's *HSD* tests to help clarify the findings.
- **d** Interpret the findings.
- **e** Are there any methodological issues to be addressed?

Participant	Incentive *A*	Incentive *B*	Incentive *C*
P_1	8	7	4
P_2	6	6	4
P_3	9	5	3
P_4	7	4	3
P_5	7	5	5

15 A psychologist uses a repeated-measures design to examine the effects of subliminal messages on problem-solving. Simple arithmetic problems are presented on a computer screen and participants are told to work as quickly as possible. In the positive condition, the phrase "Good Work" is flashed every 30 seconds just below recognition threshold. In the negative condition, the phrase "Don't Fail" is flashed. In the control condition, no subliminal phrase is projected. Conduct a repeated-measures ANOVA on the following data.

a Find the condition means and summarize the results in an ANOVA summary table.

b Calculate a measure of effect size (either one), if appropriate.

c If necessary, run Fisher's *LSD* tests to help clarify the findings.

d Interpret the findings.

e Are there any methodological issues that need to be addressed?

Participant	Positive	Negative	Control
P_1	45	20	30
P_2	56	18	29
P_3	59	10	24
P_4	48	15	25

16 We observe that people seem to be happier when they are wearing a new article of clothing. We would also like to see if this happiness depends on the *type* of new clothing worn. A random sample of five students is given new T-shirts and new shoes. With order counterbalanced, they are instructed to wear each article of clothing for one day. At the end of each day, participants rate their happiness. On another day, when they are not wearing new clothing, a happiness rating is taken. Conduct a repeated-measures ANOVA on the following data.

a Find the level means and summarize the results in an ANOVA summary table.

b Calculate a measure of effect size (either one), if appropriate.
c If necessary, run Fisher's *LSD* tests to help clarify the findings.
d Interpret the findings.
e Are there any methodological issues that need to be addressed?

Participant	New T-shirt	New shoes	Control
P_1	6	8	4
P_2	6	7	6
P_3	7	9	5
P_4	5	7	3
P_5	8	10	5

17 A psychologist is interested in the effects of distraction on pain tolerance. Three different slide shows, which vary in distraction level, are projected on a screen while participants have their hands immersed in ice-cold water. A 20-minute interval is used between experimental conditions to allow participants to recover from each hand immersion. The number of seconds participants kept their hands in the water is recorded. Conduct a repeated-measures ANOVA on the following data.

a Find the condition means and summarize the results in an ANOVA summary table.
b Calculate a measure of effect size (either one), if appropriate.
c If necessary, run Fisher's *LSD* tests to help clarify the findings.
d Interpret the findings.
e Are there any methodological issues that need to be addressed?

	Distraction		
Participant	**Low**	**Medium**	**High**
P_1	56	32	120
P_2	76	65	90
P_3	60	55	69
P_4	72	70	100
P_5	50	57	111

18 An educational psychologist is interested in comparing three visual scanning techniques on reading speed. The reading speeds of six participants are recorded (in seconds) after training in each technique. Conduct a repeated-measures ANOVA on the following data.

a Find the condition means and summarize the results in an ANOVA summary table.
b Calculate a measure of effect size (either one), if appropriate.
c If necessary, run Fisher's *LSD* tests to help clarify the findings.
d Interpret the findings.
e Are there any methodological issues that need to be addressed?

Participant	Technique *A*	Technique *B*	Technique *C*
P_1	450	250	500
P_2	426	300	456
P_3	399	170	300
P_4	400	227	310
P_5	420	225	250
P_6	350	270	350

19 Where students study may be important. To explore this idea, a random sample of six freshman psychology students were asked to study their course material for one hour every day, first in a special quiet room in the library, second in the dining hall, and last rotating among a classroom, dining hall, and library quiet room. Students spent two weeks in each condition; order was counterbalanced. They were tested with a 25-point quiz at the end of each condition. Conduct a repeated-measures ANOVA on the following data.

a Find the condition means and summarize the results in an ANOVA summary table.
b Calculate a measure of effect size (either one), if appropriate.
c If necessary, run Tukey's *HSD* tests to help clarify the findings.
d Interpret the findings.
e Are there any methodological issues that need to be addressed?

Participant	Quiet room	Dining hall	Various sites
P_1	12	10	9
P_2	20	16	18
P_3	19	16	20
P_4	25	20	22
P_5	18	15	13
P_6	14	10	14

20 Jackson and Howton (2008) looked at the health benefits of electronic pedometers. Ten participants were told to record the number of steps they took in a week. The study lasted 12 weeks. Constructed data from weeks 1, 6, and 12 are below. Conduct a repeated-measures ANOVA. Use Fisher's *LSD* to perform follow-up comparisons if warranted.

Number of steps (×1000)			
	Week		
Participant	**1**	**6**	**12**
P_1	6	7	8
P_2	3	4	7
P_3	6	3	6
P_4	1	6	5
P_5	1	5	6
P_6	5	8	9
P_7	4	6	8
P_8	5	10	7
P_9	3	5	10
P_{10}	4	7	5

Part 4 Review

Analyses of Variance

Review of Concepts Presented in Part 4

The purpose of this brief review section is to revisit both the similar concepts from Chapters 11–13 as well as those that distinguish them one from another. First let us look at the similarities. All three of the statistical tests presented in these chapters (one-way ANOVA, two-way ANOVA, repeated-measures ANOVA) are grounded on the same basic logic. That is, each one is designed to test a null hypothesis of no difference between population means by comparing a measure of variance *between* the samples based on both primary variance (or "treatment effect") and random factors with another measure of variance *within* the samples based on random factors alone. This comparison is done in ratio form such that if there is no primary variance present, the measures of variance due to random factors should roughly equate, yielding a resultant value close to 1. If, however, primary variance is present, the resulting ratio will increase corresponding to the relative amount of primary variance present. The statistic generated in all ANOVAs is an F. As with the t tests presented earlier, a table of critical values based on null distributions of various design forms can be found in Table A.5. If the observed F equals or exceeds the critical F, statistical evidence has been found suggesting the null hypothesis of no differences between population means is false. (Of course, the risk of making Type I and Type II decision errors is still present.)

Another point of similarity between the chapters concerns what can be done if an overall null hypothesis is rejected. In all three chapters we are introduced to two different ways to estimate the size of an effect (using omega-squared, ω^2, or eta-squared, η^2). The formulas vary slightly depending on the features of the design,

Essentials of Behavioral and Social Science Statistics, Third Edition.
K. Paul Nesselroade, Jr. and Laurence G. Grimm.
© 2025 John Wiley & Sons, Inc. Published 2025 by John Wiley & Sons, Inc.
Companion website: https://www.wiley.com/go/Essentials_of_Behavioral_and_Social_Science_Statistics3e

but each one creates a ratio of primary variance to total variance in the system. Additionally, to locate the source(s) of statistical evidence, all three ANOVAs introduce tools for the comparison of means between levels of conditions and cells, namely Tukey's *HSD* and Fisher's *LSD*. The formulas between the three chapters vary slightly depending on the design form, but each one allows the investigator to see if the mean difference between any two conditional means is large enough to suggest a difference at the population level.

The differences between ANOVAs result from the type of research design and the number of factors. Both the one-way and repeated-measures ANOVA allow only one factor. Each can accommodate numerous conditions, but these conditions must vary across a single dimension. The two-way ANOVA is used for designs where two factors are being used. Because of this, two different types of effects are introduced: main effects and interactions. Main effects are found when complex designs are viewed from the perspective of only one factor. Since two-way designs have two factors, there are two potential main effects for each two-way design. The interaction is an effect caused by the combination of factors and is not reducible to either main effect. Interactions are considered higher-order effects than main effects and, if inferred, must be prioritized.

One-way and repeated-measures ANOVAs are distinguished by the way participants are assigned to the various conditions. One-way ANOVAs are used to analyze between-participants (or independent-group) designs, where each participant is assigned to one condition and measured only once. Repeated-measures ANOVAs are used to measure within-group (or repeated-measures) designs where each participant is exposed to each level of the factor (or independent variable). Because the same participants are in each condition, the numerator of a repeated-measures F ratio does not include variance due to individual differences. To restore the F ratio logic, individual differences variance is partitioned out of the denominator. This creates a more statistically efficient mechanism for detecting primary variance; it is more powerful than the one-way ANOVA.

The problems with repeated-measures designs are methodological. Confounding variance can be introduced if carryover effects from the repeated measurement of participants is not controlled. Also, some treatment types are irreversible or leave long-lasting effects on participants. However, for treatments that are reversible, if the methodological issues can be addressed, the statistical advantage gained when using a repeated-measures ANOVA can be dramatic.

Since real-world research problems do not come with labels informing the researcher of which test to use for analysis, it is important for us to work on our diagnostic skills. Understandably, to solve the exercises at the end of each chapter requires the use of only the tests found within that chapter. These exercises are not designed to challenge our diagnostic skills (i.e. knowing which test to use for a given situation). With that in mind, the following review section has been created to help us develop these skills.

The exercises below will help us review the statistical differences between the various ANOVAs introduced in Chapters 11–13 as well as the *t* tests introduced in the preceding chapters. The hypothesis testing exercises will not identify which test is appropriate. We will need to use the available information presented in the exercise to make that determination.

Part 4 Review Exercises

1 The independent-samples *t* test is to the one-way ANOVA as the dependent-samples *t* test is the __________?

2 Which two types of ANOVAs are needed to understand the term "mixed design"?

3 A psychologist who works with alcoholics is interested in the effects of therapy and medication in preventing relapse. Sixteen patients at a treatment center volunteer to participate in a study. Participants are randomly assigned to either a "talk" or "relaxation" therapy condition and to either an antidepressant or vitamin B_1 medication condition. Total alcohol-free days in a one-month period are used as the dependent measure. Select the appropriate statistical analysis.

4 A friend wants to see if background color influences unconscious perceptions of female attractiveness for biological males. Our friend has male participants look at a series of 200 pictures of individuals; many different types of backgrounds are used. Unbeknown to the participants, the picture of one individual is repeated in the series, once with a red background (believing this color to be unconsciously associated with sexuality) and once with an off-white background. The dependent variable is the amount of time (in seconds) the participants look at the target images before moving on to the next one. The presentation order of the two images is counterbalanced across the participants. Select the appropriate statistical analysis.

5 Another friend does not believe the effect found in Exercise #4 is due to the supposed sexual nature of red but rather to the fact that it is bright in comparison to the boring off-white color. The friend constructs another similar study, but this time introduces a bright green background as well. So, now the target individual is shown three times, once each with a red, green, and off-white background. Counterbalancing is once again used. Select the appropriate analysis.

6 A preschool teacher would like to make sure students rest during quiet time. The teacher wonders if children will relax more quickly if a story is read, soft music is played, or a glass of milk is drunk. Children are randomly assigned to one treatment condition. For one week, the teacher records the average number of minutes it takes each child to fall asleep. Select the appropriate statistical analysis.

7 Where students choose to study may be important. To study this, a random sample of 18 freshman psychology students were asked to study their course material for one hour every day, some in a special quiet room in the university library, some in the dining hall, and some rotating among a classroom, dining hall, and library quiet room. Students spent two weeks studying in this manner. They were tested with a 25-point quiz at the end of the two weeks. Select the appropriate statistical analysis.

8 A frequently pondered topic of many university students concerns the broad area of biological sex differences (i.e., male versus female). Suppose, in particular, we are interested in exploring differences in preparation time for a formal social gathering (e.g. a campus social). Timing devices are randomly given to 15 classmates. They are asked to start the device when they begin to get themselves ready for the social and to stop the device when they are ready to leave their dorm room. The times, in minutes, are recorded. Select the appropriate statistical analysis.

9 A creative writing course instructor wonders if creativity changes with the weather, specifically the outdoor temperature. Since the instructor teaches a yearlong course, student poems can be sampled from when the weather was hot, warm, and cold. The works of seven students are randomly selected and rated by other professors for their creativity across all three conditions. Select the appropriate statistical analysis. Are there any potential methodological issues?

10 Over the years, much research has been conducted on the psychology of negotiations (see Loschelder et al. 2014; for an example). A couple variables of particular interest to many researchers are the first mover (i.e. whether the seller or buyer initially states the conditions of transaction) and the biological sex of the seller. In this experiment a raffle ticket for a new mountain bike is initially gifted to each participant. An acquaintance of the researcher who is ignorant of the hypothesis and is only told to buy back each ticket using as little money as possible engages in a negotiation with each participant to buy back the ticket prior to the raffle. In one condition the participant (seller) is told to make the opening bid; in another condition the buyer is asked to make the opening bid. The biological sex of the seller is also noted. Select the appropriate statistical analysis.

Part 5

Inferential Statistics

Bivariate Data and Chi-Square Tests

14

Linear Correlation

14.1 The Research Context

This chapter discusses correlational analysis. A **correlation coefficient** is a measure of the strength of association between two variables. The mathematics of calculating a correlation results in a single number that can range from −1 to +1. The larger the absolute value of the correlation, the stronger the association. Correlations from samples use the symbol *r*, while those from populations use the Greek symbol ρ (*rho*). Correlations have a very broad and useful function for scientific investigations, not just for the behavioral and social sciences, but across the breadth of human investigation; wherever variables are measured and presumed to relate.

Correlational analyses require the measurement of two variables – in the social and behavioral sciences, these are often either two attributes or an attribute and behavior. (Recall, when we introduced the correlation concept back in Chapter 1, we considered the relationship between age and vacation type; restful or active.) Together, the two measures define a *pair* of scores. The distribution of pairs of scores is called a **bivariate distribution** (*bi* meaning two, *variate* meaning variable). This chapter addresses the correlation of data measured on an interval or ratio scale. Chapter 16 briefly discusses two other correlational analyses to be used when data come from ordinal or nominal scales.

The statistical methods for calculating the correlation were first invented by Sir Francis Galton (1822–1911). However, the precise formula for the statistic introduced in this chapter, the **Pearson correlation** (more properly, the *Pearson product-moment correlation coefficient*), was derived by and named after Karl Pearson (1857–1936). (See Spotlight 14.1) This form of the statistic is the most powerful and frequently used version.

Essentials of Behavioral and Social Science Statistics, Third Edition.
K. Paul Nesselroade, Jr. and Laurence G. Grimm.
© 2025 John Wiley & Sons, Inc. Published 2025 by John Wiley & Sons, Inc.
Companion website: https://www.wiley.com/go/Essentials_of_Behavioral_and_Social_Science_Statistics3e

Spotlight 14.1 Karl Pearson

Karl Pearson, the father of modern statistics, was born in London in 1857, two years before Darwin published *Origin of Species,* a work that would shape Pearson's entire academic life. He is most known for completing the work Galton had started on the correlation concept, culminating in the coefficient that bears his name. Subsequently, he devised formulas for computing correlations for noncontinuous variables as well as many of the concepts and terms introduced earlier (e.g. the histogram, mode, and standard deviation). He also invented the chi-square test (see Chapter 16).

After earning a degree in mathematics at King's College, Pearson studied law and then established a private practice. In 1884, he abandoned law, became a professor of mathematics at University College London, and began his illustrious career. Pearson, like many scholars of that day, was fascinated by Darwin's theories about the role of heredity, variation, and natural selection in evolution. Additionally, he was influenced by a book written by Darwin's cousin, Sir Francis Galton, entitled, *Natural Inheritance*. Pearson eventually became a disciple and defender of Galton, spending the bulk of his professional life developing statistical tools designed to test Darwin's and Galton's theories (e.g. Pearson 1896, 1903).

Pearson became a strong advocate for one of his mentor's most notorious ideas, eugenics (See Spotlight 15.1). He could be quite accurately described as a social Darwinist, imperialist, nationalist, and racist (Grosskurth 1980). For instance, he believed that war was necessary to eliminate "inferior stock." He also opposed most forms of humanitarian legislation once writing, "No degenerate and feeble stock will ever be converted into healthy and sound stock by the accumulated effects of education, good laws, and sanitary surroundings" (Semmel 1958). Unfortunately, his nationalistic and racist beliefs even influenced his scientific conclusions (e.g. Delzell and Poliak 2013). Several of his research projects serve as an example of how personal bias and preconceived notions can influence the methods used and even the conclusions drawn from scientific investigations (see Box 2.2).

The eugenics movement, for which Pearson advocated in the late 1800s, spiraled out across the globe, coming of age in the United States (where forced sterilizations were legalized in the 1920s), and adopted as national policy by Nazi Germany in the 1930s (e.g. Kühl 1994). The death of millions of physically, mentally, racially, and socially "inferior" people was the result.

Pearson died in 1936, just a few years before the outbreak of World War II. Moving forward we must take only the best of Karl Pearson, namely, his insistence on the importance of quantifying social phenomena and the numerous correlational techniques he developed to achieve this goal.

14.1.1 The Correlational Design and the Correlation Coefficient

We have probably heard the motto "correlation does not imply causation." This phrase can be misleading if we do not keep in mind the difference between how data are collected, on one hand, and the statistical analysis used to interpret the data, on the other. If we conduct a study in which the researcher manipulates one variable and the other is observed and measured, the study is called an experiment. Under these conditions, we may be warranted in making causal statements about the relationship between independent and dependent variables, even if we use the correlation statistic for analysis. However, if we conduct a study and do not exert control over an independent variable, then we are using a correlational design, and a causal analysis is inappropriate. Whether causal language can be used to explain a relationship is entirely determined by the nature of the research design, *not* the statistical analysis.[1] A much better motto would be, "correlational *design* does not imply causation." Let's consider a few research scenarios.

Some people believe that a student's academic achievement is affected by their global self-esteem. And correlational data has been found to suggest that those who have high global self-esteem perform better in a variety of academic measures as compared to those with low global self-esteem. Has a causal connection been established? Not necessarily. It could be that academic success influences how one feels about one's value, or it could also be that some third variable, for instance, the nature of one's social support network, influences both academic success and global self-esteem.

If we wanted to determine if there is a direct, causal relationship between self-esteem and academic achievement, we would have to find a way to systematically manipulate global self-esteem and then wait to see what sort of grades and accolades a student earned. If this could be done, the research design would then be experimental, and a causal interpretation would be allowed. However, with only a moment's thought, we can see that there are several important and varied reasons why an experimental approach cannot be taken to study this relationship.

It is also helpful to note that the statistical test used in the analysis does not clarify whether the design is experimental or correlational. Consider the following two examples. Suppose that we want to test the hypothesis that anxious individuals finish their exams faster than calm individuals. Based on a standard assessment technique, we classify everyone in the class as either anxious or calm. Then, we time how long it takes each student to complete their exam. An independent-sample t test is used to analyze the data. Can we use causal language to explain a rejected null hypothesis? No, because we have not manipulated any variables. Whenever a participant variable is used to create groups, the design is correlational.

It is true that correlational designs often use a correlation coefficient (r) for analysis and experiments often use t tests and F tests for analysis. However, this is not

1 For review, see Section 1.6.

always the case. For example, suppose we are interested in understanding the relationship between the use of practice imagery and bowling proficiency. To do this, we randomly assign participants to various amounts of imagery practice time where they visualize themselves using proper bowling form. Subsequently, bowling scores are recorded. In this situation, a correlational analysis could be used to identify the relationship between the amount of time visualizing proper bowling form and bowling scores. If we found a strong correlation between these variables, could we state that visualization technique *caused* an increase in bowling performance? Yes, because we manipulated an independent variable. We directed participants to practice imagery for differing stretches of time. In this context, we would be justified in connecting a mathematical correlation with causal language. The basis for making causal statements *never* resides with the type of statistical analysis, but rather with the methodology used to collect the data.

14.2 The Correlation Coefficient and Scatter Diagrams

This section addresses the statistical aspects of correlation and hypothesis testing. Recall that a correlation coefficient is represented by a number that ranges from -1 to $+1$; the higher the coefficient's absolute value, the stronger the association between the two variables. An r of $-.80$ reflects an association as strong as an r of $+.80$. A correlation of 0 reflects no relationship between the variables. If higher values of one variable are paired with higher values of another (as with IQ and academic performance), the correlation is said to be *positive*. If higher values of one variable are paired with lower values of another, the correlation is said to be *negative*. For instance, educational achievement and number of children are negatively correlated. That is, the more educated a person is, the fewer children they tend to have.

Chapter 2 showed a few ways in which a distribution of scores can be displayed. For instance, polygons and histograms are common ways of displaying *univariate* data. However, when a *bivariate* distribution is plotted on a graph, it is called a **scatterplot** (or *scatter diagram*). Consider a child psychologist who is interested in developing a measure of pre-pubescent aggression. A rating scale is given to each of ten children in a third-grade class, who are then asked to rate the aggressiveness of their peers (from 1, not aggressive, to 10, extremely aggressive). Independently, the psychologist conducts behavioral observations of the children during recess. In Table 14.1, the X column lists the average peer ratings gathered for each child in the class. The Y column lists the number of observed instances of aggression, over three days, for each child. The designation of the X and Y variables is arbitrary. The X scores for all the children constitute one univariate distribution, and the Y scores constitute another. Since 2 scores are recorded for each

Table 14.1 Hypothetical data depicting peer ratings of aggression (X) and observed aggression (Y).

	Peer ratings of aggression: X	Observed aggression: Y
P_1	8	14
P_2	10	12
P_3	4	9
P_4	1	4
P_5	5	11
P_6	6	10
P_7	3	1
P_8	9	12
P_9	7	10
P_{10}	2	4

child (peer ratings and behavioral observations), the 10 pairs of scores define a bivariate distribution.

The distribution can be represented visually by plotting participant's X and Y scores on a graph. For this data set, there are 10 points; each point corresponds to a participant's X and Y score. Figure 14.1 is the scatterplot of the data in Table 14.1.

To draw a scatterplot, place the X variable on the horizontal axis and the Y variable on the vertical axis. To plot the data point for participant 1's score, follow the X axis to the number 8. Then, imagine a line drawn vertically, parallel to the Y axis, up to the number 14. This is the location where we plot the participant's data point. As Figure 14.1 is examined, make sure that all the pairs of scores in Table 14.1 have been accurately plotted.

14.2.1 Interpreting the Scatter Diagram

The scatterplot provides a wealth of information about the relationship between two variables. Figure 14.2 shows the scatterplot diagrams for several correlations. The magnitude of the correlation can be estimated by looking at the general shape formed by the data points. The narrower the width of the data, the stronger the correlation. The more the data takes the shape of a circle, the weaker the correlation. Compare the correlation of +.70 in Figure 14.2a with the correlation of +.40 in Figure 14.2c. Figure 14.2d illustrates a plot of unrelated variables.

A scatterplot indicates not only the magnitude of correlations but also their direction. If an imagined oval encircling most of the data slopes from the lower left to the upper right, the correlation is positive. In other words, as the X scores increase, the Y scores also tend to increase (see Figure 14.2a and c). If the encircled data slopes from

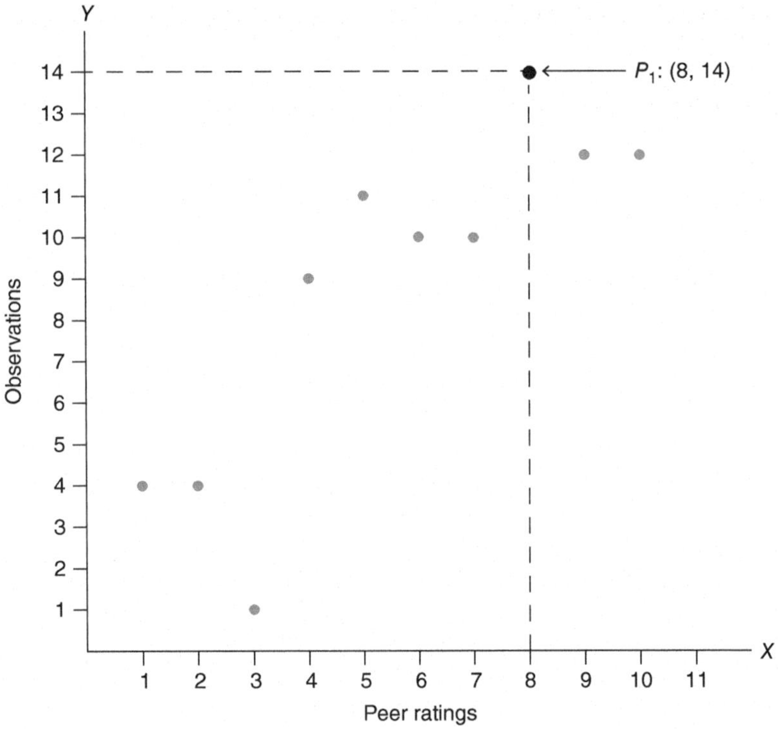

Figure 14.1 The scatterplot of the data from Table 14.1.

the upper left to the lower right, then the correlation is negative. In this case, as the X scores increase, the Y scores tend to decrease (see Figure 14.2b). Please note that the degree of the slope (e.g. gradual vs. steep) is *not* indicative of the strength of the correlation. The scatterplots of two correlations of the same magnitude, but with different signs, are shown in Figure 14.2a and b.

14.2.2 Linear and Nonlinear Correlations

The scatterplot provides a graphic representation of the relationship between the distributions of both variables. Viewing the plot's approximate shape allows us to make a crude estimate of the strength of association of the variables. However, a researcher would *never* report a correlation based solely on the shape and direction of the data in a scatterplot. So why construct one? Well, a visual analysis can be helpful, especially for detecting nonlinear relationships.

In the correlation examples provided so far, the higher values of one variable are associated with either higher values of the second variable (positive correlation) or

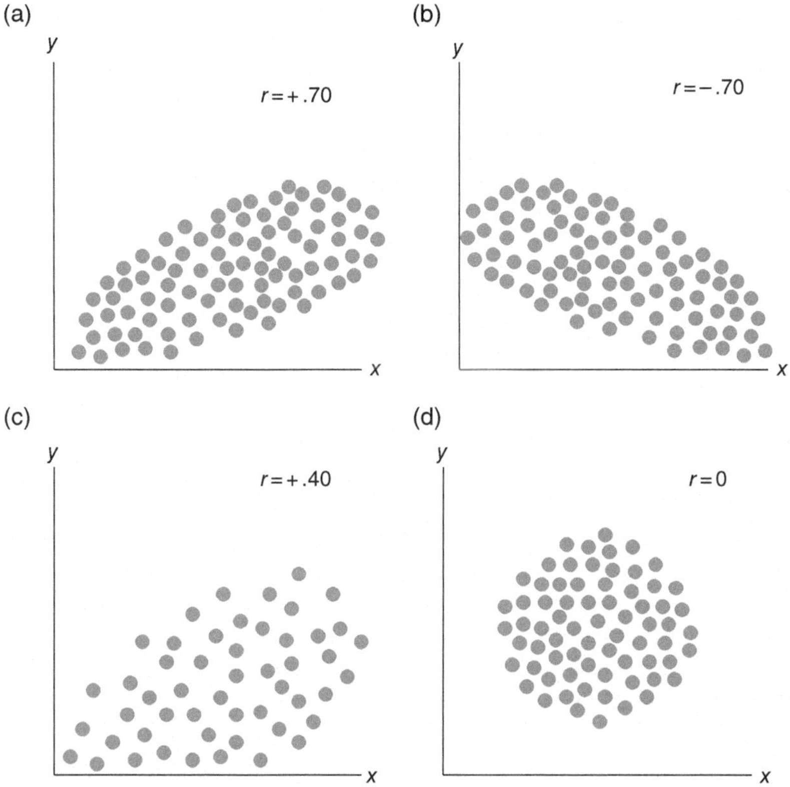

Figure 14.2 Different scatterplots depicting the magnitude and direction of a correlation (a) strong positive; (b) strong negative; (c) weak positive; (d) no correlation.

with lower values of the second variable (negative correlation). Whether the correlation is positive or negative, the scatterplots illustrated in Figure 14.2a–c show linear relationships between *X* and *Y*. When a single straight line running from one end of the data to the other does a fairly good job of reflecting the basic nature of the relationship, a **linear relationship** is exhibited. On the other hand, what if we observed a pattern similar to the one displayed in Figure 14.3? Here, lower scores on *X* are associated with lower scores on *Y*, medium *X* scores are associated with medium *Y* scores, *but* higher *X* scores are associated with *lower Y* scores. The data are distributed like an arch (∩). The variables are obviously associated (notice the narrowness of the arch), but any straight line fails to capture this association. This is an example of a **curvilinear relationship.** How would we interpret this relationship between arousal and performance? Given a certain level of task difficulty, performance is optimal when the person performing a task is experiencing

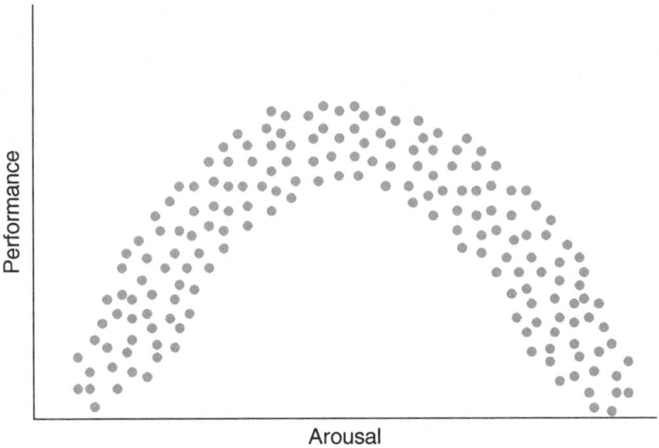

Figure 14.3 A plot of a curvilinear relationship between arousal and performance.

a moderate level of arousal. Either a low or high level of arousal is associated with poor performance.

The formulas for calculating linear and curvilinear correlations are different. In fact, if a linear formula were applied to the data of Figure 14.3, the correlation would be close to 0, suggesting *no* relationship. However, a view of the scatterplot shows a strong correlation between the variables. Indeed, we should *always* inspect a scatterplot *before* we interpret a correlation coefficient. If the plot shows an arch or an "s" shape, the formula for a linear correlation should not be used. This chapter only addresses linear correlations since they are the most basic and common type. Consult an advanced statistics resource to understand the analysis of curvilinear relationships.

Now that we are familiar with how to graph bivariate data and interpret a scatterplot, we can turn to the statistical basis and computational methods of the correlation coefficient.

14.2.3 The Computational Formula for the Correlation Coefficient

Formula 14.1 is the *computational formula* for computing the Pearson correlation. The symbol r indicates that the correlation is being derived from a sample of scores, not a population.

Computational formula for Pearson r

$$r = \frac{n_p(\Sigma XY) - (\Sigma X)(\Sigma Y)}{\sqrt{\left[n_p(\Sigma X^2) - (\Sigma X)^2\right]\left[n_p(\Sigma Y^2) - (\Sigma Y)^2\right]}} \quad (14.1)$$

where

ΣX^2 = the sum of all the squared X scores
ΣY^2 = the sum of all the squared Y scores
XY = the cross product of an X and Y score
ΣXY = the sum of the cross products
$(\Sigma X)^2$ = the sum of all the X scores, quantity squared
$(\Sigma Y)^2$ = the sum of all the Y scores, quantity squared
n_p = number of *pairs* of observations

At first glance, this appears to be an imposing formula. However, it is not nearly as difficult as it might appear. If we need to calculate a Pearson correlation by hand, the worked problem below will walk us through the steps.

■ **Question** *An investigator wants to know the correlation between subjective feelings of discomfort and the length of time participants can keep their hands submerged in ice water. The X variable is a measure of discomfort; higher scores indicate greater discomfort. The Y variable is the number of minutes participants kept their hands submerged. What is the correlation for this sample of participants?*

Solution

X	Y	X^2	Y^2	XY
3	2	9	4	6
5	3	25	9	15
4	4	16	16	16
7	5	49	25	35
10	6	100	36	60
$\Sigma X = 29$	$\Sigma Y = 20$	$\Sigma X^2 = 199$	$\Sigma Y^2 = 90$	$\Sigma XY = 132$

Use the given X and Y values to create X^2, Y^2, and XY columns. After we compute the summary statistics shown at the bottom of each column, proceed as follows:

$$r = \frac{5(132) - (29)(20)}{\sqrt{[5(199) - (29)^2]\,[5(90) - (20)^2]}} = +\mathbf{.91}$$

■

14.3 The Coefficient of Determination, r^2

Recall from the ANOVA chapters the effect size measures of omega-squared (ω^2) and eta-squared (η^2). They reflect the amount of variation among the scores of the dependent measure that are accounted for by the factor. The **coefficient of determination**, calculated by squaring r and symbolized as r^2, performs a similar service for the correlation coefficient. An r^2 is a measure of the amount of variation of the Y variable accounted for by variation in the X variable; a measure of **shared (or common) variance.**

Shared variance is usually stated as a percentage. If a correlation between two measures is .80, then the amount of shared variance is $.80^2 \times 100 = 64\,\%$. What is the coefficient of determination when the correlation is −.30? Square the correlation and multiply by 100 to arrive at 9%. Researchers use different phrases when referring to shared variance, for example:

1) Sixty percent of Y score variance is *explained* by the variation of X scores.
2) Fifteen percent of Y *is due to* X.
3) Twenty-eight percent of the variance of Y is *accounted for* by X.

It is important to note that shared variance is bidirectional. This means the relationship can be stated from either the perspective of X or Y. In addition, do not be confused by the terms "explained" or "accounted for." Strictly speaking, the coefficient of determination does not *explain* or *account for* anything. In other words, the coefficient of determination does not tell us *why* there is a relationship between two variables. This can be a difficult concept to understand. The following examples are designed to help us better understand it.

Suppose we administer two *different* tests of anxiety to a group of individuals. Since both scales measure anxiety, would we expect the correlation between them to be +1? Probably not. Even though both tests are measuring the same concept, the correlation between them will likely not be perfect. The score of *any* measure always contains some degree of *measurement error*. For example, in addition to the presence of anxiety, a person's score could be affected by fatigue or distractions during testing.[2] Also, one test may focus on the physical experience of anxiety (e.g. "How often do you feel your heart pounding?"), while the other looks at the cognitive aspect of anxiety (e.g. "How often do you find yourself worrying?"). Different scales may tap into different aspects of the same trait. In Figure 14.4, the shaded areas are what the two measures have in common: the shared variance. Note how the amount of shared variance is larger as r^2 increases. The nonoverlapping area is the *unshared variance*.

Since the concept of r^2 is important, and because it can be difficult to grasp, let us consider another more detailed example. The Wechsler Intelligence Scale for

---

2 *Measurement error* is similar to the concept of *experimental error* discussed in Chapters 11–13.

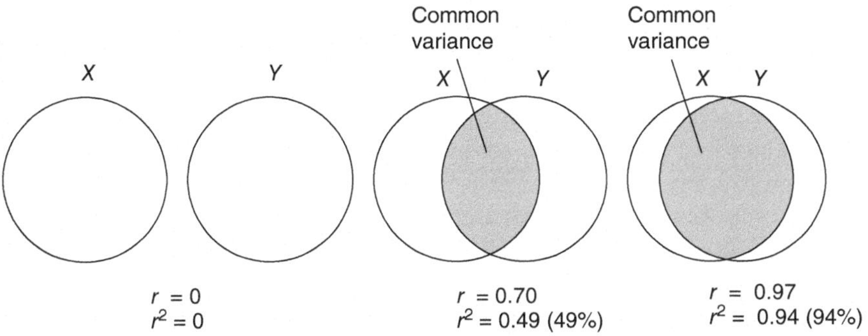

Figure 14.4 The shaded overlapping areas depict shared variance. As r increases so does r^2.

Children (WISC-R) includes dozens of questions and tasks that are used to assess various aspects of intelligence. The "block design" task, for instance, requires the child to look at a picture of a design (e.g. a diamond). Several wooden blocks, some white, some red, and some half white and half red are given to the child who must then arrange the blocks to reproduce the design in the picture. Another task, "object assembly," presents the pieces of a puzzle; the child must fit them together to form the correct figure (e.g. a horse). It is reasonable to think that some of the abilities that would lead a child to do well on the "block design" task would also lead to good performance on the "object assembly" task. This common ability is what makes up the shared variance of the two tasks. Table 14.2 lists some of these common sources of variance. However, each test is also influenced by unique abilities, introducing unshared variance (also listed in Table 14.2). Suppose the correlation between performance on "block design" and "object assembly" is +.70. Refer to the middle illustration in Figure 14.4. If the correlation is +.70, r^2 is approximately 50%. (See Box 14.1 for more information about how the correlation concept is used in the development and assessment of measurement tools.)

Table 14.2 Shared and unshared abilities between two tasks of the WISC-R.[a]

Block design and object assembly	
Shared abilities	**Abilities not shared**
(Common variance; shared variance)	**(Uncommon variance; unshared variance)**
Visual–motor coordination	Analysis of whole into parts (BD)
Spatial relations	Reproduction of models (BD)
Perceptual organization	Learn from sensory–motor feedback (OA)
Working under time pressure	

[a] Kaufman (1979)/with permission of John Wiley & Sons. BD, block design; OA, object assembly.

Box 14.1 Next Steps with Correlations: Scale Development

A common activity for many academic psychologists is the construction of measuring tools. There are literally hundreds of different psychological traits and abilities that psychologists are interested in measuring, from ubiquitous concepts like extroversion and self-esteem to more targeted concepts like humility (e.g. Rowatt et al. 2006) and right-wing authoritarianism (e.g. Mirels and Dean 2006). All these scales, however, need to be created. They do not appear out of thin air.

Scale development is an extensive process. First, the concept is carefully defined, with a lot of thought given to identifying various subcomponents of the concept (e.g. is extroversion marked by being very talkative, striving to be the center of attention, enjoying meeting new people, being very physically demonstrative, or all four?). Second, researchers gather data to see if their understanding of the concept fits well with how people answer questions about themselves. This usually involves the generation of numerous response items, oftentimes presented in the form of a question or a statement to be agreed or disagreed with by participants. These responses are then statistically analyzed.

Here, the correlation concept is useful. By looking at the size and nature of the relationships between pairs of items, researchers can gain feedback information regarding the nature and scope of the concept they are studying. For instance, if high correlations were found between individuals' responses regarding their talkativeness, how much they enjoy being the center of attention, and how much they like meeting new people, this interrelatability (or shared variance), suggests that the concept of "extroversion" encapsulates all these subcomponents. If, on the other hand, the responses to questions about being physically demonstrative do not tend to correlate highly with these other components, this might suggest that this component is not part of the concept of "extroversion." Once data has been gathered and analyzed, this two-step process of concept definition/clarification and data gathering/analysis can repeat itself, often several times over.

The process, as one might imagine, is actually much more sophisticated than what has been presented here. It is described in simple terms to show the relationship between the correlation concept and this important professional activity. Correlations are also the base concept behind other sophisticated analytical techniques (see Box 15.3).

Table 14.3 The relationship between r and r^2.

	r (±)	r^2
	.00	.00
Small r's	.10	.01
	.20	**.04**
	.30	**.09**
Moderate r's	.40	.16
	.50	.25
	.60	.36
Large r's	.70	.49
	.80	**.64**
	.90	**.81**
	1.00	1.00

Correlations of small, moderate, and large are indicated on the left.

14.3.1 Comparing r and r^2

If we compare an r of +.50 with an r of +.25, *numerically speaking,* one is twice the size of the other. However, the *worth* of correlations is better captured by the r^2 concept, the shared variance. Here we see that an r of .50 reflects four times more shared variance than an r of .25.

$$(.50)^2 \times 100 = 25\% \qquad \frac{25\%}{6.25\%} = 4$$
$$(.25)^2 \times 100 = 6.25\%$$

Table 14.3 allows us to further explore this relationship. Here, we see that correlations of .20 and .30 produce a mere 5% change in the amount of shared variance (.04 as compared to .09). Yet, the correlations of .80 and .90 yield a variance difference of 17% (.64 as compared to .81). In other words, the difference between a correlation of .20 and .30 is not as meaningful as the difference between a correlation of .80 and .90.

14.4 Using the Pearson r for Hypothesis Testing

14.4.1 The Null and Alternative Hypotheses

The statistic, r, is based on a *sample* of bivariate scores. It is an estimate of ρ, a *population* of bivariate scores. The statistic, r, can be used to make a number of inferences about ρ. For example, two r's can be compared to see if they have been drawn from populations having different ρ's. This section only addresses the most common form of hypothesis testing: Is ρ different from 0? When asking this question, the null and alternative hypotheses are stated as:

$$H_0 : \rho = 0$$
$$H_1 : \rho \neq 0$$

14.4.2 Sampling Distributions

A statistical test of significance asks the question, "What is the probability that we would obtain this sample statistic by chance when the null hypothesis is true?" If the probability is low (e.g. less than five times out of 100), then the null hypothesis is rejected, and we cautiously suggest $\rho \neq 0$. To test the null hypothesis, we need to compute r from a sample of bivariate scores and then determine how likely it is the r we obtained comes from a bivariate population with no correlation.

A direct method for testing the null hypothesis is to use a table of critical r's (see Table A.7). This table is derived in the same manner as t tables, by using the sampling distribution concept. It allows us to evaluate the likelihood that an observed r came from a bivariate population with no correlation. As with t tests, we will need to determine both the size and the placement of the rejection region. The same issues discussed in Chapter 8 present themselves here. The size of the rejection region determines the degree of Type I error risk we wish to tolerate. In terms of placement, it is usually considered prudent to split the rejection region evenly between the two tails of the distribution; that is, to use a two-tailed test.

The following worked problem illustrates how to use the table of critical values to test the null hypothesis of no correlation. Note: The degrees of freedom for a Pearson r corresponds to the number of paired scores minus two (or $n_p - 2$).

■ **Question** *An educational psychologist hypothesizes a relationship between trait anxiety and GPA. A sample of 20 students is randomly selected; the correlation between anxiety and GPA is found to be −.50. Is there statistical evidence that ρ is different from 0?*

Solution

Step 1. State the null and alternative hypotheses. In this case, H_0: $\rho = 0$ and H_1: $\rho \neq 0$.
Step 2. Establish an alpha level. Use $\alpha = .05$.
Step 3. Compute r_{obt}. It is given as −.50.
Step 4. Locate r_{crit} in Table A.7, using $df = n_p - 2 = 18$, two-tailed test. The critical value equals ±.444.
Step 5. Compare r_{obt} with r_{crit}. Since −.50 falls outside of ±.444, reject H_0: $\rho = 0$.
Step 6. Interpret the findings. "There is statistical evidence to suggest a negative correlation exists between trait anxiety and GPA, $r(18) = -.50$, $p < .05$."

Note that we cannot make a causal statement that trait anxiety *leads* to lower grades. It could be the case that having lower grades leads to a chronic feeling of anxiety. Moreover, there is always the possibility that a third, unmeasured variable, accounts for this correlation. ■

14.4.3 General Considerations in Testing $H_0: \rho = 0$

Hypothesis testing is an inferential procedure. In the context of correlational analyses, a decision is made as to whether the population correlation differs from 0. As with any test of a null hypothesis, two types of decision errors can be made. Any rejection of a null hypothesis carries with it a chance of making a Type I error, and any failure to reject a null runs the risk of making a Type II error.

Recall that the power of a statistical test is the probability that a null will be correctly rejected. In the context of this test, power is the ability to detect a nonzero population correlation. As ρ departs from 0, it is easier to obtain an r that leads to rejecting the null. Moreover, increasing the sample size decreases r_{crit}. Indeed, with very large samples, extremely small population correlations can be detected. Whether a researcher would want to detect such small correlations, however, is questionable. (This issue is further discussed below.)

14.4.4 Rejecting Null Hypotheses and r^2

On the left-hand column of Table A.7, we will see that critical values are once again linked to the degrees of freedom (*df*) associated with the inferential test. Notice that as we look down any column, r_{crit} becomes *smaller* as the *df* becomes *larger*. This indicates that with very large samples it is possible for a correlation to prompt the rejection of the null hypothesis even when that correlation value is small. Under these circumstances, it is especially important that we keep the proportion of shared variance, r^2, in mind. A *statistically* significant correlation does not necessarily mean it is *theoretically* or *practically* significant.

For example, Table 14.4 reports some published correlational information reflecting the relationship between prejudicial attitudes and antismoking attitudes.

Table 14.4 Correlations between political prejudice and antismoking attitudes.[a]

Prejudice	r	p	r^2	Percent shared variance (%)
Anti-Semitic	.06	.001	.0036	.36
Anti-Arab	.05	.001	.0025	.25
Racist	.11	.001	.0121	1.21
Anti-American	.05	.001	.0025	.25

[a] Grossarth-Maticek et al. (1988)/with permission of SAGE Publications.

Notice, for instance, the correlation between anti-Semitism and antismoking prejudice is .06, significant at the .001 level. The r^2 is .0036, meaning that well under 1% of the variance in antismoking attitudes is accounted for by variation in the anti-Semitic scores! How could such small correlations lead to a rejection of the null? Well, the sample size was 5977! With such a large sample, the statistical test was exceptionally powerful. Situations like this remind us that we should always look at the size of the correlation and the amount of shared variance.

As with other inferential tests, the Pearson r rests on assumptions namely representativeness, independent observations, interval or ratio scale of measurement, and normality.

14.5 Misleading Correlation Coefficients

The definition of a correlation coefficient can be deceptively simple; it is a measure of the strength of association between two variables. However, several factors can affect this measure and obscure the real nature of the relationship. This section addresses some of the issues that arise when measuring and interpreting correlations.

14.5.1 Restricted Range

One problem concerns the difference between the *potential* range and the *obtained* range of scores. For instance, a scale that measures kindness might range from 20 to 60. If, for instance, the scores from a given sample only range from 45 to 60, the entire span of the concept is not reflected in the sample. If we correlate kindness with some other measure, a problem might arise. *Samples with restricted ranges tend to underrepresent population correlations.* This problem is visually represented in Figure 14.5, where the smaller square depicts a sample with a restricted range. Within this smaller grid, X and Y do not look particularly well corrected.

One example is the relationship between the Graduate Record Examination (GRE) scores and the GPAs of students in prestigious graduate programs. Since only students with high GRE scores tend to be admitted into these programs, and since the grades earned by these students tend to be strong, the range of both measures is restricted. An analysis of this relationship might produce a rather low correlation. Failure to understand this issue might result in school officials falsely concluding that GRE scores are not predictive of graduate school success.

14.5.2 A Nonlinear Relationship Between X and Y

The formula for the Pearson correlation discussed in this chapter is only appropriate when there is a linear relationship between X and Y. The Pearson formula

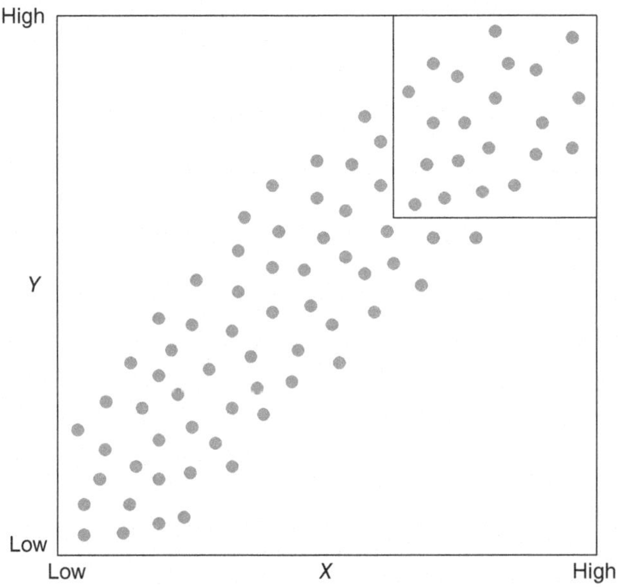

Figure 14.5 Strength of association is underestimated when range of scores is restricted.

becomes less appropriate as the relationship between *X* and *Y* departs from linearity; it underestimates the correlation.

Recall Figure 14.3 which depicted a curvilinear relationship between arousal and task performance. The Pearson formula would underrepresent this correlation. We need to protect ourselves from misapplying the Pearson formula; fortunately, to do so is not difficult. Simply examine the scatterplot. If the data is balanced and reasonably takes the form of an oval, we are safe in using the Pearson formula. Figure 14.6 illustrates some nonlinear relations between *X* and *Y*. Note that nonlinearity is a matter of degree. The data does not need to take a dramatic shape like ∪ or ∩. Nonlinearity is reflected by any plot that shows a curve.

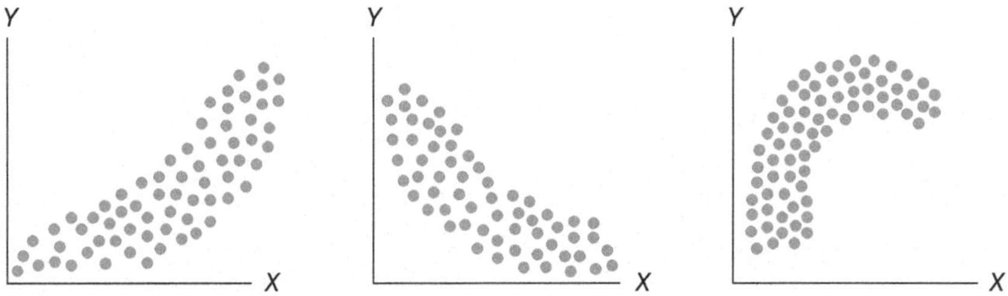

Figure 14.6 Some nonlinear relations between *X* and *Y*.

14.5.3 The *X* and *Y* Distributions are Skewed

Distributions that are terribly skewed can also create correlation computations that are artificially low, especially if the X and Y distributions are skewed in opposite directions. There are several potential remedies, the simplest being to transform the entire data set (see Section 4.6). Mathematical steps like these can help to normalize the shape of a distribution. An in-depth discussion of transformation techniques is beyond the scope of this introductory-level textbook; consult an advanced statistics resource for more information.

14.5.4 The Use of Extreme Samples

A common type of correlational design, particularly in the field of personality research, is to compare extreme groups with respect to some other variable. For example, suppose we are interested in the relationship between narcissism and aggressiveness. To explore it, we first use a scale to measure narcissism. We later administered an aggressiveness questionnaire to only those individuals who scored in the top and bottom 10% of the narcissism scale.

This methodological approach will typically yield a correlation that is larger than would be found if the entire population were used. If, instead of selecting only those with very high and very low narcissism scores, we include *all* participants who were given the questionnaire, the resulting correlation will most likely be lower. The association between *X* and *Y* is oftentimes weaker in the midrange of the distributions. Figure 14.7 depicts this point by showing a

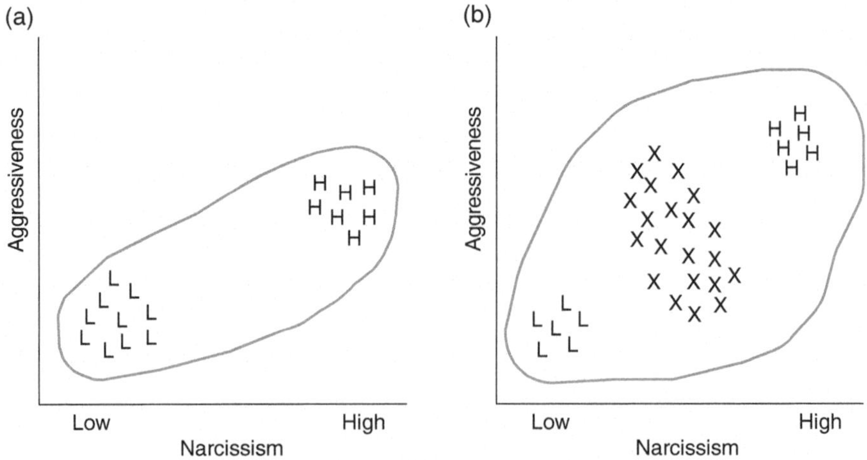

Figure 14.7 Use of only extreme groups (a) can artificially increase the magnitude of a correlation (b).

narrower oval when using only extreme groups (a) and a more circular oval when the entire range is included (b). For this reason, we should use caution when interpreting correlations based on data drawn only from the ends of distributions.

14.5.5 The Effect of an Extreme Score

A data point that stands off by itself, whether it be abnormally high or low, is called an outlier. Outliers can impact the calculation of a correlation, especially for small samples. The only difference between Sample *A* and Sample *B* below is the tripling of the *Y* score of the last participant.

Sample *A*		Sample *B*	
X	*Y*	*X*	*Y*
2	4	2	4
3	4	3	4
5	6	5	6
9	8	9	8
3	**3**	3	**9**

The *r* between *X* and *Y* in sample *A* is +.94. Higher *X* values strongly correspond with higher *Y* values. The creation of one outlier in sample *B* has dropped *r* to +.49.

Outliers present a dilemma for researchers. Researchers should not simply discard data because it is inconsistent with their expectations. This guards against biasing the outcome of a study to fit preconceived notions. Without this rule, confidence in the findings of empirical research would drop dramatically. The conventional approach is to analyze the data with and without the outlier; noting the difference it makes in the conclusions. Unusual data, although often perplexing, can serve a useful purpose. (For example, an interview of the participant responsible for the outlier may lead to another hypothesis for a subsequent study, or it may reveal that the participant misunderstood what was being asked of them.) In all cases, outliers need to be handled carefully and with transparency.

The correlation coefficient is one of the most useful measures for assessing the degree to which two variables are related. However, since there are numerous factors that affect the interpretation of the correlation, it is imperative that we carefully "dig into the data." Keep r^2 in mind when *r* is statistically significant and pay close attention to the scatterplot to check for linearity, skewness, restricted ranges, extreme samples, and the presence of outliers.

14.6 Formal Presentation of Findings

The proper reporting of Pearson r findings is similar to the proper reporting of t test findings. When reporting a significant Pearson r, we must include the degrees of freedom, the value of the obtained r (with the proper valence and without an "0" preceding the decimal), and the alpha value used. For instance, "Statistical evidence suggests a relationship exists between shoe size and height, $r(6) = +.94$, $p < .05$." A failure to reject the null might read, "There was no statistical evidence found to suggest a relationship between shoe size and height, $r(6) = +.04$, *n.s.*" A measure of shared variance (r^2) can be added to a rejected null hypothesis if needed.

Many other principles common to the proper reporting of all types of statistical findings were first laid out in Section 8.8. Please consult this portion of the text for more general information about the proper reporting of statistical findings.

Summary

A correlation is a measure of the strength of association between two variables. A data set of pairs of scores is called a bivariate distribution. A correlation based on a sample is symbolized as r and a population coefficient is symbolized as ρ.

Be careful not to imply causality when interpreting correlations. This is a common interpretive mistake. Some correlational coefficients may imply causality but only if the data has been gathered experimentally. The basis for making causal statements never resides with the type of analysis, but rather with the methodology used to collect the data. "Correlational *design* does not imply causation" is a more accurate motto than "correlation does not imply causation."

A correlation coefficient is represented by a number that ranges from +1 to −1; the higher the coefficient's absolute value, the stronger the association between the two variables. An r of +.60 reflects a strength of association as strong as an r of −.60. If higher values of one variable are associated with higher values of the other, then the correlation is "positive." If higher values of one variable are associated with lower values of the other, then the correlation is "negative."

When a bivariate distribution is plotted on a graph, it is called a scatterplot or scatter diagram. The scatterplot provides a great deal of information about the relationship between two variables. The magnitude of the correlation can be estimated by looking at the general shape formed by the points. The size of the correlation is estimated by examining the width of an oval used to envelop the data: the narrower the oval, the higher the correlation. The closer the oval takes the shape of a circle, the smaller the correlation.

Not only does the scatterplot indicate the strength of association but it also reveals the nature of the correlation. If the enveloping oval slopes from the lower left to the upper right, the correlation is positive. If it slopes from the upper left to the lower right, the correlation is negative. The degree of the slope (e.g. gradual vs. steep) is *not* indicative of correlation strength.

The scatterplot can also reveal nonlinear data. When the displayed data curves, the data is nonlinear. The formulas for calculating nonlinear correlations are different from those used to calculate a linear relationship. Only the computation of linear correlations is covered in this chapter.

The coefficient of determination, r^2, is a measure of the shared variance between two variables and is usually stated as a percentage. This tool is a better indicator of the relative strength of relationships between sets of variables than the correlation coefficient.

One null hypothesis test for r asks, "Is the magnitude of r sufficiently large to conclude that ρ is not 0?" The hypotheses are typically represented as: $H_0 : \rho = 0$ and $H_1 : \rho \neq 0$.

Several factors can affect the size of the correlation; these factors can hide the real nature of the relationship between the variables being correlated. When samples are restricted in terms of range, the correlation is likely to be artificially low. The formula presented in this chapter should only be used for linear relationships. When distributions are skewed, correlation computations can be underrepresented. The use of samples with extreme data can overstate the strength of the relationship. Outliers can create a misleading r and must be handled with care.

Microsoft® Excel and SPSS®

Information regarding how to use Microsoft® Excel and SPSS® to assist in performing Pearson correlations can be found in Appendix C.

Key Formula

Computational formula for Pearson *r*

$$r = \frac{n_p(\Sigma XY) - (\Sigma X)(\Sigma Y)}{\sqrt{\left[n_p(\Sigma X^2) - (\Sigma X)^2\right]\left[n_p(\Sigma Y^2) - (\Sigma Y)^2\right]}} \qquad (14.1)$$

Key Terms

Correlation coefficient
Bivariate distribution
Pearson correlation
Scatterplot
Linear relationship
Curvilinear relationship
Coefficient of determination
Shared (or common) variance

Exercises

1 Distinguish between correlational and experimental designs. Why does an experimental design allow for causal interpretations?

2 Which statement is most accurate?
a Correlational design does not imply causation.
b Correlation does not imply causation.
c Correlation implies causation.
d Correlation equals causation.

3 Describe what is meant by the term "bivariate distribution."

4 Provide examples of variable pairs that are positively correlated.

5 Provide examples of variable pairs that are negatively correlated.

6 Which of the following is the strongest legitimate correlation? Why?
a −1.14
b −.69
c 1.09
d .58

7 Describe what a scatterplot of a +1 or −1 correlation would look like.

8 What information can be gained from a scatterplot?

9 Draw a scatterplot reflecting the following correlations.
a A somewhat strong positive correlation.
b A very strong negative correlation.
c A very weak positive correlation.
d A curvilinear correlation.

10 Estimate an r for each of these scatterplots.

(a)

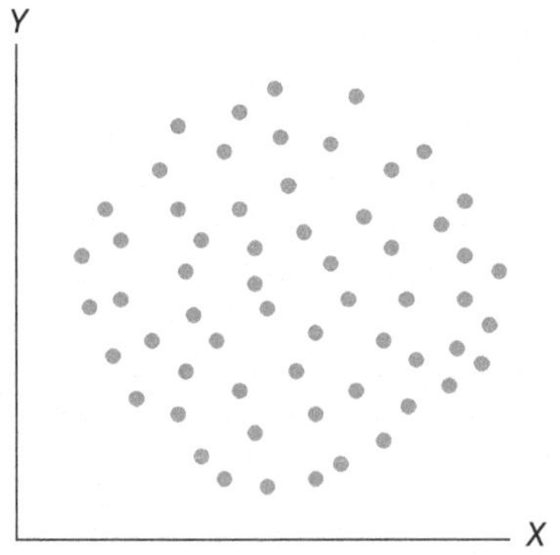

(b)

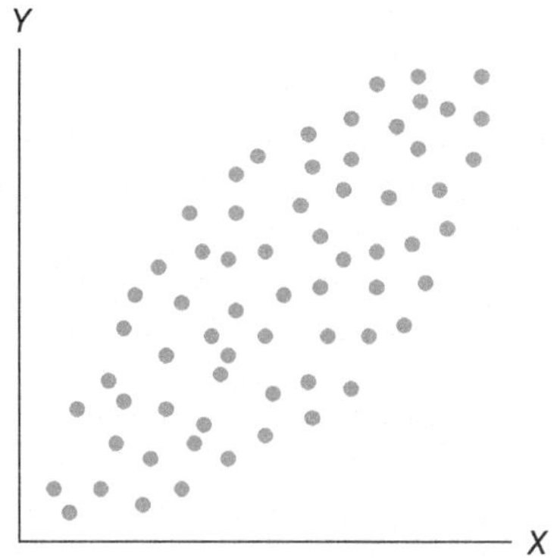

(c)

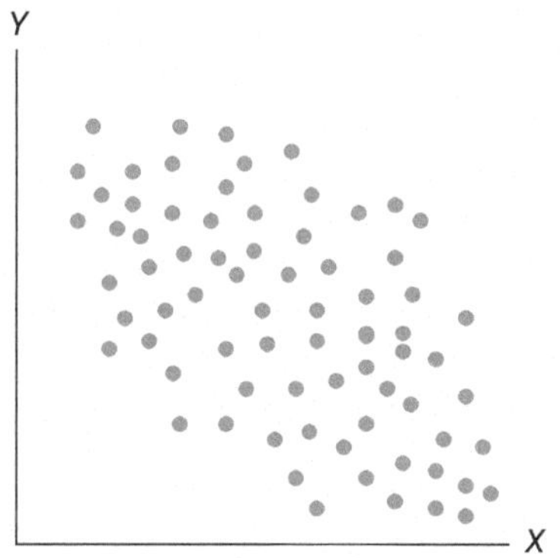

(d)

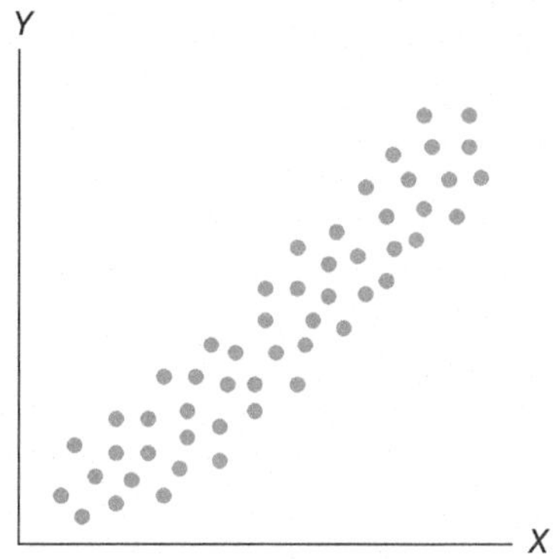

11 For the following correlations and df, what are the critical r's when using a 5 and a 1% alpha value (two-tailed test). (Note: Use an online table where Table A.7 is incomplete.) In each case, should we reject or fail to reject the null hypothesis at the 5% level? At the 1% level?

a $r = .67, df = 24$
b $r = .39, df = 24$
c $r = .89, df = 12$
d $r = .74, df = 18$
e $r = .45, df = 29$
f $r = .95, df = 7$
g $r = .62, df = 13$
h $r = .24, df = 100$

12 If X accounts for 49% of the Y variance, Y accounts for how much of the X variance?

13 Given $r^2 = .36$ or 36%, what is the value of r?

14 When rejecting a null, why should one also look at the percent of shared variance?

15 How might sampling a restricted range of scores affect a correlation?

16 How might use of only scores in the extreme of a distribution affect a correlation?

17 If we use the Pearson r to measure a nonlinear relationship, what is the likely result?

Work Problems

18 For the following data set:

a Calculate r and conduct a two-tailed test of the null hypothesis that $\rho = 0$ ($\alpha = .05$).
b Specify the hypotheses.
c What is r_{crit}?
d Reject the null hypothesis?
e What is the shared variance, r^2?

X	*Y*
6	4
7	5
7	6
4	6

19 For the following data set:

a Calculate r and conduct a two-tailed test of the null hypothesis that $\rho = 0$ ($\alpha = .05$).
b Specify the hypotheses.
c What is r_{crit}?
d Reject the null hypothesis?
e What is the shared variance, r^2?

X	Y
6	3
9	7
7	6
10	9

20 For the following data set:

a Calculate r and conduct a two-tailed test of the null hypothesis that $\rho = 0$ ($\alpha = .05$).
b Specify the hypotheses.
c What is r_{crit}?
d Reject the null hypothesis?
e What is the shared variance, r^2?

X	Y
9	12
8	6
10	10
13	13

21 For the following data set:

a Calculate r and conduct a two-tailed test of the null hypothesis that $\rho = 0$ ($\alpha = .05$).
b Specify the hypotheses.
c What is r_{crit}?
d Reject the null hypothesis?
e What is the shared variance, r^2?

X	Y
1	8
3	7
3	6
5	5
6	4

22 Draw the scatterplots for the data of Problems 18, 19, 20, and 21.

23 A sports psychologist is interested in the relationship between how many weeks people exercise and their resting heart rate. Using the data below to answer these questions.

a What is r?
b What are the hypotheses?
c What is r_{crit} ($\alpha = .05$, two-tailed)?
d Reject the null hypothesis?
e What is the shared variance, r^2?

Weeks of exercise	Resting heart rate
2	82
4	78
8	72
14	66
10	66
9	70
9	69

24 A school psychologist hypothesizes a relationship between IQ and number of siblings. Use the data below to answer these questions.

a What is r?
b What are the hypotheses?
c What is r_{crit} ($\alpha = .05$, two-tailed)?
d Reject the null hypothesis?
e What is the shared variance, r^2?

Number of siblings	IQ
8	123
3	100
1	90
4	111
2	102
0	95

25 A psychologist is interested in the strength of association between age and performance on a certain task requiring motor skills. Plot the scatter diagram of the following data and decide on the most reasonable course of action for testing the hypothesis.

Age in years	Number of errors
5	23
6	19
7	17
9	16
10	16
11	18
12	18
14	19
15	20
16	22

15

Linear Regression

15.1 The Research Context

Regression is a set of statistical procedures, built on the concepts of correlation presented in Chapter 14, that allow a researcher to use information about one variable to *predict* the value of a second variable. The idea of using statistical techniques for prediction purposes is new territory for this text, and is uniquely associated with this chapter. On many occasions, social and behavioral scientists would like to make predictions. Graduate admissions committees, for example, would like to accept applicants who are likely to do well at their institution. If Graduate Record Exam (GRE) scores are found to correlate well with future academic success, then an applicant's GRE scores can aid in the selection process.

A generation ago, researchers Zullow and Seligman (1990) showed that the outcome of presidential elections could be predicted by examining the content of campaign speeches. They found that the more candidates dwelled on negative events, the less likely they were to win.[1] However, recent findings suggest limitations to this theory. For instance, Combs and others (2009), showed that a person's level of political partisanship predicts the type of emotional reactions they will have to negative news events. These authors note that an economic collapse should be bad news for everyone, republican or democrat. However, when a negative event is interpreted as being caused by a rival party, their model predicts that those with high levels of political partisanship will experience *increased* levels of happiness. Sometimes, complex real-world relationships can be difficult for traditional experimental designs to address. In these situations, correlational designs and regression analyses are welcome tools for the researcher.

1 Technically, Zullow and Seligman (1990) used more sophisticated analyses than simple regression. However, their dataset and the type of question they were researching are consistent with regression.

Essentials of Behavioral and Social Science Statistics, Third Edition.
K. Paul Nesselroade, Jr. and Laurence G. Grimm.
© 2025 John Wiley & Sons, Inc. Published 2025 by John Wiley & Sons, Inc.
Companion website: https://www.wiley.com/go/Essentials_of_Behavioral_and_Social_Science_Statistics3e

15.2 Overview of Regression

A prediction is accomplished by using a **regression equation**. For regression to be of use to an investigator, the two variables *must* be related (correlated). This precondition makes perfect sense. If high school GPA is uncorrelated with university GPA, then there is no way to use high school GPA to predict the subsequent GPA. Additionally, *linear* regression analyses require the nature of the relationship to be *linear*, where changes in one variable are matched with a constant change in the other. If the change in the second variable is not constant, the relationship between X and Y is nonlinear. This quality will be reflected in the scatter plot. Chapter 14 included some scatter plots in which the spread of data points was curved, indicating a nonlinear relationship. In these situations, the Pearson formula should not be used. Likewise, the regression methods presented in this chapter should not be used on nonlinear relationships. Inspecting a scatterplot is very helpful when deciding whether linearity exists.

This chapter considers simple regression, where information about only one variable (called the predictor or independent variable) is used to predict a second variable (called either the predicted, criterion, or dependent variable). The terms *independent* and *dependent* variable have different meanings in the context of regression. In this context, the researcher typically does *not* manipulate the independent variable. It is simply the variable used for prediction purposes; it is the X. The dependent variable, that is, the value of Y that is predicted for a given X, is said to *depend* on the value of X. This should not be understood as implying cause and effect.[2] Despite this confusion, many resources for statistical analysis choose to employ the terms "independent" and "dependent" variables. To avoid confusion, we will not use these terms.

Regression analyses are bidirectional. For instance, given a correlation between stubbornness and empathy, regression can be used to predict a person's stubbornness based on their empathy. Conversely, a person's empathy level can be predicted based on their stubbornness. This bidirectionality reminds us that causal interpretations are typically not warranted.

The word "prediction" implies a temporal sequence between variables; this is potentially misleading. We might be tempted to conclude that a *current* predictor variable is used to predict a *future* predicted variable. However, no futurity is implied with regression. To say that empathy can be predicted from stubbornness is not to suggest that stubbornness occurs first and then empathy later. In fact, the temporal relationship can even be known to run in the opposite direction. For example, we could use resting heart rate to predict time spent exercising, even though it is obvious

---

2 Regression techniques can be used with experimental designs. Interested readers are referred to advanced statistics texts (e.g. Ryan, 2008) for a discussion of regression analyses in the context of true experiments.

that exercising is needed to reduce resting heart rates. To say "X predicts Y" is merely to say that if a Y value is unknown, X can be used to predict it.

In simple regression, only one predictor variable is used. **Multiple regression** employs more than one predictor variable. The inclusion of additional predictor variables usually increases prediction accuracy. This is attractive to researchers. For example, multiple regression was used to determine that a student's level of sexual identity-related distress could be predicted by combining information about their public and private sexual identity, organizational religiosity, intrinsic religiosity, and level of same-sex attraction (Yarhouse et al. 2018). Although multiple regression is beyond the scope of this text, it is based on the concepts discussed in this chapter. For more information about different types of analyses related to linear regression, see Box 15.1.

Box 15.1 Next Steps with Regression Analyses

This is an introductory statistics textbook, as a result, we should not be surprised to learn that there is much more to know about each one of the statistical procedures that is presented. We have already learned that there are many versions of follow-up t tests and ANOVAs that can be applied to complex research designs. The analytical tool of regression is no different. In fact, there are numerous concepts that build upon the basis of the linear regression idea. This box will briefly identify several of them.

Recall that linear regression uses a known relationship between two variables to predict unknown values. As was just mentioned, *multiple regression* uses more than one predictor variable to construct predictor equations for unknown values. If one variable (X) can account for some of the variance in another variable (Y), perhaps a third variable (Z), also shown to relate to Y, can account for even more. And, if the use of two predictor variables is better than one, might not three be better than two? What about four? And so on. The addition of other predictor variables introduces new problems and makes the analysis more complicated, but the basic idea remains; use known relationships between variables to help make more accurate predictions of unknown values for a variable of interest.

Factor analysis is a statistical technique designed to take a collection of interrelated variables and shrink them down to the fewest number of actual core concepts. Oftentimes, these are inferred variables that are not themselves being measured (they are hidden or latent). For instance, we may find that dozens of medically interesting variables seem to be correlated. A factor analysis might show that the dozens of correlated variables may be reducible down to just a handful of core variables, like activity level, social-support network, diet type, and lifestyle. Factor analysis finds hidden patterns in data by helping to identify a few core variables (also called "dimensions" or "factors") underlying the host of measured variables. These factors can then be listed according to their *factor loadings*; that is, how much variation they can account for or explain. For instance, we may learn that "diet-type" explains a large amount of the shared variance among a set of different correlated medical variables.

Path analysis is often described as a straightforward extension of multiple regression. The purpose of the procedure is to generate size and significance estimates of the hypothesized causal connections between sets of variables. As the name implies, path analysis organizes the predictor variables into a causal sequence of variables where one variable leads to another and so forth. However, rarely are the paths simple relationships. Oftentimes, the paths determined by the analysis are complex. The path analysis diagram below reflects the idea that two variables (X_1 and X_2) seem to be causing, both directly, and indirectly (through D_1), an effect on a given dependent variable (D_2). We can think of path analysis as a form of multiple regression that attempts to develop a causal model between the correlated variables.

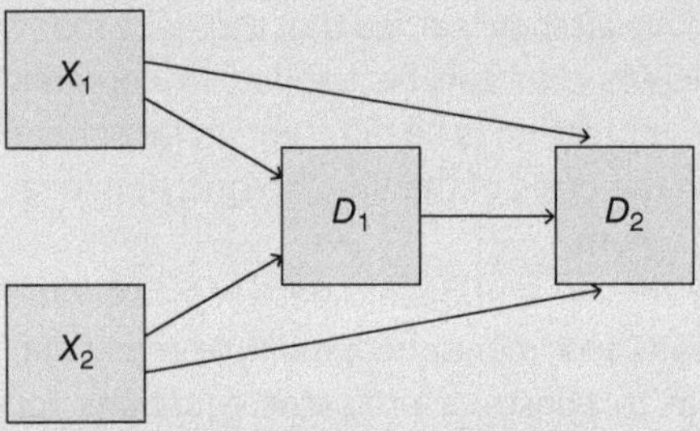

Structural equation modeling combines several regression-related tools, including factor analysis and path analysis, to "model" very complex relationships between variables. Its use is growing because it can suggest predictive (and even causal) relationships amid a collection of hidden (latent) variables while relying only on data gathered from observable variables. As computing power has increased, a sophisticated collection of analytical techniques has emerged to help researchers test their theories.

15.2.1 Using Correlated Information to Make Predictions

In this section, our working example will be a variable that was explored by Francis Galton (see Spotlight 15.1) when the mathematical expressions of correlation and regression were first taking shape in the late 1800s. If asked to predict the height of a randomly selected biological male, our best guess would be the mean height for all males. Assuming the data are normally distributed, most heights will be clustered around this value. It will probably be wrong, but it will most likely be close. It is the best guess we could make. Suppose the mean height is known to be 1.75 meters. Now, imagine we were asked to make a prediction for a series of randomly selected males. Each time we are asked, we should guess 1.75 meters. A few times we would be correct, but most of the time we would be incorrect. However, when incorrect, we would usually be only a little bit incorrect. Some of the time we would be very incorrect. Ultimately, using the mean height for all males would lead to a lot of wrong predictions, but at least our errors would be mostly small and unbiased; that is, overpredicting about as often as underpredicting.

Now let us change the game. Suppose we are told there is a correlation between the heights of fathers and sons. Furthermore, we are given a table that states the mean height of sons for fathers of a given height. We might learn that fathers who are 1.65 meters tall have sons who average 1.70 meters tall; fathers who are 2 meters tall have sons who average 2.05 meters tall; and so on. Now, if asked to predict the height of a biological male whose father is 1.65 meters tall, we now have a more specific dataset to use. Instead of predicting the mean height of all males, we should predict 1.70 meters, the mean height of males who have fathers who are 1.65 meters tall. To predict the height of a son, we should first obtain the father's height and then consult the distribution of son's heights associated with fathers of that size.

Our prediction will still be wrong most of the time, but the error will likely be smaller than if we had ignored the correlated information. If this strategy seems obvious, then we already have an intuitive understanding of regression. Although regression does not entail using a set of tables to make predictions, an equation accomplishes the same thing. Some score (e.g. height of father) is entered into the regression equation, and the predicted score (e.g. height of son) is computed.

15.2.2 Regression as a Two-Step Process

The regression procedure involves two steps. The first is to identify two variables that are correlated and to gather the bivariate data. From this data, we construct a regression equation for later use. The second step involves applying this regression equation to data from participants *not* included in the original sample – participants for which we only have an X value, not a Y. In this way, a regression equation constructed from bivariate data can be applied to new situations.

For example, if the goal is to use the height of fathers to predict the height of sons, a sample of height measurements from fathers and their sons is required. Then, a regression equation is built. This equation is then used to predict the unknown height of sons from the known height of fathers.

15.3 Establishing the Regression Line

In linear regression, the regression equation is used to plot a prediction line that goes through the middle of the scatterplot. This is called the **regression line**. The formulas for the line and the slope of that line (under certain criteria, to be explained later) are found below.

Linear regression equation

$$Y_p = M_Y + b(X - M_X) \quad (15.1)$$

where

Y_p = the predicted Y value
M_Y = the mean for the Y values
b = the slope
M_X = the mean of the X values

Computational formula for the slope

$$b = \frac{n_p(\sum XY) - (\sum X)(\sum Y)}{[n_p(\Sigma X^2)] - (\Sigma X)^2} \quad (15.2)$$

These formulas are for predicting Y given X (also called Y on X). These equations *cannot* be used for predicting X given Y. To ensure this is not a problem, be sure to label the variable to be predicted as Y and the variable that is being used as a predictor as X.

A regression equation chooses a particular line from among an infinite number of straight lines that could be drawn through the scatterplot (see Figure 15.1).

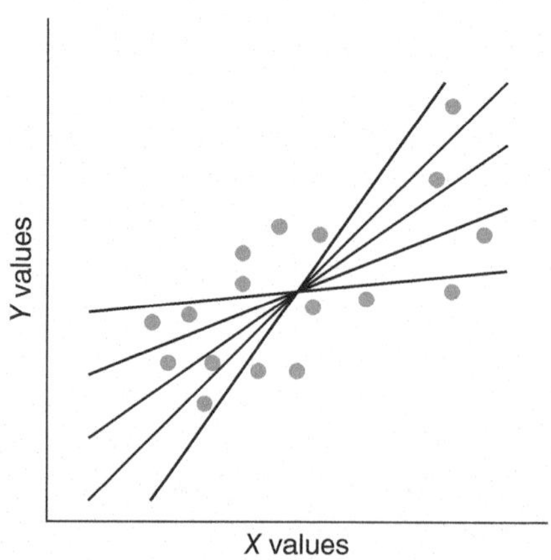

Figure 15.1 A regression equation finds the best predictor line from among many possible ones.

Before discussing how a particular regression line is chosen, it is important that we understand how to use the line.

15.3.1 The Regression Line

In regression, an X score is used to predict the value of a Y score. There is a different predicted Y for each value of X. Every predicted Y (symbolized Y_p) lies on the regression line. To find Y_p, find the corresponding X value on the horizontal axis, and draw a line parallel to the vertical axis. When that line meets the regression line, another line is drawn at a right angle (parallel to the X axis) until it meets the vertical (Y) axis. The value at which this line intersects the Y axis is Y_p. In Figure 15.2, a Y of 6 would be predicted for anyone who obtained an X of 4.

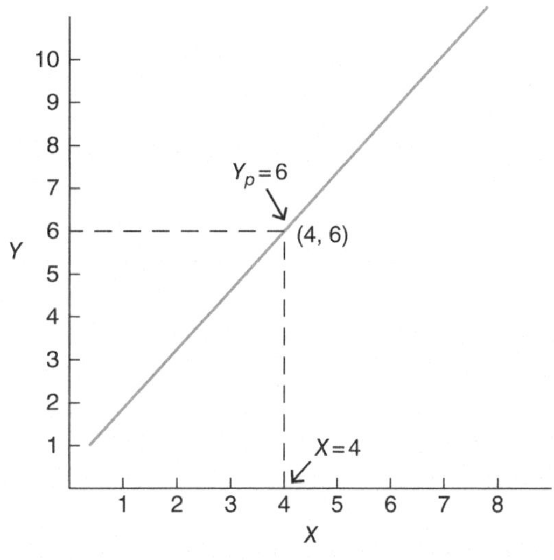

Figure 15.2 Using the regression line to predict Y given $X = 4$.

The **slope** (b) of the regression line measures the "rise over the run"; how many units the line rises on the Y axis for every one unit moved to the right on the X axis. For instance, if $b = 0.93$, the regression line ascends 0.93 Y units with each successive unit of X. In addition, a positive slope corresponds to a positive correlation, and a negative slope corresponds to a negative correlation. Larger absolute values of b indicate steeper slopes of the regression line. Figure 15.3 illustrates lines with three different slopes. In Figure 15.3a, the line "rises" one unit on the Y axis for every "run" of two units on the X axis. The slope is 0.5. In Figure 15.3b, the slope of the line neither ascends nor descends; this is a slope of 0. In Figure 15.3c, the line descends one unit for every one unit of increase on the X axis. The slope is -1.

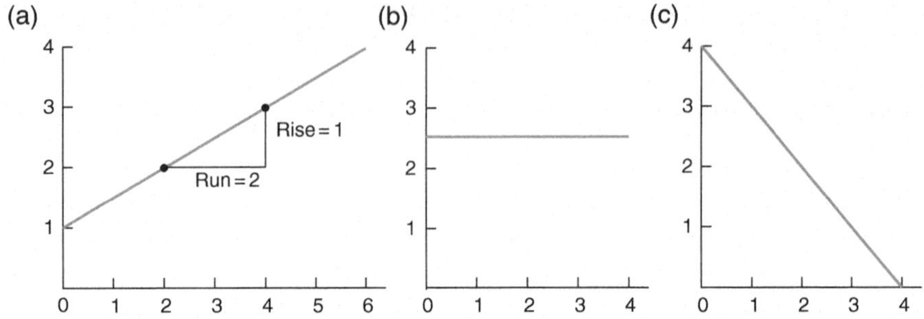

Figure 15.3 Regression lines with different slopes. (a) Slope of 0.5; (b) slope of 0; (c) slope of −1.

15.3.2 The Least Squares Criterion

The criterion most often used to select the best straight line through the scatterplot is called the **least squares criterion.** It assures that the regression line chosen has the least amount of prediction error possible.

If the relationship between X and Y is perfect ($r = \pm 1$), then all data points will lie along a straight line. The regression line will have all the points of the scatterplot on it. Since a correlation of ± 1 is extremely rare, there is typically a spread of points surrounding the line; every point not on the line reflects error. "Error" is the difference between the actual Y score obtained by an individual and Y_p, the score predicted for that person by the regression line.

Figure 15.4 shows a hypothetical regression line for the relationship between GRE scores and GPA. (For visual clarity purposes, the dataset used to create the regression line is not depicted.) Dan received a GRE score of 650. The regression line predicts a GPA of 3.62. If Dan's GPA turns out to be 3.00, we have erred by −0.62 GPA points. Marcy also scored 650 on the GRE. However, Marcy achieved a 4.00 GPA. In this case, the regression line was off by 0.38 GPA points.

Of all the straight lines that could be drawn through the scatterplot, we want the line that will serve as the best unbiased predictor of Y. Every data point in the dataset not falling on the regression line has a measure of error, defined as $Y - Y_p$ (symbolized as e; $Y - Y_p = e$). Sometimes $Y - Y_p$ will be positive, and sometimes it will be negative, depending on whether a given Y falls above or below the regression line. By *unbiased* we mean that we want the regression line to be perfectly balanced with regard to positive and negative error, such that $\Sigma(Y - Y_p) = 0$. Furthermore, we want the regression line to be fitted to the scatterplot in such

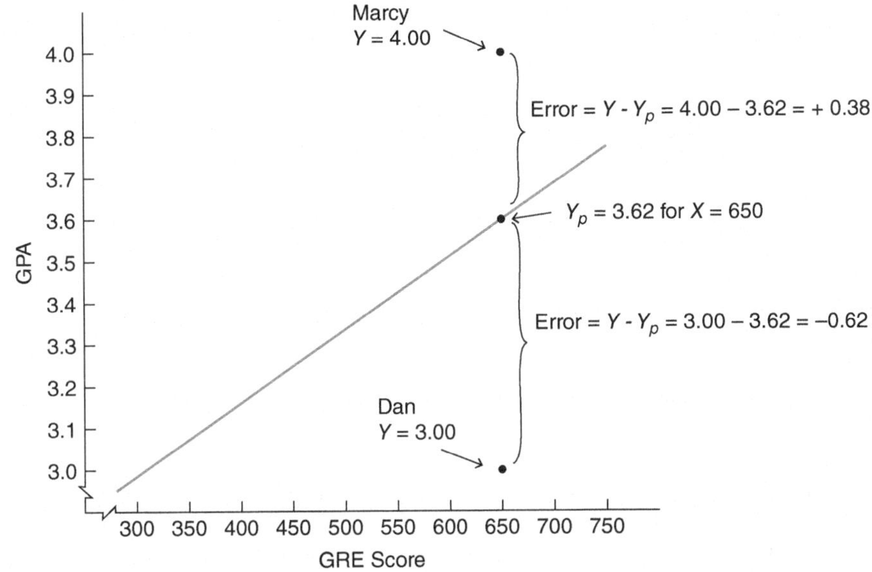

Figure 15.4 An error is the deviation of an actual Y from the predicted Y (Y_p).

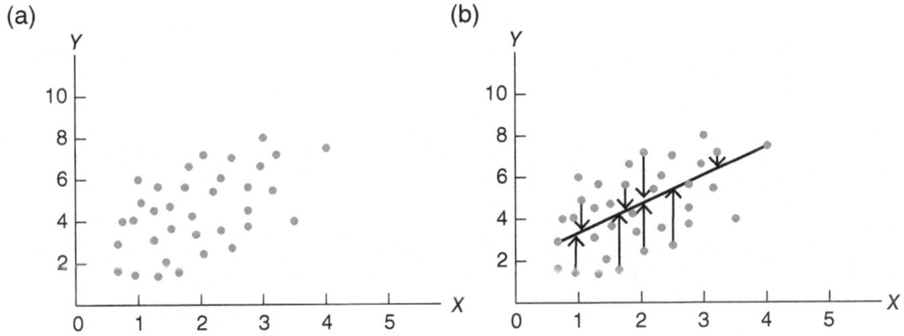

Figure 15.5 A scatterplot (a) has been fitted with a regression line (b) reflecting a minimized Σe^2.

a way that the sum of the *squared* errors (Σe^2) is as small as possible. Thus, this is the **least squares method.**

In Figure 15.5b, a regression line has been fitted to the scatter plot depicted in Figure 15.5a. Each data point not on the line possesses some amount of error. No other line drawn through the plot will yield a smaller value for Σe^2.

■ **Question** *Use data below to create a regression equation and find Y_p for $X = 9$.*

X	X^2	Y	Y^2	XY
9	81	11	121	99
6	36	8	64	48
5	25	6	36	30
7	49	9	81	63
4	16	7	49	28

Solution

$$M_X = 6.20; M_Y = 8.20; \Sigma X^2 = 207; \Sigma X = 31; \Sigma Y = 41; \Sigma XY = 268; n_p = 5$$

$$b = \frac{n_p(\Sigma XY) - (\Sigma X)(\Sigma Y)}{\left[n_p(\Sigma X)^2\right] - (\Sigma X)^2} = +.93$$

$$Y_p = M_Y + b(X - M_X)$$

$Y_p = 8.20 + 0.93(X - 6.20)$ is the regression equation for Y given X. Now, any X value can be placed in the equation to yield the Y_p for that given X. The question asks for the predicted Y when $X = 9$. When $X = 9$, $Y_p =$ **10.80**. ■

15.3.3 Plotting the Regression Line

To plot the regression line, two points are required. Simply solve the regression equation for any two values of X (ideally, the two points will be a good distance apart). If we make $X = 0$, the Y_p will fall on the Y axis – the ***Y* intercept**. There is nothing particularly special about the intercept. However, for purposes of plotting the line, it is oftentimes considered a helpful reference point.

For this example, we will select *X's* of 4 and 8 to draw the regression line.

$$Y_p = 8.20 + 0.93(\mathbf{4} - 6.20) = \mathbf{6.15}; Y_p = 8.20 + 0.93(\mathbf{8} - 6.20) = \mathbf{9.87}$$

The two coordinates are $X = 4$, $Y_p = 6.15$ and $X = 8$, $Y_p = 9.87$. Figure 15.6 shows the regression line for Y given X.

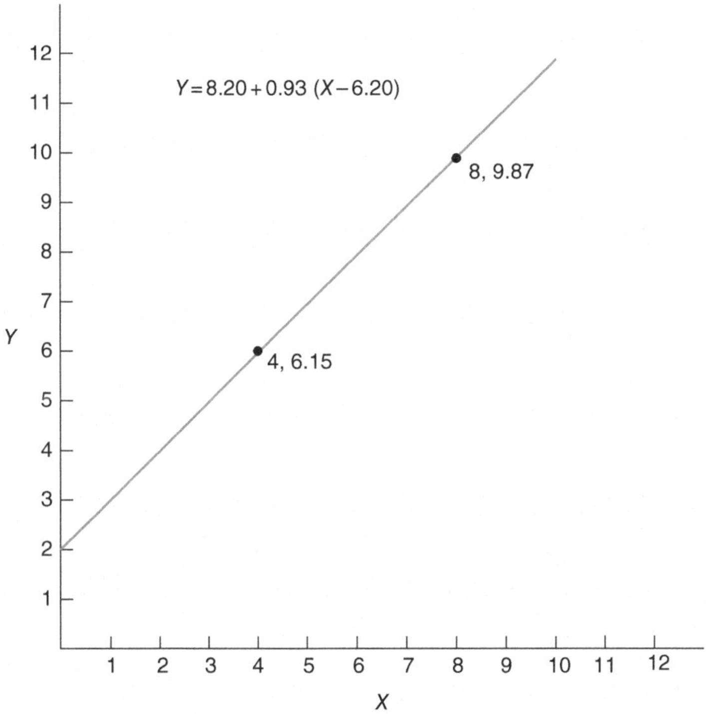

Figure 15.6 Drawing the regression line using two points: (4, 6.15) and (8, 9.87).

15.3.4 Correlation Formula for the Slope

Formula 15.2 is used to compute the slope from raw data. Formula 15.3 can be used if s_Y, s_X (the standard deviations of X and Y), and r (the correlation) are provided. Formula 15.3 is referred to as the correlation formula for the slope.

Correlation formula for the slope

$$b = r\left(\frac{s_Y}{s_X}\right) \quad (15.3)$$

15.3.5 Analysis of Regression

We do not have a separate section of the text dealing with the assumptions behind the regression analysis, but they are very similar to the assumptions for other inferential statistical procedures, namely, representative samples, independent observations, the use of interval or ratio data, and normally distributed data. (There are two additional assumptions discussed later in the chapter.)

A regression analysis can only be performed when X and Y are correlated. This can be mathematically represented by examining Formula 15.3. If $r = 0$, the slope (b) is also zero. A slope of 0 has no predictive power. This requirement is checked by most statistical analysis software programs. In deference to brevity, we will not walk through the associated ANOVA formulas or summary table. However, virtually all statistical software programs, including Microsoft® Excel and SPSS®, generate an ANOVA to test the null hypothesis that $b = 0$ when a regression analysis is performed.

15.3.6 How Accurate Is the Regression Equation?

Regression equations look very scientific. There is a tendency to equate "looking scientific" with "accurate." Regression equations, however, are merely statistical tools, and just as some carpentry tools are more refined than others, some regression equations are more accurate than others. For regression equations, the measure of accuracy is embodied in the concept of *prediction error*. The measure of prediction error is called the **standard error of the estimate**, which is often symbolized as s_e. An understanding of the conceptual basis of the standard error of the estimate first requires a discussion of conditional distributions.

15.3.7 Conditional Distributions

A bivariate distribution is based on two related univariate distributions; one univariate distribution is composed of all the observations of X, and the other is composed of all the observations of Y. Imagine we have conducted a massive study and recorded each participant's X and Y scores. Then someone asks to see all the Y scores that correspond to an X score of 5. Do we think every Y score associated with an X of 5 will be the same? Unless the correlation between X and Y is $+1$ or -1, there will be an array of Y values associated with an X of 5. In other words, for that given value of X, there exists a corresponding distribution of Y scores. The distribution of Y scores for a given X is called a **conditional distribution**. Each conditional distribution has all the characteristics of any distribution of scores: a mean, standard deviation, and so forth.

Figure 15.7 represents different conditional distributions for four different values of X. The figure is idealized such that each conditional distribution is normal. (This figure is also a bit misleading since flat surfaces force us to illustrate the concept of conditional distributions using only two dimensions. More visually appropriate would be to imagine the gray conditional distributions protruding up off from the surface into a third dimension.)

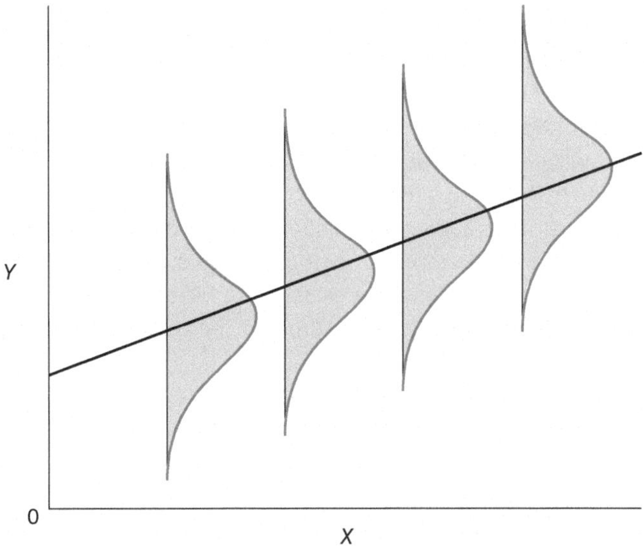

Figure 15.7 Normally distributed conditional distributions of four values of *X*.

Recall from Chapter 14 that the best prediction of an unknown value is the mean. Well, when presented with correlated information the mean is still the best prediction; but a regression analysis allows us to make this *the mean of a particular conditional distribution*. Each *X* score has an associated *Y* distribution of scores. When asked the question, "What would we predict for a given value of *X*?" the regression equation answers, "The *mean* of the conditional distribution of *Y* scores associated with that *X*." The regression line connects all the means of the conditional distributions (see Figure 15.7). For this reason, it can be referred to as the "line of moving means."

Francis Galton has the distinction of originating the concept of regression. Spotlight 15.1 describes the person in the context of his discoveries.

Spotlight 15.1 Sir Francis Galton

Francis Galton (1822–1911) is credited with developing the concepts of correlation and regression. He and his understudy, Karl Pearson (see Spotlight 14.1), are responsible for much of the conceptual content of this textbook. Born into a wealthy English family, Galton was afforded the luxury of indulging his scientific curiosities. In the mid-1800s, European explorers were mapping the interior of Africa. Perhaps inspired by the travels of his prodigious cousin, Charles Darwin, Galton departed for Africa at the age of 28. His maps of uncharted regions of the continent were well-received, and he was valorized by the British Geographical Society.

Galton was obsessed with measurement. Throughout his life, he zealously counted things. Once, when attending a lecture, he counted the number of fidgets per minute by members of the audience and correlated it with their attentiveness (Galton 1885). While his portrait was being painted, he counted the number of brush strokes made in an hour. He extrapolated it by the number of hours needed to complete the portrait and estimated the total number of strokes to be 20 000.

Without question, the greatest influence on Galton's work was Darwin's *Origin of Species*. The bulk of Galton's career was almost entirely devoted to the quantification of heredity. This was the pressing concern that spurred on his many statistical insights and inventions. In addition to the development of statistical concepts, he applied them to gathered data, collecting thousands of observations on a variety of physical and mental attributes, many from parents and their offspring. Indeed, Galton is viewed as the father of biometrics. He is also seen as the father of mental testing. His book, *Hereditary Genius* (1869), forcefully argued that intelligence is measurable, and largely inherited.

Galton's abiding belief in the inheritance of physical, mental, and, in fact, moral attributes led him to begin a movement to better humanity through selective breeding, which he called *eugenics*, meaning "good birth." To be fair, Galton could not have foreseen where his ideas might lead. At the time, eugenics held great appeal for many of the brightest scientists, writers, and politicians of the era. Karl Pearson, Sir Ronald Fisher, George Bernard Shaw, George Orwell, Teddy Roosevelt, and Winston Churchill were just some of the most prominent and ardent supporters of the movement. However, the sorrowful story of eugenics in America includes forced sterilizations, marriage laws for the poor and marginalized, and immigration laws against non-Europeans. Even more horrific are the drastic measures undertaken in the name of eugenics by German scientists (e.g. Lifton 2000) and politicians (e.g. Kühl 1994) in Nazi Germany. There is simply no positive spin that can throw this darkest chapter of modern scientific reasoning into a better light. Nevertheless, we can say that despite the evils that emerged in the wake of Galton's utopian ideal, the *techniques* he invented for exploring the world around us have unquestionably revolutionized many disciplines.

15.3.8 Formulas for the Standard Error of the Estimate

To recap, the regression line connects the means of all the conditional distributions, and Y_p is the mean of the conditional distribution associated with a given X value. We can build onto this foundation by introducing an error value, s_e, that captures the average standard deviation of the Y scores around a given X value. With two additional assumptions, we can suggest that one s_e value can be used as a measure of prediction error for any value of X. The first assumption is that each conditional distribution is normally distributed. If the regression line runs through the mean of each conditional distribution, then it is important these means fall at the center of each conditional distribution. The second assumption is that each conditional distribution has the same standard deviation. This assumption is termed **homoscedasticity** (*homo* meaning same, *scedastic* meaning scatter).

In terms of calculation, the following computational formula can be used.

Computational formula for s_e

$$s_e = \sqrt{\left[\frac{1}{n_p(n_p-2)}\right]\left[\left(n_p\Sigma Y^2-(\Sigma Y)^2\right)-\left(\frac{\left[n_p\Sigma XY-(\Sigma X)(\Sigma Y)\right]^2}{n_p\Sigma X^2-(\Sigma X)^2}\right)\right]} \tag{15.4}$$

15.3.9 The Size of the Correlation and Prediction Error

An equivalent formula for s_e is presented that allows a researcher to examine the relationship between r and s_e. Formula 15.5 is the correlation formula for the standard error of the estimate. It can be used to find s_e if s_Y, n_p, and r are provided.

Correlation formula for s_e

$$s_e = s_Y\sqrt{(1-r^2)\left[\frac{n_p}{(n_p-2)}\right]} \tag{15.5}$$

Suppose the correlation between X and Y is perfect. Substituting r in the formula with either $+1$ or -1 yields $s_e = 0$. A perfect correlation means no prediction error.

15.3.10 Interpreting the Standard Error of the Estimate

With a normal distribution, recall that approximately 68% of the scores fall within ±1 standard deviation of the mean. In regression, Y_p is the mean of a conditional distribution; approximately 68% of the actual Y scores associated with a given X value will fall within $\pm1s_e$ of the mean (Y_p). Further, recall that for a given X score,

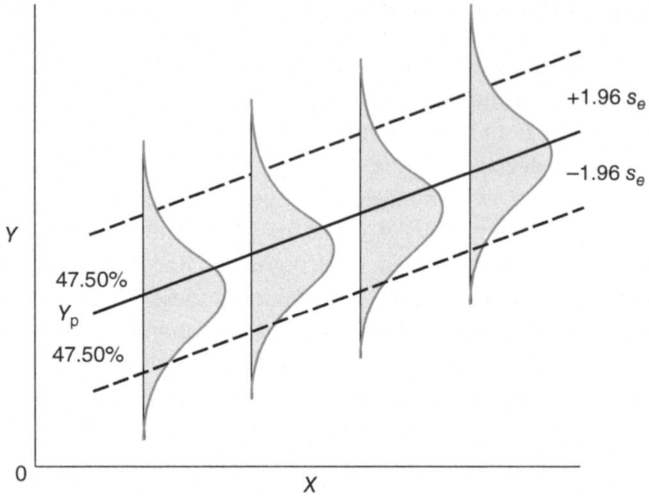

Figure 15.8 Ninety-five percent of the actual Y scores fall between $Y_p \pm 1.96s_e$.

95% of the actual Y scores fall between $Y_p \pm 1.96s_e$ (Shavelson 1996). Figure 15.8 shows conditional distributions for four values of X. The dotted lines mark the cut-offs between which 95% of the Y values fall.

15.4 Putting It All Together: A Worked Problem

A team of researchers is interested in the relationship between the marital satisfaction of couples and the subsequent marital satisfaction of children from those marriages. The researchers intend to use this data to create a regression equation that will predict the future marital satisfaction of adolescents, 15 years later when they are married. The following table and list of steps show the calculation of the correlation, regression equation, and the standard error of the estimate. Figure 15.9 shows the scatterplot of the pairs of scores. The Marital Satisfaction Questionnaire (MSQ) scores can range from 0 (very dissatisfied) to 10 (very satisfied). An interpretation of the findings is also provided.

Step 1. Construct a scatterplot to see if a linear relationship exists between X and Y. Figure 15.9 indicates the relationship is linear.

Step 2. Compute the correlation between X and Y. Although not required, a regression equation is useless if there is no correlation. The correlation coefficient is +.80 ($r_{crit} = \pm.75$).

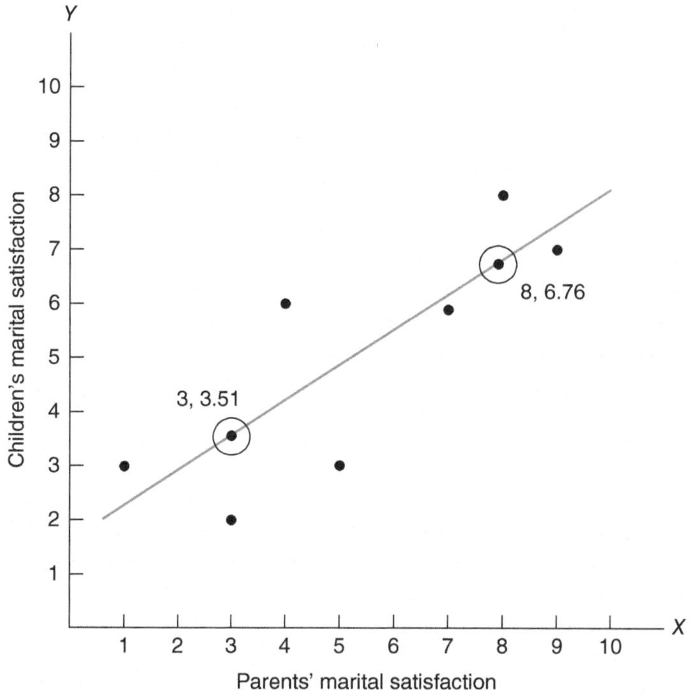

Figure 15.9 The scatterplot and regression line for the data on marital satisfaction.

Marital Satisfaction Questionnaire (MSQ) results		
Parents: X		***Children: Y***
1		3
3		2
7		6
9		7
8		8
4		6
5		3
$M_X = 5.29$		$M_Y = 5.00$
	$n_p = 7$	
$\Sigma X = 37;$		$\Sigma Y = 35;$
$\Sigma X^2 = 245;$		$\Sigma Y^2 = 207;$
$s_X = 2.87$		$s_Y = 2.31$
	$\Sigma XY = 217$	

$$r = \frac{n_p(\Sigma XY) - (\Sigma X)(\Sigma Y)}{\sqrt{\left[n_p(\Sigma X^2) - (\Sigma X)^2\right]\left[n_p(\Sigma Y^2) - (\Sigma Y)^2\right]}} = +\mathbf{.80}$$

Step 3. Compute the slope. Since r has been computed and s_X and s_Y are provided, Formula 15.3 could be used. Nonetheless, the computational formula is used for illustrative purposes.

$$b = \frac{n_p(\Sigma XY) - (\Sigma X)(\Sigma Y)}{\left[n_p(\Sigma X^2)\right] - (\Sigma X)^2} = +\mathbf{.65}$$

Step 4. All values needed to establish the regression equation are now available. Pick two discrepant values of X and use the regression equation to compute Y_p for each one, plot them, and draw a straight line between them. X values of 3 and 8 have been chosen. See Figure 15.9.

$$Y_p = M_Y + b(X - M_X)$$

$$Y_{p(X=3)} = 5.00 + 0.65(3 - 5.29) = \mathbf{3.51};\ Y_{p(X=8)} = 5.00 + 0.65(8 - 5.29) = \mathbf{6.76}$$

Step 5. Compute the standard error of the estimate Formula 15.4 will be used.

$$s_e = \sqrt{\left[\frac{1}{n_p(n_p - 2)}\right]\left[\left(n_p\Sigma Y^2 - (\Sigma Y)^2\right) - \left(\frac{\left[n_p\Sigma XY - (\Sigma X)(\Sigma Y)\right]^2}{n_p\Sigma X^2 - (\Sigma X)^2}\right)\right]}$$

$$s_e = \sqrt{\left[\frac{1}{7(7-2)}\right]\left[\left(7(207) - (35)^2\right) - \left(\frac{\left[7(217) - (37)(35)\right]^2}{7(245) - (37)^2}\right)\right]} = \mathbf{1.51}$$

Suppose for a specific adolescent, we want to predict the level of marital satisfaction. We administer the MSQ to the parents, find that their score is 7, and, using the regression equation, find $Y_p = 5.00 + 0.65(7 - 5.29) = 6.11$. Therefore, we predict that they will report a level of marital satisfaction of 6.11 as a married person 15 years late in life. However, what about the accuracy of our prediction? We can state that approximately 68% of all adolescents who have parents who scored a 7 on the MSQ will report marital satisfaction scores between 4.60 and 7.62 (6.11 ± 1.51), and about 95% will report between 3.15 and 9.07 ($6.11 \pm 1.96\ (1.51)$).

15.5 The Pitfalls of Linear Regression

Chapter 14 (Section 14.5) discussed several factors that lead to the misinterpretation of correlations. Since regression is closely connected to correlation, it should not surprise us to learn that some of the same issues adversely affect the usefulness of regression analyses. This section discusses some of the factors that can undermine the accuracy of the prediction equation.

15.5.1 Restricted Range

Figure 15.10 reveals a relationship between X and Y. What if the sampled values of X are confined to the area between the vertical lines? The scatterplot between the lines fails to capture the true relationship between X and Y. (In Chapter 14, we explored this problem within the context of using GRE scores to predict GPA scores in prestigious graduate programs.) Not only will a correlation likely underestimate this relationship, but also the accuracy of a regression equation will diminish.

Examining scatterplots will not help to detect a range restriction. We will need to know the range of *possible* values for each variable and then examine the X and Y sample distributions to see if our scores are an accurate representation of the population from which they are drawn.

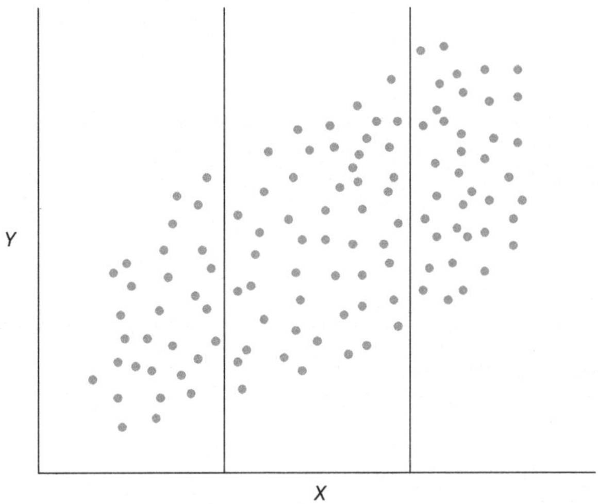

Figure 15.10 Sampling from a restricted range can lead to mismeasuring a relationship.

15.5.2 Extreme Scores

An extreme score can create a false impression of a relationship between X and Y. Figure 15.11 shows a scatterplot in which there is one outlier in the upper right corner. Were it not for this extreme point, a regression analysis might indicate no relationship between X and Y. However, one extreme point could lead to a regression analysis that suggests a relationship. Thankfully, this pitfall is mitigated as the sample size increases. This is another reason to create a scatterplot; it reveals the presence of extreme scores.

15.5.3 Overgeneralization

It is a mistake to apply a regression equation to a different population from the one used to establish it. Recall the example study in which parents' marital satisfaction

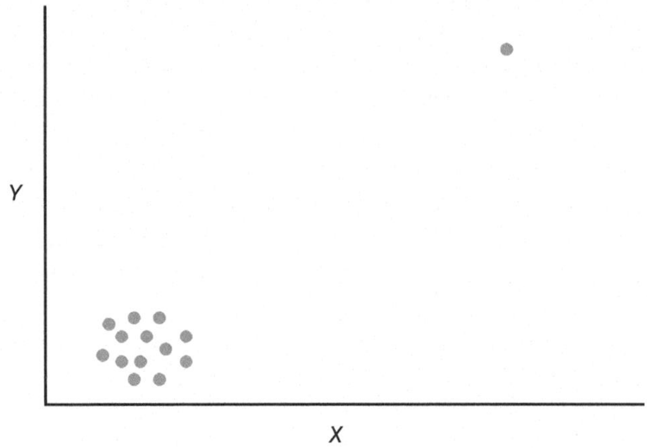

Figure 15.11 One extreme score can create the impression of a relationship.

was used to predict the future marital satisfaction of children. All the children in the sample were adolescents when the predictor variable (parents' MSQ score) was gathered. It is appropriate to apply the regression equation to predict the future marital satisfaction of other adolescents. However, it is dubious to think we could use the equation to predict the future marital satisfaction of 2-year-old children or 22-year-old children. The regression equation was not constructed using data from parents of children with these ages. To apply the regression equation to a different population is to overgeneralize the results obtained from one sample.

15.5.4 Violating Homoscedasticity

What happens to prediction error when the standard deviations of the conditional distributions are dissimilar (called **heteroscedasticity**)? Figure 15.12 shows a heteroscedastic plot for Y given X. There is a positive linear relationship between the

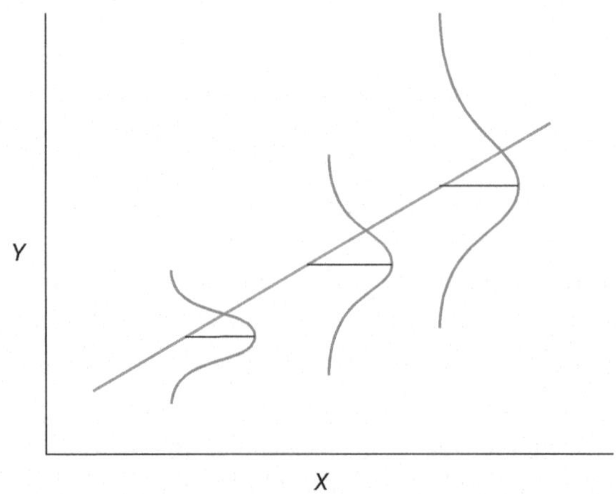

Figure 15.12 Conditional distributions with increasing error as X increases; heteroscedasticity.

variables, but the amount of prediction error increases as the value of X increases. The standard error of the estimate would be different for every value of X. The magnitude of the prediction error for each predicted Y would fluctuate accordingly. Prediction, in this case, would be worse for higher values of X as compared with lower values of X. As a result, we would not be justified in assuming that 68% of the actual Y scores for a given X score lie between $Y_p \pm 1s_e$.

15.6 Formal Presentation of Findings

The proper reporting of regression equations is a bit different from the reporting of t, F, and r tests. When reporting a linear regression, we must identify the predictor and predicted variables, register the rejected F testing the null that $b = 0$, include a quantitative measure of the relationship between them (slope), and, typically, report the r^2 value as well. For instance, "A simple linear regression was calculated to predict height based on shoe size. Statistical evidence for a regression equation was found, $F(1, 6) = 41.63$, $p < .05$, with an r^2 of .87. The resulting equation suggests height increases .46 inches for each unit increase in shoe size." A failure to reject the null might read, "A simple linear regression was calculated to predict height based on shoe size. No statistical evidence was found for a regression equation, $F(1, 6) = 1.63$, *n.s.*"

Summary

Linear regression is a statistical method using bivariate data to predict the value of one variable with information from a related other variable. Regression proceeds in two steps. First, a prediction equation is established from a representative sample of participants. Second, the prediction equation is applied to new cases where the value of only one variable is known.

The regression equation defines a straight line along which each predicted Y for any given X value lies. The slope is the angle of the regression line and is quantified by the number of units Y changes as X changes by one unit. A positive slope indicates a positive correlation, whereas a negative slope indicates a negative correlation.

A correlation of zero will yield a slope (b) of zero, neither ascending nor descending with increasing values of X. An analysis of regression can be run to test the null hypothesis that $b = 0$ for the population of bivariate scores. If this null hypothesis cannot be rejected, the findings of a regression analysis are meaningless.

An error in prediction is the difference between the actual score obtained by the participant and the predicted score. The prediction line must be drawn such that the sum of all errors will equal zero, indicating it is unbiased. Additionally, the regression line is to be drawn so that the sum of the squared errors, Σe^2, is at a minimum. This is called the least squares method.

Each X score is associated with a distribution of Y scores. The sets of Y scores for given values of X are called conditional distributions. The regression analysis assumes conditional distributions are normal and have similar standard deviations across the range of possible X values (the assumption of homoscedasticity). For any given X score, it is the mean of the corresponding conditional distribution of Y's that lies on the regression line and is the predicted Y (Y_p).

The standard error of the estimate is a measure of prediction accuracy. Roughly 68% of the actual Y scores will fall within $\pm 1s_e$ of a predicted Y, and 95% will fall within $\pm 1.96s_e$.

The viability of a regression analysis is based on several factors. It should only be applied to individuals who are represented by the sample from which the regression equation was established. A restricted range of X and/or Y scores, the presence of extreme scores, overgeneralization, or heteroscedasticity will also invalidate regression analyses.

Microsoft® Excel and SPSS®

Information regarding how to use Microsoft® Excel and SPSS® to assist in finding values related to a regression analysis can be found in Appendix C.

Key Formulas

Linear regression equation

$$Y_p = M_Y + b(X - M_X) \quad (15.1)$$

Computational formula for the slope

$$b = \frac{n_p(\Sigma XY) - (\Sigma X)(\Sigma Y)}{[n_p(\Sigma X^2)] - (\Sigma X)^2} \quad (15.2)$$

Correlation formula for the slope

$$b = r\left(\frac{s_Y}{s_X}\right) \quad (15.3)$$

Computational formula for s_e

$$s_e = \sqrt{\left[\frac{1}{n_p(n_p - 2)}\right]\left[(n_p\Sigma Y^2 - (\Sigma Y)^2) - \left(\frac{[n_p\Sigma XY - (\Sigma X)(\Sigma Y)]^2}{n_p\Sigma X^2 - (\Sigma X)^2}\right)\right]} \quad (15.4)$$

Correlation formula for s_e

$$s_e = s_Y\sqrt{(1 - r^2)\left[\frac{n_p}{(n_p - 2)}\right]} \quad (15.5)$$

Key Terms

Regression
Regression equation
Multiple regression
Regression line
Slope
Least squares criterion
Least squares method
***Y* intercept**
Standard error of the estimate
Conditional distribution
Homoscedasticity
Heteroscedasticity

Exercises

1 If guessing a value from a normal distribution, what value should be chosen? Why?

2 Can we use causal language to interpret the findings from regression analyses?

3 What does it mean to claim a regression line has been drawn by the *least squares method*?

4 What is the name of the place on the *Y* axis where the regression line passes through?

5 What concept does s_e represent?

6 What does the size of the standard error of the estimate mean?

7 If $r = 0$, what must b equal?

8 What does the slope tell us? What does it not tell us?

9 When is it not acceptable to run a regression analysis on bivariate data? Why?

10 Why is the assumption of normality for conditional distributions important?

11 Why is the assumption of homoscedasticity important?

12 Why does multiple regression explain more variability than linear regression?

13 Suppose the regression line between the *price of gas* and *miles driven for vacation* has a slope of −0.40. How might an increase in gas prices influence a family's vacation plans?

Work Problems

(Answer all questions below, even if we suspect the null that $r = 0$ cannot be rejected.)

14 Use Formula 15.3 to determine the slope in the following bivariate distributions.

a $r = -.51$, $s_Y = 3.75$; $s_X = 8.54$.
b $r = .33$, $s_Y = 1.71$; $s_X = 2.17$.
c $r = .82$, $s_Y = 1.03$; $s_X = 1.54$.

15 Suppose the relationship between "GPA" (Y) and "hours studied/week" (X) can be expressed as: $Y_p = 2.92 + 0.08(X - 18)$. What is the predicted GPA for these study plans?

a 10 hours a week.
b 22 hours a week.
c 30 hours a week.

16 Provide the requested statistics based on the following dataset.

X	Y
2	4
7	12
3	7
7	14
5	11
3	5

a Calculate the slope of the regression line for predicting Y from X.
b Establish the regression equation for predicting Y from X.
c Calculate the corresponding s_e using the computational formula.

17 Provide the requested statistics based on the following dataset.

X	Y
12	4
4	11
13	7
6	14
9	3
8	9

a Calculate the slope of the regression line for predicting Y from X.
b Establish the regression equation for predicting Y from X.
c Calculate the corresponding s_e using the computational formula.
d Determine Y_p when $X = 1$.

18 In a study on pain tolerance, a researcher is interested in predicting the amount of time participants can keep their hands in ice-cold water. Based on previous research, vitamin E intake over the past 12 hours is correlated with tolerating pain.

Vitamin E units: (X)	Tolerance Times (in seconds): (Y)
5	23
9	32
22	65
12	40
16	42

a What is the b?
b What is the Y intercept?
c What is the s_e?
d What is Y_p for 16 units of Vitamin E?
e Within what time interval should we be 68% confident the right answer lies?

19 A sociologist wants to predict yearly income (Y) based on education level (X). The following data were collected from six individuals.

Education: (X)	Income × 1000: (Y)
10	15
14	29
9	14
14	37
12	20
13	23

a What is the b?
b What is the Y intercept?
c What is the s_e?
d What is Y_p for 10 years of education?
e Within what pay range should we be 95% confident the right answer lies?

20 An admissions committee needs to predict student success for their graduate program. To survive the first year, students must have a 3.00 graduate GPA. The following data are from former students.

Undergraduate GPA: (X)	Graduate GPA: (Y)
3.50	3.33
3.98	3.63
3.10	3.40
2.90	3.41
3.40	3.40

a What is the b?
b Should they admit a student with an undergraduate GPA of 3.00?
c For an incoming student with an undergraduate GPA of 3.67, within what range are we 68% sure their graduate GPA will be found?

21 A psychologist is interested in the relationship between intelligence and keyboard typing speed. Run a regression analysis predicting typing speed and answer the following questions.

Intelligence	Typing speed
109	48
94	36
100	45
117	59
104	39
90	32
116	57
87	37
122	53
88	46
126	59
101	45

a What is the slope?
b What is the Y intercept?
c What is the s_e?
d What Y_p for an IQ of 100?
e What is the 95% confidence range of a person with an IQ of 100?

22 Gerson et al. (2016) examined the relationship between social media use and happiness (also known as "subjective well-being"). The manufactured data below reflect only one part of the study. Higher subjective well-being numbers mean greater happiness. Answer the following questions.

Using social media for social comparison purposes	
Use (hours/day)	**Subjective well-being**
3.1	3
2.8	4
1.8	6
3.0	4
3.3	2
1.5	7
0.8	6
2.2	4
1.8	6
0.6	8

a What is the slope?
b What is the regression equation?
c What is the s_e?
d What is Y_p if $X = 3.5$ hours/day?
e What is the 95% confidence range of subjective well-being for this person?
f Any methodological concerns?

16

Chi-Square Tests and Other Nonparametrics

16.1 The Research Context

Starting with Chapter 8, we have been discussing a variety of different inferential tools. However, all these tools and tests require an interval or ratio scale of measurement and involve means as well as other mean-based statistics (e.g. the variance). This last chapter will look at tests designed to analyze data sets measured on nominal or ordinal scales, or data sets that grossly violate statistical assumptions; in short, data sets where means have dubious meaning. Tests that make assumptions about population parameters are called **parametric tests** while those that do not are called **nonparametric tests.** For example, the *F* test assumes that the population distributions are normally distributed and have equal variances. If these assumptions are grossly violated, interpretations of the test results can be misleading. Nonparametric tests do not make assumptions about the shape of population distributions; for this reason, they are sometimes referred to as *distribution-free tests*. Although nonparametrics are often less powerful, they are important tools because they extend an opportunity to test hypotheses for data sets that could not otherwise be investigated.

The first pair of nonparametrics to be explored are called **chi-square tests** (pronounced *kigh* square, symbolized as χ^2). Some studies, rather than asking *how much* the means of one condition differs from another, ask questions about *how many* data points fall into different categories. In these situations, frequency counts within conditions are what matters. The following are some examples:

► **Example 16.1** A developmental psychologist hypothesizes that fear of strangers occurs at a certain age. A random sample of children, ages 2 through 6, is measured for this fear. Children from each age group are determined to be either afraid or not afraid. ◄

Essentials of Behavioral and Social Science Statistics, Third Edition.
K. Paul Nesselroade, Jr. and Laurence G. Grimm.
© 2025 John Wiley & Sons, Inc. Published 2025 by John Wiley & Sons, Inc.
Companion website: https://www.wiley.com/go/Essentials_of_Behavioral_and_Social_Science_Statistics3e

► **Example 16.2** A political scientist hypothesizes that students grow more politically liberal over their four years of undergraduate education. A random sample of freshmen and seniors is measured regarding political attitudes, and each is classified as either liberal or conservative. ◄

► **Example 16.3** A social psychologist is interested in the relationship between obedience and service in the armed forces. A random sample of former service personnel as well as those who never served is obtained. An experimental situation examines whether participants follow an unpleasant order and then classifies them as either obedient or disobedient. ◄

In each of the preceding research contexts, the data comes in the form of a frequency count tabulated for each category. Note also that the data have a discontinuous either–or quality. For example, the child is either afraid of strangers or not. This is data measured on a nominal scale. Data sets in this format are oftentimes referred to as **categorical data**.

16.2 The Goodness-of-Fit Chi-Square Test

The goodness-of-fit test is the categorical counterpart to the one-way ANOVA. It is designed to detect differences among a set of categories falling along a single factor. In these methodological situations, an ANOVA is typically used if the measuring scale is interval or ratio; the chi-square goodness-of-fit test is used if the measuring scale is nominal.

The chi-square test uses the frequency distribution of a sample to make an inference about the frequency distribution of a population. With nominal data, the frequency distribution is the number of observations per category. The **goodness-of-fit test** uses the chi-square statistic to analyze how well a sample distribution "fits" (corresponds) with a hypothesized population distribution. To do this, we will need to indicate what the population data would look like if there were *no* effect. The projected frequencies if there is no effect are called **expected frequencies**, symbolized as f_e. The obtained frequencies from a sample are called **observed frequencies**, symbolized as f_o.

16.2.1 The Hypotheses for the Goodness-of-Fit Test

The goodness-of-fit test requires us to specify the set of null frequencies. What would a null hypothesis expect? Null frequencies can be determined either *rationally* or *empirically*. First, consider the rational approach. Suppose we would like to find out if people prefer cola A or B. We decided to conduct a blind taste test. If there is no preference among participants, what percentage of people are expected to choose cola *A* and cola *B*? On average, we would expect 50% of the participants to select each one. This null hypothesis can be represented as follows:

	A	*B*
H_0	50%	50%

Now, suppose we add a third drink, cola *C*. What would be the expected frequency distribution for no preference? The answer is:

	A	*B*	*C*
H_0	33.33%	33.33%	33.33%

If we add a fourth beverage, the expected frequency would be 25% for each category. A null hypothesis of no preference states that the expected frequencies are equally distributed across the categories. The alternative hypothesis, conversely, states that the population frequencies are not distributed equally across the categories.

Another rational approach to specifying a null would be to use a theory to generate expected frequencies. A genetic theory of ulcer susceptibility might predict that after four generations of inbreeding, 40% of offspring will show stomach ulcers. This null set can be represented as:

	Yes	**No**
H_0	40%	60%

The alternative hypothesis states that frequencies are not distributed in this manner.

An *empirical* approach to specifying expected frequencies requires access to existing data. For example, suppose the voting preference in the last gubernatorial election was 60% Democrat, 30% Republican, and 10% Other. Since the last election, however, a popular Republican president campaigned for the Republican gubernatorial candidate. Before the election, a sample of voters is polled to see if this previous distribution of votes will change. Using data from the last election, the null hypothesis would be stated as:

	Democrat	**Republican**	**Other**
H_0	60%	30%	10%

The alternative hypothesis states that frequencies are not distributed in this manner.

16.2.2 Computing the Chi-Square Statistic for the Goodness-of-Fit Test

The purpose of the goodness-of-fit test is to determine if the observed frequencies, obtained from a sample of participants, meaningfully differ from the expected frequencies derived from an understanding of the null hypothesis. Recall from previous discussions that it is highly unlikely, *even when* the H_0 is true, that sample means will be identical to population expectations. In much the same way, simply finding that observed frequencies differ from expected frequencies is not a sufficient reason to reject a null. The chi-square goodness-of-fit test, however, can be used to help make that determination. The following example uses the goodness-of-fit test.

A psychology student, while talking with a friend, makes the somewhat provocative statement that business majors are most interested in making money. The friend, a business major, insists that making money, while important, is not the most important motive for entering the business world. They decide to conduct a study in which business majors are asked to respond to the statement: "Making money is the most important reason for majoring in business." Participants can either agree, disagree, or be undecided about the statement. One version of a rationally derived null hypothesis predicts the percentage of category responses to be the same.

Agree	Disagree	Undecided
33%	33%	33%

Although null hypotheses are typically stated as percentages, the chi-square test requires counts. This requires knowledge of the total number of participants in the study. Assume 90 participants were used. To determine the expected frequency for a given category, f_e, each percentage (or proportion) is multiplied by n, the number of participants.

33% of 90 = .33 (90) = 30 Agree responses
33% of 90 = .33 (90) = 30 Disagree responses
33% of 90 = .33 (90) = 30 Undecided responses

If the expected frequencies for each cell are the same, then the f_e values can be found by dividing the number of participants (observations) by the number of categories: $N/C = 90/3 = 30$. If not, the expected frequencies will differ from cell to cell. Familiarity with the null hypothesis is critical for the goodness-of-fit test.

The next step is to find the observed frequencies, f_o. This is accomplished by counting the number of participants in the sample who agreed, disagreed, and were undecided. With the set of expected and observed frequencies calculated, Formula 16.1 is used to compute χ^2.

Formula for χ^2

$$\chi^2 = \Sigma \frac{(f_0 - f_e)^2}{f_e} \quad (16.1)$$

Assume the observed frequencies for each category are as indicated below.

Agree	Disagree	Undecided
$f_e = 30$	$f_e = 30$	$f_e = 30$
$f_o = 60$	$f_o = 20$	$f_o = 10$

Formula 16.1 requires the following arithmetic operation to be performed for each category:

$$\frac{(f_o - f_e)^2}{f_e}$$

Afterward, the Σ symbol directs us to sum the values across each category.

16.2.3 Computational Steps

Step 1. For the first category, subtract the expected frequency from the observed, $f_o - f_e$.
Step 2. Square the difference, $(f_o - f_e)^2$. This removes any negative signs.
Step 3. Divide the number found in step 2 by the f_e for that category, $(f_o - f_e)^2/f_e$.
Step 4. Repeat the first three steps for each cell.
Step 5. Sum all the quantities from all the categories.

$$\chi^2 = \frac{(60-30)^2}{30} + \frac{(20-30)^2}{30} + \frac{(10-30)^2}{30} = \mathbf{46.66}$$

16.2.4 Deciding Whether to Reject the Null Hypothesis

This section will give us a deeper understanding of how hypotheses are tested in chi-square analyses. To run these tests, we will need to use the sampling distribution of the χ^2 statistic.

The formula indicates that as the differences between the set of expected and observed frequencies grow, the χ^2 value will increase, while a closer fit will result in smaller χ^2 values. As with other statistical tests, the larger the χ^2 statistic, the less likely the data which generated it came from a null distribution. To see if the null can be rejected, the obtained chi-square (χ^2_{obt}, or just χ^2) is compared with a critical value, χ^2_{crit}. If χ^2_{obt} is equal to or greater than χ^2_{crit}, the null hypothesis is to be rejected; there is statistical evidence suggesting a population consistent with the null hypothesis did not generate our observed frequencies. This means the differences between the obtained and expected values are probably *not* due to chance.

The critical value for χ^2 is found in Table A.8, a portion of which is shown in Table 16.1. The degrees of freedom for the goodness-of-fit test are the number of columns in the design minus one. Our example problem has three columns. Therefore, $df = C - 1 = 3 - 1 = 2$. Using Table 16.1, we see that $\chi^2_{crit} = 5.99$. Since the obtained $\chi^2 = 46.66$ and $46.66 > 5.99$, we would reject the null.

We will need to wait until later in the chapter to learn about follow-up analyses, but for now, we can state that statistical evidence was found suggesting the opinions of business majors were not equally distributed across the three options, $\chi^2(2, n = 90) = 46.66$, $p < .05$.

Table 16.1 A portion of the table of critical values for the chi-square distribution.

	Proportion of critical region				
	Alpha Level				
df	***.10***	***.05***	***.02***	***.01***	***.001***
1	2.71	3.84	5.41	6.64	10.83
2	4.60	5.99	7.82	9.21	13.82
3	6.25	7.82	9.84	11.34	16.27
4	7.78	9.49	11.67	13.28	18.46
5	9.24	11.07	13.39	15.09	20.52
6	10.64	12.59	15.03	16.81	22.46

Consider another problem, this time where the expected frequencies are not all the same.

■ **Question** *A professor claims that turned-in papers are graded graciously and offers the following distribution of percentages as an estimate of how the grades are distributed.*

Excellent	Above average	Average	Below average
25%	35%	25%	15%

A group of students suspect the professor is not as generous as claimed. All students who took this professor's course over the last three years are available, and amazingly, all still have their papers. A random sample of 100 of these former students yields the following distribution.

Excellent	Above average	Average	Below average
$f_o = 20$	$f_o = 25$	$f_o = 30$	$f_o = 25$

Is there a reason to conclude that the professor's claim is mistaken?

Solution

Before the formula for χ^2 can be used, the f_e's need to be calculated. Since the observed data sum to 100 papers, the calculation of the f_e's is straight forward.

Placing the expected and observed frequencies in a table shows:

Excellent	Above average	Average	Below average
$f_e = 25$	$f_e = 35$	$f_e = 25$	$f_e = 15$
$f_o = 20$	$f_o = 25$	$f_o = 30$	$f_o = 25$

Computing χ^2,

$$\chi^2 = \Sigma \frac{(f_0 - f_e)^2}{f_e}$$

$$\chi^2 = \frac{(20-25)^2}{25} + \frac{(25-35)^2}{35} + \frac{(30-25)^2}{25} + \frac{(20-15)^2}{15} = \mathbf{11.53}$$

$$df = C - 1 = 4 - 1 = 3; \alpha = .05; \text{therefore } \chi^2_{crit} = 7.82$$

Since 11.53 > 7.82, the H_0 is rejected.

This result offers statistical evidence suggesting the student's papers come from a distribution that is different than the one claimed by the professor.

A closer look at the frequencies casts suspicion on the professor's claim to be a gracious grader. However, rejecting the null does not mean we know *which cells* are responsible. More work is needed. (A follow-up test will be introduced later in the chapter.) ■

16.3 The Chi-Square Sampling Distribution

Chi-square sampling distributions are theoretical, formed by taking a vast number of samples from a null population and computing the chi-square statistic for each sample. This vast set of chi-square values is plotted to show a chi-square sampling distribution. A separate distribution can be created for each *df* value, thereby establishing a family of sampling distributions. The shapes of the distributions are positively skewed but vary based on the *df* value. Note that *df* values for χ^2 tests are *not* based on the number of participants but rather the number of categories.

16.3.1 Rejecting the Null Hypothesis

The χ^2 test measures the degree to which the observed frequencies correspond to the expected frequencies. If the null hypothesis is true, the observed frequencies and expected frequencies for each cell will be similar, and the value of χ^2 will be small. If the null hypothesis is false, the set of expected and observed frequencies will be discrepant from one another. As with previous inferential tests, the alpha level is used to determine the Type I error rate one is willing to tolerate.

Figure 16.1 shows chi-square distributions for 1, 5, and 8 degrees of freedom. The rejection region for each of the distributions when $\alpha = .05$ is also shown. An obtained value of χ^2 that falls in the rejection region would allow us to reject the null hypothesis.

16.3.2 Characteristics of the Chi-Square Distribution

The characteristics of the chi-square distribution are as follows:

1) Since the numerator of the χ^2 statistic is squared, all values of χ^2 are positive.
2) Chi-square distributions are unimodal and typically positively skewed. However, as the *df* increases, the chi-square distribution approximates the shape of a normal distribution.
3) As the *df* increases, the critical value of χ^2 likewise increases, reflecting the growing number of categories with expected and observed frequencies.

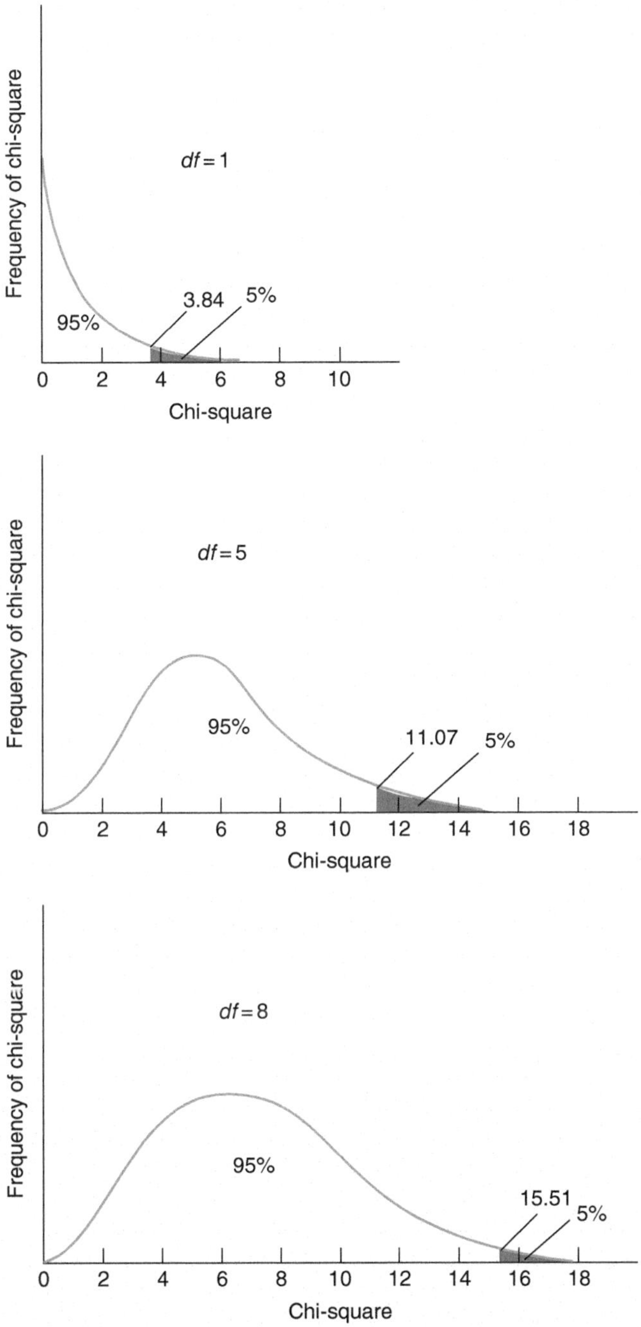

Figure 16.1 Chi-square distributions for degrees of freedom of 1, 5, and 8.

16.4 The Chi-Square Test for Independence

The chi-square test can also be used to determine if there is a relationship between two factors. Recall that the two-way ANOVA in Chapter 12 was also used to analyze two-factor designs. Similar to the goodness-of-fit test, the chi-square **test for independence** also uses the frequency counts from a set of observations. Below is an example research question:

► **Example 16.4** A clinical psychologist hypothesizes that birth complications correspond with subsequent diagnoses of schizophrenia. Each participant is categorized by *two* variables: diagnosis (schizophrenic, depressed, normal) and history of birth complications (complications or no complications). This design is represented in the following table, called a **contingency table** (also called a *frequency* or *cross-tabulation table*). Since there are two rows and three columns, the table is called a 2 × 3 contingency table.

	Schizophrenic	Depressed	Normal
Birth complications	20	6	8
No birth complications	8	20	22

The numbers in the cells refer to the number of participants that meet the classification criteria for both variables. For example, 20 schizophrenics had birth complications, 6 depressed patients had birth complications, and so on. ◄

16.4.1 The Null Hypothesis and the Concept of Independence

The null hypothesis, H_0, applied to a two-way design states that the two variables are independent. The alternative hypothesis, H_1, states that the two variables are not independent; that is, they are related. This is analogous to testing for evidence of an interaction in a two-way ANOVA analysis (see Chapter 12). Figure 16.2 shows bar graphs of the hypothetical data from the schizophrenia and birth complications example. The illustration on the left shows the relative number of schizophrenics, depressed, and normal participants who had complications associated with their births. Compare this graph with the one to the right which depicts the frequency counts for those who do *not* have a history of birth complications. Just by "eyeballing" the two graphs, we can see that the patterns of data are different. These pattern differences indicate that the data distributions for one variable *depend* on the level of the second variable.[1] In this example, the two variables do not appear to be independent.

1 The word "depends" in this context does not imply a causal relationship between the variables.

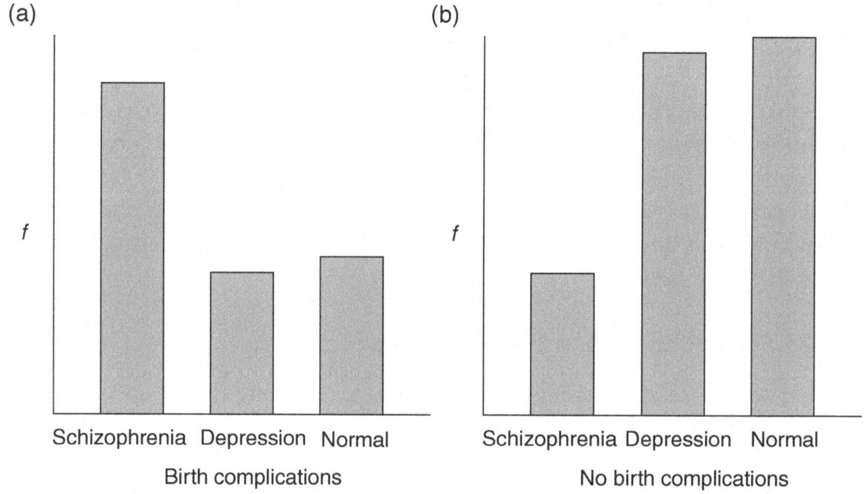

Figure 16.2 The pattern of bars in (a) differs from the pattern of bars in (b) indicating a relationship between diagnosis and birth complications.

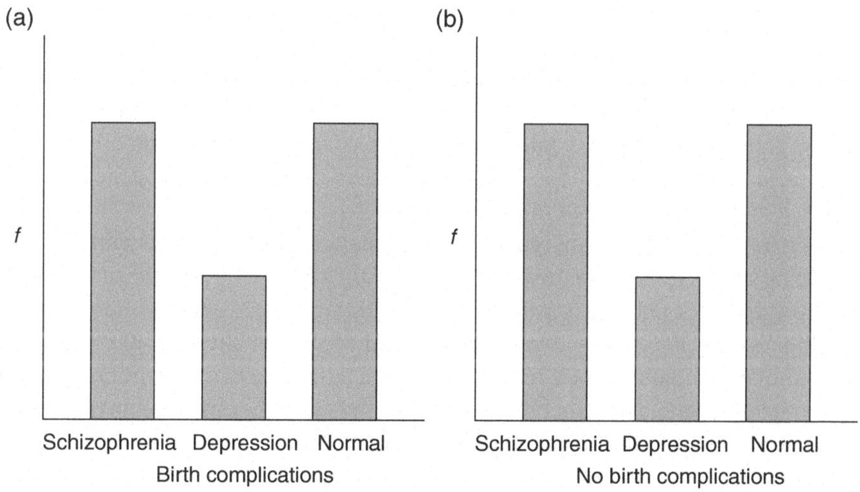

Figure 16.3 Unlike Figure 16.2, the pattern of bars is similar in (a) and (b), suggesting diagnosis and birth complications are independent; that is, unrelated.

Now, consider Figure 16.3. New data sets have been constructed that reflect independence between the two categories. The data set on the left is now quite similar to the data set on the right. In other words, the likelihood of someone being classified as schizophrenic, depressed, or normal does not appear to be related to the presence or absence of birth complications. Stated differently, the frequency distribution for one variable has the same pattern for each level of the second variable. The chi-square test for independence checks to see if data distributions for one variable differ based on the level of the second variable. If the distributions differ, then the null hypothesis that the variables are independent is rejected.

Recognize that the chi-square test for independence is restricted to hypotheses *between* categories (i.e. interactions). It does not test hypotheses *within* a category (i.e. main effects). To explore hypotheses within a category, the chi-square goodness-of-fit test is used.

16.4.2 Computing χ^2 for a Two-Way Design

Whether we are conducting a goodness-of-fit test or testing the independence of two variables, the formula for χ^2 is the same:

$$\chi^2 = \Sigma \frac{(f_0 - f_e)^2}{f_e}$$

Moreover, observed frequencies are still obtained from sample data. Specifying the expected frequencies for a test of the null, however, is not straightforward.

16.4.3 Computing Expected Frequencies in the Two-Way Design

Computing the proper f_e value for each cell requires a bit of arithmetic. Let us use the frequency count data from our working example to illustrate how to find these values. The observed data should be organized, counted, and presented in a two-factor grid such as Table 16.2. From this observed data, we can generate the expected data to test the null of independence.

Using the observed data, marginal frequency counts need to be determined. This is done by simply adding up the frequency counts in all conditions associated with each row and each column (see Table 16.3). Additionally, the total number of participants can be placed in the lower right-hand corner. In our example, the total number of participants is 84. From the marginal tallies, we can determine various values such as the total number of participants having birth complications (34) and the total number of participants with depression (26).

Table 16.2 Observed frequencies for a worked problem.

	Schizophrenic	Depressed	Normal
Birth complications	$f_o = 20$	$f_o = 6$	$f_o = 8$
No birth complications	$f_o = 8$	$f_o = 20$	$f_o = 22$

Table 16.3 Observed, expected, and marginal frequencies for a worked problem.

	Schizophrenic	Depressed	Normal	
Birth complications	$f_o = 20$ $f_e = 11.33$	$f_o = 6$ $f_e = 10.52$	$f_o = 8$ $f_e = 12.14$	**34**
No birth complications	$f_o = 8$ $f_e = 16.67$	$f_o = 20$ $f_e = 15.48$	$f_o = 22$ $f_e = 17.86$	**50**
	28	**26**	**30**	**84**

Table 16.3 also shows the expected frequencies for each cell of the matrix. To compute expected frequencies, see Formula 16.2. This formula uses the column and row frequencies, as well as the total number of cases, to find the expected frequency for each cell. For instance, for the uppermost left cell, schizophrenia and birth complications, we need the frequency for the schizophrenia column ($f_c = 28$) and the birth complications row ($f_r = 34$). When multiplied together and divided by N, the expected frequency for that cell is found to be 11.33. This can be done for each of the six cells in this design. Given that the column and row frequencies in this data set are not identical, the f_e's for the six cells will not be the same value.

Formula for computing f_e

$$f_e = \frac{f_c f_r}{N} \quad (16.2)$$

where

f_c = the frequency total for the relevant *column*
f_r = the frequency total for the relevant *row*
N = the total number of participants

16.4.4 Computing χ^2 and Testing the Null Hypothesis

The task of computing χ^2 is easier if we work from a frequency summary table. In Table 16.4, the observed and expected frequencies are ordered in the first two columns. The next two columns are interim calculations necessary for computing χ^2. The value of χ^2 computes to 16.77.

The degrees of freedom are needed to determine χ^2_{crit}. For a two-way design, the $df = (R - 1)(C - 1)$; that is the number of rows (R) minus one multiplied by the number of columns (C) minus one. For this example, $df = (2 - 1)(3 - 1) = 2$.

The critical values for the chi-square tests are found in Table A.8. If alpha is set at .05, the critical value associated with $df = 2$ is 5.99. Since 16.77 is larger than 5.99, we have statistical evidence to reject the null hypothesis. Therefore, we have

Table 16.4 Computing χ^2.

f_o	f_e	$(f_o - f_e)$	$(f_o - f_e)^2$	$(f_o - f_e)^2/f_e$
20	11.33	8.67	75.17	6.63
6	10.52	−4.52	20.43	1.94
8	12.14	−4.14	17.14	1.41
8	16.67	−8.67	75.17	4.51
20	15.48	4.52	20.43	1.32
22	17.86	4.14	17.14	0.96
				$\chi^2 = 16.77$

$$\chi^2 = \Sigma \frac{(f_0 - f_e)^2}{f_e} = \mathbf{16.77}$$

evidence suggesting that a relationship exists between psychiatric diagnosis and birth complications.

The steps involved in conducting a chi-square analysis for a two-way design are:

Step 1. Specify the hypotheses. H_0 states the two variables are independent. H_1 states that the two variables are not independent.
Step 2. Specify alpha; it is typically .05 or .01.
Step 3. Use Formula 16.2, $f_e = f_c f_r, /N$, to compute all expected cell frequencies.
Step 4. Place the observed and expected frequencies in a summary table and for χ^2.
Step 5. Compute *df by* $(R - 1)(C - 1)$.
Step 6. Find χ^2_{crit} using Table A.8 and make a decision regarding the null hypothesis.
Step 7. Interpret the findings.

16.5 The Chi-Square Test for a 2 × 2 Contingency Table

We have just completed an example of how to compute χ^2 for a 2×3 design. For 2×2 designs, a shortcut method can be used that does not involve computing expected frequencies for each cell. Formula 16.3 requires us to place the letters *A*, *B*, *C*, and *D* in the cells as indicated in the following diagram. In this case, only the observed cell frequencies are needed to compute χ^2.

A	B	$A + B$
C	D	$C + D$
$A + C$	$B + D$	N

Chi-square formula for a 2 × 2 contingency table

$$\chi^2 = \frac{N(AD - BC)^2}{(A + B)(C + D)(A + C)(B + D)} \quad (16.3)$$

where

A, B, C, D = the observed frequencies, f_o, in each cell
$AD = f_o$ for cell $A \times f_o$ for cell D
$BC = f_o$ for cell $B \times f_o$ for cell C
N = total number of participants

16.6 A Measure of Effect Size for Chi-Square Tests

Recall that hypothesis tests only indicate the degree of certainty associated with rejecting the null hypothesis; that is, how certain are we that there is an effect? However, the *certainty* of an effect and the *size* of an effect are not the same thing. For instance, with large sample sizes comes great statistical power; power to detect even very small effects. As with other inferential tests, a measure of effect size can be helpful when interpreting the meaning of a rejected null hypothesis.

For chi-square tests, the standard effect size measure is **Cramér's *V*** (or Cramér's *phi* – pronounced "fie," rhymes with *fly*). It is oftentimes symbolized as ϕ (although V can also be used). If a chi-square null has been rejected, this measure of effect size can be calculated.

Formula for Cramér's *V*

$$\phi = \sqrt{\frac{\chi^2}{(N)\left(df_{row/column}\right)}} \quad (16.4)$$

where

χ^2 = the obtained chi-square value
N = the total number of participants in the study
$df_{row\ or\ column}$ = the smaller degrees of freedom value, either rows or columns

This procedure generates a number that proportions the obtained chi-square value to the size of the sample. Larger numbers reflect greater effect sizes. Of course, Cramér's V should only be found if the null hypothesis is rejected. For our example problem, Cramér's V would be calculated as follows:

$$\phi = \sqrt{\frac{16.77}{84(1)}} = \mathbf{0.45}$$

16.7 Major Contributors to a Significant Chi-Square

In the chapters covering the analysis of variance, we learned that a significant *F* ratio could be further analyzed with follow-up comparisons designed to locate the source(s) of a significant *F* ratio. Within the context of the chi-square analysis, a follow-up test answers the question, "Which cells are *major contributors* to a significant chi-square value?" Unlike previously introduced follow-up tests, this analysis does *not* make contrasts between cells. Rather, each cell is analyzed separately to determine which cells make a *major* contribution to the χ^2 value. It requires the calculation of **cell residuals**, a ratio measure of observed to expected frequencies within a cell.

Formula for the standardized residual

$$R = \frac{f_o - f_e}{\sqrt{f_e}} \quad (16.5)$$

An adjustment to this measure that better considers the overall size of the sample is generated by many statistical software packages like SPSS®. For introductory purposes, however, only the standardized residual formula will be presented here.

How large must a cell residual be before it is considered a major contributor to the significant χ^2? The conventional standard suggests an absolute value that *equals or exceed* 2.00 (i.e. $R \geq |2.00|$). Table 16.5 presents the standardized residuals for each category of our example problem. The valence of *R* is important for interpreting the meaning of any cell designated to be a major contributor. A positive *R* indicates the number of observations in that cell is more than would be expected by the null, whereas a negative *R* means fewer observations than would be

Table 16.5 Standardized residuals for the schizophrenia and birth complications (BC) study.

Cell	f_o	f_e	$\sqrt{f_e}$	R
Cell 1 (Schizophrenia/BC)	20	11.33	3.37	**2.57**
Cell 2 (Depressed/BC)	6	10.52	3.24	−1.40
Cell 3 (Normal/BC)	8	12.14	3.48	−1.19
Cell 4 (Schizophrenia/No BC)	8	16.67	4.08	**−2.13**
Cell 5 (Depressed/No BC)	20	15.48	3.93	1.15
Cell 6 (Normal/No BC)	22	17.86	4.23	0.98

expected were found. In Table 16.5, note that two cells have R values $\geq |2.00|$. Cell 1, Schizophrenic/Birth Complications, has a large positive R (2.57), suggesting *more* of this pairing than expected. The other, Cell 4, the No Birth Complications/Schizophrenic, has a large negative R (−2.13), suggesting *fewer* cases than expected. Since none of the other cells yield an R that equals or exceeds $|2.00|$, these are the only two cells considered to have made a major contribution to the significant chi-square finding.

The analysis of the standardized residuals is a useful technique. It allows us to make a more specific interpretation of a significant χ^2. Standardized residuals can also be used to analyze significant χ^2 goodness-of-fit tests. The cell residuals are calculated the same way, and $\geq |2.00|$ is the conventional standard for any cell to be considered a major contributor.

16.8 Using the Chi-Square Test with Quantitative Variables

Although chi-square tests are usually performed using discrete variables (e.g. citizen or foreigner; homeowner or renter; single, married, divorced, or widowed), it is possible to run the test with quantitative variables that are treated as categorical variables. For example, extroversion could be measured using a scale that ranges from 0 to 30, but participants could be reclassified as either high or low extroversion. In fact, the chi-square test for independence can be used when the underlying scale for both variables is continuous. For example, Table 16.6 assigns participants to one of three "need for achievement" categories and one of four "annual income" categories.

If there is reason to suspect that a variable is *not* continuous, but rather a set of discrete categories, using a chi-square test is a good idea. Unlike parametric tests, a chi-square test does not assume the underlying scale to be interval or ratio.

Table 16.6 A 3 × 4 contingency table.

	Annual income ($1000)			
Need for achievement	**<40**	**41–60**	**61–80**	**>80**
>40	39	44	57	85
21–40	45	40	53	62
<20	76	52	40	30

16.9 The Assumptions

One of the advantages of the chi-square test is the limited number of assumptions.

1) As with any inferential test, samples should be representative of populations.
2) Each observation must be *independent* of every other observation.
3) The summative data is in the form of a frequency count, not means.
4) Expected cell frequencies need to be of sufficient size; "5" is used as a rule of thumb. Chi-square analyses with smaller expected frequencies can underrepresent the Type I error rate.

16.10 Formal Presentation of Findings

The basic information and formatting needed when presenting nonparametric findings in professional writings are not much different than those needed for the more typical analytical techniques of *t*'s, *F*'s, and *r*'s. However, there are a few differences. A typical sentence structure for a rejected null hypothesis for a chi-square goodness-of-fit test might take this general form: "A chi-square goodness-of-fit test was performed which found statistical evidence suggesting major preference was not equally preferred by undergraduates, χ^2 $(2, N = 108) = 6.22$, $p < .05$." A rejected null from a chi-square test of independence might take this general form: "A chi-square test of independence comparing biological sex and type of undergraduate major found statistical evidence suggesting a relationship, χ^2 $(2, N = 22) = 6.05$, $p < .05$." Notice how the sample size is included in the parentheses. With other tests, the degrees of freedom are sufficient to communicate sample size. The degrees of freedom for a chi-square, however, only reflect the number of cells in the design. For this reason, *N* is typically included. For both tests, a second sentence is used to describe secondary analyses. For example, "Follow-up analyses found evidence that biological females ($R = 2.75$) were disproportionately more likely than biological males to major in the humanities ($R = -3.25$)."

A typical sentence for a null that is not rejected might take the general form: "A chi-square test of independence was calculated comparing biological sex and type of undergraduate major. No statistical evidence of a relationship was found, χ^2 $(2, N = 22) = 3.05$, *n.s.*"

16.11 Other Nonparametric Tests

There are many other nonparametric tools that can be used to test hypotheses for data sets that cannot be analyzed by the standard *t*, *F*, and *r* tests (see, for example, Corder and Foreman 2014; Siegal and Castellan 1988). This text restricts itself to briefly identifying four of them.

In addition to the Pearson r from Chapter 14, there are two other measures of linear correlation. The **Spearman rank correlation** can be used for ordinal scale data or when statistical assumptions like normality or homogeneity of variance have not been met. To run a Spearman, the data from both variables is converted into ranks and then a simplified (albeit less powerful) correlation value can be calculated.

The **point-biserial correlation** is used when one variable is continuous, but the second variable is dichotomous. A **dichotomous variable** is an either/or variable, such as student/nonstudent, married/single, theist/nontheist, or resident/alien.

Two other frequently used nonparametrics are the **Mann–Whitney *U* test** and the **Wilcoxon signed-ranks test**. The Mann–Whitney U is the nonparametric alternative to the independent-samples t test and the Wilcoxon is the nonparametric alternative to the dependent-samples t test. Both tests transform data onto an ordinal scale by converting either the raw scores (Mann–Whitney U) or the difference scores between the dependent measures (Wilcoxon) into ranks.

This is not an exhaustive list of nonparametric tests; however, these are some of the most frequently used ones. It is important to recall that nonparametrics, with the notable exception of the chi-square tests, are typically less powerful compared to the more standard tests. However, these "Plan B" tools demonstrate their value by expanding the opportunity to test null hypotheses. Many statistical analysis software programs (e.g. SPSS®) can run these tests.

As we conclude the text, below is one more Box 16.1 dealing with the scientific method. It is important for all students of science to consider carefully both the strengths and limitations of scientific investigation.

Box 16.1 Is the Scientific Method Broken? The Limitations of Science

Throughout this book we have periodically stopped to look more closely at some commonly discussed problems in the world of science; in particular, we have tried to understand the current reproducibility crisis that is afflicting many of the sciences. Let us finish this series by stepping back to look a bit more philosophically at the scientific endeavor as a whole. What can we hope to accomplish with the help of science, and what, if anything, lies on the outside? This has sometimes been referred to as science's *demarcation* problem – drawing the line between what questions are addressable by the scientific method and what ones are not.

In Chapter 1, we were told that the scientific method addresses a limited set of questions. Causal explanations, when methodologically warranted, are to be understood in a limited way, not as complete, final explanations.

Unfortunately, not all claims made by scientists reflect this modesty. In other words, not all pronouncements made by scientists are, in fact, statements of *science*. Some claims made in the name of science are clearly outside this demarcation line. Perhaps the human desire to win arguments tempts us to misuse scientific authority, to declare matter-of-factly something true that we merely *want* to be true that sits outside its domain. These over-reaches, when exposed, can leave the public, who are listening in on the conversation, with the impression that science might be broken. A lack of proper modesty by some popularizers of science and the zealotry of others whose primary concern is advocacy for a particular political or philosophical position can severely damage science's public reputation.

A related problem is the claim made by some that science is the *only* avenue to truth. For instance, the famed philosopher of science, Bertrand Russell (1936/1997), once famously wrote, "... what science cannot discover, mankind cannot know." This position has come to be known as "scientism." This view, however, does not withstand scrutiny. In fact, the statement itself is circular. After all, this claim itself is not a statement of science, so, if it is true, by its own pronouncement, it cannot be known to be true.

Of course, the scientific method does an excellent job of exploring the physical machinery of reality, including social and behavioral reality. However, it is powerless to answer even simple existential questions like "why am I here?" and "what is the meaning of life?" When discussing the demarcation problem, British biologist/theologian Alister McGrath (2015) refers to Frank Rhodes famous question regarding a boiling kettle. Rhodes asks us to imagine that we find a kettle sitting on a gas ring; and upon closer inspection, we see that the kettle is boiling. "Why is the kettle boiling?" we may ask. Well, one answer, addressing the mechanics of this event, would be that there is a heat transfer taking place between the underlying burner and the bottom of the copper kettle. This transfer then continues on to the water inside the kettle. The additional energy excites the liquid water molecules and eventually changes their physical state and is released into the air as steam. This explanation has been revealed to us over the past few hundred years through careful scientific analysis. Rhodes, however, offers a second explanation. He says, "It is boiling because I want a cup of tea." Rhodes asks, "Which answer is right?" He goes on to say, "[n]ow these are different answers...But both are true, both are complementary and not competitive. One answer is appropriate within a particular frame of reference, the other within another frame of reference. There is a sense in which each is incomplete without the other." Indeed, the answers work together to give a richer explanation. One further observation is that we have been asking the *metaphysical* question of "why" much longer than we have been asking the *mechanical* question of "how." Yet, our recent advances in understanding a mechanical explanation, while interesting and helpful in many ways, have not served to bring us any closer to answering the deeper metaphysical question.

The scientific method, once given the time needed to work through human foibles and limitations, ends up doing a very good job of describing the interworking parts of much of physical reality. However, even here, at the foundation level of physical reality, we find limitations. For instance, no one understands what gravity actually *is*. We know how to measure it, and we have learned in careful detail how it works, but that is where our understanding stops. What exactly is it? We also do not know what energy is, or time, life, and consciousness. Many secondary questions related to these fundamental features can be asked and seem to have been sufficiently answered by using careful scientific investigation. However, this gained information, as helpful as it is, only serves to deepen the mystery around these basic features of our reality. It is interesting to note that these fundamental questions have been around long before the scientific method was established; in fact, they give every indication they will be perennially tied to the human experience. The famed NASA astronomer Robert Jastrow (1992), when speaking of those who wish to give too much credibility to science, once wrote, "For the scientist who has lived by his faith in the power of reason, the story ends like a bad dream. He has scaled the mountain of ignorance; he is about to conquer the highest peak; as he pulls himself over the final rock, he is greeted by a band of theologians who have been sitting there for centuries." Perhaps science is not broken after all; perhaps its' boundaries just need to be more properly considered.

Summary

Tests that make statistical assumptions are called parametric tests. The chi-square test does not make statistical assumptions and does not use means or standard deviations to infer population parameters; therefore, they are termed nonparametric tests. A chi-square analysis examines the correspondence between a hypothesized population frequency distribution and an observed sample frequency distribution. The null hypothesis states there are no differences between these distributions. The alternative hypothesis states the distributions are different.

The design of a chi-square goodness-of-fit test is analogous to a one-way ANOVA with two or more groups (categories) distributed along a single factor. The design of a chi-square test for independence is analogous to a two-way ANOVA with two or more groups (categories) distributed along two different factors. In a two-way design, the chi-square analysis tests whether two variables are independent. This is analogous to the two-way ANOVA test for an interaction. Both chi-square tests generate an observed statistic that can be compared with a critical value derived from a null sampling distribution.

In situations where the null hypothesis can be rejected, a measure of effect size, Cramér's V, can be found. Furthermore, a residual analysis can be used as a follow-up test to explore which cell or cells of the design are major contributors to the large chi-square value. The assumptions of the chi-square test include sample representativeness, frequency count data, independent observations, and sufficient large observed and expected frequencies.

Four other nonparametric tests are briefly introduced; the Spearman rank correlation, the point-biserial correlation, the Mann–Whitney U test, and the Wilcoxon signed-ranks test. Nonparametric tests demonstrate their value by expanding the opportunity to test null hypotheses.

Microsoft® Excel and SPSS®

Information regarding how to use Microsoft® Excel and SPSS® to assist in performing chi-square tests can be found in Appendix C. Neither program, however, are particularly well-suited for the work problems below. To save time, it is recommended to do them by hand.

Key Formulas

Formula for χ^2

$$\chi^2 = \sum \frac{(f_o - f_e)^2}{f_e} \quad (16.1)$$

Formula for computing f_e

$$f_e = \frac{f_c f_r}{N} \quad (16.2)$$

Chi-square formula for a 2 × 2 contingency table

$$\chi^2 = \frac{N(AD - BC)^2}{(A + B)(C + D)(A + C)(B + D)} \quad (16.3)$$

Formula for Cramér's *V*

$$\phi = \sqrt{\frac{\chi^2}{(N)\left(df_{row/column}\right)}} \quad (16.4)$$

Formula for the standardized residual

$$R = \frac{f_o - f_e}{\sqrt{f_e}} \quad (16.5)$$

Key Terms

Parametric tests
Nonparametric tests
Chi-square test
Categorical data
Goodness-of-fit test
Expected frequencies
Observed frequencies
Test for independence
Contingency table
Cramér's *V*
Cell residual
Spearman rank correlation
Point-biserial correlation
Dichotomous variable
Mann–Whitney U test
Wilcoxon signed-ranks test

Exercises

1 What assumptions are not needed for a chi-square analysis?

2 Explain the difference between the two chi-square tests presented in this chapter?

3 How are degrees of freedom for chi-square tests different from other inferential tests?

4 Why is there only one χ^2_{crit} value when cell counts can be either too large or too small?

5 A one-way ANOVA is to a two-way ANOVA as: (select the best option below)
a A nonparametric test is to a parametric test.
b A parametric test is to a nonparametric test.
c A chi-square is to correlation.
d A goodness-of-fit test is to a test for independence.

6 Which pairing below is most similar? Why?
- **a** Residuals and omega-squared
- **b** Cramér's *V* and Fisher's *LSD*
- **c** Omega-squared and Cramér's *V*
- **d** Fisher's *LSD* and r^2

7 Specify the correct *df* for each of these designs.
- **a** 2 × 2
- **b** 3 × 4
- **c** 4 × 5
- **d** 1 × 3

8 Supply the requested information for each of the following designs. Assume $\alpha = .05$.

	Design	χ^2_{obt}	*df*	χ^2_{crit}	Reject H_0?
a	2 × 2	4.5			
b	3 × 3	9.0			
c	1 × 5	17.22			
d	2 × 4	5.55			

9 For each matrix, fill in the missing frequency counts, then compute the f_e for each cell.

a

30	?	?	20	120
?	?	40	?	100
?	80	60	40	$N = ?$

b

7	?	14
?	?	
	18	30

Work Problems

10 Three different drug treatments are used to control hypertension. After treatment, the investigator classifies patients as having either a favorable or an unfavorable response to the medication. Set alpha at .05 and conduct a chi-square test for no relationship. If necessary, use the R statistic to determine which cells are major contributors to the χ^2. Interpret the findings. If there is evidence of an effect, what is the effect size?

	Treatment			
Response	**I**	**II**	**III**	
Favorable	70	160	168	398
Unfavorable	30	40	32	102

11 A psychologist hypothesized that biological males are more likely than biological females to accumulate trivial objects. All students were loaned No. 2 pencils for a multiple-choice exam. A box labeled "pencils" was positioned next to the exit. The number of males and females who returned the pencils after the test were counted. Conduct a chi-square test to analyze the data. If there is evidence of an effect? If so, what is the effect size?

	Kept pencil	**Returned pencil**
Males	15	40
Females	38	17

12 Two social scientists investigate police officer bias for traffic violations. They enlisted the aid of 50 drivers. Twenty-five drivers place a "Stop Police Brutality!" sticker on their bumper while the other 25 use a bumper sticker that reads, "Smile!" Assume no difference in driving habits between the groups. Over the next year, police stops are recorded. Those with the brutality sticker are stopped 18 times while those with the smile sticker are stopped 5 times. No driver is stopped more than once. (Use $\alpha = .05$ when running a test.)

a State the hypotheses.
b Specify f_o for each cell.
c Compute χ^2 and test the null.
d Interpret the findings.
e If an effect is found, what is the effect size?

13 A marketing psychologist is hired to consult an association of recreational vehicle dealers. The dealers are considering seasonally altering their advertising. The market researcher collects data on the number of RVs sold per season. Test a null of no selling differences between seasons, $\alpha = .05$. Perform follow-up tests if appropriate. Interpret the findings.

Spring	Summer	Fall	Winter
160	190	170	130

14 Below are data regarding driving conditions and automobile accidents. Run the appropriate chi-square test to see if a relationship exists.

	Accident	No accident
Rain	29	35
No rain	31	48

15 Both the number of people getting tattoos and the reasons they get them have changed dramatically in the last few decades. Suppose a researcher, with the help of a local tattoo parlor, organizes the reasons people give into the following four categories:

a To memorialize meaningful family event (e.g. marriage): 1724
b To memorialize meaningful nonfamily event (e.g. first-time skydiving): 1031
c Conformity reasons (e.g. friends or family are doing it together): 635
d Personal expression of an important idea (e.g. freedom, peace, etc.): 879

Test the null that each category is equally likely. Perform follow-up analyses if needed.

16 Are younger mothers more likely to give birth to physically immature babies? In this study, mothers under 20-years-old are compared with 30–35-year-old mothers. An immature baby has a birth weight equal to or less than 2500 grams. Set $\alpha = .05$ and test a null of no relationship. Perform follow-up analyses, including effect size, if needed.

	Birth weight		
Age	**≤ 2500 grams**	**> 2500 grams**	
Under 20	45	20	65
30–35	10	39	49
	55	59	114

17 A researcher is interested in seeing if diabetics show prolonged healing of wounds. Conduct a chi-square analysis on the data below and interpret any findings. Set alpha at .05.

Patient	Healing Normal	Prolonged	
Diabetic	125	329	454
Nondiabetic	245	111	356
	370	440	810

18 A dermatologist is interested in comparing dandruff treatments. After six weeks of treatment, an independent colleague decides if a patient has improved. Set alpha at .05 test the null of no difference between treatments. Perform follow-up analyses if needed.

Preparation	No improvement	Satisfactory improvement	
A	22	24	46
B	19	17	36
C	23	28	51
D	17	22	39
	81	91	172

Part 5 Review

Bivariate Data and Chi-Square Tests

Review of Concepts Presented in Part 5

The purpose of this review is to revisit the concepts introduced in Chapters 14–16. It is important to remember the major conceptual distinction separating the first two chapters which explored bivariate data (14 and 15) and the last chapter which introduced nonparametric statistic tools, namely two chi-square tests (16).

Although previous chapters dealt with the repeated measurement of participants, only one variable was measured. In Chapters 14 and 15, two different variables are measured for each participant. The purpose of bivariate data is to explore the nature and strength of the relationship between the measured variables. Both correlation and regression concepts help us realize that goal.

Unfortunately, when describing relationships between variables, it is tempting to think of, if not state, a specific causal structure for the relationship. However, the same cautions introduced in Chapter 1 are also in effect here. The language of causality is restricted to data gathered experimentally, where a hypothesized causal variable is controlled and manipulated. Typically, bivariate data is gathered in nonexperimental situations and so causal language is not warranted.

Another commonality between Chapters 14 and 15 is the notion of shared variance, the degree to which the variance associated with one variable corresponds to variance found in a second variable. It is captured as the coefficient of determination (r^2) in Chapter 14 and developed into a measure of prediction error (s_e) in Chapter 15.

There are also limitations and common pitfalls associated with analyzing bivariate data. Although advanced techniques associated with both correlation and

Essentials of Behavioral and Social Science Statistics, Third Edition.
K. Paul Nesselroade, Jr. and Laurence G. Grimm.
© 2025 John Wiley & Sons, Inc. Published 2025 by John Wiley & Sons, Inc.
Companion website: https://www.wiley.com/go/Essentials_of_Behavioral_and_Social_Science_Statistics3e

regression allow for the analysis of nonlinear relationships between variables, the introductory concepts presented in these chapters can only be used on bivariate data that is linearly related. Furthermore, bivariate data sets that do not capture the entire range of scores for each variable can lead to a mischaracterization of their relationship. Finally, the question of including or excluding outlier data can be problematic, either underrepresenting or overrepresenting the strength of the relationship between the variables.

A final area of commonality concerns the visual representation of the data, typically in the form of a scatterplot on a two-dimensional graph. Although not meant to replace the requirement for mathematical analysis, especially when making decisions regarding null hypotheses, a visual representation of the relationship can help make sense of the data in several nonquantifiable ways. Most notably, visual representation is necessary to avoid many of the pitfalls mentioned above.

There are also several areas of distinction between these two chapters. First, we should recognize that the regression concept is built upon the correlation concept. Regression takes the idea of quantifying the shared variance, extends it, and applies it to a specific purpose – the prediction of unknown values. Another distinction concerns variable designation. Correlations are bidirectional. The assignment of X and Y to the variables is inconsequential to the outcome; the r value will be the same either way, as will the decision regarding a null hypothesis. However, with regression, the mathematics used to make predictions require the proper specification of a predictor (X) and predicted (Y) variable; otherwise, the conclusions will be in error. Each technique also has a slightly different approach to testing a null hypothesis, an important objective if the data is to be used either as statistical evidence for a relationship or as a predictor variable. Correlational analyses test null hypotheses of no relationship ($\rho = 0$), while regression analyses postulate no slope ($b = 0$). The regression analysis also introduces two new assumptions on which the analysis and secondary judgments of error rest, namely the normality of conditional distributions and homoscedasticity.

The last chapter features nonparametric tests; these are statistical tools that do not make population assumptions or require interval or ratio data. The nonparametrics featured in Chapter 16 are the chi-square tests. These tools are applied to categorical data. In these analytical situations, participants are not measured in terms of "how much" but rather "how many;" that is, how many participants fit into one category versus another. If there is only one dimension or factor, the goodness-of-fit chi-square is the test to run. If, however, conditions vary across two dimensions or factors, then the chi-square test for independence is the analytical tool needed.

The null hypothesis for a goodness-of-fit test depends on what the "nothing" means for a particular investigation. It may mean *no difference* between the various conditions, or it may mean *no change* from the last time the data was collected. This needs to be carefully determined as the expected frequencies needed for chi-square calculation are derived from an understanding of the null. The null hypothesis for a chi-square test for independence, however, is always the same – no relationship between the two factors. (This is akin to investigating an interaction effect in a two-way factorial design.) The procedure for both chi-square tests is similar; compare the observed frequency counts with the expected frequency counts derived from a null hypothesis. The greater the cumulative disparity across the various conditions, the more likely the null is false.

Other nonparametrics like the Spearman rank correlation and the Mann–Whitney U test are briefly mentioned. Nonparametrics expand a researcher's ability to analyze data sets that cannot be otherwise analyzed with the standard t, F, or r tests. With this added flexibility, however, there is a trade-off. Typically, nonparametric tests are less powerful than their counterparts.

Since real-world data sets do not come with a label telling us which test to use, it is important to work on diagnostic skills. The exercises below will challenge us to recognize the appropriate diagnostic cues for *all* tests presented in the text. For the hypothesis testing exercises, we will need to use the given information to determine which test best fits the described situation.

Part 5 Review Exercises

1 Select the proper statement.
- **a** Correlation is to regression as relationship is to prediction.
- **b** Correlation is to regression as prediction is to relationship.
- **c** Correlation is to regression as independent variable is to dependent variable.
- **d** Correlation is to regression as dependent variable is to independent variable.

2 What does regression add to the concept of correlation?

3 Which type of test (parametric or nonparametric) is to be preferred and why?

4 Describe how r is related to prediction error.

5 If a researcher wants to study the relationship between a person's political conservatism and their philanthropic giving, what statistical tool seems most appropriate? Why?

6 A researcher studies the relationship between political attitudes and philanthropy by classifying people as liberals, conservatives, or independents and measuring the percent of charitable giving as noted on tax returns. What statistical seems most appropriate? Why?

7 A researcher wants to predict charitable giving based on the level of political conservatism espoused by an individual. What statistical tool seems most appropriate? Why?

8 A researcher studies the relationship between political attitudes, philanthropy, and biological sex by classifying people politically (liberals, conservatives, or independents) and biologically (male or female) and then measuring the percent of charitable giving as noted on tax returns. What statistical tool seems most appropriate? Why?

9 A school psychologist would like to determine if there are differences in reading preferences among students in a particular middle school. The number of each type of book checked out of the school library is tallied over a six-month period. Which test should be used?

Nonfiction	Sports	Romance	Science fiction	Classics
68	75	55	50	52

10 Imagine the following data set is gathered regarding biological male–female sibling pairs and their average daily time spent doing leisurely activities. Answer the questions below.

Male	Female
3.9	4.2
1.6	1.9
2.9	1.6
3.5	2.3
3.3	2.1
4.9	4.0
5.5	3.5
4.4	5.3
3.7	2.1
3.5	2.0

a What should we run to test a null of no difference between biological males and females?

b How would we test the null of no relationship between sibling pairs for this data set?

c What if our friend, Jimmy, spends two hours every day playing video games every day in an otherwise highly structured life, what analysis would we run to predict how many hours Jenny, his sibling, will engage in leisurely activity?

11 Suppose we are interested in the relationship between the number of siblings a child has and the age at which they learn to walk. We ask mothers of seven two-year-olds how many older siblings the child has and the age the child began to walk. The data are below.

Number of siblings	Age of walking (months)
1	16.0
2	8.5
0	16.5
3	10.0
1	11.0
2	10.0
4	9.0

a What statistical tool would we use to see if a relationship exists?

b Suppose a new child enters the study and has five older siblings. What statistical tool would help us predict when they would begin to walk?

12 A biology professor thinks caffeinated sodas cause more burping than non-caffeinated sodas. A large sample of students are gathered, split evenly into two groups, and randomly assigned to drink sodas with or without caffeine. The number of burps for each group over the next ten minutes is recorded. What test should be used?

13 Suppose a lot of burping is going on in both groups. So, the professor had each student count the number of burps they generated as their individual score. Would the analytical tool change? If so, to what?

14 Some research suggests early retirement may lead to memory decline (Rohwedder and Willis, 2010). The researchers gathered data based on a memory test given to numerous individuals aged 60–64 from a variety of countries throughout the world. An average memory score for each country cohort was generated. National statistics noting the percentage of retired people in that age window was also recorded. So, for each country we have an average memory score and a percentage of retired people at age 60–64. What test would be used to determine if a relationship exists? Is the causal relationship between the variables clear?

15 A soft drink company believes their new energy drink helps users lose weight responsibly. To test this hypothesis, 15 adult consumers are gathered, their current weight is measured, and they are asked to use the soft drink company's product at least twice a day. After six months, the participants are weighed again and the change in weight is recorded. What test should be used to evaluate the null hypothesis?

Appendix A

Statistical Tables

Table A.1 *z* table.

(A) z	(B) Area between mean *z*	(C) Area beyond and *z*	(A) z	(B) Area between mean and *z*	(C) Area beyond *z*	(A) z	(B) Area between mean and *z*	(C) Area beyond *z*
.00	.0000	.5000	.17	.0675	.4325	.34	.1331	.3669
.01	.0040	.4960	.18	.0714	.4286	.35	.1368	.3632
.02	.0080	.4920	.19	.0753	.4247	.36	.1406	.3594
.03	.0120	.4880	.20	.0793	.4207	.37	.1443	.3557
.04	.0160	.4840	.21	.0832	.4168	.38	.1480	.3520
.05	.0199	.4801	.22	.0871	.4129	.39	.1517	.3483
.06	.0239	.4761	.23	.0910	.4090	.40	.1554	.3446
.07	.0279	.4721	.24	.0948	.4052	.41	.1591	.3409
.08	.0319	.4681	.25	.0987	.4013	.42	.1628	.3372
.09	.0359	.4641	.26	.1026	.3974	.43	.1664	.3336
.10	.0398	.4602	.27	.1064	.3936	.44	.1700	.3300
.11	.0438	.4562	.28	.1103	.3897	.45	.1736	.3264
.12	.0478	.4522	.29	.1141	.3859	.46	.1772	.3228
.13	.0517	.4483	.30	.1179	.3821	.47	.1808	.3192
.14	.0557	.4443	.31	.1217	.3783	.48	.1844	.3156
.15	.0596	.4404	.32	.1255	.3745	.49	.1879	.3121
.16	.0636	.4364	.33	.1293	.3707	.50	.1915	.3085

(*Continued*)

Essentials of Behavioral and Social Science Statistics, Third Edition.
K. Paul Nesselroade, Jr. and Laurence G. Grimm.
© 2025 John Wiley & Sons, Inc. Published 2025 by John Wiley & Sons, Inc.
Companion website: https://www.wiley.com/go/Essentials_of_Behavioral_and_Social_Science_Statistics3e

Table A.1 (Continued)

(A) z	(B) Area between mean z	(C) Area beyond and z	(A) z	(B) Area between mean and z	(C) Area beyond z	(A) z	(B) Area between mean and z	(C) Area beyond z
.51	.1950	.3050	.80	.2881	.2119	1.09	.3621	.1379
.52	.1985	.3015	.81	.2910	.2090	1.10	.3643	.1357
.53	.2019	.2981	.82	.2939	.2061	1.11	.3665	.1335
.54	.2054	.2946	.83	.2967	.2033	1.12	.3686	.1314
.55	.2088	.2912	.84	.2995	.2005	1.13	.3708	.1292
.56	.2123	.2877	.85	.3023	.1977	1.14	.3729	.1271
.57	.2157	.2843	.86	.3051	.1949	1.15	.3749	.1251
.58	.2190	.2810	.87	.3078	.1922	1.16	.3770	.1230
.59	.2224	.2776	.88	.3106	.1894	1.17	.3790	.1210
.60	.2257	.2743	.89	.3133	.1867	1.18	.3810	.1190
.61	.2291	.2709	.90	.3159	.1841	1.19	.3830	.1170
.62	.2324	.2676	.91	.3186	.1814	1.20	.3849	.1151
.63	.2357	.2643	.92	.3212	.1788	1.21	.3869	.1131
.64	.2389	.2611	.93	.3238	.1762	1.22	.3888	.1112
.65	.2422	.2578	.94	.3264	.1736	1.23	.3907	.1093
.66	.2454	.2546	.95	.3289	.1711	1.24	.3925	.1075
.67	.2486	.2514	.96	.3315	.1685	1.25	.3944	.1056
.68	.2517	.2483	.97	.3340	.1660	1.26	.3962	.1038
.69	.2549	.2451	.98	.3365	.1635	1.27	.3980	.1020
.70	.2580	.2420	.99	.3389	.1611	1.28	.3997	.1003
.71	.2611	.2389	1.00	.3413	.1587	1.29	.4015	.0985
.72	.2642	.2358	1.01	.3438	.1562	1.30	.4032	.0968
.73	.2673	.2327	1.02	.3461	.1539	1.31	.4049	.0951
.74	.2704	.2296	1.03	.3485	.1515	1.32	.4066	.0934
.75	.2734	.2266	1.04	.3508	.1492	1.33	.4082	.0918
.76	.2764	.2236	1.05	.3531	.1469	1.34	.4099	.0901
.77	.2794	.2206	1.06	.3554	.1446	1.35	.4115	.0885
.78	.2823	.2177	1.07	.3577	.1423	1.36	.4131	.0869
.79	.2852	.2148	1.08	.3599	.1401	1.37	.4147	.0853

(*Continued*)

Table A.1 (Continued)

(A) z	(B) Area between mean z	(C) Area beyond and z	(A) z	(B) Area between mean and z	(C) Area beyond z	(A) z	(B) Area between mean and z	(C) Area beyond z
1.38	.4162	.0838	1.67	.4525	.0475	1.96	.4750	.0250
1.39	.4177	.0823	1.68	.4535	.0465	1.97	.4756	.0244
1.40	.4192	.0808	1.69	.4545	.0455	1.98	.4761	.0239
1.41	.4207	.0793	1.70	.4554	.0446	1.99	.4767	.0233
1.42	.4222	.0778	1.71	.4564	.0436	2.00	.4772	.0228
1.43	.4236	.0764	1.72	.4573	.0427	2.01	.4778	.0222
1.44	.4251	.0749	1.73	.4582	.0418	2.02	.4783	.0217
1.45	.4265	.0735	1.74	.4591	.0409	2.03	.4788	.0212
1.46	.4279	.0721	1.75	.4599	.0401	2.04	.4793	.0207
1.47	.4292	.0708	1.76	.4608	.0392	2.05	.4798	.0202
1.48	.4306	.0694	1.77	.4616	.0384	2.06	.4803	.0197
1.49	.4319	.0681	1.78	.4625	.0375	2.07	.4808	.0192
1.50	.4332	.0668	1.79	.4633	.0367	2.08	.4812	.0188
1.51	.4345	.0655	1.80	.4641	.0359	2.09	.4817	.0183
1.52	.4357	.0643	1.81	.4649	.0351	2.10	.4821	.0179
1.53	.4370	.0630	1.82	.4656	.0344	2.11	.4826	.0174
1.54	.4382	.0618	1.83	.4664	.0336	2.12	.4830	.0170
1.55	.4394	.0606	1.84	.4671	.0329	2.13	.4834	.0166
1.56	.4406	.0594	1.85	.4678	.0322	2.14	.4838	.0162
1.57	.4418	.0582	1.86	.4686	.0314	2.15	.4842	.0158
1.58	.4429	.0571	1.87	.4693	.0307	2.16	.4846	.0154
1.59	.4441	.0559	1.88	.4699	.0301	2.17	.4850	.0150
1.60	.4452	.0548	1.89	.4706	.0294	2.18	.4854	.0146
1.61	.4463	.0537	1.90	.4713	.0287	2.19	.4857	.0143
1.62	.4474	.0526	1.91	.4719	.0281	2.20	.4861	.0139
1.63	.4484	.0516	1.92	.4726	.0274	2.21	.4864	.0136
1.64	.4495	.0505	1.93	.4732	.0268	2.22	.4868	.0132
1.65	.4505	.0495	1.94	.4738	.0262	2.23	.4871	.0129
1.66	.4515	.0485	1.95	.4744	.0256	2.24	.4875	.0125

(*Continued*)

Table A.1 (Continued)

(A) z	(B) Area between mean z	(C) Area beyond and z	(A) z	(B) Area between mean and z	(C) Area beyond z	(A) z	(B) Area between mean and z	(C) Area beyond z
2.25	.4878	.0122	2.54	.4945	.0055	2.83	.4977	.0023
2.26	.4881	.0119	2.55	.4946	.0054	2.84	.4977	.0023
2.27	.4884	.0116	2.56	.4948	.0052	2.85	.4978	.0022
2.28	.4887	.0113	2.57	.4949	.0051	2.86	.4979	.0021
2.29	.4890	.0110	2.58	.4951	.0049	2.87	.4979	.0021
2.30	.4893	.0107	2.59	.4952	.0048	2.88	.4980	.0020
2.31	.4896	.0104	2.60	.4953	.0047	2.89	.4981	.0019
2.32	.4898	.0102	2.61	.4955	.0045	2.90	.4981	.0019
2.33	.4901	.0099	2.62	.4956	.0044	2.91	.4982	.0018
2.34	.4904	.0096	2.63	.4957	.0043	2.92	.4982	.0018
2.35	.4906	.0094	2.64	.4959	.0041	2.93	.4983	.0017
2.36	.4909	.0091	2.65	.4960	.0040	2.94	.4984	.0016
2.37	.4911	.0089	2.66	.4961	.0039	2.95	.4984	.0016
2.38	.4913	.0087	2.67	.4962	.0038	2.96	.4985	.0015
2.39	.4916	.0084	2.68	.4963	.0037	2.97	.4985	.0015
2.40	.4918	.0082	2.69	.4964	.0036	2.98	.4986	.0014
2.41	.4920	.0080	2.70	.4965	.0035	2.99	.4986	.0014
2.42	.4922	.0078	2.71	.4966	.0034	3.00	.4987	.0013
2.43	.4925	.0075	2.72	.4967	.0033	3.01	.4987	.0013
2.44	.4927	.0073	2.73	.4968	.0032	3.02	.4987	.0013
2.45	.4929	.0071	2.74	.4969	.0031	3.03	.4988	.0012
2.46	.4931	.0069	2.75	.4970	.0030	3.04	.4988	.0012
2.47	.4932	.0068	2.76	.4971	.0029	3.05	.4989	.0011
2.48	.4934	.0066	2.77	.4972	.0028	3.06	.4989	.0011
2.49	.4936	.0064	2.78	.4973	.0027	3.07	.4989	.0011
2.50	.4938	.0062	2.79	.4974	.0026	3.08	.4990	.0010
2.51	.4940	.0060	2.80	.4974	.0026	3.09	.4990	.0010
2.52	.4941	.0059	2.81	.4975	.0025	3.10	.4990	.0010
2.53	.4943	.0057	2.82	.4976	.0024	3.11	.4991	.0009

(Continued)

Table A.1 (Continued)

(A) z	(B) Area between mean z	(C) Area beyond and z	(A) z	(B) Area between mean and z	(C) Area beyond z	(A) z	(B) Area between mean and z	(C) Area beyond z
3.12	.4991	.0009	3.19	.4993	.0007	3.40	.4997	.0003
3.13	.4991	.0009	3.20	.4993	.0007	3.50	.4998	.0002
3.14	.4992	.0008	3.21	.4993	.0007	3.60	.4998	.0002
3.15	.4992	.0008	3.22	.4994	.0006	3.70	.4999	.0001
3.16	.4992	.0008	3.23	.4994	.0006	3.80	.49993	.00007
3.17	.4992	.0008	3.24	.4994	.0006	3.90	.49995	.00005
3.18	.4993	.0007	3.30	.4995	.0005	4.00	.49997	.00003

Column *A* lists the *z* score values. Column *B* provides the proportion of area between the mean and the *z* score value. Column *C* provides the proportion of area beyond the *z* score.
Note: Because the normal distribution is symmetrical, areas for negative *z* scores are the same as those for positive *z* scores.

Table A.2 *t* table.

	α values for two-tailed test					
	.20	*.10*	*.05*	*.02*	*.01*	*.001*
	α values for one-tailed test					
df	*.10*	*.05*	*.025*	*.01*	*.005*	*.0005*
1	3.078	6.314	12.706	31.821	63.657	636.619
2	1.886	2.920	4.303	6.965	9.925	31.598
3	1.638	2.353	3.182	4.541	5.841	12.924
4	1.533	2.132	2.776	3.747	4.604	8.610
5	1.476	2.015	2.571	3.365	4.032	6.869
6	1.440	1.943	2.447	3.143	3.707	5.959
7	1.415	1.895	2.365	2.998	3.499	5.408
8	1.397	1.860	2.306	2.896	3.355	5.041
9	1.383	1.833	2.262	2.821	3.250	4.781
10	1.372	1.812	2.228	2.764	3.169	4.587
11	1.363	1.796	2.201	2.718	3.106	4.437
12	1.356	1.782	2.179	2.681	3.055	4.318
13	1.350	1.771	2.160	2.650	3.012	4.221
14	1.345	1.761	2.145	2.624	2.977	4.140
15	1.341	1.753	2.131	2.602	2.947	4.073
16	1.337	1.746	2.120	2.583	2.921	4.015
17	1.333	1.740	2.110	2.567	2.898	3.965
18	1.330	1.734	2.101	2.552	2.878	3.922
19	1.328	1.729	2.093	2.539	2.861	3.883
20	1.325	1.725	2.086	2.528	2.845	3.850
21	1.323	1.721	2.080	2.518	2.831	3.819
22	1.321	1.717	2.074	2.508	2.819	3.792
23	1.319	1.714	2.069	2.500	2.807	3.767
24	1.318	1.711	2.064	2.492	2.797	3.745
25	1.316	1.708	2.060	2.485	2.787	3.725
26	1.315	1.706	2.056	2.479	2.779	3.707
27	1.314	1.703	2.052	2.473	2.771	3.690
28	1.313	1.701	2.048	2.467	2.763	3.674
29	1.311	1.699	2.045	2.462	2.756	3.659
30	1.310	1.697	2.042	2.457	2.750	3.646
40	1.303	1.684	2.021	2.423	2.704	3.551
60	1.296	1.671	2.000	2.390	2.660	3.460
120	1.289	1.658	1.980	2.358	2.617	3.373
∞	1.282	1.645	1.960	2.326	2.576	3.291

To be significant, the *t* obtained from the data must be equal to or larger than the value shown in the table.
Source: Adapted from Table III, Fisher et al.

Table A.3 Power table (finding power).

	One-tailed test (α)					One-tailed test (α)			
	.05	*.025*	*.01*	*.005*		*.05*	*.025*	*.01*	*.005*
	Two-tailed test (α)					Two-tailed test (α)			
δ	*.10*	*.05*	*.02*	*.01*	δ	*.10*	*.05*	*.02*	*.01*
.0	.10*	.05*	.02	.01	2.5	.80	.71	.57	.47
.1	.10*	.05*	.02	.01	2.6	.83	.74	.61	.51
.2	.11*	.05	.02	.01	2.7	.85	.77	.65	.55
.3	.12*	.06	.03	.01	2.8	.88	.80	.68	.59
.4	.13*	.07	.03	.01	2.9	.90	.83	.72	.63
.5	.14	.08	.03	.02	3.0	.91	.85	.75	.66
.6	.16	.09	.04	.02	3.1	.93	.87	.78	.70
.7	.18	.11	.05	.03	3.2	.94	.89	.81	.73
.8	.21	.13	.06	.04	3.3	.96	.91	.83	.77
.9	.23	.15	.08	.05	3.4	.96	.93	.86	.80
1.0	.26	.17	.09	.06	3.5	.97	.94	.88	.82
1.1	.30	.20	.11	.07	3.6	.97	.95	.90	.85
1.2	.33	.22	.13	.08	3.7	.98	.96	.92	.87
1.3	.37	.26	.15	.10	3.8	.98	.97	.93	.89
1.4	.40	.29	.18	.12	3.9	.99	.97	.94	.91
1.5	.44	.32	.20	.14	4.0	.99	.98	.95	.92
1.6	.48	.36	.23	.16	4.1	.99	.98	.96	.94
1.7	.52	.40	.27	.19	4.2	.99	.99	.97	.95
1.8	.56	.44	.30	.22	4.3	**	.99	.98	.96
1.9	.60	.48	.33	.25	4.4		.99	.98	.97
2.0	.64	.52	.37	.28	4.5		.99	.99	.97
2.1	.68	.56	.41	.32	4.6		**	.99	.98
2.2	.71	.59	.45	.35	4.7			.99	.98
2.3	.74	.63	.49	.39	4.8			.99	.99
2.4	.77	.67	.53	.43	4.9			.99	.99
					5.0			**	.99
					5.1				.99
					5.2				**

* Values inaccurate for *one-tailed* test by more than .01.
** The power at and below this point is greater than .995.
Source: Welkowitz et al. (1988)/with permission from Harcourt Brace Jovanovich.

Table A.4 Power table (finding delta).

	One-tailed test (α)			
	.05	*.025*	*.01*	*.005*
	Two-tailed test (α)			
Power	*.10*	*.05*	*.02*	*.01*
.25	.97	1.29	1.65	1.90
.50	1.64	1.96	2.33	2.58
.60	1.90	2.21	2.58	2.83
.67	2.08	2.39	2.76	3.01
.70	2.17	2.48	2.85	3.10
.75	2.32	2.63	3.00	3.25
.80	2.49	2.80	3.17	3.42
.85	2.68	3.00	3.36	3.61
.90	2.93	3.24	3.61	3.86
.95	3.29	3.60	3.97	4.22
.99	3.97	4.29	4.65	4.90
.999	4.37	5.05	5.42	5.67

Source: Welkowitz et al. (1988)/with permission from Harcourt Brace Jovanovich.

Table A.5 *F* table.

df: denominator	Degrees of freedom: numerator														
	1	*2*	*3*	*4*	*5*	*6*	*7*	*8*	*9*	*10*	*11*	*12*	*14*	*16*	*20*
1	161	200	216	225	230	234	237	239	241	242	243	244	245	246	248
	4052	**4999**	**5403**	**5625**	**5764**	**5859**	**5928**	**5981**	**6022**	**6056**	**6082**	**6106**	**6142**	**6169**	**6208**
2	18.51	19.00	19.16	19.25	19.30	19.33	19.36	19.37	19.38	19.39	19.40	19.41	19.42	19.43	19.44
	98.49	**99.00**	**99.17**	**99.25**	**99.30**	**99.33**	**99.34**	**99.36**	**99.38**	**99.40**	**99.41**	**99.42**	**99.43**	**99.44**	**99.45**
3	10.13	9.55	9.28	9.12	9.01	8.94	8.88	8.84	8.81	8.78	8.76	8.74	8.71	8.69	8.66
	34.12	**30.82**	**29.46**	**28.71**	**28.24**	**27.91**	**27.67**	**27.49**	**27.34**	**27.23**	**27.13**	**27.05**	**26.92**	**26.83**	**26.69**
4	7.71	6.94	6.59	6.39	6.26	6.16	6.09	6.04	6.00	5.96	5.93	5.91	5.87	5.84	5.80
	21.20	**18.00**	**16.69**	**15.98**	**15.52**	**15.21**	**14.98**	**14.80**	**14.66**	**14.54**	**14.45**	**14.37**	**14.24**	**14.15**	**14.02**
5	6.61	5.79	5.41	5.19	5.05	4.95	4.88	4.82	4.78	4.74	4.70	4.68	4.64	4.60	4.56
	16.26	**13.27**	**12.06**	**11.39**	**10.97**	**10.67**	**10.45**	**10.27**	**10.15**	**10.05**	**9.96**	**9.89**	**9.77**	**9.68**	**9.55**
6	5.99	5.14	4.76	4.53	4.39	4.28	4.21	4.15	4.10	4.06	4.03	4.00	3.96	3.92	3.87
	13.74	**10.92**	**9.78**	**9.15**	**8.75**	**8.47**	**8.26**	**8.10**	**7.98**	**7.87**	**7.79**	**7.72**	**7.60**	**7.52**	**7.39**
7	5.59	4.47	4.35	4.12	3.97	3.87	3.79	3.73	3.68	3.63	3.60	3.57	3.52	3.49	3.44
	12.25	**9.55**	**8.45**	**7.85**	**7.46**	**7.19**	**7.00**	**6.84**	**6.71**	**6.62**	**6.54**	**6.47**	**6.35**	**6.27**	**6.15**
8	5.32	4.46	4.07	3.84	3.69	3.58	3.50	3.44	3.39	3.34	3.31	3.28	3.23	3.20	3.15
	11.26	**8.65**	**7.59**	**7.01**	**6.63**	**6.37**	**6.19**	**6.03**	**5.91**	**5.82**	**5.74**	**5.67**	**5.56**	**5.48**	**5.36**
9	5.12	4.26	3.86	3.63	3.48	3.37	3.29	3.23	3.18	3.13	3.10	3.07	3.02	2.98	2.93
	10.56	**8.02**	**6.99**	**6.42**	**6.06**	**5.80**	**5.62**	**5.47**	**5.35**	**5.26**	**5.18**	**5.11**	**5.00**	**4.92**	**4.80**

(*Continued*)

Table A.5 (Continued)

df: denominator	Degrees of freedom: numerator														
	1	*2*	*3*	*4*	*5*	*6*	*7*	*8*	*9*	*10*	*11*	*12*	*14*	*16*	*20*
10	4.96	4.10	3.71	3.48	3.33	3.22	3.14	3.07	3.02	2.97	2.94	2.91	2.86	2.82	2.77
	10.04	**7.56**	**6.55**	**5.99**	**5.64**	**5.39**	**5.21**	**5.06**	**4.95**	**4.85**	**4.78**	**4.71**	**4.60**	**4.52**	**4.41**
11	4.84	3.98	3.59	3.36	3.20	3.09	3.01	2.95	2.90	2.86	2.82	2.79	2.74	2.70	2.65
	9.65	**7.20**	**6.22**	**5.67**	**5.32**	**5.07**	**4.88**	**4.74**	**4.63**	**4.54**	**4.46**	**4.40**	**4.29**	**4.21**	**4.10**
12	4.75	3.88	3.49	3.26	3.11	3.00	2.92	2.85	2.80	2.76	2.72	2.69	2.64	2.60	2.54
	9.33	**6.93**	**5.95**	**5.41**	**5.06**	**4.82**	**4.65**	**4.50**	**4.39**	**4.30**	**4.22**	**4.16**	**4.05**	**3.98**	**3.86**
13	4.67	3.80	3.41	3.18	3.02	2.92	2.84	2.77	2.72	2.67	2.63	2.60	2.55	2.51	2.46
	9.07	**6.70**	**5.74**	**5.20**	**4.86**	**4.62**	**4.44**	**4.30**	**4.19**	**4.10**	**4.02**	**3.96**	**3.85**	**3.78**	**3.67**
14	4.60	3.74	3.34	3.11	2.96	2.85	2.77	2.70	2.65	2.60	2.56	2.53	2.48	2.44	2.39
	8.86	**6.51**	**5.56**	**5.03**	**4.69**	**4.46**	**4.28**	**4.14**	**4.03**	**3.94**	**3.86**	**3.80**	**3.70**	**3.62**	**3.51**
15	4.54	3.68	3.29	3.06	2.90	2.79	2.70	2.64	2.59	2.55	2.51	2.48	2.43	2.39	2.33
	8.68	**6.36**	**5.42**	**4.89**	**4.56**	**4.32**	**4.14**	**4.00**	**3.89**	**3.80**	**3.73**	**3.67**	**3.56**	**3.48**	**3.36**
16	4.49	3.63	3.24	3.01	2.85	2.74	2.66	2.59	2.54	2.49	2.45	2.42	2.37	2.33	2.28
	8.53	**6.23**	**5.29**	**4.77**	**4.44**	**4.20**	**4.03**	**3.89**	**3.78**	**3.69**	**3.61**	**3.55**	**3.45**	**3.37**	**3.25**
17	4.45	3.59	3.20	2.96	2.81	2.70	2.62	2.55	2.50	2.45	2.41	2.38	2.33	2.29	2.23
	8.40	**6.11**	**5.18**	**4.67**	**4.34**	**4.10**	**3.93**	**3.79**	**3.68**	**3.59**	**3.52**	**3.45**	**3.35**	**3.27**	**3.16**
18	4.41	3.55	3.16	2.93	2.77	2.66	2.58	2.51	2.46	2.41	2.37	2.34	2.29	2.25	2.19
	8.28	**6.01**	**5.09**	**4.58**	**4.25**	**4.01**	**3.85**	**3.71**	**3.60**	**3.51**	**3.44**	**3.37**	**3.27**	**3.19**	**3.07**
19	4.38	3.52	3.13	2.90	2.74	2.63	2.55	2.48	2.43	2.38	2.34	2.31	2.26	2.21	2.15
	8.18	**5.93**	**5.01**	**4.50**	**4.17**	**3.94**	**3.77**	**3.63**	**3.52**	**3.43**	**3.36**	**3.30**	**3.19**	**3.12**	**3.00**

20	4.35	3.49	3.10	2.87	2.71	2.60	2.52	2.45	2.40	2.35	2.31	2.28	2.23	2.18	2.12
	8.10	**5.85**	**4.94**	**4.43**	**4.10**	**3.87**	**3.71**	**3.56**	**3.45**	**3.37**	**3.30**	**3.23**	**3.13**	**3.05**	**2.94**
21	4.32	3.47	3.07	2.84	2.68	2.57	2.49	2.42	2.37	2.32	2.28	2.25	2.20	2.15	2.09
	8.02	**5.78**	**4.87**	**4.37**	**4.04**	**3.81**	**3.65**	**3.51**	**3.40**	**3.31**	**3.24**	**3.17**	**3.07**	**2.99**	**2.88**
22	4.30	3.44	3.05	2.82	2.66	2.55	2.47	2.40	2.35	2.30	2.26	2.23	2.18	2.13	2.07
	7.94	**5.72**	**4.82**	**4.31**	**3.99**	**3.76**	**3.59**	**3.45**	**3.35**	**3.26**	**3.18**	**3.12**	**3.02**	**2.94**	**2.83**
23	4.28	3.42	3.03	2.80	2.64	2.53	2.45	2.38	2.32	2.28	2.24	2.20	2.14	2.10	2.04
	7.88	**5.66**	**4.76**	**4.26**	**3.94**	**3.71**	**3.54**	**3.41**	**3.30**	**3.21**	**3.14**	**3.07**	**2.97**	**2.89**	**2.78**
24	4.26	3.40	3.01	2.78	2.62	2.51	2.43	2.36	2.30	2.26	2.22	2.18	2.13	2.09	2.02
	7.82	**5.61**	**4.72**	**4.22**	**3.90**	**3.67**	**3.50**	**3.36**	**3.25**	**3.17**	**3.09**	**3.03**	**2.93**	**2.85**	**2.74**
25	4.24	3.38	2.99	2.76	2.60	2.49	2.41	2.34	2.28	2.24	2.20	2.16	2.11	2.06	2.00
	7.77	**5.57**	**4.68**	**4.18**	**3.86**	**3.63**	**3.46**	**3.32**	**3.21**	**3.13**	**3.05**	**2.99**	**2.89**	**2.81**	**2.70**
26	4.22	3.37	2.98	2.74	2.59	2.47	2.39	2.32	2.27	2.22	2.18	2.15	2.10	2.05	1.99
	7.72	**5.53**	**4.64**	**4.14**	**3.82**	**3.59**	**3.42**	**3.29**	**3.17**	**3.09**	**3.02**	**2.96**	**2.86**	**2.77**	**2.66**
27	4.21	3.35	2.96	2.73	2.57	2.46	2.37	2.30	2.25	2.20	2.16	2.13	2.08	2.03	1.97
	7.68	**5.49**	**4.60**	**4.11**	**3.79**	**3.56**	**3.39**	**3.26**	**3.14**	**3.06**	**2.98**	**2.93**	**2.83**	**2.74**	**2.63**
28	4.20	3.34	2.95	2.71	2.56	2.44	2.36	2.29	2.24	2.19	2.15	2.12	2.06	2.02	1.96
	7.64	**5.45**	**4.57**	**4.07**	**3.76**	**3.53**	**3.36**	**3.23**	**3.11**	**3.03**	**2.95**	**2.90**	**2.80**	**2.71**	**2.60**
29	4.18	3.33	2.93	2.70	2.54	2.43	2.35	2.28	2.22	2.18	2.14	2.10	2.05	2.00	1.94
	7.60	**5.42**	**4.54**	**4.04**	**3.73**	**3.50**	**3.33**	**3.20**	**3.08**	**3.00**	**2.92**	**2.87**	**2.77**	**2.68**	**2.57**
30	4.17	3.32	2.92	2.69	2.53	2.42	2.34	2.27	2.21	2.16	2.12	2.09	2.04	1.99	1.93
	7.56	**5.39**	**4.51**	**4.02**	**3.70**	**3.47**	**3.30**	**3.17**	**3.06**	**2.98**	**2.90**	**2.84**	**2.74**	**2.66**	**2.55**

(Continued)

Table A.5 (Continued)

df: denominator	Degrees of freedom: numerator														
	1	*2*	*3*	*4*	*5*	*6*	*7*	*8*	*9*	*10*	*11*	*12*	*14*	*16*	*20*
32	4.15	3.30	2.90	2.67	2.51	2.40	2.32	2.25	2.19	2.14	2.10	2.07	2.02	1.97	1.91
	7.50	**5.34**	**4.46**	**3.97**	**3.66**	**3.42**	**3.25**	**3.12**	**3.01**	**2.94**	**2.86**	**2.80**	**2.70**	**2.62**	**2.51**
34	4.13	3.28	2.88	2.65	2.49	2.38	2.30	2.23	2.17	2.12	2.08	2.05	2.00	1.95	1.89
	7.44	**5.29**	**4.42**	**3.93**	**3.61**	**3.38**	**3.21**	**3.08**	**2.97**	**2.89**	**2.82**	**2.76**	**2.66**	**2.58**	**2.47**
36	4.11	3.26	2.86	2.63	2.48	2.36	2.28	2.21	2.15	2.10	2.06	2.03	1.98	1.93	1.87
	7.39	**5.25**	**4.38**	**3.89**	**3.58**	**3.35**	**3.18**	**3.04**	**2.94**	**2.86**	**2.78**	**2.72**	**2.62**	**2.54**	**2.43**
38	4.10	3.25	2.85	2.62	2.46	2.35	2.26	2.19	2.14	2.09	2.05	2.02	1.96	1.92	1.85
	7.35	**5.21**	**4.34**	**3.86**	**3.54**	**3.32**	**3.15**	**3.02**	**2.91**	**2.82**	**2.75**	**2.69**	**2.59**	**2.51**	**2.40**
40	4.08	3.23	2.84	2.61	2.45	2.34	2.25	2.18	2.12	2.07	2.04	2.00	1.95	1.90	1.84
	7.31	**5.18**	**4.31**	**3.83**	**3.51**	**3.29**	**3.12**	**2.99**	**2.88**	**2.80**	**2.73**	**2.66**	**2.56**	**2.49**	**2.37**
42	4.07	3.22	2.83	2.59	2.44	2.32	2.24	2.17	2.11	2.06	2.02	1.99	1.94	1.89	1.82
	7.27	**5.15**	**4.29**	**3.80**	**3.49**	**3.26**	**3.10**	**2.96**	**2.86**	**2.77**	**2.70**	**2.64**	**2.54**	**2.46**	**2.35**
44	4.06	3.21	2.82	2.58	2.43	2.31	2.23	2.16	2.10	2.05	2.01	1.98	1.92	1.88	1.81
	7.24	**5.12**	**4.26**	**3.78**	**3.46**	**3.24**	**3.07**	**2.94**	**2.84**	**2.75**	**2.68**	**2.62**	**2.52**	**2.44**	**2.32**
46	4.05	3.20	2.81	2.57	2.42	2.30	2.22	2.14	2.09	2.04	2.00	1.97	1.91	1.87	1.80
	7.21	**5.10**	**4.24**	**3.76**	**3.44**	**3.22**	**3.05**	**2.92**	**2.82**	**2.73**	**2.66**	**2.60**	**2.50**	**2.42**	**2.30**
48	4.04	3.19	2.80	2.56	2.41	2.30	2.21	2.14	2.08	2.03	1.99	1.96	1.90	1.86	1.79
	7.19	**5.08**	**4.22**	**3.74**	**3.42**	**3.20**	**3.04**	**2.90**	**2.80**	**2.71**	**2.64**	**2.58**	**2.48**	**2.40**	**2.28**
50	4.03	3.18	2.79	2.56	2.40	2.29	2.20	2.13	2.07	2.02	1.98	1.95	1.90	1.85	1.78
	7.17	**5.06**	**4.20**	**3.72**	**3.41**	**3.18**	**3.02**	**2.88**	**2.78**	**2.70**	**2.62**	**2.56**	**2.46**	**2.39**	**2.26**

55	4.02	3.17	2.78	2.54	2.38	2.27	2.18	2.11	2.05	2.00	1.97	1.93	1.88	1.83	1.76
	7.12	**5.01**	**4.16**	**3.68**	**3.37**	**3.15**	**2.98**	**2.85**	**2.75**	**2.66**	**2.59**	**2.53**	**2.43**	**2.35**	**2.23**
60	4.00	3.15	2.76	2.52	2.37	2.25	2.17	2.10	2.04	1.99	1.95	1.92	1.86	1.81	1.75
	7.08	**4.98**	**4.13**	**3.65**	**3.34**	**3.12**	**2.95**	**2.82**	**2.72**	**2.63**	**2.56**	**2.50**	**2.40**	**2.32**	**2.20**
65	3.99	3.14	2.75	2.51	2.36	2.24	2.15	2.08	2.02	1.98	1.94	1.90	1.85	1.80	1.73
	7.04	**4.95**	**4.10**	**3.62**	**3.31**	**3.09**	**2.93**	**2.79**	**2.70**	**2.61**	**2.54**	**2.47**	**2.37**	**2.30**	**2.18**
70	3.98	3.13	2.74	2.50	2.35	2.23	2.14	2.07	2.01	1.97	1.93	1.89	1.84	1.79	1.72
	7.01	**4.92**	**4.08**	**3.60**	**3.29**	**3.07**	**2.91**	**2.77**	**2.67**	**2.59**	**2.51**	**2.45**	**2.35**	**2.28**	**2.15**
80	3.96	3.11	2.72	2.48	2.33	2.21	2.12	2.05	1.99	1.95	1.91	1.88	1.82	1.77	1.70
	6.96	**4.88**	**4.04**	**3.56**	**3.25**	**3.04**	**2.87**	**2.74**	**2.64**	**2.55**	**2.48**	**2.41**	**2.32**	**2.24**	**2.11**
100	3.94	3.09	2.70	2.46	2.30	2.19	2.10	2.03	1.97	1.92	1.88	1.85	1.79	1.75	1.68
	6.90	**4.82**	**3.98**	**3.51**	**3.20**	**2.99**	**2.82**	**2.69**	**2.59**	**2.51**	**2.43**	**2.36**	**2.26**	**2.19**	**2.06**
125	3.92	3.07	2.68	2.44	2.29	2.17	2.08	2.01	1.95	1.90	1.86	1.83	1.77	1.72	1.65
	6.84	**4.78**	**3.94**	**3.47**	**3.17**	**2.95**	**2.79**	**2.65**	**2.56**	**2.47**	**2.40**	**2.33**	**2.23**	**2.15**	**2.03**
150	3.91	3.06	2.67	2.43	2.27	2.16	2.07	2.00	1.94	1.89	1.85	1.82	1.76	1.71	1.64
	6.81	**4.75**	**3.91**	**3.44**	**3.14**	**2.92**	**2.76**	**2.62**	**2.53**	**2.44**	**2.37**	**2.30**	**2.20**	**2.12**	**2.00**
200	3.89	3.04	2.65	2.41	2.26	2.14	2.05	1.98	1.92	1.87	1.83	1.80	1.74	1.69	1.62
	6.76	**4.71**	**3.88**	**3.41**	**3.11**	**2.90**	**2.73**	**2.60**	**2.50**	**2.41**	**2.34**	**2.28**	**2.17**	**2.09**	**1.97**
400	3.86	3.02	2.62	2.39	2.23	2.12	2.03	1.96	1.90	1.85	1.81	1.78	1.72	1.67	1.60
	6.70	**4.66**	**3.83**	**3.36**	**3.06**	**2.85**	**2.69**	**2.55**	**2.46**	**2.37**	**2.29**	**2.23**	**2.12**	**2.04**	**1.92**
1000	3.85	3.00	2.61	2.38	2.22	2.10	2.02	1.95	1.89	1.84	1.80	1.76	1.70	1.65	1.58
	6.66	**4.62**	**3.80**	**3.34**	**3.04**	**2.82**	**2.66**	**2.53**	**2.43**	**2.34**	**2.26**	**2.20**	**2.09**	**2.01**	**1.89**
∞	3.84	2.99	2.60	2.37	2.21	2.09	2.01	1.94	1.88	1.83	1.79	1.75	1.69	1.64	1.57
	6.64	**4.60**	**3.78**	**3.32**	**3.02**	**2.80**	**2.64**	**2.51**	**2.41**	**2.32**	**2.24**	**2.18**	**2.07**	**1.99**	**1.87**

Table entries in lightface type are critical values for the .05 level of significance. Boldface type values are for the .01 level of significance.
Source: Snedecor (1956)/with permission of The Iowa State College Press.

Table A.6 The critical values for studentized range statistic (q), α = .05.

	k = number of treatments								
df: error term	*2*	*3*	*4*	*5*	*6*	*7*	*8*	*9*	*10*
5	3.64	4.60	5.22	5.67	6.03	6.33	6.58	6.80	6.99
	5.70	**6.98**	**7.80**	**8.42**	**8.91**	**9.32**	**9.67**	**9.97**	**10.24**
6	3.46	4.34	4.90	5.30	5.63	5.90	6.12	6.32	6.49
	5.24	**6.33**	**7.03**	**7.56**	**7.97**	**8.32**	**8.61**	**8.87**	**9.10**
7	3.34	4.16	4.68	5.06	5.36	5.61	5.82	6.00	6.16
	4.95	**5.92**	**6.54**	**7.01**	**7.37**	**7.68**	**7.94**	**8.17**	**8.37**
8	3.26	4.04	4.53	4.89	5.17	5.40	5.60	5.77	5.92
	4.75	**5.64**	**6.20**	**6.63**	**6.96**	**7.24**	**7.47**	**7.68**	**7.86**
9	3.20	3.95	4.41	4.76	5.02	5.24	5.43	5.59	5.74
	4.56	**5.43**	**5.96**	**6.35**	**6.66**	**6.91**	**7.13**	**7.33**	**7.51**
10	3.15	3.88	4.33	4.65	4.91	5.12	5.30	5.46	5.60
	4.48	**5.27**	**5.77**	**6.14**	**6.43**	**6.67**	**6.88**	**7.05**	**7.21**
11	3.11	3.82	4.26	4.57	4.82	5.03	5.20	5.35	5.49
	4.39	**5.15**	**5.62**	**5.97**	**6.25**	**6.48**	**6.67**	**6.84**	**6.99**
12	3.08	3.77	4.20	4.51	4.75	4.95	5.12	5.27	5.39
	4.32	**5.05**	**5.50**	**5.84**	**6.10**	**6.32**	**6.51**	**6.67**	**6.81**
13	3.06	3.73	4.15	4.45	4.69	4.88	5.05	5.19	5.32
	4.26	**4.96**	**5.40**	**5.73**	**5.98**	**6.19**	**6.37**	**6.53**	**6.67**
14	3.03	3.70	4.11	4.41	4.64	4.83	4.99	5.13	5.25
	4.21	**4.89**	**5.32**	**5.63**	**5.88**	**6.08**	**6.26**	**6.41**	**6.54**
15	3.01	3.67	4.08	4.37	4.59	4.78	4.94	5.08	5.20
	4.17	**4.84**	**5.25**	**5.56**	**5.80**	**5.99**	**6.16**	**6.31**	**6.44**
16	3.00	3.65	4.05	4.33	4.56	4.74	4.90	5.03	5.15
	4.13	**4.79**	**5.19**	**5.49**	**5.72**	**5.92**	**6.08**	**6.22**	**6.35**
17	2.98	3.63	4.02	4.30	4.52	4.70	4.86	4.99	5.11
	4.10	**4.74**	**5.14**	**5.43**	**5.66**	**5.85**	**6.01**	**6.15**	**6.27**
18	2.97	3.61	4.00	4.28	4.49	4.67	4.82	4.96	5.07
	4.07	**4.70**	**5.09**	**5.38**	**5.60**	**5.79**	**5.94**	**6.08**	**6.20**
19	2.96	3.59	3.98	4.25	4.47	4.65	4.79	4.92	5.04
	4.05	**4.67**	**5.05**	**5.33**	**5.55**	**5.73**	**5.89**	**6.02**	**6.14**
20	2.95	3.58	3.96	4.23	4.44	4.62	4.77	4.90	5.01
	4.02	**4.64**	**5.02**	**5.29**	**5.51**	**5.69**	**5.84**	**5.97**	**6.09**

(Continued)

Table A.6 (Continued)

	k = number of treatments								
df: error term	*2*	*3*	*4*	*5*	*6*	*7*	*8*	*9*	*10*
24	2.92	3.53	3.90	4.17	4.37	4.54	4.68	4.81	4.91
	3.96	**4.55**	**4.91**	**5.17**	**5.37**	**5.54**	**5.69**	**5.81**	**5.92**
30	2.89	3.49	3.85	4.10	4.30	4.46	4.60	4.72	4.82
	3.89	**4.45**	**4.80**	**5.05**	**5.24**	**5.40**	**5.54**	**5.65**	**5.76**
40	2.86	3.44	3.79	4.04	4.23	4.39	4.52	4.63	4.73
	3.82	**4.37**	**4.70**	**4.93**	**5.11**	**5.26**	**5.39**	**5.50**	**5.60**
60	2.83	3.40	3.74	3.98	4.16	4.31	4.44	4.55	4.65
	3.76	**4.28**	**4.59**	**4.82**	**4.99**	**5.13**	**5.25**	**5.36**	**5.45**
120	2.80	3.36	3.68	3.92	4.10	4.24	4.36	4.47	4.56
	3.70	**4.20**	**4.50**	**4.71**	**4.87**	**5.01**	**5.12**	**5.21**	**5.30**
∞	2.77	3.31	3.63	3.86	4.03	4.17	4.29	4.39	4.47
	3.64	**4.12**	**4.40**	**4.60**	**4.76**	**4.88**	**4.99**	**5.08**	**5.16**

Table entries in lightface type are critical values for the .05 level of significance. Boldface type values are for the .01 level of significance.

Table A.7 Pearson *r* table.

	α **levels for two-tailed test**				
	.10	***.05***	***.02***	***.01***	***.001***
	α **levels for one-tailed test**				
($df = n_p - 2$)	***.05***	***.025***	***.01***	***.005***	***.0005***
1	.98769	.99692	.999507	.999877	.9999988
2	.90000	.95000	.98000	.990000	.99900
3	.8054	.8783	.93433	.95873	.99116
4	.7293	.8114	.8822	.91720	.97406
5	.6694	.7545	.8329	.8745	.95074
6	.6215	.7067	.7887	.8343	.92493
7	.5822	.6664	.7498	.7977	.8982
8	.5494	.6319	.7155	.7646	.8721

(Continued)

Table A.7 (Continued)

	α levels for two-tailed test				
	.10	*.05*	*.02*	*.01*	*.001*
	α levels for one-tailed test				
($df = n_p - 2$)	*.05*	*.025*	*.01*	*.005*	*.0005*
9	.5214	.6021	.6851	.7348	.8371
10	.4973	.5760	.6581	.7079	.8233
11	.4762	.5529	.6339	.6835	.8010
12	.4575	.5324	.6120	.6614	.7800
13	.4409	.5139	.5923	.6411	.7603
14	.4259	.4973	.5742	.6226	.7420
15	.4124	.4821	.5577	.6055	.7246
16	.4000	.4683	.5425	.5897	.7084
17	.3887	.4555	.5285	.5751	.6932
18	.3783	.4438	.5155	.5614	.6787
19	.3687	.4329	.5034	.5487	.6652
20	.3598	.4227	.4921	.5368	.6524
25	.3233	.3809	.4451	.4869	.5974
30	.2960	.3494	.4093	.4487	.5541
35	.2746	.3246	.3810	.4182	.5189
40	.2573	.3044	.3578	.3932	.4896
45	.2428	.2875	.3384	.3721	.4648
50	.2306	.2732	.3218	.3541	.4433
60	.2108	.2500	.2948	.3248	.4078
70	.1954	.2319	.2737	.3017	.3799
80	.1829	.2172	.2565	.2830	.3568
90	.1726	.2050	.2422	.2673	.3375
100	.1638	.1946	.2301	.2540	.3211

To be significant, the *r* obtained from the data must be equal to or larger than the value shown in the table.
Source: Adapted from Table VII, Fisher et al.

Table A.8 Chi-square table.

	α levels				
df	*.10*	*.05*	*.02*	*.01*	*.001*
1	2.71	3.84	5.41	6.64	10.83
2	4.60	5.99	7.82	9.21	13.82
3	6.25	7.82	9.84	11.34	16.27
4	7.78	9.49	11.67	13.28	18.46
5	9.24	11.07	13.39	15.09	20.52
6	10.64	12.59	15.03	16.81	22.46
7	12.02	14.07	16.62	18.48	24.32
8	13.36	15.51	18.17	20.09	26.12
9	14.68	16.92	19.68	21.67	27.88
10	15.99	18.31	21.16	23.21	29.59
11	17.28	19.68	22.62	24.72	31.26
12	18.55	21.03	24.05	26.22	32.91
13	19.81	22.36	25.47	27.69	34.53
14	21.06	23.68	26.87	29.14	36.12
15	22.31	25.00	28.26	30.58	37.70
16	23.54	26.30	29.63	32.00	39.25
17	24.77	27.59	31.00	33.41	40.79
18	25.99	28.87	32.35	34.80	42.31
19	27.20	30.14	33.69	36.19	43.82
20	28.41	31.41	35.02	37.57	45.32
21	29.62	32.67	36.34	38.93	46.80
22	30.81	33.92	37.66	40.29	48.27
23	32.01	35.17	38.97	41.64	49.73
24	33.20	36.42	40.27	42.98	51.18
25	34.38	37.65	41.57	44.31	52.62
26	35.56	38.88	42.86	45.64	54.05
27	36.74	40.11	44.14	46.96	55.48
28	37.92	41.34	45.42	48.28	56.89
29	39.09	42.56	46.69	49.59	58.30
30	40.26	43.77	47.96	50.89	59.70

To be significant, the χ^2 obtained from the data must be equal to or larger than the value shown in the table.
Source: Adapted from Table IV, Fisher et al.

Appendix B

Answers to Exercises and Work Problems

Chapter 1

1 Here are some examples:
Description (1) How frequently are tiny homes being constructed? (2) How many different types of tiny homes are there? (3) What is the ratio of different types of tiny homes? (4) What are the defining features of a tiny home? (5) What is the square foot range of a tiny home?
Correlation (1) Is there a relationship between tiny homeowners and various socioeconomic variables? (2) Is there a relationship between parts of the country and tiny home prevalence? (3) Do tiny homeowners tend to have similar political viewpoints?
Understanding (1) Are tiny home designs more attractive as space is dedicated to specific functionality (e.g. bathroom, bedroom, kitchen) or to general living space? (2) Are tiny homes more attractive if they are mobile or anchored?
Comment: Because the topic of interest is a major life decision (what kind of home to buy), it is hard to manipulate and study it experimentally, therefore it is hard to approach the goal of understanding. Some topics of study are limited because manipulation is either ethically dubious, logically impossible, or logistically impractical.

2 Quantitative: Answers will vary. Some examples of different quantitative independent variables and amounts include: (1) Gummy Bears: 0 gummy bears, 2 gummy bears, 4 gummy bears; (2) Verbal Praise: no verbal praise, one statement of praise, two statements of praise; (3) Screen Time: 10 minutes of screen time, 20 minutes of screen time. (Studies can have different numbers of levels of the independent variable.)
Qualitative: Answers will vary. Some examples of different qualitative independent variables and kinds include: (1) Reward Type: verbal praise, sugar

Essentials of Behavioral and Social Science Statistics, Third Edition.
K. Paul Nesselroade, Jr. and Laurence G. Grimm.
© 2025 John Wiley & Sons, Inc. Published 2025 by John Wiley & Sons, Inc.
Companion website: https://www.wiley.com/go/Essentials_of_Behavioral_and_Social_Science_Statistics3e

snack, salty snack; (2) Emotion: verbal praise with emotion, verbal praise without emotion; (3) Reward Type: gifted toy, screen time, snack. (Studies can have different numbers of levels of the independent variable.)

In terms of holding an extraneous variable constant, answers will vary. One example would be the "time spent administering the reward." If some rewarding procedures took longer than others, it could be that the time of the interaction with the child and not the reward itself would be responsible for some or all the differences found between conditions.

3 Below are some example designs.

a *Correlational design*: Have all participants experience the same painful stimulus. Do not manipulate the level of pain. Measure each participant's pain perception and anxiety level.

Experiment: First, select a method for inducing pain. Next, randomly assign participants to at least two levels of pain induction that differ in intensity (e.g. placing arms in buckets of water with different temperatures – both cold, but one colder than the other). For the dependent variable, select a known method for measuring anxiety.

b *Correlational design*: Randomly select a group of participants. Find out how much participants exercise and **measure** their resting heart rate. Note that participants are not randomly assigned to different exercise conditions.

Experiment: Randomly assign participants to a "high frequency of exercise" condition and a "low frequency of exercise" condition. After a predetermined length of time, say, three months, obtain a measure of resting heart rate. If we wanted three conditions, we could also have a group that is asked not to exercise.

c *Correlational design*: Take a random sample of participants and have them record the number of hours they work per week **over** a one-month period. Also, measure each participant's need for achievement.

Experiment: Need for achievement is a personality (participant) variable. There is no way to manipulate it and use it as an independent variable in an experiment. Hours worked per week are also a major component of a person's life. This relationship would be hard to study experimentally.

d *Correlational design*: Take a random sample of children, making sure that the sample **includes** some children who will attend preschool and some children who will not attend preschool. Measure their social skills in first grade. Another approach is to take a random sample of first-grade students, measure social skills, and identify which students attended preschool and which did not.

Experiment: Preschool children would be randomly assigned to attend or not attend day care. Measure all children's level of social skills when they are in first grade. Obviously, there are real-world problems with implementing this design. What if the parent of a child assigned not to attend day care wants their child to attend day care, and vice versa?

4 **a** Observation
b Hypothesis
c Theory
d Theory
e Observation
f Hypothesis

5 **d** It is the only one of the four definitions that reflects a concrete way of measuring the concept for the purposes of a scientific study.

6 **a** *Independent variable:* the amount of Vitamin E (20, 60, placebo). *Dependent variable:* Time participants spent riding the bicycle.
b *Independent variable:* Educational programs. Dependent variable: There are two dependent variables – comprehension and reading speed.
c *Independent variable:* Type of justification prompted for behaving counter-attitudinally (voluntary or forced). Dependent variable: Amount of attitude change.
d *Independent variable:* Days of natural light. Dependent variable: Number of widgets made.

7 **a** The letter on the can is confounded with the types of root beer. Morgan's is always the letter *A* and Rose's is always the letter *B*. Is it the taste that participants are responding to, or are they simply showing a preference for the letter *A* over *B*? This is admittedly a "stretch," but sometimes differences can be traced to seemingly innocuous differences such as this – see Box 1.1.
b Higher evaluations might be due to ownership and not mere association. A way needs to be found to associate the target object with the participant that does not involve ownership.
c The presence of the radar units alone could have caused motorists to drive more carefully. Without using the radar, it is possible that reducing the speed limit (even if motorists obey the new speed limit) would not have influenced traffic accidents.
d Participants are not randomly assigned to the two conditions. It is possible that those who select the headphones are different in some way from those who select novocaine. Perhaps those selecting "headphones" are less anxious about dental work in general, and therefore reported less anxiety and pain.

8 b

9 Since participants were not randomly assigned to the "title page" vs. "no title page" conditions, it is possible that higher grades are associated with "title page" papers not because the professor is biased in favor of them, but because students with better writing skills tend to include them.

10 "a" is for sure. "b" is most likely, unless there is some way of assigning ID numbers that would suggest that different types of students get different types of numbers. "c" is probably not a good method – there are too many reasons to think meaningful differences between students might correspond with different sign-up preferences. "d" is probably not a good system either – color preference could be associated with personality and temperament differences. "e" also has potential problems – if we are mainly using collegiate freshman, the younger 18-year-olds are going to end up in one group, and the older 19-year-olds will end up in the other.

11 "a" is problematic in many ways, one being that not everyone is likely to eat at the cafeteria. "b" might be a good method, but there are likely privacy violations constraining the registrar. "c" is problematic because participants are selecting themselves; the resulting sample could misrepresent the larger population in terms of extroversion, amount of free time, helpfulness, etc. "d" is problematic for the same reasons as "c." "e," although awkward and time-consuming, would be an excellent way to generate a random sample of the student population.

Chapter 2

1 **a** *Nominal* scale examples include "pass/no pass" or "good student" and "bad student."

Ordinal scale examples include grades in each class (A, B, C, D, F) or a set of categories like great, good, so-so, poor, and terrible.

Interval scale examples are hard to think of – debatably GPA is an interval measure since a "0" GPA is more easily thought of as a grouping of very poor students as opposed to a single place on a scale. For instance, one student may fail every class but barely, while another fails every class dreadfully. Both get a "0" GPA. Also, a student with a GPA of "1" is not necessarily half as successful as a student with a GPA of "2." Furthermore, an argument can be made that the spacing between integers is conserved, at least in terms of the number of academic points needed to move from one notch on the scale up to another.

Ratio scale examples include the number correct on a given measure or perhaps the number of credits earned (10 credits is half of 20).

b *Nominal* scale examples include a "student–athlete/non-student–athlete," a self-reported "athlete/nonathlete," and an independent judge report of "athlete/nonathlete."

Ordinal scale examples include a Likert scale question like "how athletic are you?" (very, somewhat, not really, not at all), "what place did you earn in the tournament?" (first, second...), or "what ranking does one have on their tennis team?"

Interval scale examples are hard to think of – but one might include Likert scale questions, depending upon how one argues Likert scale data are to be interpreted. Golf scores if stated above or below par might work. A stroke is a stroke is a stroke, but the zero point (par) is determined on other grounds. In other words, a person who scores 2 below par is not half as good as a player who scores 4 below par.

Ratio scale examples include the number of trophies won, how far one can throw a javelin, and how fast one can run a mile.

c *Nominal* scale examples include categorizations like "creative/not creative."

Ordinal scale examples include a ranking by independent judges of creative products, a grade in a class involving creativity, and a Likert scale question like "how creative are you?" (very, somewhat, so-so, not very, not at all).

Interval scale examples are hard to think of – but one might include Likert scale questions, depending upon how one argues the Likert scale is to be interpreted.

Ratio scale examples include how many art awards a person has won and the number of judges out of 10 who classify a person as "artistic."

d *Nominal* scale examples include categorizations like "high amount" vs. "low amount."

Ordinal scale examples include Likert scale questions like "how much food did you eat today?" (a lot, an average amount, a little) or an independent judge rankings.

Interval scale examples are hard to think of – but one might include Likert scale questions, depending upon how one argues the Likert scale is to be interpreted; weighing the plate might be considered interval if one does not take out the weight of the plate itself.

Ratio scale examples include the weight of food (minus the plate), the number of calories eaten, or even the difference between pre-meal weight and after-meal weight.

2 Here are two examples of likely normal distributions: pounds of garbage per week for a given family (it makes sense to think that most weeks would produce garbage amounts that are roughly similar to most other weeks – with some weeks being a bit less and some weeks being a bit more) and the total inches of snowfall for a given northern American city (it makes sense to think that most years produce a total snowfall that is very similar to most other years – with some years being a bit less and some being a bit more).

3 Here are two examples of likely negatively skewed distributions: GPAs at a given university (it makes sense to suggest that most GPAs will cluster around 3.25 or 3.3 with some being greater, but many more will trail off down the scale to about a 1.75 or 1.5 and free-throw percentages of professional basketball players (it makes sense to think that most have high percentages, but since the ceiling of 100% cannot be exceeded, the dispersion is much more likely to stretch out toward the lower percentages).

4 Here are two examples of likely positively skewed distributions: yards rushed by an NFL running back (it makes sense that most rushing attempts produce just a few yards gained, sometimes even the loss of a few yards; but there will be a good number of rushes that will amount to significant yardage gains) and number of books read per year (it makes sense that a few read voraciously, but most people consume just a few.

5 **a** Nominal
b Ordinal
c Ratio
d Ratio
e Interval or ordinal – how to interpret Likert scale data?
f Ratio
g Nominal
h Nominal
i Nominal
j Ordinal

6

	Width	*LL*	*Midpoint*	*UL*
a	3	0.5	2	3.5
b	5	5.5	8	10.5
c	5	−8.5	−6	−3.5
d	5	−2.5	0	2.5

7 and 8 **Part a:** Only the top and bottom four numbers of the distribution are provided here (and also in **Part b**).

LL	*X*	*UL*	*f*	*cf*
97.5	98	98.5	1	36
96.5	97	97.5	0	35
95.5	96	96.5	0	35
94.5	95	95.5	0	35
⋮	⋮	⋮	⋮	⋮
43.5	44	44.5	0	3
42.5	43	43.5	2	3
41.5	42	42.5	0	1
40.5	41	41.5	1	1

Part b

LL	*X*	*UL*	*f*	*cf*
95.5	96–98	98.5	1	36
92.5	93–95	95.5	0	35
89.5	90–92	92.5	3	35
86.5	87–89	89.5	3	32
⋮	⋮	⋮	⋮	⋮
47.5	48–50	50.5	1	5
44.5	45–47	47.5	1	4
41.5	42–44	44.5	2	3
38.5	39–41	41.5	1	1

Part c

LL	*X*	*UL*	*f*	*cf*
89.5	90–99	99.5	4	36
79.5	80–89	89.5	11	32
69.5	70–79	79.5	9	21
59.5	60–69	69.5	5	12
49.5	50–59	59.5	2	7
39.5	40–49	49.5	5	5

Part d

LL	*X*	*UL*	*f*	*cf*
79.5	80–99	99.5	15	36
59.5	60–79	79.5	14	21
39.5	40–59	59.5	7	7

9 Refer to graph.

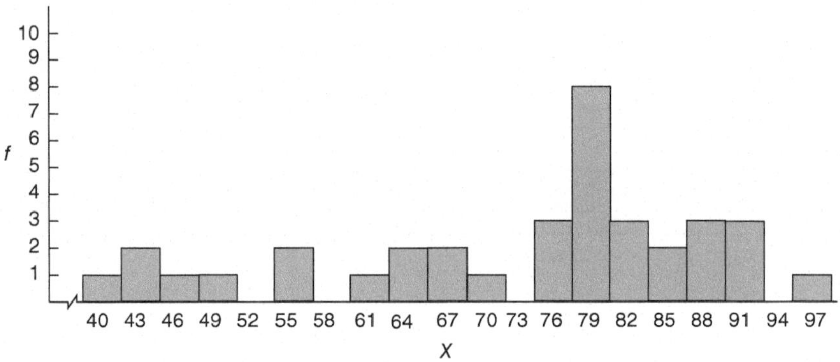

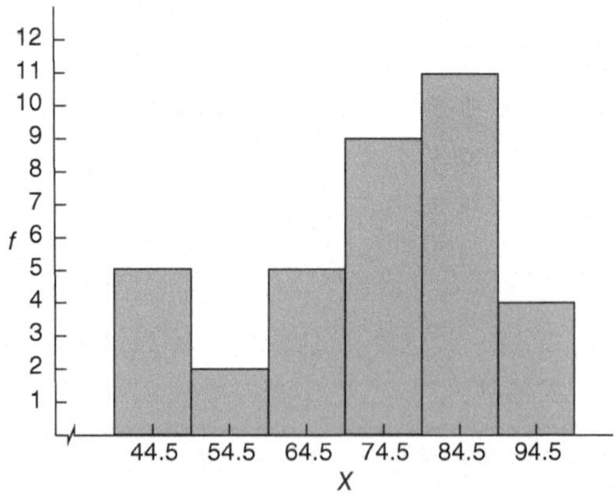

10 Refer to graph.

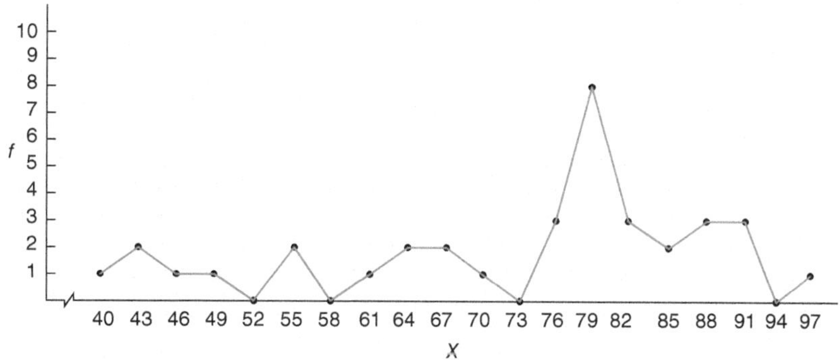

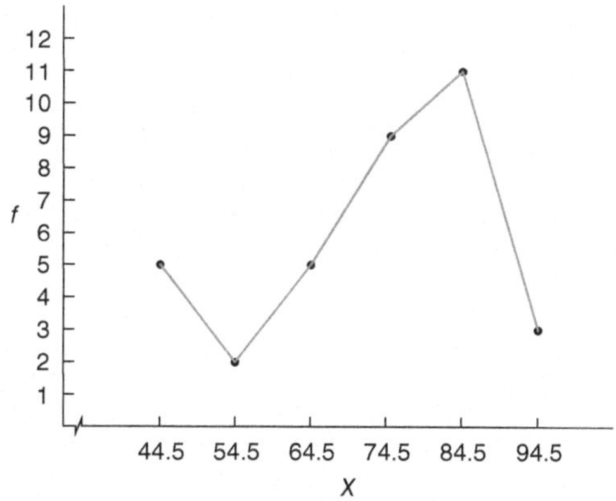

11 Here is a picture of a graph drawn in Excel that faithfully represents the relationship between the cereal types in terms of sugar per serving.

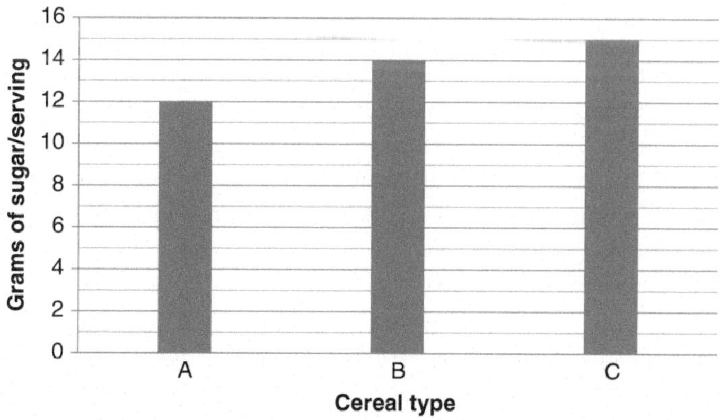

Here is a graph drawn in Excel that truncates the *Y* axis without labeling it, potentially leaving the viewer with a false understanding of the relationship between the cereal types in terms of sugar per serving.

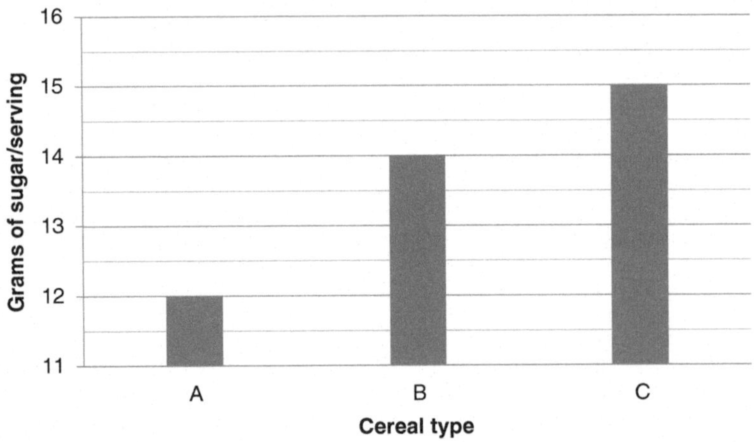

Chapter 3

1 Answers will vary. Here are two possibilities. (1) Grade Point Average: perhaps it is more important to think of a set of grades in terms of the median. This downplays the effect of an extreme score where perhaps a life event interfered with schooling. Also, perhaps a case can be made that the mode is a good way to capture a student's typical academic abilities. (2) Digital step counter: perhaps it is more important to know how many days one exceeds 10 000 steps in a week than it is to know a weekly average. A median might be more beneficial.

2 Answers will vary. Here are a couple of possibilities. (1) Height measurements for a basketball team composed of only guards and centers. (The shorter guards would all be clustered around a smaller height measure, and the taller centers would cluster around a larger number.) (2) The running times for Olympians running the 100-m dash. (The men's times would be clustered around 10 seconds, while the women's times would be clustered around 11 seconds.)

3 It is almost certainly symmetrical.

4 The mean will still be 15. The formula for both means is functionally equivalent.

5 **a** 2
b −1
c −10
d 0
e −11
f −0.5

6

a	$M = 6$	Median = 6	Mode = 8
b	$M = 4.86$	Median = 4	Mode = 4
c	$M = 8.71$	Median = 9	Mode = 10
d	$M = 3.86$	Median = 4	Mode = 1 and 4
e	$M = 5.67$	Median = 6.5	Mode = 8
f	$M = 8.22$	Median = 8	Mode = 5

For "**f**," remember to rearrange the numbers from lowest to highest.

7 Median = 7 (two sevens fall on each side of the middle – because of this, it does not matter if the data are declared to be discrete or discontinuous)

8 Median = 18.83. One of the three 19's needs to be below the median. We need to add .33 to the start of the interval in which 19 is the midpoint (i.e. 18.5 – 19.5). So, 18.5 + .33 = 18.83.

9 **a** Positively skewed
b Negatively skewed
c Symmetrical, unimodal
d Symmetrical, bimodal
e Negatively skewed
f Positively skewed

10 **a** $M = 104.80$
b Median = 101
c Yes, there is a mode. It is also 101.

11 The mean has the biggest difficulty with extreme scores. Because the mean considers the distance from each score to the middle, extreme scores, especially for small datasets, can generate a mean that does not sit amidst the bulk of the scores.

12 The median. Neither an ordinal scale nor the concept of the median makes any assumptions about the uniformity of the intervals between values.

13 The mode. Whenever data are in the form of *how many* (i.e. a nominal scale) rather than *how much* (i.e. an interval or ratio scale), the mode is the only appropriate measure of centrality.

14 101. If the mean of the original distribution is 100, the sum of the 10 scores must be 1000. If one of those numbers goes from 80 to 90, the new sum must be 1010. There are still a total of 10 numbers, so the new mean is now 101.

15 27. The first set of 6 scores must sum to 150, since the mean is 25. If 1 score is removed, we now have 5 scores summing to 135, which means the value of the new mean is 27.

16 The best guess is 52 minutes. As sample size increases, sample statistics better approximate population statistics. The sample including 500 people is most likely the most accurate.

17 $M = 109.73$; median $= 100.50$; mode $= 100$
The measures of centrality suggest a positively skewed distribution.

Chapter 4

1 One cannot say which would have the *larger* variance, but the sample with $n = 60$ would likely give a more accurate estimate of the population variance.

2 The variance cannot be estimated based on the size of M.

3 The denominator in the sample formula has a correction factor $(n - 1)$. This helps to make the resulting value and unbiased estimate of the population variance. Otherwise, a sample variance would most likely be too small. Reducing the numerator by one helps to correct this problem.

4 All scores are 50. There is no deviation from that mean value.

5 The variance is the average squared deviation score.

6 **a** σ – all the scores for the population (one season) are present
b s – a sample is being taken to evaluate something much bigger
c σ – all scores for one class's test are available, this is not a sample
d s – a subset from a much larger collection of light bulbs is sampled.

7 No. The 68-95-99.7 rule assumes a normal distribution.

8 The range is entirely determined by two extreme scores – the highest and lowest.

9 The standard deviation is usually considered the better descriptor statistic. The squaring used to create the variance has been removed by taking the square root.

10 The coach would need to transform the data by dividing each value recorded by three, thereby turning feet measurements into yards.

11 25 minutes. As sample size increases, sample statistics better approximate population statistics. The sample including 500 people is most likely the most accurate.

12 **a** $IQR = 111 - 81 = 30$
b $SIQR = (111 - 81)/2 = 15$

13 **a** 3.5
b 12.8
c 2.58

14 The sample mean (M) is 103.50; the sample standard deviation (s) is 5.24.

15 The range is 7; the variance (s^2) is 4.27; the standard deviation (s) is 2.07.

16 $\sigma^2 = 23.35$ and $\sigma = 4.83$

17 $s_A^2 = 7.125$ $s_B^2 = 5.90$ Distribution A has the larger variance.

18 Using the 68-95-99.7 rule, values 40–60 would encapsulate the middle 68% of scores, 30–70 would encapsulate the middle 95%, and 20–80 would encapsulate the middle 99.7%.

19 Using the 68-95-99.7 rule, values 48–52 would encapsulate the middle 68% of scores, 46–54 would encapsulate the middle 95%, and 44–56 would encapsulate the middle 99.7%.

20 According to the 68-95-99.7 rule, the standard deviation must be 12. This means the variance of this distribution must be 144.

21 A variance of 100 means a standard deviation of 10. According to the 68-95-99.7 rule, the values 160 and 120 must be ± 2 standard deviations away from the mean. This means the distribution must have a mean of 140.

22 **a** 15
b 250
c 30
d 500

23 $M = 14.50$; $s^2 = 1.60$

24

$X + 10$	$X - 10$	$X(10)$	$X/10$
$\mu = 60$	$\mu = 40$	$\mu = 500$	$\mu = 5$
$\sigma^2 = 25$	$\sigma^2 = 25$	$\sigma^2 = 2500$	$\sigma^2 = .25$

25 Range = 15 $IQR = 6$ $SIQR = 3$

26 $M = 965.24$ $s^2 = 13\,856.20$ $s = 117.71$

27 $M = 2895.72$ $s^2 = 124\,705.80$ $s = 353.14$

28 $M = 321.75$ $s^2 = 1539.58$ $s = 39.24$

Chapter 5

1. Percentile rank notes the percentage of scores falling at or bclow a given score in a distribution.

2. A z score represents the number of standard deviations a raw score is away from the mean.

3. A positive z score means the corresponding raw score is larger than the mean, while a negative z score means the corresponding raw score is smaller than the mean.

4. Answers will vary. Variables that may be normally distributed include the number of slices of pizza eaten in a month by university students, the body weight of biological male students, the level of extroversion of all students, etc.

5. Answers will vary. Variables that might not be normally distributed include the miles away from home for university students (positively distributed), the driving speed of cars on an interstate relative to the speed limit (positively

distributed), the number of credits taken per semester by university students (negatively distributed), the number of miles on car odometers (positively distributed), etc.

6

a $PR = \frac{23 + .5(12)}{49}(100) = .59(100) = 59$

b $PR = \frac{42 + .5(4)}{49}(100) = .90(100) = 90$

c $PR = \frac{13 + .5(10)}{49}(100) = .37(100) = 37$

d $PR = \frac{6 + .5(7)}{49}(100) = .19(100) = 19$

7 $\mu = 7.67; \sigma = 2.56$

X	*z*
4	−1.43
5	−1.04
7	−0.26
9	+0.52
10	+0.91
11	+1.30

8 Recall that a variance of 16 equals a standard deviation of 4.

$$z = \frac{11 - 14}{4} = \frac{-3}{4} = -0.75$$

9 $X = 25 + .36(3) = 25 + 1.08 - 26.08$

10 $z = \frac{140 - 130}{13} = \frac{10}{13} = 0.77$ *Answer*: .22 or 22%

11 $z = \frac{27 - 34}{3} = -2.33$ *Answer*: .99% (less than 1%)

12 0.025 + .025 = 2.5% + 2.5% = 5%

13 0.40 + .40 = 40% + 40% = 80%

14 Column B for z of .56 = .2123; Column B for z of -1.2 = .3849. Sum = .5972 or 59.72%.

15 (z's = ± 1) *Answer*: $.3413 + .3413 = .6826$

16 z's = ± 1.28 (from table); $X = 70 \pm 1.28(7) = 61$ and 79; 79 and above = best students while 61 or below = worst students.

17

a $z = \dfrac{38 - 56}{5} = -3.6$ *Answer*: .02%

b $z = \pm 0.39$; X's $= 56 \pm .39(5) = 54.05$ and 57.95, so "C" category is roughly $54 - 58$.

c $z = 1.28$; $X = 56 + 1.28(5) = 62.4$. The "A" category is (rounding) 62 and up.

18 If a number that is 10 units above a mean has a z score of 2.5, then $10 = 2.5\,s$, $s = 4$.

19 If a raw score of 51 corresponds to a z score of -1.00, then 51 is 1 standard deviation below the mean; therefore, the mean must be 66. $X = \mu + (z\sigma)$ or $51 = 66 + (-1 * 15)$.

20 If a raw score of 31 corresponds to a z score of 2.00, then 31 is two standard deviations above the mean; therefore, the mean must be 21. $X = M + (zs)$ or $31 = 21 + (2 * 5)$.

21 If the mean is 75 and a raw score of 60 corresponds to a z of -2.00, then the standard deviation must be 7.5. $X = M + zs$ or $60 = 75 + (-2 * 7.5)$.

22 If a raw score of 35 corresponds to a z score of -1.00 (which means it is 1 standard deviation below the mean) and a raw score of 40 corresponds to a z score of -0.50 (which means it is one-half of a standard deviation below the mean), then s must be 2 times the distance between 35 and 40, that is, 10. And if 35 is 1 s below the mean and 40 is .50 s below, $M = 45$. (Hint: if we are having trouble, draw it out.)

23 If a raw score of 77 corresponds to a z score of 2.50 (which means it is 2.5 standard deviation above the mean) and a raw score of 41 corresponds to a z score of -5.00 (which means it is 5 standard deviations below the mean), the distance between 77 and 41 must be 7.5 σ away from each other.

Therefore, $\sigma = 36/7.5 = 4.8$. Further, this means $\mu = 65$. (Hint: if we are having trouble, draw it out.)

24 Bottom 20% corresponds to a z score of -0.84; $X = \mu + z\sigma$ or
$X = 25\,000 + (-0.84 * 6000) = 19\,960$.
Bottom 40% corresponds to a z score of -0.25; $X = 25\,000 + (-0.25 * 6000) = 23\,500$.
Bottom 60% corresponds to a z score of .25; $X = 25\,000 + (0.25 * 6000) = 26\,500$.
Bottom 80% corresponds to a z score of .84; $X = 25\,000 + (0.84 * 6000) = 30\,040$.
The highest value cannot be determined since theoretically the corresponding z score would be infinite. (Recall that normal distributions are asymptotic.)

25 The top 15% corresponds to a z score of 1.04;
$X = 45 + (1.04 * 11) = 56.44$ lbs.
The bottom 28% corresponds to a z score of -0.58;
$X = 45 + (-0.58 * 11) = 38.62$ lbs.

26 $X = M + zs$ so, $X = 25 + (-1.75 * 4) = 18$; the z table suggests that 95.99% of the raw scores will be greater than 18. $(50 + 45.99 = 95.99\%)$

27 There are several ways to find this area under the curve. Here is one:

The raw score of .7 corresponds to a z score of -2.00; 2.28% of the area under the curve is below a raw score of .7, including area we do not want to include. The raw score of .6 corresponds to a z score of -2.67; there is 0.38% of the area under the curve below .6. We can subtract .38% from 2.28%, leaving us with 1.90% between raw scores of .6 and .7.

28 Andrew's 54 completed passes from a distribution centered on 44 with a standard deviation of 6 produce a z score of 1.67. Lisa's 48 completed passes from a distribution centered on 38 with a standard deviation of 7 produce a z score of 1.43. Both did well, but Andrew's passing performance was stronger than Lisa's relative to their respective teams.

29 Sarah's 20 minutes deciding what to wear from a distribution centered on 15 with a standard deviation of 4 produces a z score of 1.25. Justine's 90 minutes on social media from a distribution centered on 65 with a standard deviation of 20 produces a z score of 1.25. Relative to their respective activities, the time wasted by each person is the same.

Chapter 6

1 **b**

2 **a** No, some countries allow dual citizenship.
b No
c Yes
d No
e No
f Yes
g Yes
h Yes (well, I suppose one could be both, but I have never met one)
i Uncertain(!) – cheap shot, I know. Sorry about that Trekkies! We love you.

3 **a** 13/52 or ¼ or .25 or 25%
b 1/10 or .1 or 10%
c 16/32 or ½ or .5 or 50%
d 1/50 or .02 or 2% (assuming all states are equally likely to be drawn)
e Impossible to determine without knowing more information

4 **a** $\frac{40}{100} = .40$ or 40%
b $\frac{60}{100} = .60$ or 60%
c $0.40 + .60 = 1.00$ or 100%
d $(0.40)(0.40) = .16$ or 16%
e $(0.60)(0.60) = .36$ or 36%

5 **a** Dependent (getting one number necessarily changes the likelihood of getting another).
b Independent (the numbers selected the previous week do not change the likelihood of the numbers selected the subsequent week).
c Dependent (once one 5 card is selected, the probability of drawing the next 5 decreases).
d Independent (the first draw does not change the chance of getting a 6 on the second draw).

6 **a** .3
b .4
c .7
d .6 (mutually exclusive – so $.3 + .3$)
e .55 (not mutually exclusive – so $.3 + .35 - .1$)

f .1 (uncertain relationship – so $P(Red|X)\ P(X) = 10/35\ (.35)$)
g .29 (conditional – so .1/.35)
h .33 (conditional – so .1/.3)
i No
j $P(red)P(X)$ does not equal $P(Red|X)P(X)$.

7 **a** .35
b 0.4
c 0.65 (mutually exclusive – so .25 + .4)
d 1 (mutually exclusive – so .35 + .25 + .4)
e 0.75 (not mutually exclusive – so .25 + .6 − .1)
f .65
g 1 (mutually exclusive – so .6 + .4)
h 0.3 (uncertain relationship – so $P(Yellow|X)\ P(X) = .5\ (.6)$)
i 0 (uncertain relationship – so $P(Yellow|Red)\ P(Red) = 0\ (.25)$)
j .375 or .38 (conditional – so $P(Red\ and\ Y)/P(Y) = .15/.4$)
k .6 (conditional – so $P(Y\ and\ Red)/P(Red) = .15/.25$)
l No, because $P(X)\ P(Green) \neq P(X|Green)\ P(Green)$
m No, because $P(Yellow)\ P(Y) \neq P(Yellow|Y)\ P(Y)$

8 **a** .8 or 80%
b 0.4 (not mutually exclusive – so .2 + .3 − .1)
c 0.33 (conditional – so .1/.3)
d 0.5 (conditional – so .1/.2)
e No because $P(Red)\ P(X) \neq P(Red|X)\ P(X)$; [$.2(.3) \neq .33(.3)$].

9 **a** .3
b .5
c .7 (mutually exclusive – so .3 + .4)
d .7 (not mutually exclusive – so .3 + .5 − .1)
e .7
f 1 (mutually exclusive – so .5 + .5)
g .25 (uncertain relationship – so $P(Yellow|X)\ P(X) = (.5)\ (.5)$)
h 0 (dependency – so $P(Yellow|Red)\ P(Red) = (0)\ (.3)$)
i 0.3 (conditional – so $P(Yellow\ and\ Y)/P(Y) = (0.15)/(0.5)$)
j 0.38 (conditional – so $P(Y\ and\ Yellow)/P(Yellow) = (0.15)/(0.4)$)
k Yes, because $P(Green)\ P(X) = P(Green|X)\ P(X)$
l No, because $P(Yellow)\ P(Y) \neq P(Yellow|Y)\ P(Y)$

10 0.22. Use the basic probability formula (number of favorable events divided by the total number of events; 8/36). There are 2 ways to get an 11 (5 and 6; 6 and 5) and 6 ways to get a 7 (1 and 6; 2 and 5; 3 and 4; 4 and 3; 2 and 5; 1 and 6).

Chapter 7

1 Choose a population of scores. Decide on a sample size, *n*. Take a random sample of size *n*. Compute the mean and replace the scores back into the population. Repeat the sampling procedure, always using the same sample size, until all possible samples have been drawn. Plot the relative frequency distribution of the means.

2 It is directly related. The standard error of the mean increases as the population standard deviation increases, and vice versa. The relationship is $\sigma_M = \sigma/\sqrt{n}$.

3 It is inversely related. The standard error of the mean decreases as the sample size (n) increases, and vice versa. The relationship is $\sigma_M = \sigma/\sqrt{n}$.

4 There is no relationship between a sample mean value and the standard error.

5 Point estimation estimates a parameter as a single value. Confidence intervals establish a range of values within which the population parameter is expected to lie.

6 Estimation is an inferential procedure that uses data from a sample to infer the value of a population parameter. Hypothesis testing is a set of inferential procedures that uses data from samples to establish the credibility of a hypothesis about population parameters.

7 Single-sample designs use one sample to test a hypothesis about the mean of a population. Two-sample research uses two samples to test a hypothesis about the difference between two population means.

8 A research hypothesis is a formal statement or expectation about the outcome of a study, often specifying the expectation of a relationship between an independent and dependent variable. A statistical hypothesis is a numerical statement about the outcome of a study. The null and alternative hypotheses are statistical hypotheses.

9 The null hypothesis states that there is no effect or no relationship between variables. The alternative hypothesis states that there is an effect or there is a relationship between variables.

10 **a** H_0: $\mu = 3$ hours; H_1: $\mu \neq 3$ hours
b Finding a sample mean that was very similar to three hours.
c Finding a sample mean that was much less than three hours.
d Finding a sample mean that was much greater than three hours.

11 **a** $H_0: \mu_A = \mu_B$; $H_1: \mu_A \neq \mu_B$
b If the two-sample means are very similar.
c Failure to reject the null means that we are uncertain about whether one drug works better than another. (It does not mean that we are certain there is no difference between the drugs.)
d If the sample mean for Drug A outperforms the Drug B sample mean.
e If the sample mean for Drug B outperforms the Drug A sample mean.

12 $\mu_M = 100$; $\sigma_M = 10/\sqrt{9} = 3.33$. Its shape is likely to be normal or approximating normal, because even if the population of scores is not normal, n is of decent size (9).

13 Standard error formula is $\sigma_M = \sigma/\sqrt{n}$, so $\sigma_M = 10/\sqrt{20} = 2.24$.

14 Standard error formula is $\sigma_M = \sigma/\sqrt{n}$, so $\sigma_M = 0.5/\sqrt{100} = 0.05$.

15 Well, $\mu = \mu_M$, so $\mu_M = 20$ kg. The rest of the information is not necessary.

16 **a** Sample means are unbiased estimates of pop. means, so $M = 17 \approx \mu$. Sample standard deviations are unbiased estimates of pop. standard deviations, so $s = 2 \approx \sigma$.
b $s_M = s/\sqrt{n} = 2/\sqrt{20} = 0.45$.

17 **a** The research hypothesis would be that squirrels that eat the genetically modified nuts would grow to become larger squirrels.
b The null hypothesis would be that $\mu_{modified\ diet} = 17$ ounces; the alternative hypothesis would be that $\mu_{modified\ diet} \neq 17$ ounces.
c Finding the sample mean to be significantly larger than 17 ounces.
d Finding the sample mean to be significantly smaller than 17 ounces.
e Finding the sample mean is very close to 17 ounces.

Chapter 8

1 A statistical test helps ascertain if the difference between a sample mean and a population mean can be explained by sampling error (chance) or if there is evidence the sample came from a different population. Since the judgment is probabilistic, we can never be certain.

2 Answers will vary. The research hypothesis should be stated in terms of concepts and relationships – without the use of numbers but with a clear direction (more, less, stronger, weaker, etc.); the statistical hypotheses should formulate the idea into a null and alternative hypothesis (including both directions).

3 **a** H_0: $\mu = 8$ breaths per minute for smokers; H_1: $\mu \neq 8$ breaths per minute for smokers
b H_0: $\mu = 12$ on the BDI; H_1: $\mu \neq 12$ on the BDI
c H_0: $\mu = 20$ mpg; H_1: $\mu \neq 20$ mpg

4 We should use the *t* distribution when the population standard deviation is unknown.

5 The researcher should use the *z* test. The ability to determine the actual standard error should always be preferred to estimating it by using a sample standard deviation.

6 The relative consequences of a Type I vs. a Type II error. If a Type I error is not a particular concern, the researcher may move alpha to .10 (or 10%) to help avoid making a Type II error. If a Type I error is a major concern, the researcher may move alpha to .01 (or 1%) or even less.

7 Type I error; Type II error

8 Type II error

9 Type I error

10 We can set the risk rate for Type I errors (usually .05).

11 Cohen's $d = \textit{mean difference}/\sigma = 5/10 = .5$.

12 Cohen's $d = \textit{mean difference}/\sigma = 3/4 = .75$.

13 Cohen's $d = \textit{mean difference}/\sigma$. So, if $.4 = 12/\sigma$, then $\sigma = 30$.

14 Cohen's $d = \textit{mean difference}/\sigma$. So, if $.2 =$ mean difference/20, then the mean difference must $= 4$. If μ is 100, then the publisher is claiming new textbook users will average 104.

15 **a** Use the z test since σ is known.
b H_0: $\mu = 60$; H_1: $\mu \neq 60$
c $z_{crit} = \pm 1.96$
d $z_{obt} = \dfrac{65 - 60}{5/\sqrt{50}} = \dfrac{65 - 60}{5/7.07} = 7.07$
e Reject the null hypothesis.

f Type I

g Yes. Statistical evidence suggests typing speed has improved with the new keyboard.

h Effect size (Cohen's d) = 5/5 = 1.

16 **a** Use the t test since σ is unknown.

b $H_0: \mu = 90$; $H_1: \mu \neq 90$

c $t_{crit} = \pm 2.021$ ($df = 40$)

d $t_{obt} = \dfrac{110 - 90}{30/\sqrt{41}} = 4.27$

e Reject the null hypothesis.

f Effect size (Cohen's d) = 20/30 = .67

g Type I

h Statistical evidence suggests the hormone produces heavier golden retrievers.

17 **a** The single-sample z test. We know σ (the standard deviation of the population).

b $H_0: \mu = 7.5$; $H_1: \mu \neq 7.5$

c No; the sample size is so large that we are robust to the assumption of normality.

d Since we are using a z test, we can use the standard z_{crit} *of* ± 1.96 for an α of .05.

e $z_{obt} = \dfrac{7.2 - 7.5}{2.4/\sqrt{200}} = \dfrac{-0.3}{0.17} = -1.76$

f No, fail to reject.

g Not applicable – the null was not rejected.

h Type II

i There is no statistical evidence to suggest that students at this university sleep different amounts than university students in general.

18 **a** The t distribution because we do not know σ.

b $H_0: \mu = 50$; $H_1: \mu \neq 50$

c ± 2.093

d $t_{obt} = \dfrac{M - \mu}{s/\sqrt{n}} = \dfrac{63 - 50}{17/\sqrt{20}} = \dfrac{13}{3.80} = 3.42$

e Yes

f The estimate of the effect size (Cohen's d) = *mean difference*/*s* = 13/17 = .76.

g Type I

h Statistical evidence suggests people who work at home are more satisfied with their jobs than workers in general.

19 **a** Use t distribution since σ is unknown.
b H_0: $\mu = 72.40$; H_1: $\mu \neq 72.40$
c $t_{crit} = \pm 2.306\ (df = 8)$
d $t_{obt} = \dfrac{77 - 72.4}{3.1/\sqrt{9}} = 4.47$
e Reject the null hypothesis.
f The estimate of the effect size (Cohen's d) = *mean difference/s* = 4.6/3.1 = 1.48.
g Type I
h Statistical evidence suggests that biological males overestimate their life expectancy.

20 For males, $t(7) = 4.52$, $p < .05$; for females, $t(7) = 3.08$, $p < .05$. Statistical evidence suggests that the mean height for physically stressed biological males ($M = 69.25$) is greater than the average height for biological males, $t(7) = 4.52$, $p < .05$. Statistical evidence suggests that biological females who have been physically stressed ($M = 61.88$) also are taller than the average biological female, $t(7) = 3.08$, $p < .05$.

21 $LL = M - t_\alpha s_M = 24.50 - 2.11[(2.61/\sqrt{18})] = 23.20$
$UL = M + t_\alpha s_M = 24.50 + 2.11[(2.61/\sqrt{18})] = 25.80$

22 $LL = M - t_\alpha s_M = 4.2 - 2.18[(.86/\sqrt{13})] = 3.68$
$UL = M + t_\alpha s_M = 4.2 + 2.18[(.86/\sqrt{13})] = 4.72$

23 $LL = M - t_\alpha s_M = 56\,000 - 2.00[(4300/\sqrt{60}) = 54\,889.74$
$UL = M + t_\alpha s_M = 56\,000 + 2.00[(4300/\sqrt{60}) = 57\,110.26$

24 There is no statistical evidence that the average number of days to process a claim is different from 15, $t(39) = -0.83$, *n.s.*

25 If we can use 24.5 minutes as a population mean for lateness prior to the addition of extra trains, statistical evidence suggests that the addition of extra trains during rush hour reduced the amount of time the train is late, $t(29) = -5.18$, $p < .05$.

Chapter 9

1 sample; population

2 **a**

3 b

4 a

5 Because dependent-samples t tests use the difference between a participant's two responses.

6 c

7 d

8 d

9 a

10 Sampling error

11 No. However, 1 and 2 are typically used, but researchers are free to use other subscripts as well; for instance, $\mu_{control}$ and μ_{exp} or μ_{drug} and $\mu_{placebo}$ or even letters such as μ_A and μ_B.

12 b

13 s_D is the symbol for the standard deviation of the difference scores; $s_{\overline{D}}$ is the symbol for the estimate of the standard error of the differences between means. The first is a measure of dispersion for the sample of difference scores found from the actual data gathered; the second is an estimate of dispersion for the theoretical sampling distribution used by the dependent-samples t test. They are not the same, although s_D is used to calculate $s_{\overline{D}}$.

14 Representativeness.

15 b

16 As the sample size increases, df increases, and t_{crit} decreases. As the sample size increases, the t distribution approaches the standard normal curve, which means that the tails pull in toward 0. Consequently, the values marking the outermost 5%, (t_{crit}) are closer to 0.

17

	Critical values	Decision
a	±2.160	Fail to reject
b	±2.663 (online table)	Reject
c	±1.746	Reject
d	±2.064	Fail to reject

18 **d**

19 **b**

20 **a** H_0: $\mu_1 = \mu_2$; H_1: $\mu_1 \neq \mu_2$

b The appropriate inferential test is the independent-samples *t* test.

No siblings	Siblings
$M_1 = 7.33$; $s_1 = 2.16$; $n_1 = 6$	$M_2 = 3.17$; $s_2 = 2.14$; $n_2 = 6$

Using the computational formula,

$$t_{obt} = \frac{7.33 - 3.17}{\sqrt{\{[(346 - (44)^2/6) + (83 - (19)^2/6)]/(6+6-2)\}(1/6+1/6)}} = 3.36$$

c $t_{crit} = \pm 2.228$ $(df = 10)$

d Reject the null hypothesis.

e First, we need to pool together the two-sample standard deviations to get the best estimate of the population standard deviation.

$$\sqrt{\frac{s_1^2(n_1-1) + s_2^2(n_2-1)}{n_1 + n_2 - 2}} = \sqrt{\frac{2.16^2(5) + 2.14^2(5)}{6 + 6 - 2}} = \sqrt{\frac{23.22 + 22.9}{10}}$$

$$= \sqrt{4.61} = 2.15.$$

The effect size (Cohen's *d*) = *estimated mean difference*/*estimated s* = 4.16/2.15 = 1.93.

f Type I

g Statistical evidence suggests that two-year-olds with siblings have less fear than two-year-olds with no siblings, $t(10) = 3.36$, $p < .05$.

21 **a** $H_0: \mu_1 = \mu_2$; $H_1: \mu_1 \neq \mu_2$

b The appropriate test is the independent-samples t test.

$$t_{obt} = \frac{4.2 - 2.2}{\sqrt{[(0.5(9) + 0.7(9))/(10 + 10 - 2)](1/10 + 1/10)}}$$

$$t_{obt} = 5.71$$

c $t_{crit} = \pm 2.101$ $(df = 18)$

d Reject the null hypothesis.

e First, we need to pool together the two-sample standard deviations to get the best estimate of the population standard deviation.

$$\sqrt{\frac{s_1^2(n_1 - 1) + s_2^2(n_2 - 1)}{n_1 + n_2 - 2}} = \sqrt{\frac{0.5(9) + 0.7(9)}{10 + 10 - 2}} = \sqrt{\frac{4.5 + 6.3}{18}} = \sqrt{0.6} = 0.77.$$

The estimate effect size (Cohen's d) = *estimated mean difference/ estimated* $s = 2/0.77 = 2.60$.

f Type I

g Statistical evidence suggests that a high level of anxiety leads to greater attraction than a low level of anxiety, $t(18) = 5.71$, $p < .05$.

22 **a** $$t_{obt} = \frac{4.2 - 2.2}{\sqrt{[(5.2(9) + 5.4(9))/(10 + 10 - 2)](1/10 + 1/10)}}$$

$$t_{obt} = 1.94$$

b With $t_{crit} = \pm 2.101$, the t_{obt} of 1.94 does not lead to rejection of the null hypothesis.

c The increased variability has increased the denominator and reduced the size of t_{obt}.

23 **a** $$t_{obt} = \frac{4.2 - 2.2}{\sqrt{[(5.2(29) + 5.4(29))/(30 + 30 - 2)](1/30 + 1/30)}}$$

$$t_{obt} = 3.51$$

b With $t_{crit} = \pm 2.002$ (use an online t table to find t_{crit}), the t_{obt} of 3.51 leads to a rejection of the null hypothesis.

c Increasing the sample size decreases t_{crit} and, more importantly, increases t_{obt} by shrinking the estimate of the standard error (the denominator).

24 **a** $H_0: \mu_1 = \mu_2$; $H_1: \mu_1 \neq \mu_2$

b The appropriate inferential test is the independent-samples t test.

Males	Females
$M_1 = 15.60$	$M_2 = 9.0$
$s_1 = 4.16$	$s_2 = 3.61$
$n_1 = 5$	$n_2 = 5$

Using the computational formula,

$$t_{obt} = \frac{15.6 - 9.0}{\sqrt{\left\{\frac{[(1286 - 1216.8) + (457 - 405)]}{5 + 5 - 2}\right\}\left(\frac{1}{5} + \frac{1}{5}\right)}} = 2.68$$

c $t_{crit} = \pm 2.306$ (with $df = 8$)

d Yes, reject the null.

e First, we need to pool together the two-sample standard deviations to get the best estimate of the population standard deviation.

$$\sqrt{\frac{s_1^2(n_1 - 1) + s_2^2(n_2 - 1)}{n_1 + n_2 - 2}} = \sqrt{\frac{4.16^2(4) + 3.61^2(4)}{5 + 5 - 2}} = \sqrt{\frac{69.22 + 52.13}{8}} =$$

$\sqrt{15.17} = 3.89$. The estimate effect size (Cohen's d) = *estimated mean difference/estimated s* $= 6.6/3.89 = 1.70$.

f Type I

g Statistical evidence suggests that biological male college students report more anger reactions than biological female college students, $t(8) = 2.68$, $p < .05$.

25 **a** $H_0: \mu_{Gouda} = \mu_{Swiss}$; $H_1: \mu_{Gouda} \neq \mu_{Swiss}$

b $M_x = 7.0$; $M_y = 5.60$; $\Sigma D = 7$; $\Sigma D^2 = 35$

$$t_{obt} = \frac{7.0 - 5.60}{1.12} = \frac{1.40}{1.12} = 1.25$$

c $df = n_p - 1 = 4$; $t_{crit} = \pm 2.776$

d No, fail to reject.

e No need for an effect size analysis – the null was not rejected.

f Type II

g There is no statistical evidence that college students have a preference between Gouda and Swiss cheese, $t(4) = 1.25$, *n.s.*

h The researcher should counterbalance exposure to the cheeses.

26 Here is the data according to Formula 9.10, plugging in the proper values for $s_{\overline{D}}$ and using n_p to find the right t_{crit} score:

$$LL = (7.0 - 5.60) - 2.776\left(\frac{2.51}{\sqrt{5}}\right) = 1.4 - 3.11 = -1.71$$

$$UL = (7.0 - 5.60) + 2.776\left(\frac{2.51}{\sqrt{5}}\right) = 1.4 + 3.11 = 4.51$$

Note: the negative difference for the lower limit suggests the possibility that students actually prefer Swiss cheese to Gouda. The amount of variability reflected in this confidence interval analysis properly reflects the failure to reject the null hypothesis in the previous question.

27 **a** *Males:*

$$t_{obt} = \frac{23 - 16}{\sqrt{[(61.47(9) + 41.34(14))/(10 + 15 - 2)](1/10 + 1/15)}}$$

$$t_{obt} = 2.42$$

$$t_{crit} = \pm 2.07 \text{ (with } df = 23)$$

Statistical evidence suggests that firstborn biological males are more narcissistic than laterborn biological males, $t(23) = 2.42$, $p < .05$.

b *Females:*

$$t_{obt} = \frac{17 - 12}{\sqrt{[(42.51(18) + 43.16(27))/(19 + 28 - 2)](1/19 + 1/28)}}$$

$$t_{obt} = 2.55$$

$$t_{crit} = \pm 2.01 \text{(used Internet source to find complete } t \text{ table)}$$

Statistical evidence suggests that firstborn biological females are more narcissistic than laterborn biological females, $t(45) = 2.55$, $p < .05$.

28 Here is the data according to Formula 9.5:

For biological males:

$$LL = (23 - 16) - 2.07(2.89) = 7 - 5.98 = 1.02$$

$$UL = (23 - 16) + 2.07(2.89) = 7 + 5.98 = 12.98$$

For biological females:

$$LL = (17 - 12) - 2.01(1.96) = 5 - 3.94 = 1.06$$

$$UL = (17 - 12) + 2.01(1.96) = 5 + 3.94 = 8.94$$

29 **a** $H_0: \mu_1 = \mu_2$; $H_1: \mu_1 \neq \mu_2$

b The appropriate inferential test is the independent-samples t test:

$$t_{obt} = \frac{41 - 47.2}{\sqrt{\frac{4^2(15-1) + 5^2(14-1)}{15 + 14 - 2}\left(\frac{1}{15} + \frac{1}{14}\right)}}$$

$$= \frac{-6.2}{\sqrt{\frac{224 + 325}{27}(0.14)}} = \frac{-6.2}{\sqrt{20.33(0.14)}}$$

$$t_{obt} = \frac{-6.2}{1.69} = -3.68$$

c $t_{crit} = \pm 2.052$ (with $df = 27$)

d Reject the null hypothesis.

e First, we need to pool together the two-sample standard deviations to get the best estimate of the population standard deviation.

$$\sqrt{\frac{s_1^2(n_1-1) + s_2^2(n_2-1)}{n_1 + n_2 - 2}} = \sqrt{\frac{4^2(14) + 5^2(13)}{15 + 14 - 2}} = \sqrt{\frac{224 + 325}{27}} = \sqrt{20.33} =$$

4.51. The estimate effect size (Cohen's d) = *estimated mean difference/estimated* $s = -6.2/4.51 = -1.37$ or just 1.37 (recall that negative effect sizes do not need to be reported as negatives).

f Type I

g Statistical evidence suggests that people with several negative life experiences have higher measures of subjective well-being than those with few, $t(27) = -3.68$, $p < .05$.

30 **a** $H_0: \mu_1 = \mu_2$; $H_1: \mu_1 \neq \mu_2$

b $t_{obt} = \frac{17.0 - 13.5}{2} = 1.75$

c $t_{crit} = \pm 2.002$ (used online table)

d No, fail to reject the null.

e There is no statistical evidence suggesting children who receive training performed differently than those who do not, $t(58) = 1.75$, *n.s.*

31 **a** $H_0: \mu_E > \mu_C$; $H_1: \mu_E \leq \mu_C$

b (Same as Problem 30b) $t_{obt} = 1.75$

c $t_{crit} = 1.672$ (used online table). Now, using a one-tailed test, the null hypothesis would be rejected. Statistical evidence suggests children who are training to start conversations spend more time interacting with peers than those who do not, $t(58) = 1.75$, $p < .05$.

d No. There is no compelling reason for a one-tailed test. Indeed, a finding that children who receive training engage in *less* peer interaction would be important to know.

32 **a** $H_0: \mu_1 = \mu_2$; $H_1: \mu_1 \neq \mu_2$

b The appropriate inferential test is the independent-samples t test:

$$t_{obt} = \frac{80 - 69}{\sqrt{\frac{106.09(15-1) + 156.25(15-1)}{15 + 15 - 2}\left(\frac{1}{15} + \frac{1}{15}\right)}}$$

$t_{obt} = 2.63$

c $t_{crit} = \pm 2.048$ (with $df = 28$)

d Reject the null hypothesis.

e First, we need to pool together the two-sample standard deviations to get the best estimate of the population standard deviation.

$$\sqrt{\frac{s_1^2(n_1-1) + s_2^2(n_2-1)}{n_1 + n_2 - 2}} = \sqrt{\frac{106.09(14) + 156.25(14)}{15 + 15 - 2}} =$$

$\sqrt{\frac{1485.26 + 2187.5}{28}} = \sqrt{131.17} = 11.45$. The estimate effect size (Cohen's d) $=$ *estimated mean difference/estimated s* $= 11/11.45 = .96$.

f Type I

g Statistical evidence suggests that students who study in one place perform better than students who rotate their study location, $t(28) = 2.63$, $p < .05$.

h One potential confound might be noise. The degree of ambient noise present when studying might compete with the number of locations as an explanation for the statistical difference.

33 **a** $H_0: \mu_x = \mu_y$; $H_1: \mu_x \neq \mu_y$ (X = Pre; Y = Post)

b $M_x = 92.50$; $M_y = 45.0$; $\Sigma D = 190$; $\Sigma D^2 = 12\,500$

$$t_{obt} = \frac{92.50 - 45.0}{17.015} = 2.79$$

c $df = n_p - 1 = 3$; $t_{crit} = \pm 3.182$

d No, fail to reject.

e No need for an effect size analysis – the null was not rejected.

f Type II

g There is no statistical evidence that the new drug for insomnia decreases the amount of time needed to fall asleep, $t(3) = 2.79$, *n.s.*

34 **a** $H_0: \mu_{story} = \mu_{music}$; $H_1: \mu_{story} \neq \mu_{music}$

b $M_{story} = 8.43$; $M_{music} = 6.57$; $\Sigma D = 13$; $\Sigma D^2 = 85$

$$t_{obt} = \frac{8.43 - 6.57}{1.204} = 1.54$$

c $df = n_p - 1 = 6$; $t_{crit} = \pm 2.447$

d No, fail to reject.

e No need for an effect size analysis – the null was not rejected.

f Type II

g No statistical evidence was found to suggest that a difference exists between being read a story and listening to music as sleep inducers for preschoolers, $t(6) = 1.54$, *n.s.*

35 $t_{obt} = 2.06$, t_{crit} for 22 degrees of freedom $= \pm 2.07$ for a two-tailed test. There is no statistical evidence to suggest that a new t-shirt generates more happiness in students, $t(22) = 2.06$, *n.s.* With every study, there are always several issues, but one that stands out for this study is the use of a rating scale. Might it be ordinal data? Perhaps another way of measuring happiness should be considered, or a different test could be run (see Chapter 16).

Chapter 10

1 d

2 The lenses of a microscope also vary by power. Just as low-powered lenses may not see small objects (e.g. germs) that high-powered lenses can see, so studies with low power may not be able to detect small treatment effect sizes as well as those studies with high power.

3 b

4 Nothing. The Type I error rate does not influence power; it can only occur when the null is true.

5 Power is inversely related to the Type II error rate. As it increases, Type II error rates decrease.

6 The hypothesized treatment effect (γ) is found by dividing the difference between the null and hypothesized mean by the standard error. It is represented mathematically as: $\gamma = \frac{\mu_{alt} - \mu_0}{\sigma}$.

7 The size of the treatment effect influences power directly. As it increases, so does power.

8 One way would be to try and increase the treatment effect. The researcher may try to do this by extending the sleep deprivation from three hours to something more than three hours.

9 The sample size influences power directly. As it increases, so does power.

10 **b** Sampling error. The variability of raw scores for any measure cannot usually be influenced without altering the sampling method (e.g. only accepting participants into the study whose score falls into a prescribed range based on a premeasure).

11 Increasing alpha increases power by increasing the rejection region of the hypothesis test. Decreasing alpha correspondingly decreases power by decreasing the rejection region.

12 Compared with a two-tailed test, a one-tailed test increases alpha for the predicted end of the sampling distribution. This, in effect, is analogous to increasing alpha.

13 **a** $\gamma = \dfrac{345 - 300}{70} = 0.64$

b $\gamma = \dfrac{345 - 300}{20} = 2.25$

c $\gamma = \dfrac{310 - 300}{20} = 0.50$

d $\gamma = \dfrac{310 - 300}{50} = 0.20$

14 **a** $n = \left(\dfrac{2.8}{0.64}\right)^2 = 19$

b $n = \left(\dfrac{2.8}{2.25}\right)^2 = 2$

c $n = \left(\dfrac{2.8}{0.5}\right)^2 = 31$

d $n = \left(\dfrac{2.8}{0.2}\right)^2 = 196$

15 **a** $\gamma = \dfrac{120-130}{15} = -0.67$ $\quad\delta = -0.67\left(\sqrt{10}\right) = -2.12$ $\quad$ Power $= 0.56$

b $\gamma = -0.67$ $\quad\delta = -0.67\left(\sqrt{40}\right) = -4.24$ $\quad$ Power $= 0.99$

c $\gamma = \dfrac{52-50}{10} = 0.20$ $\quad\delta = 0.20\left(\sqrt{15}\right) = 0.77$ $\quad$ Power $= 0.13$

d $\gamma = 0.20$ $\quad\delta = 0.20\left(\sqrt{100}\right) = 2.00$ $\quad$ Power $= 0.52$

16 The value of δ, for a desired power of .80 with $\alpha = .05$ and two-tailed test, = 2.8

a $n = \left(\dfrac{2.8}{0.67}\right)^2 = 17.47$; about 17 participants. (We can remove the negative sign before calculating. Effect sizes can be thought of as positive for calculation purposes.)

b $n = \left(\dfrac{2.8}{0.2}\right)^2 = 196$; so, 196 participants.

Part 3 Review

1 A power analysis. Although power analyses can be run after data have been gathered or after a pilot study has been run, they can also be performed before any data have been gathered.

2 **a** $t_{crit} = \pm 2.093$
b $t_{crit} = \pm 1.746$
c $z_{crit} = \pm 2.33$
d $t_{crit} = \pm 1.895$
e $t_{crit} = \pm 2.626$
f $t_{crit} = \pm 3.335$
g $t_{crit} = \pm 2.776$
h t_{crit} = either 1.86 or −1.86, depending on the predicted direction.
i $z_{crit} = \pm 1.96$
j $t_{crit} = \pm 1.699$
k $t_{crit} = \pm 2.771$
l $t_{crit} = \pm 9.925$

3 The single-sample t test. One sample mean is compared with a population mean; σ is not given.

4 The dependent-samples t test. Two measures from one sample; a repeated-measures design.

5 The single-sample z test. One sample mean is compared with a population mean; σ is given.

6 The single-sample t test. One sample mean is compared with a population mean; σ is not given.

7 The independent-samples t test. Two measures from two different and independent samples.

8 The dependent-samples t test. Two measures from one sample; a repeated-measures design.

9 The single-sample z test. One sample mean is compared with a population mean; σ is given.

10 The independent-samples t test. Two measures from two different and independent samples.

11 This question involves data from three conditions – we have not yet learned what statistical tools can be used to analyze data in this type of research design. See Chapter 11, question 21, in the next part of the text for the answers.

Chapter 11

1 ANOVA

2 two; independent of

3 An ANOVA avoids the inflated alpha problem that comes with running multiple t tests.

4 Treatment variance or primary variance.

5 Treatment effect (or primary variance), individual differences, and experimental error.

6 Secondary variance or error variance.

7 Individual differences and experimental error (when combined; secondary or error variance).

8 MS_{BG}; MS_W

9 One measurement of variance, MS_{BG}, incorporates primary variance as well as secondary (error) variance. The other measurement of variance, MS_W, only incorporates secondary (error) variance. By comparing them in ratio form (MS_{BG} over MS_W), primary variance is evidenced by a larger than 1 result.

10 They are a family of positively skewed distributions, each peaking at the value of 1.

11 Independent-samples *t*.

12 Normality and homogeneity of variance, as long as the *ns* are large and roughly equal.

13 ANOVA summary table

14 When H_0 is correct, the numerator of the *F* ratio only contains error variance (random factors). When H_0 is incorrect, the numerator includes error variance plus primary variance.

15 After a null has been rejected, and there is an interest in making certain group comparisons.

16 Least significant difference.

17 Honestly significant difference.

18 Is; group mean differences which equal or exceed the *HSD* suggest different populations.

19

Source	*SS*	*df*	*MS*	*F*	*p*
Between groups	280.3	3	93.43	12.83	<.05
Within groups	247.68	34	7.28		
Total	527.98	37			

Reject the null: F_{obt} of 12.83 exceeds F_{crit} of 2.88.

20

Source	SS	f	MS	F	p
Between groups	5.88	4	1.47	2.59	n.s.
Within groups	6.30	11	0.57		
Total	12.18	15			

Fail to reject the null: F_{obt} of 2.59 does not equal or exceed F_{crit} of 3.36.

21 **a** H_0: $\mu_1 = \mu_2 = \mu_3$; H_1: at least two of the means are different.
Below are summative values needed for hand calculations.

Aerobics	Circuit	Control	Summary values
$\Sigma X_1 = 243$	$\Sigma X_2 = 273$	$\Sigma X_3 = 313$	$\Sigma X = 829$
$n_1 = 4$	$n_2 = 4$	$n_3 = 4$	$\Sigma X^2 = 58\,115$
$M = 60.75$	$M = 68.25$	$M = 78.25$	$(\Sigma X)^2 = 687\,241$
			$N = 12,\ k = 3$

b $SS_{BG} = \frac{243^2}{4} + \frac{273^2}{4} + \frac{313^2}{4} - \frac{829^2}{12} = 57\,886.75 - 57\,270.08 = \mathbf{616.67}$

c $SS_W = 58\,115 - \frac{243^2}{4} + \frac{273^2}{4} + \frac{313^2}{4} = \mathbf{228.25}$

d $df_{BG} = k - 1 = 3 - 1 = \mathbf{2}$

e $df_W = N - k = 12 - 3 = \mathbf{9}$

f $MS_{BG} = \frac{616.67}{2} = \mathbf{308.34}$

g $gMS_W = \frac{228.25}{9} = \mathbf{25.36}$

h $SS_T = 58\,115 - \frac{829^2}{12} = \mathbf{844.92}$

i $df_T = N - 1 = df_{BG} + df_W = \mathbf{11}$

j $F = \frac{308.34}{25.36} = \mathbf{12.16}$

k $F_{\text{crit}}(2,9) = \mathbf{4.26}$ (for $\alpha = .05$)

l Reject H_0

m

Source of variation	SS	df	MS	F	p
Between groups	616.17	2	308.34	12.16	<.05
Within groups(error)	228.25	9	25.36		
Total	844.92	11			

n $\varpi^2 = \frac{616.67 - 2(25.36)}{844.92 + 25.36} = 0.65$ or 65%

o (t_{crit} for all ts 2.262, $df = 9$, $\alpha = .05$)

Aerobics vs. circuit:

$$t = \frac{68.25 - 60.75}{\sqrt{25.36\left(\frac{1}{4} + \frac{1}{4}\right)}} = 2.11\ n.s.$$

Aerobics vs. control:

$$t = \frac{78.25 - 60.75}{3.56} = 4.92(p < .05)$$

Circuit vs. control:

$$t = \frac{78.25 - 68.25}{3.56} = 2.81(p < .05)$$

There is statistical evidence that both the aerobic and circuit training conditions are superior to the control, $t(11) = 4.92$, $p < .05$ and $t(11) = 2.81$, $p < .05$, respectively. There is no statistical evidence of a difference between aerobic and circuit training, $t(11) - 2.11$, *n.s.*

p Yes. The participants were randomly assigned, and their experience was directed by the researcher. This is an experiment. Evidence has been found suggesting that aerobic and circuit methods of training significantly reduce resting heart rate.

22 **a** H_0: $\mu_1 = \mu_2 = \mu_3$; H_1: at least two of the means are different. Below are summative values needed for hand calculations.

West	Midwest	East	Summary values
$\Sigma X_1 = 18$	$\Sigma X_2 = 43$	$\Sigma X_3 = 19$	$\Sigma X = 80$
$n_1 = 5$	$n_2 = 5$	$n_3 = 5$	$\Sigma X^2 = 624$
$M_1 = 3.60$	$M_2 = 8.60$	$M_3 = 3.80$	$(\Sigma X)^2 = 6400$
			$N = 15$, $k = 3$

b $$SS_{BG} = \frac{(\Sigma X_1)^2}{n_1} + \frac{(\Sigma X_2)^2}{n_2} + \ldots + \frac{(\Sigma X_k)^2}{n_k} - \left[\frac{(\Sigma X)^2}{N}\right] = 506.80 - 426.67$$

$$= \mathbf{80.13}$$

c $$SS_W = \Sigma X^2 - \left[\frac{(\Sigma X_1)^2}{n_1} + \frac{(\Sigma X_2)^2}{n_2} + \ldots + \frac{(\Sigma X_k)^2}{n_k}\right] = 624 - 506.80 = \mathbf{117.20}$$

d $df_{BG} = 3 - 1 = \mathbf{2}$

e $df_W = 15 - 3 = \mathbf{12}$

f $$MS_{BG} = \frac{80.13}{2} = \mathbf{40.07}$$

g $MS_W = \dfrac{117.20}{12} = \mathbf{9.77}$

h $SS_T = 624 - 426.67 = \mathbf{197.33}$

i $df_T = 15 - 1 = 2 + 12 = \mathbf{14}$

j $F = \dfrac{40.07}{9.77} = \mathbf{4.10}$

k $F_{crit}\,(2{,}12) = \mathbf{3.88}$ (for $\alpha = .05$)

l Reject H_0.

m

Source of variation	*SS*	*df*	*MS*	*F*	*p*
Between groups	80.13	2	40.07	4.10	<.05
Within groups(error)	117.20	12	9.77		
Total	197.33	14			

n $\eta^2 = \dfrac{80.13}{197.33} = 0.4061$ or 40.61%

o Tukey's HSD for all comparisons equals

$$HSD = q\sqrt{\frac{MS_W}{n}} = 3.77\sqrt{\frac{9.77}{5}} = 3.77(1.40) = 5.27$$

West vs. Midwest: $8.60 - 3.60 = 5$ (*n.s.*)
East vs. Midwest: $8.60 - 3.80 = 4.8$ (*n.s.*)
West vs. East: $3.60 - 3.80 = -0.2$ (*n.s.*)

This is an unusual case where the one-way ANOVA registers significance, but the follow-up test does not find evidence of a difference between any pair of groups. Tukey's test is rather conservative.

If Fisher's *LSD* had run, we would have found evidence of two differences:

(t_{crit} for all t's $= 2.179$, $df = 12$, $\alpha - .05$)

$$t = \frac{8.60 - 3.60}{\sqrt{9.77\left(\frac{1}{5} = \frac{1}{5}\right)}} = 2.53\ (p < 0.05)$$

East vs. Midwest:

$$t = \frac{8.60 - 3.80}{1.98} = 2.42\ (p < 0.05)$$

West vs. East:

$$t = \frac{3.60 - 3.80}{1.98} = -0.10\ (n.s.)$$

p No. This is a correlational design. The independent variable (geographical region) is not manipulated.

23 **a** H_0: $\mu_1 = \mu_2 = \mu_3$; H_1: at least two of the means are different.
Below are summative values needed for hand calculations.

Breathing	Medication	Control	Summary values
$\Sigma X_1 = 75$	$\Sigma X_2 = 70$	$\Sigma X_3 = 65$	$\Sigma X = 210$
$n_1 = 5$	$n_2 = 5$	$n_3 = 5$	$\Sigma X^2 = 3126$
$M_1 = 15$	$M_2 = 14$	$M_3 = 13$	$(\Sigma X)^2 = 44\,100$
			$N = 15,\ k = 3$

b $$SS_{BG} = \frac{(\Sigma X_1)^2}{n_1} + \frac{(\Sigma X_2)^2}{n_2} + \ldots + \frac{(\Sigma X_k)^2}{n_k} - \left[\frac{(\Sigma X)^2}{N}\right]$$
$$= 2950 - 2940 = \mathbf{10}$$

c $$SS_W = \Sigma X^2 - \left[\frac{(\Sigma X_1)^2}{n_1} + \frac{(\Sigma X_2)^2}{n_2} + \ldots + \frac{(\Sigma X_k)^2}{n_k}\right] = 3126 - 2950 = \mathbf{176}$$

d $df_{BG} = 3 - 1 = \mathbf{2}$

e $df_W = 15 - 3 = \mathbf{12}$

f $MS_{BG} = \frac{10}{2} = \mathbf{5}$

g $MS_W = \frac{176}{12} = \mathbf{14.67}$

h $SS_T = 3126 - 2940 = \mathbf{186}$

i $df_T = 15 - 1 = 2 + 12 = \mathbf{14}$

j $F = \frac{5}{14.67} = \mathbf{0.34}$

k $F_{crit}(2,12) = \mathbf{3.88}$ (for $\alpha = 0.05$)

l Fail to reject the H_0.

m

Source of variation	*SS*	*df*	*MS*	*F*	*p*
Between groups	10	2	5	0.34	*n.s.*
Within groups (error)	176	12	14.67		
Total	186	14			

n There is no statistical evidence of a difference among treatment conditions in the alleviation of panic attacks, $F(2,12) = .34$, *n.s.*

o Since the F ratio is nonsignificant, ω^2 is meaningless.

p Conducting post hoc comparisons is unwarranted since the F ratio is nonsignificant.

q It would have, if evidence of a difference had been found. The methodology is experimental. However, since no evidence of a difference was found, no causal claims can be made.

24 **a** H_0: $\mu_1 = \mu_2 = \mu_3$; H_1: at least two of the means are different.
Below are summative values needed for hand calculations.

New T-shirt	New shoes	Control	Summary values
$\Sigma X_1 = 32$	$\Sigma X_2 = 41$	$\Sigma X_3 = 23$	$\Sigma X = 96$
$n_1 = 5$	$n_2 = 5$	$n_3 = 5$	$\Sigma X^2 = 664$
$M_1 = 6.4$	$M_2 = 8.2$	$M_3 = 4.6$	$(\Sigma X)^2 = 9216$
			$N = 15, k = 3$

b $SS_{BG} = \dfrac{(\Sigma X_1)^2}{n_1} + \dfrac{(\Sigma X_2)^2}{n_2} + \ldots + \dfrac{(\Sigma X_k)^2}{n_k} - \left[\dfrac{(\Sigma X)^2}{N}\right] = 646.8 - 614.4$
$= \mathbf{32.4}$

c $SS_W = \Sigma X^2 - \left[\dfrac{(\Sigma X_1)^2}{n_1} + \dfrac{(\Sigma X_2)^2}{n_2} + \ldots + \dfrac{(\Sigma X_k)^2}{n_k}\right] = 664 - 646.8 = \mathbf{17.2}$

d $df_{BG} = 3 - 1 = \mathbf{2}$

e $df_W = 15 - 3 = \mathbf{12}$

f $MS_{BG} = \dfrac{32.4}{2} = \mathbf{16.2}$

g $MS_W = \dfrac{17.2}{12} = \mathbf{1.43}$

h $SS_T = 32.4 + 17.2 = 664 - 614.4 = \mathbf{49.6}$

i $df_T = 15 - 1 = 2 + 12 = \mathbf{14}$

j $F = \dfrac{16.2}{1.43} = \mathbf{11.33}$

k $F_{crit}\,(2,12) = \mathbf{3.88}$ (for $\alpha = .05$)

l Reject H_0.

m

Source of variation	*SS*	*df*	*MS*	*F*	*p*
Between groups	32.4	2	16.2	11.33	<.05
Within groups(error)	17.2	12	1.43		
Total	49.6	14			

n $\eta^2 = \dfrac{32.4}{49.6} = 0.6532$ or 65.32%.

o Tukey's HSD for all comparisons equals

$$HSD = q\sqrt{\frac{MS_W}{n}} = 3.77\sqrt{\frac{1.43}{5}} = 3.77(0.53) = 2.02$$

Shirt vs. Shoes: $6.4 - 8.2 = -1.8$ (*n.s.*)
Shirt vs. Control: $6.4 - 4.6 = 1.8$ (*n.s.*)
Shoes vs. Control: $8.2 - 4.6 = 3.6$ ($p < .05$)

p Statistical evidence suggests the type of clothing worn influences perceptions of happiness, $F(2,12) = 11.33$, $p < .05$. A Tukey *HSD* test found statistical evidence suggesting those wearing new shoes felt happier than those not wearing a new article of clothing at $p < .05$.

q Yes. This is an experimental design.

25

Source	*SS*	*df*	*MS*	*F*	*p*
Between groups	51.91	2	25.96	3.94	<.05
Within groups	276.53	42	6.58		
Total	328.44	44			

Statistical evidence suggests of a reduction in headaches as a result of treatment, $F(2,42)$ 3.94, $p < .05$. Fisher's *LSD* analysis showed a significant difference between both kinds of treatment vs. the control condition ($t_{med\ v\ control} = 2.29$, $p < .05$; $t_{bf\ v\ control} = 2.58$, $p < .05$); however, there is no significant difference between the two forms of treatment, ($t_{med\ v\ bf} = .29$, *n.s.*).

26

Source	*SS*	*df*	*MS*	*F*	*p*
Between groups	143.22	3	47.74	11.19	<.05
Within groups	273.06	64	4.27		
Total	416.28	67			

There is statistical evidence suggesting that there is at least one difference between majors in terms of the number of shoes brought to the university, $F(3,64) = 11.19$, $p < .05$. Tukey's *HSD* (1.87) found evidence with $p < .05$ that chemistry majors bring fewer shoes than history majors and theater majors, and psychology majors bring fewer shoes than theater majors.

Chapter 12

1 There must be two factors (or independent variables), each factor must have at least two conditions, participants must be only assigned to one combination of levels of the two factors (in other words, not repeatedly measured), and each of the possible combinations of conditions between the factors must have participants (no empty cells, in fact, similar *n*'s is best).

2 In the first stage of a two-way ANOVA, the total variance is partitioned into between-group and within-group variance. In the second stage, the between-group variance is partitioned into variance due to Factor *A*, variance due to Factor *B*, and variance due to the interaction.

3 A main effect occurs when there is an effect found among the conditions of one factor, independent of the influence of another factor. An interaction occurs when the effect of one factor is altered depending on the value of a second factor. This is a new and unique variance, not explained by the main effects.

4 The following are examples. Labels selected by the student will vary; the grid design will not.

a 2 × 2

	In a group	Not in a group
Peg-word System		
Acronyms		

Both factors can be independent variables.

b 3 × 2

	White collar	Blue collar
Group		
Behavioral		
Person-centered		

The factor described in rows (therapy type) can be an independent variable.

c 4 × 3

	Humanities	Sciences	Preprofessional
Dorm room			
Library			
Food commons			
Park bench			

The factor described in rows (study location) can be an independent variable.

d 3 × 6

	Mon.	Tue.	Wed.	Thur.	Fri.	Sat.
Eating						
Mood						
Dissociative						

5 If they are in the same design the researcher will be able to look for interactions.

6 Two different types (main effects and interaction). Two main effects and one interaction.

7 **a** H_0: $\mu_{A1} = \mu_{A2} = \mu_{Ak}$; there is no difference in population means of the levels of Factor *A*.

H_1: H_0 is false, or at least two means from Factor *A* are not equal.

b H_0: $\mu_{B1} = \mu_{B2} = \mu_{Bk}$; there is no difference in population means of the levels of Factor *B*.

H_1: H_0 is false, or at least two means from Factor *B* are not equal.

c H_0: There is no interaction; H_1: There is an interaction.

8 **a** $F_A = \dfrac{73.50}{2.12} = 34.67, p < .05; \quad F_{crit}(2{,}45) = 3.23$

$F_B = \dfrac{13.72}{2.12} = 6.47, p < 0.05; \quad F_{\text{crit}}(2{,}45) = 3.23$

$F_{A \times B} = \dfrac{3.05}{2.12} = 1.44, \text{n.s.}; \quad F_{\text{crit}}(4{,}45) = 2.61$

Only this information is needed since the rest of the table is already complete.

b Critical values: $F_A(1{,}30) = 4.17$; $F_B(2{,}30) = 3.32$; $F_{A \times B}(2{,}30) = 3.32$

Source	***SS***	***df***	***MS***	***F***	***p***
Factor *A*	3.36	1	3.36	0.90	*n.s.*
Factor *B*	66.67	2	33.34	8.94	<.05
$A \times B$	56.89	2	28.45	7.63	<.05
Within groups	111.83	30	3.73		
Total	238.75	35			

c All critical values are based on $df = 1{,}16$ and equal 4.49.

Source	***SS***	***df***	***MS***	***F***	***p***
Factor *A*	0.45	1	0.45	0.35	*n.s.*
Factor *B*	6.05	1	6.05	4.73	<.05
$A \times B$	84.05	1	84.05	65.66	<.05
Within groups	20.48	16	1.28		
Total	111.03	19			

9 Refer to graphs.

(a)

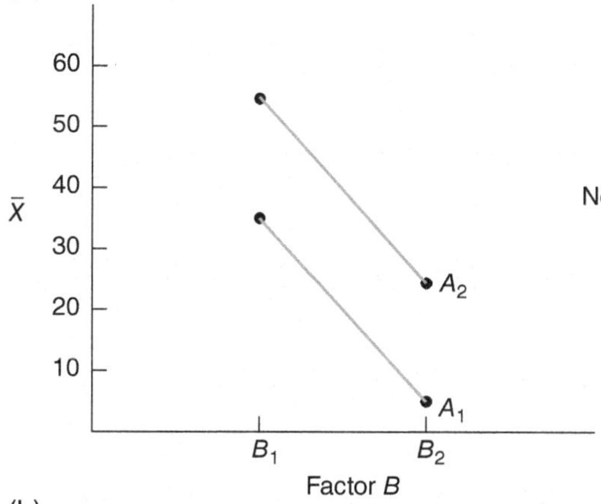

No interaction

(b)

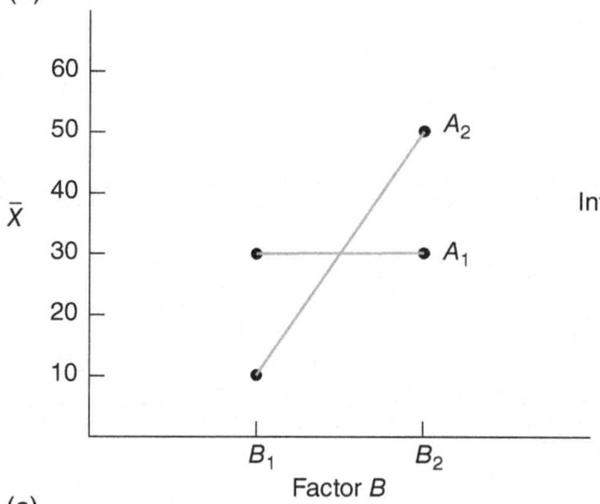

Interaction

(c)

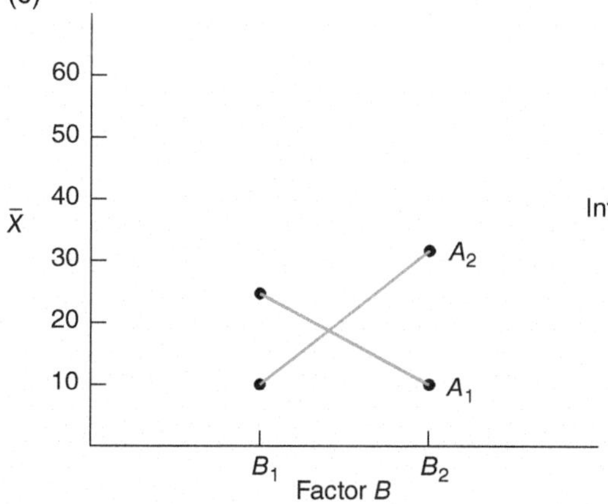

Interaction

10 The size of an F reflects the level of certainty that the null of no difference between population means can be rejected. A large treatment effect does lead to large F values, but so do small effects with high statistical power. Another statistic is needed to measure effect size alone.

11 These values reflect slightly different ways to calculate the ratio of treatment (primary) variance to total variance. For example, $\varpi_A^2 = .3$ means 30% of the total variance is associated with Factor A.

12 In a 2×2 design, there are only two conditions for each factor. A rejected main effect null means the evidence of a difference must be between the only two-cell means for that factor.

13 Fisher's *LSD*. Tukey's *HSD* requires each cell n to be the same.

14 The presence of a significant interaction qualifies a straightforward interpretation of a main effect. In other words, the interaction may be making it look as if there is a main effect, when, in reality, the variance is due to the interaction; the lower-order effect might be illusory.

15 There would be no effect on the main effects, interaction, or MS_W. Adding a constant does not affect the variance; therefore, it has no effect on an ANOVA.

16 a **Cell means**: M for Hi-Dep/Lo-Inc = 7.0; M for Hi-Dep/Hi-Inc = 4.8; M for Lo-Dep/Lo-Inc = 7.4; M for Lo-Dep/Hi-Inc = 7.6

Marginal Means: Lo-Inc = 7.2; Hi-Inc = 6.2; Hi-Dep = 5.9; Lo-Dep = 7.5

Source	*SS*	*df*	*MS*	*F*	*p*
Factor *A*	5.00	1	5.00	3.45	*n.s.*
Factor *B*	12.80	1	12.80	8.83	<.05
$A \times B$	7.20	1	7.20	4.97	<.05
Within groups	23.20	16	1.45		
Total	48.20	19			

b

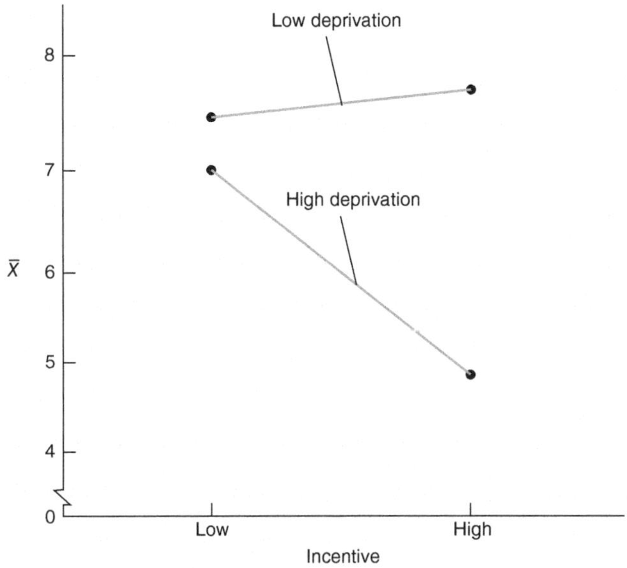

c $\varpi_B^2 = \dfrac{11.35}{49.65} = \mathbf{0.23\ or\ 23\%}$ $\varpi_{A \times B}^2 = \dfrac{5.75}{49.65} = \mathbf{0.12\ or\ 12\%}$

17 **a** **Cell means**: Freshmen/No Delay: $M = 12.20$; Freshmen/Two-Hour Delay: $M = 7.40$; Freshmen/One-Day Delay: $M = 6.20$; Seniors/No Delay: $M = 12.20$; Seniors/Two-Hour Delay: $M = 6.40$; Seniors/One-Day Delay: $M = 6.20$

Marginal means: Freshmen: $M = 8.60$; Seniors: $M = 8.27$; No Delay: $M = 12.20$; Two-Hour Delay: $M = 6.90$; One-Day Delay: $M = 6.20$

Source	*SS*	*df*	*MS*	*F*	*p*
Factor *A*	0.84	1	0.84	0.31	*n.s.*
Factor *B*	215.27	2	107.64	39.43	<.05
$A \times B$	1.66	2	0.83	0.30	*n.s.*
Within groups	65.60	24	2.73		
Total	283.37	29			

b

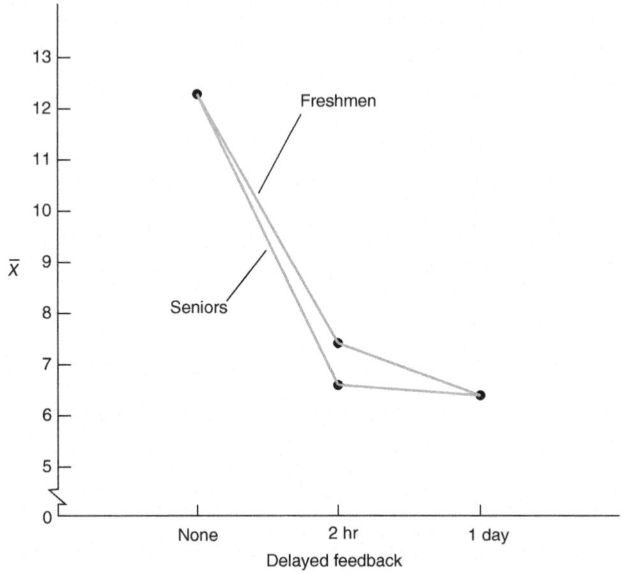

c $\eta_B^2 = \dfrac{SS_B}{SS_B + SS_W} = \dfrac{215.27}{215.27 + 65.6} = \dfrac{215.27}{280.87} = 0.77 \text{ or } 77\%$

d No Delay vs. One-Day Delay: $t = \dfrac{6}{0.74} = 8.11, p < .05$

No Delay vs. Two-Hour Delay: $t = \dfrac{5.3}{0.74} = 7.16, p < .05$

Two-Hour Delay vs. One-Day Delay: $t = \dfrac{0.7}{0.74} = 0.95$, *n.s.*

$t_{\text{crit}}(24) = \pm 2.064$

Tukey's HSD is $HSD_B = q_B\sqrt{\dfrac{MS_W}{n_B}} = 3.88\sqrt{\dfrac{2.73}{10}} = 2.03$. When applied to the three sets of means, the same findings emerge – fail to reject the null between Two-Hour Delay vs. One-Day Delay, and reject the other two.

e Statistical evidence suggests that delayed feedback influences learning [$F(2,24) = 39.43$, $p < .05$], with Two-Hour Delays [$t(24) = 7.16$, $p < .05$] and One-Day Delays [$t(24) = 8.11$, $p < .05$] both showing lower retention than No Delay. No evidence of a difference between Two-Hour and One-Day Delays, $t(24) = .95$, *n.s.* There was also no evidence of a difference for educational level, $F(1,24) = .31$, *n.s.*, or an interaction, $F(2,24) = .30$, *n.s.*

18

Source	SS	df	MS	F	p
Factor A	51.04	1	51.04	.94	*n.s.*
Factor B	260.04	1	260.04	4.81	<.05
$A \times B$	3.38	1	3.38	.06	*n.s.*
Within groups	1082.17	20	54.11		
Total	1396.63	23			

$M_{B1} = 28.42$, $M_{B2} = 21.84$. Since the F for Factor B allows us to reject the null hypothesis, we can conclude that statistical evidence suggests that the race of the defendant is associated with sentencing, with Black defendants drawing stiffer sentences.

19

Source	SS	df	MS	F	p
Factor A	34 003.34	1	34 003.34	100.24	<.05
Factor B	31.86	2	15.93	.05	*n.s.*
$A \times B$	23.03	2	11.52	.03	*n.s.*
Within groups	38 671	114	339.22		
Total	72 729.20	119			

There is statistical evidence suggesting that the biological sex of a child relates to their ability to delay gratification [$F(1,114) = 100.24$, $p < .05$], with biological females being able to delay longer ($M = 63.03$) than males ($M = 29.37$). There is no statistical evidence of differences between cognitive strategies [$F(2,114) = .05$, *n.s.*] and no statistical evidence of an interaction between biological sex and cognitive strategy [$F(2,114) = .03$, *n.s.*].

The Factor A effect size, according to η^2, is [34 003.34/(34 003.34 + 38 671)] .47 or 47%.

20

Source	SS	df	MS	F	p
Factor A	187.26	1	187.26	6.35	<.05
Factor B	224.26	1	224.26	7.61	<.05
$A \times B$	693.60	1	693.60	23.53	<.05
Within groups	1650.80	56	29.48		
Total	2755.93	59			

Marginal means: $M_{h-drive} = 21.73$; $M_{l-drive} = 18.2$; $M_{easy} = 18.03$; $M_{difficult} = 21.90$
Cell means: $M_{high/easy} = 16.4$; $M_{high/diff} = 27.07$; $M_{lo/easy} = 19.67$; $M_{lo/diff} = 16.73$
Follow-up tests: Let us use Tukey's *HSD* to compare group means from the same factor:

$$HSD_{A \times B} = q_{A \times B}\sqrt{\frac{MS_W}{n_{A \times B}}} = 3.41\sqrt{\frac{29.48}{15}} = 4.78$$

Using this value and then looking at our cell means, we see that one cell is driving all three effects, the "high-drive state/difficult task" condition. Since the other three cells are similar, it looks like the main effects are illusory. This simplifies our interpretation. We have statistical evidence to suggest that drive state and task difficulty combine to create greater arithmetic errors on the given task, $F(1{,}56) = 23.53$, $p < .05$. Further analysis shows that participants in the high-drive condition make more errors if the task is difficult ($M = 27.07$) compared with those solving easy problems ($M = 16.4$) or those in the low-drive state, regardless of task difficulty ($M_{easy} = 19.67$; $M_{diff} = 16.73$), Tukey's $HSD = 4.78$.

21

Source	*SS*	*df*	*MS*	*F*	*p*
Factor *A*	19.01	1	19.01	8.85	<.05
Factor *B*	1.19	2	0.58	0.28	*n.s.*
A × *B*	5.03	2	2.51	1.17	*n.s.*
Within groups	141.75	66	2.15		
Total	166.98	71			

Since the only null rejected is Factor *A*, and since it only has two conditions, no follow-up tests are needed. The analysis is straightforward; statistical evidence suggests students who use social media while studying ($M = 3.72$) perform poorer than students who do not ($M = 4.75$), $F(1{,}66) = 8.85$, $p < .05$. [This is not a causal interpretation.] There was no evidence of a difference for the type of material studied [$F(2{,}66) = .28$, *n.s.*] or for a social media use *x* type of material interaction [$F(2{,}66) = 1.17$, *n.s*].

We should be cautious about using Likert scale data. Means only make sense when the scale being used has conserved the quantitative space between each integer.

22

Source	*SS*	*df*	*MS*	*F*	*p*
Factor *A*	0.56	1	0.56	0.01	*n.s.*
Factor *B*	1105.56	1	1105.56	12.82	<.05
$A \times B$	3.06	1	3.06	0.04	*n.s.*
Within groups	1034.75	12	86.23		
Total	2143.93	15			

The only rejected null is Factor *B*. It only has two conditions; no follow-up tests are needed. The analysis is straightforward; statistical evidence suggests students who study with music ($M = 23.25$) performed more poorly than those who did not ($M = 39.88$), $F(1,12) = 12.82$, $p < .05$. There was no evidence to suggest a type-of-material-studied effect [$F(1,12) = .01$, *n.s.*] or an interaction between music use and the type of material [$F(1,12) = .04$, *n.s.*].

Chapter 13

1 They both analyze data from designs where participants are exposed to all conditions.

2 In a repeated-measures design, each participant is exposed to each treatment condition. In a between-groups design, each participant receives only one treatment.

3 "Order effects" is the name given to the variance introduced in repeated-measures designs when differences between treatment conditions result from the order of presentation. When the order of presentation is not being investigated, order effects introduce confounding variance.

4 "Counterbalancing" is a strategy used with repeated-measures designs in which participants differ by the order in which conditions are presented. The purpose of counterbalancing is to prevent confounding of the independent variable with order effects by distributing the carryover effects that come with repeated measuring across all experimental conditions.

5 Repeated-measures designs cannot be used when the variable is not subject to manipulation (e.g. participant variables). It should not be run when carryover effects cannot be controlled. For example, some forms of therapy result in a relatively permanent change in the participant.

6 In a repeated-measures design, between-group variation is made up of treatment effect and experimental error (there is no variation due to individual

differences); within-group variation consists of individual differences (between-participant variability) and experimental error.

7 Both contain experimental error variance. However, individual differences variance is mathematically removed from the denominator of a repeated-measures ANOVA.

8 MS_{BG} divided by MS_{Error}

9 By using the same participants in every treatment condition, it is impossible for one condition to have more or less of a participant variable than another treatment condition. This explains the absence of individual differences variance in the numerator. Individual differences variance can be partitioned out of the denominator, thus removing it from both parts of the *F* ratio.

10 For one-way, $df_{BG} = k - 1 = 2$; $df_W = N - k = 12$. For a repeated-measures, $df_{BG} = k - 1 = 2$; $df_{error} = (N - k) - (n - 1) = 28$. F_{crit} (one way) = 3.88; F_{crit} (repeated-measures) = 3.34.

11 df_T (degrees of freedom, total)

12 $df_{BG} = k - 1 = 4$; df_{error} $(N - k) - (n - 1) = 76$. The answers are 4 and 76.

13 9. Finding *n*, the number of participants, is hard. We know that $df_{Error} = 24$, but when we try to figure it out, we see that we have two missing values; *N* and *n*. Trial and error might be best. To pick our first number, think of a sample size that when it is measured 4 times (since *k* must equal 4, given that $df_{BG} = k - 1$) would be something a bit bigger than 24. Let's try 8. So, if $n = 8$, that means that *N*, given $k = 4$, must be 32. So, what would df_{Error} be? Well, $df_{Error} = (N - k) - (n - 1)$. That would be $(32 - 4) - (8 - 1)$. This equals 21. Our guess of 8 is too small. Let's try 9. That would mean that $df_{Error} = (36 - 4) - (9 - 1)$. What does that equal? 24. That is it!

14 **a** $M_A = 7.4$; $M_B = 5.4$; $M_C = 3.8$

Source	*SS*	*df*	*MS*	*F*	*p*
Between groups	32.53	2	16.27	14.79	<.05
Within groups	13.20	12			
Between participants	4.40	4			
Error	8.80	8	1.10		
Total	45.73	14			

b $\varpi^2 = \dfrac{32.53 - 2(1.10)}{45.73 + 1.10} = \dfrac{30.33}{46.83} = 0.65 \text{ or } 65\%; \eta^2 = \dfrac{32.53}{45.73} = 0.71 \text{ or } 71\%$

c $HSD = 4.04\sqrt{\dfrac{1.10}{5}} = 1.89$

Mean differences for Incentive A vs. $B = 2$, $p < .05$
Mean differences for Incentive A vs. $C = 3.6$, $p < .05$
Mean differences for Incentive B vs. $C = 1.6$, *n.s.*

d Statistical evidence suggests type-of-incentive influenced sales, $F(2,8) = 14.79$, $p < .05$. Further analyses using Tukey's *HSD* found statistical evidence suggesting that Incentive *A* worked better than Incentive *B*, $HSD = 2$, $p < .05$, and Incentive *C*, $HSD = -3.6$, $p < .05$. No evidence of a difference was found between Incentives *B* and *C*, $HSD = 1.6$, *n.s.*

e The description is not clear if the incentive programs were counterbalanced. As a result, carryover effects might be introducing confounding variance.

15 **a** $M_{pos} = 52$; $M_{neg} = 15.75$; $M_{con} = 27$

Source	SS	df	MS	F	p
Between groups	2754.17	2	1377.09	47.54	<.05
Within groups	212.75	9			
Between participants	38.92	3			
Error	173.83	6	28.97		
Total	2966.92	11			

b $\varpi^2 = \dfrac{2754.17 - (2)28.97}{2966.92 + 28.97} = \dfrac{2696.23}{2995.89} = 0.90 \text{ or } 90\%;$

$\eta^2 = \dfrac{2754.17}{2966.92} = 0.93 \text{ or } 93\%$

$t_{crit}(6) = \pm 2.447$

c Positive vs. Negative: $t = \dfrac{52 - 15.75}{3.81} = 9.51, p < 0.05$

Positive vs. Control: $t = \dfrac{52 - 27}{3.81} = 6.56, p < 0.05$

Negative vs. Control: $t = \dfrac{15.75 - 27}{3.81} = -2.95, p < 0.05$

d Statistical evidence suggests the type of subliminal message influenced problem-solving, $F(2,6) = 47.54$, $p < .05$. Further analyses found evidence suggesting positive messages worked better than negative messages, $t(6) = 9.51$, $p < .05$, and the control group, $t(6) = 6.56$, $p < .05$; and the control group outperformed the negative feedback group, $t(6) = -2.95$, $p < .05$.

e Since participants are repeatedly measured, it is hoped that each set of math problems is deemed equally difficult (and not merely the same list of problems). Furthermore, no statement was made about counterbalancing the order of presentation. If this was not performed, carryover effects may be introducing confounding variance.

16 **a** $M_{shirt} = 6.4$; $M_{shoes} = 8.2$; $M_{control} = 4.6$

Source	SS	df	MS	F	p
Between groups	32.40	2	16.20	26.27	<.05
Within groups	17.20	12			
Between participants	12.27	4			
Error	4.93	8	0.62		
Total	49.6	14			

b $\varpi^2 = \dfrac{32.40 - (2)0.62}{49.60 + 0.62} = \dfrac{31.16}{50.22} = 0.62$ or 62%; $\eta^2 = \dfrac{32.40}{49.60} = 0.65$ or 65%

c $t_{crit}(8) = \pm 2.306$

Shirt vs. Shoes: $t = \dfrac{6.40 - 8.20}{0.50} = -3.60,\ p < 0.05$

Shirt vs. Control: $t = \dfrac{6.40 - 4.60}{0.50} = 3.60,\ p < 0.05$

Shoes vs. Control: $t = \dfrac{8.20 - 4.60}{0.50} = 7.20,\ p < 0.05$

d Statistical evidence suggests that the type of new clothing influenced perceptions of happiness, $F(2,8) = 16.20$, $p < .05$. Further analyses found evidence suggesting that wearing new shoes generated more happiness than wearing a new shirt, $t(8) = 3.60$, $p < .05$, and the control group, $t(8) = 7.20$, $p < .05$; and the new shirt group seemed happier than the control group, $t(8) = 3.60$, $p < .05$. Notice the different findings compared with Problem 24 in Chapter 11 which used the same data but with an independent-groups design. The increased power of a repeated-measures ANOVA found evidence of more differences.

e It appears that the researchers did not fully employ counterbalancing by having the control condition always go last (at least this is implied). Furthermore, there is the issue of measuring happiness using a rating scale: Is this data from an interval scale?

17 **a** $M_{low} = 62.8$; $M_{medium} = 55.8$; $M_{high} = 98$

Source	SS	df	MS	F	p
Between groups	5114.80	2	2557.40	9.18	<.05
Within groups	2893.60	12			
Between participants	663.73	4			
Error	2229.87	8	278.73		
Total	8008.40	14			

b $\varpi^2 = \dfrac{5114.80 - (2)278.73}{8008.40 + 278.73} = \dfrac{4557.34}{8287.13} = 0.55$ or 55%;

$\eta^2 = \dfrac{5114.80}{8008.40} = 0.64$ or 64%

c $t_{\text{crit}}(8) = \pm 2.306$

Low vs. Medium: $t = \dfrac{62.80 - 55.80}{10.56} = 0.66$, *n.s.*

Low vs. High: $t = \dfrac{62.80 - 98}{10.56} = -3.33$, $p < 0.05$

Medium vs. High: $t = \dfrac{55.80 - 98}{10.56} = -4.00$, $p < 0.05$

d Statistical evidence suggests that the degree of distraction influenced pain tolerance, $F(2,8) = 9.18$, $p < .05$. Further analyses found evidence suggesting that high degrees of distraction worked better than both low degrees of distraction, $t(8) = -3.33$, $p < .05$, and medium degrees of distraction, $t(8) = -4.00$, $p < .05$. No evidence of a difference was found between low and medium degrees of distraction, $t(8) = .66$, *n.s.*

e It is assumed that the 20-minute interval in between the presentation of different conditions eliminated carryover effects. No statement was made in the description about counterbalancing the conditions. Carryover effects due to extreme nature of the dependent variable and/or the order of presentation may be introducing confounding variance.

18 **a** $M_A = 407.5$; $M_B = 240.33$; $M_C = 361$

Source	SS	df	MS	F	p
Between groups	89 334.77	2	44 667.39	16.24	<.05
Within groups	62 742.83	15			
Between participants	35 236.93	5			
Error	27 505.90	10	2 750.59		
Total	152 077.61	17			

b $\varpi^2 = \dfrac{89\,334.77 - 2(2750.59)}{15\,2077.61 + 2750.59} = \dfrac{83\,833.59}{154\,828.20} = 0.54 \text{ or } 54\%;$

$\eta^2 = \dfrac{89\,334.77}{152\,077.61} = 0.59 \text{ or } 59\%$

c $t_{crit}(10) = \pm 2.228$

Technique A vs. B: $t = \dfrac{407.50 - 240.33}{30.28} = 5.52, p < 0.05$

Technique A vs. C: $t = \dfrac{407.50 - 361}{30.28} = 1.54$, *n.s.*

Technique B vs. C: $t = \dfrac{240.33 - 361}{30.28} = -3.99, p < 0.05$

d Statistical evidence suggests type-of-technique influenced reading speed, $F(2,10) = 16.24$, $p < .05$. Further analyses found evidence suggesting Technique A outpaced Technique B, $t(10) = 5.52$, $p < .05$, and Technique C outpaced Technique B, $t(10) = -3.99$, $p < .05$. No evidence of a difference was found between Techniques A and C, $t(10) = 1.54$, *n.s.*

e The order of presentation is not addressed by the problem. This could be an issue if carryover effects are experienced.

19 **a** $M_{quiet} = 18$; $M_{DH} = 14.5$; $M_{Various} = 16$

Source	*SS*	*df*	*MS*	*F*	*p*
Between groups	37.00	2	18.50	9.10	<.05
Within groups	299.50	15			
Between participants	279.17	5			
Error	20.33	10	2.03		
Total	336.50	17			

b $\varpi^2 = \dfrac{37.00 - (2)2.03}{336.50 + 2.03} = \dfrac{32.94}{338.53} = 0.10 \text{ or } 10\%;\ \eta^2 = \dfrac{37.00}{336.50} = 0.11 \text{ or } 11\%$

c $HSD = 3.88\sqrt{\dfrac{2.03}{6}} = 2.26$

Mean differences for Quiet Room vs. Dining Hall = 3.5, $p < .05$
Mean differences for Quiet Room vs. Various = 2.0, *n.s.*
Mean differences for Dining Hall vs. Various = 1.5, *n.s.*

d Statistical evidence suggests that study location influences student performance, $F(2,10) = 9.10$, $p < .05$. Further analyses showed evidence suggesting that students perform better having studied in a quiet room compared with the dining hall, $HSD = 3.5$, $p < .05$. No evidence of a difference between a quiet room and a variety of locations was found, $HSD = 2.0$, *n.s.*, and no evidence of a difference between studying in the dining hall or a variety of locations was found, $HSD = 1.5$, *n.s.*

e Order of presentation is addressed. Assuming the various 25-point tests are of equal difficulty, the study seems to be in pretty good shape, methodologically speaking.

20

Source	*SS*	*df*	*MS*	*F*	*p*
Between groups	57.27	2	28.63	10.87	<.05
Within groups	91.40	27			
Between participants	44.00	9			
Error	47.40	18	2.63		
Total	148.67	29			

Statistical evidence suggests that the number of steps that participants walk changes across time as they use a pedometer, $F(2,18) = 10.87$, $p < .05$. Follow-up tests provide evidence suggesting that participants walk more after 6 and 12 weeks of using a pedometer compared with the first week, $t(18) = 3.15$, $p < .05$ and $t(18) = 4.52$, $p < .05$, respectively. No statistical evidence was found of a difference between 6 and 12 weeks, $t(18) = 1.36$, *n.s.*

Part 4 Review

1 Repeated-measures ANOVA. Just as the one-way ANOVA allows for a more expansive examination of between-participants designs (more than two conditions), the repeated-measures ANOVA allows for a more expansive examination of within-participants designs.

2 A mixed design is a form of a two-way ANOVA but one where one factor is repeatedly measured. So, both the two-way and repeated-measures ANOVAs are needed.

3 This is a 2×2 factorial design and requires a two-way ANOVA for analysis.

4 This is a within-participants design. It can be analyzed using either a dependent-samples *t* test or a repeated-measures ANOVA. Conventionally, the *t* test is used for two-cell designs.

5 This 3-condition version of Problem 4 now requires a repeated-measures ANOVA.

6 This is a between-groups single-factor design. A one-way ANOVA should be selected.

7 This is a between-groups single-factor design. A one-way ANOVA should be selected.

8 This is a between-groups single-factor design. Either an independent-samples *t* test or one-way ANOVA should be selected. Conventionally, the *t* test is used for two-cell designs.

9 This is a repeated-measures design with more than two conditions. A repeated-measures ANOVA should be selected. No counterbalancing means carryover effects might be present.

10 There are two factors involved in this design: first mover and biological sex of the seller. Each is also a between-groups factor. A two-way ANOVA should be selected.

Chapter 14

1 In a correlational design, a study is conducted without exerting control over the phenomenon under investigation; data are gathered as they present themself to the researcher. In an experimental design, some procedural variables are held constant, and others are purposely manipulated by the experimenter (e.g. drug dosages). The manipulation of an independent variable allows one to make a causal relationship from the independent variable to the dependent variable.

2 **a;** Causal interpretations are a function of methodological design, not statistical analysis.

3 "Bivariate distribution" is the term used to describe a dataset in which there are pairs of scores, each pair belonging to a given participant (or entity).

4 Answers will vary. Positive correlations are marked by higher scores on one variable being associated with higher scores on a second variable.

5 Answers will vary. Negative correlations are marked by higher scores on one variable being associated with lower scores on a second variable.

6 **b** Two of the options (a and c) are not legitimate correlations (correlations cannot exceed the absolute value of 1). Of the remaining two, −.69 is stronger in absolute value terms, which is how strength of association is measured.

7 The data would form a straight line. The slope would correspond to the valence, either positive or negative. The angle of the slope would be irrelevant.

8 An estimate of the strength of the correlation, direction of correlation (i.e. positive or negative based on the type of slope, not the particular angle), linearity, and presence of outliers.

9 **a** The scatter plot will have data points going from the lower left to the upper right. They will be gathered in an oval, not a straight line.
b The scatter plot will have data points going from the upper left to the lower right. They will be gathered in a tight oval, close to a straight diagonal line.
c The scatter plot will have data points going from the lower left to the upper right. They will be gathered in a slight oval, something close to a circle.
d The scatter plot will have data points that will gather into a shape that is not well represented by a straight line. All manner of shapes is possible (U's, S's, C's, etc.).

10 These are just rough estimates, designed to see if we understand the basic way in which scatter plots reflect correlations.
a $r = 0$
b $r = +0.50$
c $r = -0.50$
d $r = +0.90$

11

	r_{crit} 5%	r_{crit} 1%	r_{crit} 5%	r_{crit} 1%
a	0.388	0.496	Reject	Reject
b	0.388	0.496	Reject	Fail to reject
c	0.532	0.661	Reject	Reject
d	0.444	0.561	Reject	Reject
e	0.355	0.456	Reject	Fail to reject
f	0.666	0.798	Reject	Reject
g	0.514	0.641	Reject	Fail to reject
h	0.195	0.254	Reject	Fail to reject

12 It would also be 49%. Shared variance is bidirectional.

13 Since r^2 is the value of r squared, if $r^2 = .36$, r would have to be .6.

14 The rejection criteria (perhaps an $\alpha = .05$) only communicates the certainty of the relationship, not the size of one. If power is high, even trivial relationships can be found with high certainty. The shared variance represents the size of the relationship.

15 A restricted range of scores may underestimate the population correlation.

16 Using only extreme scores from a population tends to overestimate the correlation.

17 Pearson r will likely underestimate the actual correlation.

18 **a** Intermediate values, if hand calculated:

$\Sigma X = 24$	$\Sigma Y = 21$	$\Sigma XY = 125$
$(\Sigma X)^2 = 576$	$(\Sigma Y)^2 = 441$	$n_p = 4$
$\Sigma X^2 = 150$	$\Sigma Y^2 = 113$	

$$r_{obt} = \frac{-4}{16.25} = -0.25$$

b $H_0: \rho = 0$; $H_1: \rho \neq 0$
c $r_{crit}(2) = \pm 0.950$
d Do not reject the null hypothesis.
e $r^2 = (-0.25)^2 = 0.0625$ or 6.25%.

19 **a** Intermediate values, if hand calculated:

$\Sigma X = 32$	$\Sigma Y = 25$	$\Sigma XY = 213$
$(\Sigma X)^2 = 1024$	$(\Sigma Y)^2 = 625$	$n_p = 4$
$\Sigma X^2 = 266$	$\Sigma Y^2 = 175$	

$$r_{obt} = \frac{52}{54.78} = 0.95$$

b $H_0: \rho = 0$; $H_1: \rho \neq 0$
c $r_{crit}\ (2) = \pm 0.950$
d Reject the null hypothesis when r is equal to or greater than r_{crit}.
e $r^2 = 0.9025$ or 90.25%

20 **a** Intermediate values, if hand calculated:

$\Sigma X = 40$	$\Sigma Y = 41$	$\Sigma XY = 425$
$(\Sigma X)^2 = 1600$	$(\Sigma Y)^2 = 1681$	$n_p = 4$
$\Sigma X^2 = 414$	$\Sigma Y^2 = 449$	

$$r_{obt} = \frac{60}{80.25} = 0.75$$

b $H_0: \rho = 0$; $H_1: \rho \neq 0$
c $r_{crit}\ (2) = \pm 0.950$
d Do not reject the null hypothesis.
e $r^2 = 0.5625$ or 56.25%.

21 **a** Intermediate values, if hand calculated:

$\Sigma X = 18$	$\Sigma Y = 30$	$\Sigma XY = 96$
$(\Sigma X)^2 = 324$	$(\Sigma Y)^2 = 900$	$n_p = 5$
$\Sigma X^2 = 80$	$\Sigma Y^2 = 190$	
$r_{abt} = \dfrac{-60}{61.64} = -0.97$		

b H_0: $\rho = 0$; H_1: $\rho \neq 0$
c $r_{crit}\ (3) = \pm 0.878$
d Reject the null hypothesis.
e $r^2 = 0.941$ or 94.10%

22 Refer to graphs.

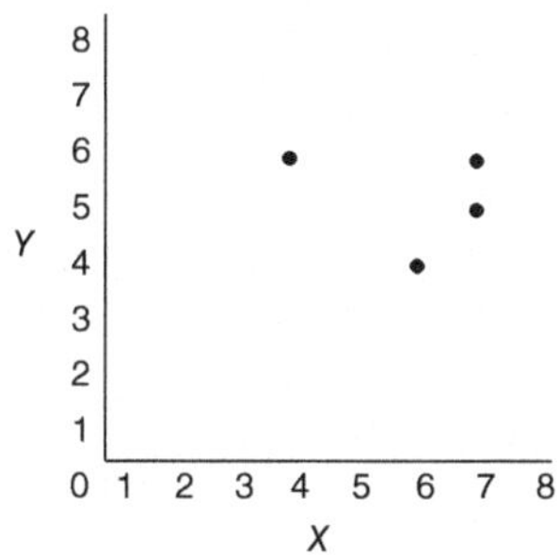

(Problem 18)

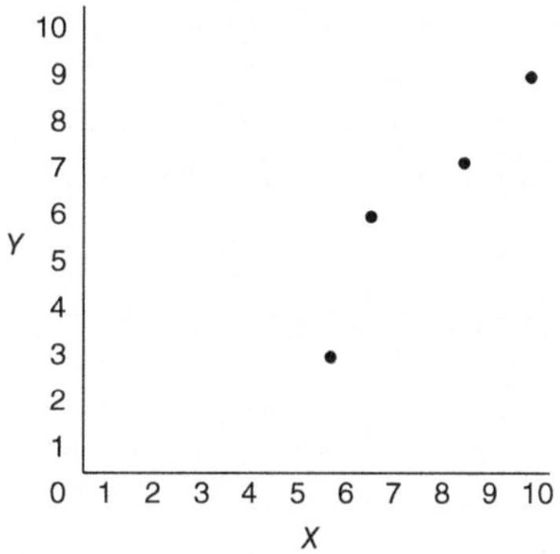

(Problem 19)

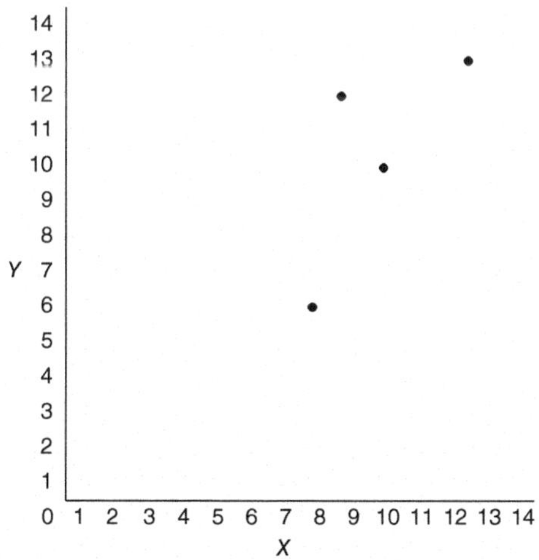

(Problem 20)

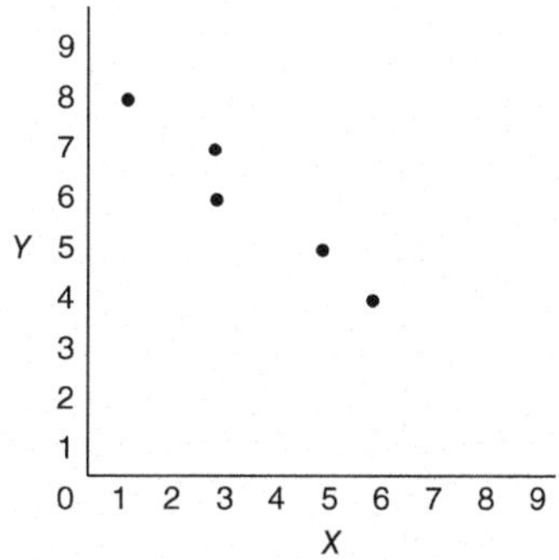

(Problem 21)

23 **a** Intermediate values, if hand calculated:

$\Sigma X = 56$	$\Sigma Y = 503$	$\Sigma XY = 3887$
$(\Sigma X)^2 = 3136$	$(\Sigma Y)^2 = 253\ 009$	$n_p = 7$
$\Sigma X^2 = 542$	$\Sigma Y^2 = 36\ 365$	

$$r_{abt} = \frac{-959}{1008.60} = -0.95$$

b $H_0: \rho = 0$; $H_1: \rho \neq 0$
c $r_{crit}\ (5) = \pm 0.754$
d Reject the null hypothesis.
e $r^2 = 0.9025$ or 90.25%

24 **a** Intermediate values, if hand calculated:

$\Sigma X = 18$	$\Sigma Y = 621$	$\Sigma XY = 2022$
$(\Sigma X)^2 = 324$	$(\Sigma Y)^2 = 385\ 641$	$n_p = 6$
$\Sigma X^2 = 94$	$\Sigma Y^2 = 64\ 979$	

$$r_{abt} = \frac{954}{1007.93} = 0.95$$

b $H_0: \rho = 0$; $H_1: \rho \neq 0$
c $r_{crit}\ (4) = \pm 0.811$
d Reject the null hypothesis.
e $r^2 = 0.9025$ or 90.25%

25 Refer to graph.

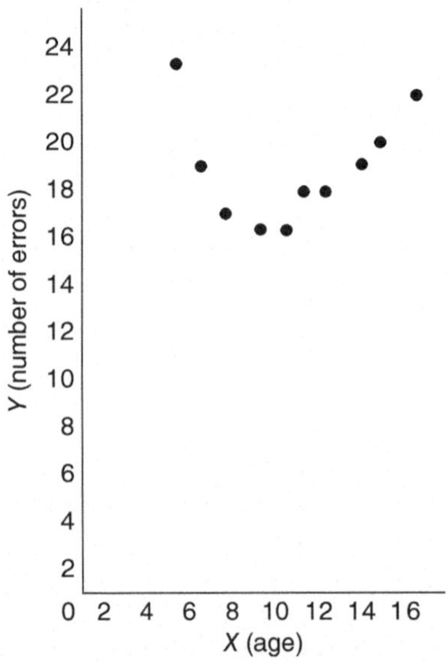

The variables appear to be related, but not in a linear fashion. The most reasonable course of action is to not run a Pearson r (it will most likely underestimate the relationship between these variables), but to run a curvilinear measure of association, something not covered in this text.

Chapter 15

1 The mean. It is the most frequent score in a normal distribution, and the error associated with it will likely be smaller than other guesses.

2 Bivariate data are typically not experimentally gathered. Data gathered from correlational designs do not imply causation. Only experimentally gathered data allow for causal analysis.

3 The least squares criterion means the regression line is fitted to the scatterplot so that the errors sum to zero and the sum of the squared errors (Σe^2) is minimized.

4 Y intercept

5 s_e is termed "the standard error of the estimate." It is the average standard deviation for a set of conditional distributions in a dataset. If the homoscedasticity assumption is met, it is a measure of error associated with predicted Y's.

6 A small standard error of the estimate means there is little error associated with a prediction; the larger the standard error, the greater the error associated with a prediction.

7 If $r = 0$, then $b = 0$. There is no relationship between X and Y.

8 The slope is a ratio of the rise over the run for bivariate data placed on a scatterplot. It tells us the degree and direction of change in Y for each unit change in X. It, however, does not communicate the strength of the relationship between the variables. (Except when $b = 0$. When this happens, the relationship is nonexistent.)

9 When the slope (b) equals 0, or when the null that $r = 0$ cannot be rejected. A regression analysis should not be run if the two variables are not correlated.

10 It is important because predicted Y (Y_p) is the mean of each conditional distribution. For skewed distributions, the mean may not represent centrality well (see Chapter 3).

11 It allows us to assume that s_e is the same for every value of X. If this were not true, the amount of error associated with different predictions would vary and be unpredictable.

12 Multiple regression uses multiple predictor variables; each one adds to the predictive power.

13 With a negative correlation, it is reasonable to conclude that higher gas prices will be associated with vacations closer to home. However, caution is needed; we must not convince ourselves that we understand the causal relationship between these two factors.

14 **a** $b = -0.22$
b $b = 0.26$
c $b = 0.55$

15 **a** 2.28
b 3.24
c 3.88

16 **a** $b = 1.81$
b $Y_p = 8.83 + 1.81(X - 4.50)$
c $s_e = 1.22$

17 **a** $b = -0.84$
b $Y_p = 8.00 - 0.84(X - 8.67)$
c $s_e = 3.39$
d $Y_p = 8.00 - 0.84(1 - 8.67) = 14.44$

18 Intermediate values, if hand calculated:

$M_X = 12.80$	$M_Y = 40.40$	
$\Sigma X = 64$	$\Sigma Y = 202$	$\Sigma XY = 2985$
$(\Sigma X)^2 = 4096$	$(\Sigma Y)^2 = 40\,804$	$n_p = 5$
$\Sigma X^2 = 990$	$\Sigma Y^2 = 9142$	

a $b = \dfrac{1997}{854} = 2.34$
b $Y_p = 40.40 + 2.34(0 - 12.80) = 10.45$

c $s_e = \sqrt{\left[\frac{1}{5(3)}\right]\left[(5)9142 - 40\,804 - \left(\frac{[5(2985) - (64)(202)]^2}{(5)990 - 4096}\right)\right]} = 3.97$

d $Y_p = 40.40 + 2.34(16 - 12.80) = 47.89$ seconds

e $47.89\ (\pm 1s_e) = 43.92$ to 51.86

19 Intermediate values if hand calculated:

$M_X = 12$	$M_Y = 23$	
$\Sigma X = 72$	$\Sigma Y = 138$	$\Sigma XY = 1739$
$(\Sigma X)^2 = 5184$	$(\Sigma Y)^2 = 19\,044$	$n_p = 6$
$\Sigma X^2 = 886$	$\Sigma Y^2 = 3560$	

a $b = \frac{10\,434 - 9936}{5316 - 5184} = 3.77$

b $Y_p = 23 + 3.77(0 - 12) = -22.24$

c $s_e = \sqrt{\left[\frac{1}{6(4)}\right]\left[(6)3560 - 19\,044 - \left(\frac{[6(1739) - (72)(138)]^2}{6(886) - 5184}\right)\right]} = 4.27$

d $Y_p = 23 + 3.77(10 - 12) = 15.46$; when converted to \$1000's = \$15 460

e $15.46\ (\pm 1.96s_e) = 7.091$ to 23.829; when converted to \$1000's = \$7 091 to \$23 829

20 Intermediate values, if hand calculated:

$M_X = 3.38$	$M_Y = 3.43$	
$\Sigma X = 16.88$	$\Sigma Y = 17.17$	$\Sigma XY = 58.09$
$(\Sigma X)^2 = 284.93$	$(\Sigma Y)^2 = 294.81$	$n_p = 5$
$\Sigma X^2 = 57.67$	$\Sigma Y^2 = 59.01$	

a $b = \frac{290.45 - 289.83}{288.35 - 284.93} = 0.18$

b $Y_p = 3.43 + 0.18(3.00 - 3.38) = 3.36$

Probably, since the predicted graduate GPA is 3.36.

c $s_e = \sqrt{\left[\frac{1}{5(3)}\right]\left[(5)59.01 - 294.81 - \left(\frac{[5(58.09) - (16.88)(17.17)]^2}{5(57.67) - 284.93}\right)\right]}$

$= 0.092$

$Y_p = 3.43 + 0.18(3.67 - 3.38) = 3.48$

$3.48 \pm 1se = 3.388$ to 3.572

21 Regression formulas is: $Y_p = 46.33 + 0.59(X - 104.5)$

a $b = 0.59$

b $Y_p = 46.33 + 0.59(0 - 104.5) = -15.33$

c $s_e = 4.99$

d $Y_p = 46.33 + 0.59(100 - 104.5) = 43.68$

e $43.68 \pm 1.96s_e = 33.9$ to 53.46

22 **a** $b = -1.81$

b $Y_p = 5.00 - 1.81\ (X - 2.09)$

c $s_e = 0.79$

d $Y_p = 5.00 - 1.81\ (3.5 - 2.09) = 2.45$

e $2.45 \pm 1.96s_e = 0.90$ to 4

f Be aware that the subjective well-being scale may not be interval or ratio.

Chapter 16

1 Statistical assumptions like normality and homogeneity of variance.

2 The goodness-of-fit test is used on categorical data spread across one dimension or factor. The test for independence is used on categorical data spread across two dimensions or factors.

3 *df* for chi-square tests are based on the number of categories in the design, not the sample size.

4 Chi-square calculations square $f_o - f_e$ differences – this makes all differences positive. Cell residuals, however, do not square this difference. They show which way cells are deviating from what is expected.

5 **d**

6 **c** Both are measures of effect size.

7 **a** $df = 1$

b $df = 6$

c $df = 12$

d $df = 2$

8

	Design	χ^2_{obt}	df	χ^2_{crit}	Reject H_0?
a	2 × 2	4.5	1	3.84	Yes
b	3 × 3	9.0	4	9.49	No
c	1 × 5	17.22	4	9.49	Yes
d	2 × 4	5.55	3	7.82	No

9 **a**

(f_o)

30	**50**	**20**	20	120
10	**30**	40	**20**	100
40	80	60	40	N = **220**

(f_e)

21.82	43.64	32.73	21.82
18.18	36.36	27.27	18.18

b

(f_o)

7	**7**	14
5	**11**	**16**
12	18	N = 30

(f_e)

5.60	8.40
6.40	9.60

10

Cell #	f_o	f_e	$(f_o - f_e)^2$	$(f_o - f_e)^2/f_e$	R
1	70	79.60	92.16	1.16	−1.08
2	160	159.20	0.64	0.004	0.06
3	168	159.20	77.44	0.49	0.70

(*Continued*)

Cell #	f_o	f_e	$(f_o - f_e)^2$	$(f_o - f_e)^2/f_e$	R
4	30	20.40	92.16	4.52	2.13
5	40	40.80	0.64	0.016	−0.13
6	32	40.80	77.44	1.90	−1.38
				$\chi^2 = \mathbf{8.09}$	
				$\chi^2_{crit} = 5.99$	

We have statistical evidence to reject the null hypothesis of no relationship, $\chi^2(2, N = 500) = 8.09$. Cell 4, *Unfavorable Response to Treatment I*, makes a significant contribution to the χ^2. There are more unfavorable responses than the null would expect. Cramér's $V = 0.13$

11 Using the contingency table formula $\chi^2 = \dfrac{110(255 - 1520)^2}{(55)(55)(53)(57)} = 19.26$

$\chi^2_{crit} = 3.84 \quad \chi^2(1, N = 110) = 19.26, p < 0.05$

There is statistical evidence of a difference, but the frequencies suggest it is one that is opposite to the hypothesis. Cell residuals were not requested. Cramér's $V = 0.42$.

12 **a** H_0: No relationship between type of bumper sticker and being stopped by police.
H_1: The variables are related (not independent).

b and c

	Stop Brutality	Smile	
Stopped	18	5	23
Not Stopped	7	20	27
	25	25	$N = 50$

f_o	f_e	$(f_o - f_e)^2$	$(f_o - f_e)^2/f_e$
18	11.50	42.25	3.67
5	11.50	42.25	3.67
7	13.50	42.25	3.13
20	13.50	42.25	3.13
			$\chi^2 = 13.60$

$\chi^2_{crit} = 3.84$

d Statistical evidence suggests drivers displaying "Stop Brutality" stickers are stopped more often than drivers displaying "Smile" stickers, χ^2 (1, $N = 50$) = 13.60, $p < .05$. Cell residuals were not requested (though none reached the ±2 threshold).

e Cramér's $V = 0.52$

13

Spring	Summer	Fall	Winter	
160	190	170	130	f_o
162.5	162.5	162.5	162.5	f_e

f_o	f_e	$(f_o - f_e)^2$	$(f_o - f_e)^2/f_e$	R
160	162.5	6.25	0.04	−0.19
190	162.5	756.25	4.65	2.16
170	162.5	56.25	0.35	0.59
130	162.5	1056.25	6.50	−2.55
			$\chi^2 = 11.54$	

There is statistical evidence to reject the null hypothesis of no seasonal selling differences, $\chi^2 = (3, N = 650)$ 11.54, $p < .05$. Follow-up analyses suggest that more people buy RVs in the summer ($R = 2.16$) and fewer in the winter ($R = -2.55$) than would be expected by the null. The advice to dealers depends on other issues. If there is a need to even out sales throughout the year, the dealers would be advised to advertise more heavily in the winter. If they want to take advantage of the peak buying season, they might be advised to advertise more heavily in the summer.

14 Since it is a 2 × 2 design, a contingency table will be a shortcut through the analysis.

	Accident	No Accident	
Rain	29	35	64
No Rain	31	48	79
	60	83	143

$$\chi^2 = \frac{143(1392 - 1085)^2}{(64)(79)(60)(83)} = \frac{13\,477\,607}{25\,178\,880} = 0.54$$

$\chi^2_{crit} = 3.84$; $\chi^2(1, N = 143) = 0.54$, n.s.

15 The null hypothesis that each reason is equally likely can be rejected, $\chi^2 = 613.65, p < .05$. A residual analysis found evidence suggesting that more people than would be expected to get tattoos for familial reasons ($R = 20.10$) and less people than would be expected to get tattoos for conformity ($R = -13.23$) or personal expression reasons ($R = -5.76$).

16 Younger mothers give birth to more physically immature ($R = 2.44$) and fewer physically mature ($R = -2.35$) babies than would be expected by chance. Older mothers give birth to more physical mature ($R = 2.71$) and fewer physically immature ($R = -2.81$) babies than would be expected by chance, χ^2 (1, $N = 114$) = 26.67, $p < .05$. Cramér's $V = 0.48$.

17 There is an association between diabetes and healing times, with diabetics showing longer healing times than the null would expect ($R = 5.25$), χ^2 (1, $N = 810$) = 137.08, $p < .05$.

18 There is no differential effect due to treatment, $\chi^2(1, N = 172) = 0.75$, *n.s.*

Part 5 Review

1 a

2 In a word, *prediction*. A regression analysis allows us to leverage a known relationship between two variables for predictive use when a given value is unknown.

3 Parametric tests are preferred because they have more power. If the null is wrong, parametric tests are more likely to generate statistical evidence to reject it. Chi-square tests, however, have power and avoid Type II errors at a rate similar to parametric tests.

4 As r (correlation) increases, prediction error decreases.

5 A correlation. Each variable can be measured quantitatively, and the amount of shared variance between them can be represented indirectly with an r or directly with an r^2.

6 A one-way ANOVA seems best, assuming charitable giving is now classified into categories. Each political category is a condition; charitable giving percentage would be the dependent variable. If donations are classified, then a chi-square test for independence is needed.

7 Regression. It is the only statistical tool that will generate a predicted value. Appropriate bivariate data will need to be gathered, however.

8 A two-way ANOVA seems best. Political attitude and biological sex are classified and serve as the two dimensions. Charitable giving percentage would be the dependent variable.

9 A chi-square goodness-of-fit test. (By the way, the $\chi^2 = 7.98$; χ^2_{crit} is 9.49. Fail to reject.)

10 **a** This one is tricky. If we decide to see these groups as being independent of each other (setting aside that each set of scores is a sibling pair), we will run an independent-samples *t* test and test the null that $\mu_{males} = \mu_{females}$. (By the way, the result is $t(18) = 1.57$, *n.s.* Fail to reject.) If, however, we see each pair of scores as a dependent set, then we can run a dependent-samples *t* test. (The result is $t(9) = 2.68$, $p < .05$. The null can be rejected.)
b Use an *r* to test the null. (By the way, reject the null, $r(8) = +0.66$, $p < .05$.)
c Regression. (By the way, $Y_p = 2.90 + 0.77(2 - 3.72) = 1.58$ hours or about 1 hour and 35 minutes. Just enough time for most movies!)

11 **a** A correlation. (By the way, reject the null of $\rho = 0$, $r(5) = -0.77$, $p < .05$.)
b Regression. (By the way, the prediction formula is: $Y_p = 11.57 - 1.88(5 - 1.86) = 5.67$. Anyone familiar with developmental milestones in children realizes this would be highly unusual. Perhaps the relationship is not linear for high values of *X*.)

12 Chi-square goodness-of-fit. This is a frequency count of categorical data with one factor (drink type).

13 In this case, the independent-samples *t* test seems best. There are two independent groups, and the dependent variable is being measured on a ratio scale, assuming all assumptions are met.

14 A correlation. No causal analysis is allowed. This data were gathered correlationally.

15 Assuming σ is not known, a single-sample *t* test seems best. (The null hypothesis would be 0, no weight gain across the six months.)

Appendix C

Instructions for Microsoft® Excel and SPSS®

Many of the calculation needs for this text can be met by using statistical processing software. This appendix will help students gain familiarity with two programs, Microsoft® Excel and SPSS®. Excel was chosen for its accessibility; it is ideal for students who do not have access to college or university software programs. SPSS was chosen because it is commonly available for students who do have access to college and university computer systems. This appendix will first provide general data input instruction, followed by particular information related to specific inferential tests. As these programs change with new versions and as the outputs are expressed differently on different platforms, some of the descriptions below may vary. However, the basic structure should be similar. (Note: The information presented below can be easily supplemented with tutorials found on the Internet.)

General Instruction for Data Entry into Microsoft® Excel and SPSS®

Microsoft® Excel

Following are some guidelines for how to enter data into Excel:

1) Open a new file (or an existing file if adding additional data).
2) Activate the cell in which the data are to be inputted.
3) Input data by highlighting a cell where data are to be placed and typing in the information.
4) Hitting the return key will activate the cell directly below. Hitting the tab key will activate the cell immediately to the right. A mouse click can also activate a cell.

Essentials of Behavioral and Social Science Statistics, Third Edition.
K. Paul Nesselroade, Jr. and Laurence G. Grimm.
© 2025 John Wiley & Sons, Inc. Published 2025 by John Wiley & Sons, Inc.
Companion website: https://www.wiley.com/go/Essentials_of_Behavioral_and_Social_Science_Statistics3e

5) Data are typically organized such that each participant is assigned one row, and each variable is assigned one column. Column names are typically added across the top row to help identify the data. Participant numbers are often not added due to the already existing numbering system down the left-hand column. However, sometimes it is important to create a specific column for participant identification purposes. This is usually the leftmost column.
6) If the data are repetitive or patterned, shortcuts can be used to input data. For instance, highlight a set of values, then grab the lower right-hand corner of the grid, and drag it downward.
7) Information can be edited while the cell is activated through standard word processing editing procedures, or, if the data have already been registered, a cell can be edited by activating it and then using the data entry block at the top of the spreadsheet.
8) There are cutting and pasting functions available. In fact, there are numerous data management tools available to the user. These are identified across the top of the file. The most useful ones for inputting data can be found under "Home," "Formulas," and "Data."
9) Under **Data** look for a **Data Analysis** option. If it is not there, it will need to be added. Use the Excel User's Manual to install the "Data Analysis ToolPak" (or some similarly worded add-on; different versions of Excel use different terms). This feature is required to run many of the statistical tests discussed in the textbook. (Some versions of Excel have a "Tools" tab. The Data Analysis function may be under this tab.)

Excel is a very flexible and sophisticated data organization tool. This is a brief introduction to data entry and organization. Please consult the user's manual or any number of online tutorials to understand further the features of this very helpful program.

SPSS®

Following are some guidelines for how to enter data into SPSS:

1) The first important thing to notice about SPSS is the two different views that can be reached by clicking boxes in the lower left of the screen: the **Data View** and the **Variable View**.
2) The Variable View should be accessed first (even though the program typically opens with the Data View activated). Here is where variables are named. They are also identified by Type (Numeric is the default value). Other specifications can also be made. For instance, one can note what various categorical values mean in the "Values" column and identify what type of scaling was used for

the measure under the "Measure" column. The default is "Scale," which means interval or ratio. "Ordinal" or "Nominal" can be selected.

3) Toggling to the Data View allows data to be inputted. Here the procedure is very similar to any other database program. Use the return key to move down a column and the tab key to move to the right along a row. The mouse can also be used to move around the grid. Data can be inputted into a cell once it is highlighted.
4) Typically, data from the same participant is placed along the same row. This is important for any repeated-measures or bivariate data analysis. In fact, data along the same row will be assumed to have come from the same participant (or to be matched in some way). For this reason, data from independent groups will need to be identified through a second variable, oftentimes labeled as "Condition." In this way, scores independently gathered across two or more conditions will populate one column, and nominal values such as "0," "1," "2," and so on will populate another column and identify different conditions.

SPSS is designed for statistical analysis. As a result, it is quite sophisticated. To better understand this program, consult a tutorial resource and/or use tutorial videos found online.

Chapter 2 Scales of Measurement and Data Display

Microsoft® Excel

Tables and graphs can be made using Excel (see Figure C.1). Tables are constructed by creating column headings and then inputting the corresponding data below the column headings. There are three different features within Excel that

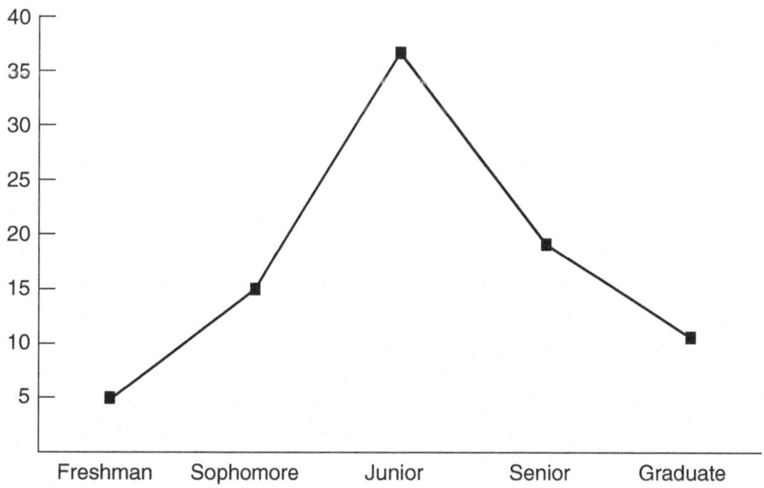

Figure C.1 A table and graph generated using Microsoft Excel.

can be used to create graphs. The Pivot Table feature is recommended when working with discrete or nominal categories (e.g. ethnicity, religion, or academic major) for the *X* axis. Either the Histogram Analysis Tool or the Frequency function is recommended when working with continuous variables (e.g. time or age) for the *X* axis. Excel has many options available to the user for generating customized graphical displays. For specific help using Excel, students are recommended to either purchase an Excel tutorial manual for statistics or use instructional videos easily found online.

SPSS®

SPSS is an extremely sophisticated but easy-to-use statistical software analysis tool. For creating tables, use the *Custom Tables* option. This feature is not standard on all forms of SPSS. If not available, users will need to export data into Excel to create tables. For creating graphs, SPSS has the *Chart Builder* feature (see Figure C.2). For specific help using SPSS, students are recommended to either purchase an SPSS tutorial manual or use available instructional videos easily found online.

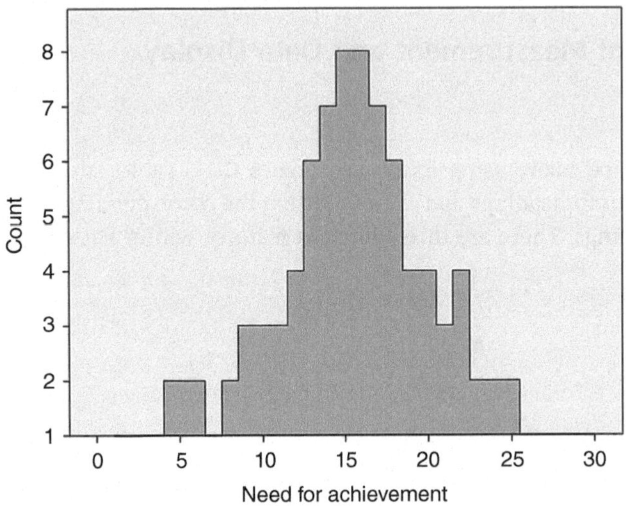

Figure C.2 A histogram generated using the Chart Builder function is SPSS.

Chapters 3 and 4 Descriptive Statistics

Microsoft® Excel

1) Once the data are entered, select **Data Analysis** and then **Descriptive Statistics**. Click **OK**.

2) Highlight the scores and put those quadrant numbers into the **Input Range.** (If labels are included, be sure to click the "Labels in first row" box.)
3) Select a location for the output. Use the **Output Range** box if needed.
4) Make sure to click **Summary Statistics** before clicking **OK**.

This will generate a table with all three measures of centrality, the sample variance and standard deviation, as well as other descriptive statistics.

SPSS®

1) Once the data have been entered, click **Analyze** on the toolbar and select **Descriptive Statistics** and then **Frequencies**.
2) Move the column label containing the data to be analyzed from the left box to the **Variable** box. Also, unclick the **Display Frequency Tables** box in the lower left corner.
3) Click the **Statistics** box in the upper right corner. Here are boxes labeled Mean, Median, and Mode. Other boxes in the lower left-hand corner under **Dispersion** are labeled **Std. deviation**, **Variance**, and **Range**. In addition, in the upper left-hand corner, find in the **Percentile Values** area a box labeled **Quartiles**. Once everything desired has been selected, click **Continue**.
4) Then click **OK**. The output is self-explanatory.

Chapter 5 *z* scores

Microsoft® Excel

1) Once a dataset has been inputted, Excel will first need the mean and standard deviation. Select **Data Analysis** and then **Descriptive Statistics**. Click **OK**.
2) Highlight all the needed scores and put those quadrant numbers into the **Input Range** box.
3) Select a location for the output. Use the **Output Range** box if needed.
4) Make sure to click **Summary Statistics** before clicking **OK**. This should generate a box of descriptive statistics, including the mean and standard deviation of the distribution.
5) Select the location for the z score computation for a given raw score number (typically this is the cell directly adjacent to the right of the raw score) and click the ***fx*** key to the immediate left of the input box at the top of the spreadsheet.
6) Excel uses the term **standardize** for z scores. Search for and select this term.
7) In the **X** box, select the raw score to be transformed.
8) In the **Mean** and **Standard-dev** box, select the appropriate values from the Summary Statistics box just constructed and click **OK**. The z score will show up in the selected box.

9) To transform all the raw scores into z scores, use the "autofill" function. However, first highlight the z score already transformed and place a $ in front of the number of the coordinate for both the mean and standard deviation components of the equation (since these values should not change). For instance, if the mean value is found in cell E5, change that to E$5; if the standard deviation value is found in cell E9, change that to E$9. This will keep this value constant for all the autofill calculations. Then highlight the cell containing the z score and move the cursor to the lower right of the cell until a "+" appears. Then drag down to a cell corresponding to the last raw score cell and release. This will produce a column of z scores corresponding to the adjacent raw scores.

SPSS®

1) Once the data have been entered, click **Analyze** on the toolbar, select **Descriptive Statistics**, and then **Descriptives.**
2) Move the column label containing the data to be converted into z scores from the left box to the **Variable** box. Also, make sure the **Save standardized values as variables** box in the lower left corner is checked.
3) Click **OK** (no other work is needed).

The output screen will generate some descriptive statistics. In the data file, there will be a new variable (same name as the selected variable, but with a "Z" in front of it) with each raw score transformed into its corresponding z score.

Chapter 8 The *t* Test

Microsoft® Excel

1) Enter the data into one column. Label the column appropriately (see Figure C.3).
2) Excel has built-in programs for several t tests; however, it does not have one for a single-sample t test. As a result, the Formula 8.3 components will need to be computed.
3) Determine the null hypothesis (μ), and record it in an open cell (label it appropriately).
4) Determine the sample mean by using the built-in Excel function "AVERAGE." Record it in an open cell (label it appropriately).
5) Determine the estimate of the standard error (s_M) by first determining the sample standard deviation using the built-in function (either STDEV or STDEV.S; both calculate the standard deviation of a sample) and the sample size (n). With these values, find the estimate of the standard error ($s_M = s/\sqrt{n}$). Record this value in an open cell (label it appropriately).

Figure C.3 A worked example of a single-sample t test using Microsoft Excel.

Number of cousins		
5	H_0=	6.5
7	M =	9.2
12	s =	4.131182
8	s_m =	1.307336
18	t_{obt} =	2.065268
3	t_{crit} =	±2.262
10		
8	Fail to reject the null hypothesis	
11		
10		

6) The t value can now be determined using Formula 8.3; $t_{obt} = (M - \mu)/s_M$.
7) Use n to find the appropriate *df*. The *df* for a single-sample t test is $n - 1$.
8) Go to Table A.2 (t Table) in the Appendix and use the *df* value and the alpha value (usually .05) to find the appropriate t_{crit} value(s).
9) Compare the t_{obt} with the t_{crit} and make a decision regarding the null (see Figure C.3).

SPSS®

1) Label the variable appropriately in Variable View. Enter all the scores from the sample into the appropriate column in Data View; one score per row (see Figure C.4).
2) Click **Analyze**, then select **Compare Means**, and then **One-Sample T Test.**
3) Highlight the appropriate column in the left box and move it into the **Test Variable(s)** box.
4) In the **Test Value** box on the bottom right-hand side, enter the hypothesized value for the population mean given a true null. The default is zero.
5) Click **OK**.

The output generates two boxes (see Figure C.5). The first will identify the number of scores in the sample (N) as well as the mean, standard deviation, and standard error. The second will identify the t value, degrees of freedom, significance level, mean difference, and the 95% confidence intervals of the actual mean difference. It will not generate t_{crit}. Either look up t_{crit}, or see if the significance level is equal to or lower than .05.

	Cousin_num
1	5
2	7
3	12
4	8
5	18
6	3
7	10
8	8
9	11
10	10

Figure C.4 An example of entered data for a single-sample t test is SPSS.

T-test

One-sample statistics

	N	Mean	Std. deviation	Std. error mean
cousin_num	10	9.20	4.131	1.306

One-sample test

	Test value = 6.5					
					95% confidence interval of the difference	
	t	*df*	Sig. (2-tailed)	Mean difference	Lower	Upper
cousin_num	2.067	9	0.069	2.700	–0.26	5.66

Figure C.5 An example of a single-sample *t* test output in SPSS.

Chapter 9 Independent- and Dependent-Samples *t* Tests
Independent-Samples *t* Tests

Microsoft® Excel

1) Enter all the scores from the samples into two adjacent columns, one sample in each column. Label the columns appropriately (see Figure C.6).
2) Excel has a built-in program for the independent-samples *t* test. To access it, click on the **Data** tab on the top menu and then click **Data Analysis**.

Control	Drug
2	16
0	20
10	2
7	22
2	3
5	17
8	23
5	4
6	23
0	14
5	22
11	9
2	22
10	20
2	5

t-Test: Two-sample assuming equal variances

	Control	Drug
Mean	5	14.8
Variance	13.28571429	64.31428571
Observations	15	15
Pooled Variance	38.8	
Hypothesized Mean Difference	0	
df	28	
t Stat	–4.308646306	
P(*T*≤*t*) one-tail	9.13838E-05	
t Critical one-tail	1.701130934	
P(*T*≤*t*) two-tail	0.000182768	
t Critical two-tail	2.048407142	

Figure C.6 A worked example of an independent-samples *t* test using Microsoft Excel.

3) With the Data Analysis box open, select ***t*-Test: Two-Sample Assuming Equal Variances**. (Do not select a similar option, ***t*-Test: Two-Sample Assuming Unequal Variances**. This function does not create a pooled variance.)
4) Input the data for one variable in box **Variable 1 Range**. Input the data for the other variable in box **Variable 2 Range**. (If labels were included, make sure to click the **Labels** box.)
5) Decide on an Output option. The default is to place it on a separate worksheet.
6) Click **OK**.

The output box (see Figure C.6) presents the means, variances, and observations (sample sizes). Also presented are the pooled variance, hypothesized mean difference (0, unless otherwise specified), degrees of freedom (labeled *df*), observed *t* value (labeled "*t* stat"), and the critical scores and probabilities for both one- and two-tailed versions of the test. Compare *t* stat with either the probability value (labeled ***P*(*T*≤*t*) two-tail**) or the critical score (labeled ***t* Critical two-tail**) associated with the two-tailed test to make a decision regarding the null.

SPSS®

1) In SPSS, each row of the data file represents a participant. Since both samples in an independent-samples *t* test have different participants, all the dependent variable data from both samples needs to be placed into one column. Within **Variable View**, label this variable appropriately. Create a second variable (typically labeled "condition") to assign the data to different groups. Then, go to **Data View** and input the data (see Figure C.7). Use a nominal variable in the "condition" column to distinguish the samples (e.g. "0" and "1").
2) Click **Analyze,** then select **Compare Means**, and then **Independent-Samples T Test.**
3) Highlight the dependent variable column label in the left box and move it into the **Test Variable(s)** box. Move the "condition" variable to the **Grouping Variable** box.
4) Click **Define Groups**. Place the nominal values used to distinguish the groups (e.g. "0" and "1") into the two group boxes, one in each. Click **Continue.**
5) Click **OK.**

The output generates two boxes (see Figure C.8). The first identifies how many scores were in the sample (*N*) as well as the mean, standard deviation, and an estimate of the standard error from the perspective of each variable (neither of these

	Heart_rate	Condition
1	84	1
2	78	1
3	67	1
4	87	1
5	80	1
6	78	1
7	78	1
8	79	1
9	82	1
10	81	1
11	88	2
12	97	2
13	74	2
14	80	2
15	87	2
16	90	2
17	90	2
18	86	2
19	84	2
20	78	2

Figure C.7 An example of entered data for an independent-samples *t* test in SPSS.

***T*-test**

Group statistics

	Condition	*N*	Mean	Std. deviation	Std. error mean
heart_rate	1	10	79.40	5.254	1.661
	2	10	85.40	6.687	2.115

Independent samples test

		Levene's test for equality of variances		*t*-test for equality of means			
		F	Sig.	*t*	*df*	Sig. (2-tailed)	Mean difference
heart_rate	Equal variances assumed	0.975	0.337	−2.231	18	0.039	−6.000
	Equal variances not assumed			−2.231	17.046	0.039	−6.000

Independent samples test

		t-test for equality of means		
			95% confidence interval of the difference	
		Std. error difference	Lower	Upper
heart_rate	Equal variances assumed	2.689	−11.650	−0.350
	Equal variances not assumed	2.689	−11.672	−0.328

Figure C.8 An example of an independent-samples *t* test output in SPSS.

is used in the *t* test). The second identifies the *t* value, the degrees of freedom, the significance level, the mean difference, and the estimate of the standard error (labeled as "Std. Error Difference"). The output will not generate t_{crit}. Use a table to find t_{crit} or look at the given significance level to see if it is equal to or lower than .05.

Dependent-Samples *t* Tests

Microsoft® Excel

1) Enter the data into two adjacent columns, one sample in each column. Label appropriately. Keep the data from each participant in the same row across the two columns. This is necessary to gain a difference score for each participant (see Figure C.9).
2) Excel has a built-in program for the dependent-samples *t* test. To access it, click on the **Data** tab on the top menu and then click **Data Analysis**.
3) With the Data Analysis box open, select ***t*-Test: Paired Two-Sample for Means**.
4) Input the data for one variable in box **Variable 1 Range** and the data for the other variable in box **Variable 2 Range**. (If labels were included, make sure to click the **Labels** box.)
5) Decide on an Output option. The default is to place it on a separate worksheet.
6) Click **OK**.

The output box (see Figure C.9) presents the means, variances, and observations (sample sizes). Also presented are the Pearson correlation (covered in Chapter 14), hypothesized mean difference (0, unless otherwise specified), degrees of freedom (labeled *df*), and observed *t* value (labeled "*t* stat"), as well as the critical scores and probabilities for both one- and two-tailed versions of the test. Compare "t stat" with the critical score (labeled ***t* Critical two-tail**) to make a decision regarding the null.

SPSS®

1) In SPSS, each row of the data file represents a participant. Since each participant is being measured twice, two columns are needed to hold the raw data. Within **Variable View**, label the two-column headings using terms that will distinguish between the two conditions (e.g. Alcohol/Placebo.). Then, go to

Alcohol	Placebo
8	4
12	8
6	2
4	6
11	8
15	9
8	5
7	4

t-Test: paired two sample for means

	Alcohol	*Placebo*
Mean	8.875	5.75
Variance	12.69 643	5.928 571
Observations	8	8
Pearson correlation	0.769 782	
Hypothesized mean difference	0	
df	7	
t Stat	3.85 104	
$P(T \leq t)$ one-tail	0.003 142	
t Critical one-tail	1.894 579	
$P(T \leq t)$ two-tail	0.006 284	
t Critical two-tail	2.364 624	

Figure C.9 A worked example of a dependent-samples *t* test using Microsoft Excel.

Data View. Input the sample data to the appropriate columns, being careful to keep the data from each participant within the same row; this will be essential for creating the proper difference score (see Figure C.10).

2) Click **Analyze**, then select **Compare Means**, and then **Paired-Samples *T* Test.**
3) Highlight one variable and move it to the **Variable1** box. Move the other variable to the **Variable2** box in the same manner. (Disregard the new row of

	Alcohol	Placebo
1	8	4
2	12	8
3	6	2
4	4	6
5	11	8
6	15	9
7	8	5
8	7	4

Figure C.10 An example of entered data for a dependent-samples *t* test in SPSS.

boxes that appear underneath. These are for running more than one dependent-samples *t* test at once.)

4) Click **OK**.

The output generates three boxes (see Figure C.11). The first identifies some descriptive statistics, including the sample means. The second runs a correlation analysis that can be disregarded. The third will identify, among other things, the mean difference (Mean), the estimate of the standard error (Std. Error Mean), the obtained *t* value (*t*), the degrees of freedom (*df*), and the probability of obtaining a *t* value of that size given a true null and a two-tailed test (Sig. [two-tailed]). It will not generate t_{crit}. Either use a table to find t_{crit}, or look at the given significance level to see if it is equal to or lower than .05.

***T*-test**

Paired samples statistics

		Mean	*N*	Std. deviation	Std. error mean
Pair 1	Alcohol	8.88	8	3.563	1.260
	Placebo	5.75	8	2.435	.861

Paired samples correlations

	N	Correlation	Sig.
Pair 1 Alcohol and Placebo	8	.770	.025

Paired samples test

	Paired differences						
				95% Confidence interval of the difference			
	Mean	Std. deviation	Std. error mean	Lower	Upper	*t*	*df*
Pair 1 Alcohol and Placebo	3.125	2.295	.811	1.206	5.044	3.851	7

Paired samples test

	Sig. (2-tailed)
Pair 1 Alcohol and Placebo	.006

Figure C.11 An example of a dependent-samples *t* test output in SPSS.

Chapter 11 One-Way ANOVA

Microsoft® Excel

1) Enter all the scores from the samples into adjacent columns (the number of columns equaling the number of conditions in the research design), with one sample in each column. Label the columns appropriately (see Figure C.12).
2) Excel has a built-in program for the one-way ANOVA test. To access it, click on the **Data** tab on the top menu and then click **Data Analysis**.
3) With the Data Analysis box open, select **Anova: Single Factor**.
4) Input the data by dragging over the entire dataset and placing those coordinates into the **Input Range** box. (If labels are included, make sure to click the **Labels** box.)
5) Decide on an Output option. The default is to place it on a separate worksheet.
6) Click **OK**.

The first output box (see Figure C.12) labeled "Summary" presents the count, sum of all values, means (average), and variance for all conditions. The second box is an ANOVA summary table (labeled "ANOVA"), similar to the one described in the chapter (an F_{crit} value is added).

Cond1	Cond2	Control
4	7	2
4	8	2
5	7	3
5	6	2
4	7	2
5	9	3

Anova: Single factor

Summary

Groups	*Count*	*Sum*	*Average*	*Variance*
Cond1	6	27	4.5	0.3
Cond2	6	44	7.333 333	1.066 667
Control	6	14	2.333 333	0.266 667

ANOVA

Source of variation	*SS*	*df*	*MS*	*F*	*P-value*	F_{crit}
Between groups	75.444 444	2	37.72 222	69.28 571	2.7E-08	3.68 232
Within groups	8.1 666 667	15	0.544 444			
Total	83.611111	17				

Figure C.12 A worked example of a one-way ANOVA using Microsoft Excel.

SPSS®

1) In SPSS, each row of the data file represents a participant. Since all samples in a one-way ANOVA test have different participants, the dependent variable data from all samples needs to be placed in one column. Within **Variable View**, label this variable appropriately. However, also create a second variable that will allow the user to identify which data goes with which group. A typical label for this variable might be "condition." Then, go to **Data View**. Input the sample data to the appropriate column and use a nominal variable in the "condition" column to distinguish between the samples (e.g. "1," "2," and "3") (see Figure C.13).
2) Click **Analyze** on the toolbar, select **Compare Means**, and then click **One-Way ANOVA**.
3) Move the dependent variable to the **Dependent List** box. Move the "condition" variable to the **Factor** box.
4) To make specific group comparisons, click on **Post Hoc** and make appropriate selections.
5) If descriptive statistics are wanted, click on **Options** and then **Descriptive**.
6) Click **OK**.

	Hoursslept	Condition
1	4	1
2	4	1
3	5	1
4	5	1
5	4	1
6	5	1
7	7	2
8	8	2
9	7	2
10	6	2
11	7	2
12	9	2
13	2	3
14	2	3
15	3	3
16	2	3
17	2	3
18	3	3

Figure C.13 An example of entered data for a one-way ANOVA in SPSS.

The output will generate an ANOVA summary table similar to the one described in the text (see Figure C.14). However, instead of a *p* column, SPSS offers a column labeled Sig. for significance. The meaning, however, is the same. If the Sig. value is .05 or less, reject the null.

One way

ANOVA

hoursslept

	Sum of squares	*df*	Mean square	*F*	Sig.
Between groups	75.444	2	37.722	69.286	.000
Within groups	8.167	15	.544		
Total	83.611	17			

Figure C.14 An example of a one-way ANOVA output in SPSS.

Chapter 12 Two-Way ANOVA

Microsoft® Excel

1) Choose one of the two factors and enter all the scores from the samples into adjacent columns (the number of columns equaling the number of conditions for that factor), with one sample in each column. Label the columns appropriately. Within the columns, organize the other factor so that all the data for the first level of this second factor comes first, then all the data for the second level, and so on. Leave the first column blank so it can be used to label the levels of the second factor (see Figure C.15).

	Behavioral	Group	Control
	4	8	2
	5	6	3
	6	4	4
Anxiety	7	7	1
	3	6	3
	4	7	4
	3	8	2
Depression	2	5	2

Anova: Two-factor with replication

Summary	Behavioral	Group	Control	Total
Count	4	4	4	12
Sum	22	25	10	57
Average	5.5	6.25	2.5	4.75
Variance	1.6666667	2.9166667	1.6666667	4.568182
Count	4	4	4	12
Sum	12	26	11	49
Average	3	6.5	2.75	4.083333
Variance	0.6666667	1.6666667	0.9166667	4.083333
Total				
Count	8	8	8	
Sum	34	51	21	
Average	4.25	6.375	2.625	
Variance	2.7857143	1.9821429	1.125	

ANOVA

Source of variation	*SS*	*df*	*MS*	*F*	*P-value*	F_{crit}
Sample	2.6666667	1	2.6666667	1.684211	0.210751124	4.413873
Columns	56.583333	2	28.291667	17.86842	5.30888E-05	3.554557
Interaction	10.083333	2	5.0416667	3.184211	0.065464527	3.554557
Within	28.5	18	1.5833333			
Total	97.833333	23				

Figure C.15 A worked example of a two-way ANOVA using Microsoft Excel.

2) Excel has a built-in program for the two-way ANOVA. To access it, click on the **Data** tab on the top menu and then click **Data Analysis**.
3) With the Data Analysis box open, select **Anova: Two Factor with Replication**. (Yes, the title is very confusing; this is not a repeated-measures ANOVA.)
4) Input the data range by dragging over the entire dataset, including the labels, and placing those coordinates into the **Input Range** box.
5) Determine how many rows there are per condition of the second factor. Excel requires an equal number of rows for each condition of both factors. It also does not allow for empty cells. Input this value into the **Rows per sample** box.
6) Decide on an alpha value. The default is .05.
7) Decide on an Output option. The default is to place it on a separate worksheet.
8) Click **OK**.

Several output tables are produced (see Figure C.15). The first contains summary data involving the counts, sums, averages, and variances for all cell and marginal means. The last is the ANOVA summary table (labeled "ANOVA"); it looks very similar to the ANOVA summary table described earlier in the chapter. However, it uses the term "sample" for the row factor and "column" for the column factor. Also, an F_{crit} value is included for each F in the design.

SPSS®

1) In SPSS, each row of the data file represents a participant. Since all samples in a two-way ANOVA test have different participants, all the dependent variable data from all samples will need to be placed in one column. Within **Variable View**, label this variable appropriately. However, also create a second and third variable that will allow the user to identify which data goes with which condition of each of the two factors. Label these factor columns appropriately. Then, go to **Data View**. Input the sample data to the appropriate column and use nominal variables in the factor columns to distinguish between the cells of the factorial design. For example, data in the first level of both factors would get a [1, 1] in the factor columns that follow; data in the second level of the first factor and the first level of the second factor would get a [2, 1] in associated factor columns; and so on (see Figure C.16).
2) Click **Analyze** on the toolbar, select **General Linear Model**, and then click **Univariate**.
3) Move the dependent variable to the **Dependent Variable** box. Move both factor variables into the **Fixed Factor(s)** box.

	Incidents	disordertype	treatmenttype
1	4	1	1
2	5	1	1
3	6	1	1
4	7	1	1
5	3	2	1
6	4	2	1
7	3	2	1
8	2	2	1
9	8	1	2
10	6	1	2
11	4	1	2
12	7	1	2
13	6	2	2
14	7	2	2
15	8	2	2
16	5	2	2
17	2	1	3
18	3	1	3
19	4	1	3
20	1	1	3
21	3	2	3
22	4	2	3
23	2	2	3
24	2	2	3

Figure C.16 An example of entered data for a two-way ANOVA in SPSS.

4) If specific group comparisons are desired, click on the **Post Hoc** tab, move the factor variables into the **Post Hoc Tests for** box, and make the appropriate selections.
5) If descriptive statistics are needed, click on **Options** and then **Descriptive**.
6) Click **OK**.

The output will generate an expanded ANOVA summary table compared with the one in the text (see Figure C.17). It is labeled **Tests of Between-Subjects Effects**. The first row (*Corrected Model*) displays the between-groups variance as if it was a one-way ANOVA. Simply drop this line from the ANOVA summary

Univariate analysis of variance

Between-subjects factors

		N
Disorder type	1	12
	2	12
Treatment type	1	8
	2	8
	3	8

Tests of between-subjects effects

Dependent variable: incidents

Source	Type III sum of squares	*df*	Mean square	*F*	Sig.
Corrected model	69.333[a]	5	13.867	8.758	.000
Intercept	468.167	1	468.167	295.684	.000
Disorder type	2.667	1	2.667	1.684	.211
Treatment type	56.583	2	28.292	17.868	.000
Disorder type * Treatment type	10.083	2	5.042	3.184	.065
Error	28.500	18	1.583		
Total	566.000	24			
Corrected total	97.833	23			

[a] *R* squared = .709 (Adjusted *R* squared = .628)

Figure C.17 An example of a two-way ANOVA output in SPSS.

table. The second line (*intercept*) can also be dropped. The following three rows reflect the three effects of interest, the two main effects, and the interaction. Once again, look to see if the calculated *F*'s falls in the outermost 5% of the null *F* distribution. If the value found under **Sig.** is .05 or less, reject the null. Please note: Use the "Corrected Total" as the total.

Chapter 13 Repeated-Measures ANOVA

Microsoft® Excel

1) Enter all the scores into adjacent columns (one column per condition), with each row assigned to a participant. For example, participant 1 will have all their data entered on the same row, then participant 2's data, and so on. Label the columns appropriately (see Figure C.18).

Participant	Therapy A	Therapy B	Therapy C
1	12	15	16
2	14	17	18
3	14	14	16
4	12	14	16
5	8	14	20
6	9	16	17
7	13	12	21
8	12	11	15

Figure C.18 An example of entered data for a repeated-measures ANOVA in Microsoft Excel.

2) Excel has a built-in program for a repeated-measures ANOVA. To access it, click on the **Data** tab on the top menu and then click **Data Analysis.**
3) With the Data Analysis box open, select **Anova: Two-Factor Without Replication**. (*Yes, this is a confusing title for a repeated-measures ANOVA.*)
4) Input the data by dragging over the entire dataset and placing those coordinates into the **Input Range** box. (If labels are included, make sure to click the **Labels** box.)
5) Decide on an Output option. The default is to place it on a separate worksheet.
6) Click **OK.**

The first output box (see Figure C.19) provides summary data: the count, sum of all values, means (average), and variance for all conditions. The second is an ANOVA summary table (labeled "ANOVA"), similar to the one found in this text. The *F* of importance (between groups) is associated with the "Columns" row. The row labeled "Rows" is the "Between-participants" row. (Ignore the *F* on this row.) The ANOVA table presented in this chapter can be fully constructed once the Within-groups line is determined by adding the data from the "Rows" row and the "Error" row. Also, note the addition of an F_{crit} value.

SPSS®

1) In SPSS, each row of the data file represents a participant. Since all samples in a repeated-measures ANOVA have the same participants, multiple columns will be needed to house the data. Within **Variable View**, create a series of variables corresponding to the various conditions in the study. Then, go to **Data View** and input the data. Be careful to keep data from each participant within a given row (see Figure C.20).
2) Click **Analyze** on the toolbar, select **General Linear Model**, and then **Repeated Measures.**
3) In the **Within-Subject Factor Name** box, assign a label for the independent variable. In the **Number of Levels** box, type the number of conditions present and click **Add**.
4) Once this is done, the **Define** box at the bottom will become active. Click it.

Anova: two-factor without replication

Summary	*Count*	*Sum*	*Average*	*Variance*
1	3	43	14.33333	4.333333
2	3	49	16.33333	4.333333
3	3	44	14.66667	1.333333
4	3	42	14	4
5	3	42	14	36
6	3	42	14	19
7	3	46	15.33333	24.33333
8	3	38	12.66667	4.333333
Therapy A	8	94	11.75	4.785714
Therapy B	8	113	14.125	3.839286
Therapy C	8	139	17.375	4.553571

ANOVA

Source of variation	*SS*	*df*	*MS*	*F*	*P-value*	F_{crit}
Rows	24.5	7	3.5	0.723247	0.655321	2.764199
Columns	127.5833	2	63.79167	13.18204	0.000604	3.738892
Error	67.75	14	4.839286			
Total	219.8333	23				

Figure C.19 An example of a repeated-measures ANOVA output in Microsoft Excel.

	Therapy A	Therapy B	Therapy C
1	12	15	16
2	14	17	18
3	14	14	16
4	12	14	16
5	8	14	20
6	9	16	17
7	13	12	21
8	12	11	15

Figure C.20 An example of entered data for a repeated-measures ANOVA in SPSS.

5) Move the names of the conditions to be analyzed into the **Within-Subjects Variable** box.
6) If post hoc (follow-up) tests are needed, select the **Options** box. Use the arrow button to move the independent variable into the **Display Means for** box. Then click **Compare Main Effects** and select the test of choice. (Tukey's *HSD* is not available.) Then click **Continue**.
7) For descriptive statistics and/or effect size, click on **Options** and then **Descriptive** and/or **Estimates of effect size**, as the case may be, and then **Continue**.

The output will generate multiple tables (see Figure C.21). A repeated-measures ANOVA summary table like the one presented in the text can be constructed from

General linear model

Tests of within-subjects effects

Measure: MEASURE_1

Source		Type III sum of squares	*df*	Mean square	*F*	Sig.
factor1	Sphericity assumed	127.583	2	63.792	13.182	.001
	Greenhouse-Geisser	127.583	1.935	65.924	13.182	.001
	Huynh-Feldt	127.583	2.000	63.792	13.182	.001
	Lower bound	127.583	1.000	127.583	13.182	.008
Error(factor1)	Sphericity assumed	67.750	14	4.839		
	Greenhouse-Geisser	67.750	13.547	5.001		
	Huynh-Feldt	67.750	14.000	4.839		
	Lower bound	67.750	7.000	9.679		

Tests of between-subjects effects

Measure: MEASURE_1

Transformed variable: Average

Source	Type III sum of squares	*df*	Mean square	*F*	Sig.
Intercept	4988.167	1	4988.167	1425.190	.000
Error	24.500	7	3.500		

Figure C.21 An example of a repeated-measures ANOVA output in SPSS.

the available information. Find the **Tests of Within-Subjects Effects** box. In the "Sphericity assumed" rows are the Sum-of-Squares, *df*, Mean Square, *F*, and probability of *F* for both the Between-groups row (factor 1) and the Error row. (The other rows reflect more sophisticated analyses which are beyond the scope of this text.) To complete the ANOVA summary table as presented in the chapter, the Between-participants line needs to be found. Go to the **Tests of Between-Subjects Effects** box (the very last box unless post hoc tests are run). The row labeled "Error" is the "Between-participants" row. Now with both the Error row (from the **Tests of Within-Subjects Effects** box) and Between-participants row (from the **Tests of Between-Subjects Effects** box), the table can be fully constructed. Recall SS_W is found by adding SS_{BP} and SS_{error}; df_W is found by adding df_{BP} and df_{error}; SS_T is found by adding SS_{BG} and SS_W; and df_T is found by adding df_{BG} and df_W.

Chapter 14 Pearson *r*

Microsoft® Excel

1) Enter the bivariate data into two adjacent columns, being sure to keep the data from each participant together in the same row. Label the columns appropriately (see Figure C.22).
2) Excel has a built-in program for a Pearson *r*. To access it, click on the **Data** tab on the top menu and then click **Data Analysis.**
3) With the Data Analysis box open, select **Correlation.**
4) Input the data by dragging over the entire dataset and placing those coordinates into the **Input Range** box. (If labels are included, make sure to click the **Labels** box.)
5) Decide on an Output option. The default is to place it on a separate worksheet.
6) Click **OK.**

A correlation grid is produced (see Figure C.22). Each variable is listed on both a column and a row. (Excel can run multiple correlations at once.) The correlation

Figure C.22 A worked example of a Pearson *r* using Microsoft Excel.

Shoe size	Height
72	10
66	9
74	13
68	10
63	7
70	10
73	12
67	9

	Shoe size	*Height*
Shoe size	1	
Height	0.93489	1

between the X and Y variables is produced. Excel does not test the null that $\rho = 0$. Use a critical value table (e.g. Table A.7 in Appendix A) to find r_{crit}.

SPSS®

1) In SPSS, each row of the data file represents a participant. Since bivariate data are used in calculating a Pearson r, create a series of variables within **Variable View** corresponding to the variables measured. Then, go to **Data View** and input the data, being careful to keep the values from each participant within a given row (see Figure C.23).
2) Click **Analyze** on the toolbar, select **Correlate**, and then click **Bivariate.**
3) Use the arrow key to move the variables of interest into the **Variables** box.
4) The default correlation coefficient calculated is the Pearson; leave this box checked. Make a selection regarding the critical r value to be calculated as well as a one or two-tailed test.
5) If descriptive statistics are needed, click the **Means and standard deviations** option.
6) Click **Ok**.

	Shoe size	Height
1	72	10
2	66	9
3	74	13
4	68	10
5	63	7
6	70	10
7	73	12
8	67	9

Figure C.23 An example of entered data for a Pearson r in SPSS.

If descriptives were asked for, the first box will present the means, standard deviations, and sample size of all selected variables. The next box is the correlation grid simply labeled **Correlations** (see Figure C.24). Each variable is listed on the left-side column and across the top of the grid. (SPSS can run multiple correlations at once.) The correlation between the X and Y variables is produced. The table is redundant showing each correlation twice. Within each correlation box can also be found the probability of getting a Pearson r of that size if $\rho = 0$ [**Sig. (2-tailed)**] as well as the number of paired scores (**N**). If the significance value is equal to or less than .05, there is statistical evidence to reject the null.

Correlations

Correlations

		Shoe size	Height
Shoe size	Pearson correlation	1	.935**
	Sig. (2-tailed)		.001
	N	8	8
Height	Pearson correlation	.935**	1
	Sig. (2-tailed)	.001	
	N	8	8

** Correlation is significant at the 0.01 level (2-tailed).

Figure C.24 An example of a Pearson *r* output in SPSS.

Chapter 15 Linear Regression

Microsoft® Excel

1) Enter the bivariate data into two adjacent columns, being sure to keep the data from each participant together in the same row. Label the columns appropriately (see Figure C.25).
2) Excel has a built-in program for linear regression. To access it, click on the **Data** tab on the top menu and then click **Data Analysis**.

Shoe size	Height
72	10
66	9
74	13
60	10
63	7
70	10
73	12
67	9

Summary output

Regression statistics	
Multiple *R*	0.9348898
R Square	0.874019
Adjusted *R* square	0.8530222
Standard Error	0.709876
Observations	8

ANOVA

	df	*SS*	*MS*	*F*	*Significance F*
Regression	1	20.97645601	20.97646	41.62623	0.000656801
Residual	6	3.02354399	0.503924		
Total	7	24			

	Coefficients	*Standard error*	*t Stat*	*P-value*
Intercept	−21.521685	4.892131999	−4.39924	.004572
X variable 1	0.4560099	0.070679057	6.451839	.000657

Figure C.25 A worked example of a regression analysis using Microsoft Excel.

3) With the Data Analysis box open, select **Regression**.
4) Input the data for each variable by dragging over the dataset and placing each variable into the corresponding **Input Range** box. Make sure the predicted variable (Y) is placed in the **Input Y Range** box and the predictor variable (X) is placed in the **Input X Range** box. Accidentally switching these around results in a faulty slope value. (If the labels are included in the data range, make sure to click the **Labels** box.) Leave the **Constant to Zero** box unchecked and leave the **Confidence Level** box at the 95% default value.
5) Decide on an **Output** option. Leave the other boxes unchecked.
6) Click **OK**. (Increase column width as necessary so the longer output labels can be read.)

Three tables are produced (see Figure C.25). **Multiple R** is the correlation between the two variables (Pearson's r). **R Square** is the coefficient of determination (r^2). **Standard Error** is the standard error of the estimate. [The **ANOVA** analysis in the middle box tests the null that $b = 0$. If the regression **F** is not large and the probability associated with it (**Significance F**) is larger than .05, then the null cannot be rejected, and the regression analysis has dubious meaning. This is akin to testing the null that $\rho = 0$ in Chapter 14 and failing to reject.] The coefficient associated with the X variable (**X Variable 1**) in the last box is the slope. Excel's regression analysis does not produce either M_X or M_Y. Use the **Descriptive Statistics** function found in the **Data Analysis** toolbox to generate those values. Once the slope, M_X, and M_Y are found, a regression line equation can be constructed, and a Y_p for any given value of X can be calculated.

SPSS®

1) In SPSS, each row of the data file represents a participant. Since bivariate data are used in calculating a regression equation, create a series of variables within **Variable View** corresponding to the variables measured. Then, go to **Data View** and input the data, being careful to keep data from each participant within a given row (see Figure C.26).
2) Click **Analyze** on the toolbar, select **Regression**, and then click **Linear**.
3) Use the arrow key to move the variable to be predicted into the **Dependent** box and the predictor variable into the **Independent(s)** box. Be careful not to get these two confused. Doing so would generate a wrong slope value.
4) Since M_X and M_Y are needed to create the regression equation, click the **Statistics** box in the upper right-hand corner, and then select **Descriptives**. Click **Continue**.
5) Click **Ok** on the main menu and run the analysis.

The first box provides both M_X and M_Y. (Not pictured in Figure C.27.) The slope can be found in the **Coefficients** box at the bottom. The slope is the

Figure C.26 An example of entered data for regression analysis in SPSS.

	Shoe size	Height
1	72	10
2	66	9
3	74	13
4	68	10
5	63	7
6	70	10
7	73	12
8	67	9

Regression

Model summary

Model	*R*	*R* square	Adjusted *R* square	Std. error of the estimate
1	.935[a]	.874	.853	.710

[a] Predictors: (constant), shoe size

ANOVA[a]

Model		Sum of squares	*df*	Mean square	*F*	Sig.
1	Regression	20.976	1	20.976	41.626	.001[b]
	Residual	3.024	6	.504		
	Total	24.000	7			

[a] Dependent variable: height

[b] Predictors: (constant), shoe size

Coefficients[a]

Model		Unstandardized coefficients		Standardized coefficients		
		B	Std. error	Beta	*t*	Sig.
1	(Constant)	–21.522	4.892		–4.399	.005
	Shoe size	.456	.071	.935	6.452	.001

[a] Dependent variable: height

Figure C.27 An example of a regression analysis output in SPSS.

unstandardized coefficients B value associated with the predictor variable. Other helpful information presented includes an ANOVA test of the null hypothesis $b = 0$. In the **ANOVA** box, if the probability (**Sig.**) associated with the **F** value is larger than .05, then the null cannot be rejected, and the analysis has dubious meaning. (This is akin to testing the null that $\rho = 0$ in Chapter 14 and failing to reject.) In the **Model Summary** box, **R** is the Pearson r, **R Square** is the coefficient of determination (r^2), and **Std. Error of the Estimate** is the standard error of the estimate (see Figure C.27).

Chapter 16 Chi-Square

Microsoft® Excel

There is no specific Excel function for the chi-square goodness-of-fit test. However, Excel, just like a calculator, can be used to generate the necessary values to perform the test.

For Excel to run a chi-square test for independence, pivot tables must be created.

1) Enter the bivariate data into two adjacent columns, being sure to keep the data from each participant in the same row. Categorical data need not be numerical (see Figure C.28).

Biological sex	Major
male	arts
male	sciences
female	humanities
male	sciences
female	arts
male	sciences
female	humanities
female	humanities
female	arts
male	humanities
female	sciences
female	arts
male	sciences
female	humanities
female	humanities
male	sciences
female	arts
female	sciences
female	humanities
male	sciences
male	humanities
female	humanities

Count of biological sex	Column labels		
Row labels	**Female**	**Male**	**Grand total**
Arts	4	1	5
Humanities	7	2	9
Sciences	2	6	8
Grand total	**13**	**9**	**22**

Figure C.28 An example of a chi-square test for independence analysis using Microsoft Excel.

2) Click on any cell that has data, then click the **Insert** tab, and select **Pivot Table**. The entire data range should become activated. Select a location for the Pivot Table and click **OK**.
3) On the right-hand side of the monitor is a Pivot Table Field List displaying the two variables. Drag one variable in the **Column Labels** box and the other in the **Row Labels** box. Either variable can be placed in the Σ **Values** box. The resulting values in the table are the observed frequencies that can be used to calculate the chi-square statistic (see Figure C.28).
4) Now, the arithmetic features of Excel can be used to calculate f_e's as well as the χ^2_{obt}.

SPSS® Goodness-of-fit Test

SPSS will run a goodness-of-fit chi-square, but the frequency counts for each category need to be calculated ahead of time. Then, create two variables within **Variable View**, one labeled "Category" and the other labeled "Frequency." Use a nominal scale to identify the categories (1, 2, 3, ...) and label them using **Values** in the **Variable View**.

1) Input the frequency count data (see Figure C.29).
2) Go to **Data** and select **Weight Cases**. Move the "Frequency" variable into the **Frequency Variable** box. This will properly weight each category.
3) Click **Analyze,** then **Nonparametric Tests**, then **Legacy Dialogues**, then **Chi-square**.
4) Move "Category" into the **Test Variable List** box.
5) Leave the default value under **Expected Values** if the null hypothesis predicts each category to be equal. (If the null hypothesis is more complex, click **Values** and put in, in order of the category number, the percent expected for each category numerical label.)
6) Click **OK**.

The first output box will present the observed frequency (**Observed N**), expected frequency (**Expected N**), and a measure of the difference between them

Figure C.29 An example of entered data for a chi-square goodness-of-fit test in SPSS.

	Category	Frequency
1	1	25
2	2	39
3	3	44

Chi-square test

Frequencies

Category

	Observed *N*	Expected *N*	Residual
Arts	25	36.0	−11.0
Sciences	39	36.0	3.0
Humanities	44	36.0	8.0
Total	108		

Test statistics

	Category
Chi-square	5.389[a]
df	2
Asymp. sig.	.068

[a] 0 cells (.0%) have expected frequencies less than 5. The minimum expected cell frequency is 36.0.

Figure C.30 An example of a chi-square goodness-of-fit test output in SPSS.

(**Residual**) for each category. The second will present the overall chi-square statistic (**Chi-Square**), the degrees of freedom (**df**), and the probability of getting a chi-square of that value if the null is true (**Asymp. Sig.**). There is statistical evidence to reject the null if this number is ≤.05 (see Figure C.30).

SPSS® Test for Independence

SPSS will run a test for independence. First, create the two categorical variables in **Variable View**. Label each numerical value used under the **Values** tab while in **Variable View**.

1) Input the data properly being sure to remember that each row represents a case; typically, that is a participant (see Figure C.31).
2) Click **Analyze** on the toolbar, select **Descriptives Statistics**, and then select **Crosstabs**.
3) Select one of the two categorical variables and place it in the **Row(s)** box. Select the other categorical variable and place it in the **Column(s)** box. Then click **OK**.
4) The second table (**Crosstabulation**) presents the observed frequencies for each combination of categories between the two variables.
5) Go back to **Analyze**, then **Descriptive Statistics**, and then **Crosstabs**. Then click on **Statistics**. Click **Chi-square**. (An effect size measure like Cramér's *V*, can also be generated.)
6) Click **Continue** to leave the **Cells** menu and then click **OK** to run the analysis.

Figure C.31 An example of entered data for a chi-square test for independence in SPSS.

	Biological_Sex	Major
1	1	1
2	1	2
3	2	3
4	1	2
5	2	1
6	1	2
7	2	3
8	2	3
9	2	1
10	1	3
11	2	2
12	2	1
13	1	2
14	2	3
15	2	3
16	1	2
17	2	1
18	2	2
19	2	3
20	1	2
21	1	3
22	2	3

The third table of the output is entitled **Chi-Square Tests** (see Figure C.32). The first line labeled "Pearson Chi-Square" is the test statistic. Along with that value, find the *df* value (**df**) and the probability of getting a chi-square value of that size if the null is true (**Asym. Sig. (2-sided)**). If the significance value is equal to or less than .05, there is statistical evidence to reject the null (assuming the alpha value is 5%). If a residual analysis is needed, the various residual options can be found under the **Cells** tab on the **Crosstabs** menu.

Chi-square test

Crosstabs

Chi-square tests

	Value	*df*	Asymp. sig. (2-sided)
Pearson chi-square	6.051[a]	2	.049
Likelihood ratio	6.231	2	.044
Linear-by-linear Association	.120	1	.729
N of valid cases	22		

[a] 5 cells (83.3%) have expected count less than 5. The minimum expected count is 2.05.

Figure C.32 An example of a chi-square test for independence output in SPSS.

References

Abelson, R.P. (1995). *Statistics as Principled Argument.* Hillsdale, NJ: Erlbaum.

Addison, W.E. (1989). Beardedness as a factor in perceived masculinity. *Perceptual and Motor Skills* 68: 921–922.

American Psychological Association (2020). *Publication Manual of the American Psychological Association*, 7e. Washington, DC: American Psychological Association.

Back, M.D., Schmukle, S.C., and Egloff, B. (2008). Becoming friends by chance. *Psychological Science* 19: 365–380.

Baddeley, A.D. and Longman, D.J.A. (1978). The influence of length and frequency of training sessions on the rate of learning to type. *Ergonomics* 21 (8): 627–635.

Bakan, D. (1966). The test of significance in psychological research. *Psychological Bulletin* 66 (6): 423–437.

Baker, M. (2015). Over half of psychology studies fail reproducibility test. *Nature News* https://www.nature.com/articles/nature.2015.18248.

Biaggio, M.K. (1989). Sex differences in behavioral reactions to provocation of anger. *Psychological Reports* 64: 23–26.

Box, J.F. (1978). *R. A. Fisher: The Life of a Scientist.* New York: Wiley.

Brady, J.V. (1958). Ulcers in executive monkeys. *Scientific American* 199: 95–100.

Byers, W. (2011). *The Blind Spot: Science and the Crisis of Uncertainty.* Princeton, NJ: Princeton University Press.

Campbell, D.T. and Stanley, J. (1963). Experimental and quasi-experimental designs for research on Teaching. In: *Handbook of Research on Teaching* (ed. N.L. Gage), 171–246. Chicago: Rand McNally.

Campbell, R.B. (2001). John Graunt, John Arbutnott, and the human sex ratio. *Human Biology* 73 (4): 605–610.

Carey, K. (2016). A peek inside the strange world of fake academia. *The New York Times, A3.* http://www.nytimes.com/2016/12/29/upshot/fake-academe-looking-much-like-the-real-thing.html?_r=1.

Carson, R. (1962). *Silent Spring.* Boston: Houghton Mifflin.

Essentials of Behavioral and Social Science Statistics, Third Edition.
K. Paul Nesselroade, Jr. and Laurence G. Grimm.
© 2025 John Wiley & Sons, Inc. Published 2025 by John Wiley & Sons, Inc.
Companion website: https://www.wiley.com/go/Essentials_of_Behavioral_and_Social_Science_Statistics3e

Cimpean, A. and David, D. (2019). The mechanisms of pain tolerance and pain-related anxiety in acute pain. *Health Psychology Open* 6 (2): https://www.ncbi.nlm.nih.gov/pmc/articles/PMC6886278/.

Cochran, W.G. (1976). Early development of techniques in comparative experimentation. In: *On the History of Statistics and Probability* (ed. D.B. Owen). New York: Marcel Dekker.

Cohen, B. (2013). *Explaining Psychological Statistics*, 4e. Hoboken, NJ: Wiley.

Cohen, J. (1962). The statistical power of abnormal psychological research: a review. *Journal of Abnormal and Social Psychology* 65: 145–153.

Cohen, J. (1977). *Statistical Power Analysis for the Behavioral Sciences* (Rev. ed.). New York: Academic Press.

Cohen, J. (1988). *Statistical Power Analysis for the Behavioral Sciences*, 2e. Hillsdale, NJ: Lawrence Erlbaum.

Cohen, J. (1990). Things I have learned (so far). *American Psychologist* 45 (12): 1304–1312.

Combs, D.J.Y., Powell, C.A.J., Schurtz, D.R., and Smith, R.H. (2009). Politics, schadenfreude, and ingroup identification: the sometimes happy thing about a poor economy and death. *Journal of Experimental Social Psychology* 45: 635–646.

Cook, T.D. and Campbell, D.T. (1979). *Quasi-Experimentation: Design and Analysis Issues for FieldSsettings*. Chicago: Rand McNally.

Corder, G.W. and Foreman, D.I. (2014). *Nonparametric Statistics: A Step-by-Step Approach*, 2e. Hoboken, NJ: Wiley.

Cowles, M. and Davis, C. (1982). On the origins of the .05 level of statistical significance. *American Psychologist* 37 (5): 553–558.

Croasmun, J.T. and Ostrom, L. (2011). Using Likert-type scales in the social sciences. *Journal of Adult Education* 40 (1): 19–22.

Darley, J.M. and Batson, C.D. (1973). From Jerusalem to Jericho: a study of situational and dispositional variables in helping behavior. *Journal of Personality and Social Psychology* 27: 100–108.

De Moivre, A. (1738/1959). On the law of normal probability. In: *A Source Book in Mathematics*, vol. 2 (ed. D.E. Smith), 566–575. New York: Dover.

Delzell, D.A.P. and Poliak, C.D. (2013). Karl Pearson and eugenics: personal opinions and scientific rigor. *Science and Engineering Ethics* 19: 1057–1070.

Devlin, H. (2013). Academic benefits of music 'a myth'. *The Times of London.* http://www.thetimes.co.uk/tto/science/article3946476.ece.

Dunnington, G.W. (1955). *Carl Friedrich Gauss: Titan of Science.* New York: The Free Press.

Ebbinghaus, H. (1885). *Memory: A Contribution to Experimental Psychology* (ed. H.A. Ruger and C.E. Bussenius), Trans. New York, NY: Columbia University.

Fisher, R.A. (1939). Student. *Annals of Eugenics* 9: 1–9.

Fisher, R.A. and MacKenzie, W.A. (1923). Studies in crop variation. II: the manurial response of different potato varieties. *Journal of Agricultural Science* 13: 311–320.

Freedman, D.H. (2010). Lies, damned lies, and medical science. *The Atlantic* 306 (11): https://www.theatlantic.com/magazine/archive/2010/11/lies-damned-lies-and-medical-science/308269/?single_page=true.

Galton, F. (1869). *Hereditary Genius: Its Laws and Consequences.* New York: Appleton.

Galton, F. (1885). The measure of fidget. *Nature* 32: 174–175.

Gerson, J., Plagnol, A., and Corr, P.J. (2016). Subjective well-being and social media use: do personality traits moderate the impact of social comparison on Facebook? *Computers in Human Behavior* 63: 813–822.

Gliozzi, M. (2008). Cardono, Girolamo. *Complete dictionary of scientific biography.* Retrieved from http://www.encyclopedia.com/people/philosophy-and-religion/other-religious-beliefs-biographies/girolamo-cardano.

Gosset, W.S. (1970). *Letters from W S. Gosset to R. A. Fisher, 1915–1936.* Issued for private circulation. Dublin: Arthur Guinness.

Grace-Martin, K. (2008). Can Likert Scale data ever be continuous? [The Analysis Factor]. Retrieved from: http://www.theanalysisfactor.com/can-likert-scale-data-ever-be-continuous/

Grossarth-Maticek, R., Eysenck, H.J., and Vetter, H. (1988). Antismoking attitudes and general prejudice: An empirical study. *Perceptual and Motor Skills* 66: 927–931.

Grosskurth, P. (1980). *Havelock Ellis: A biography.* New York: Knopf. Haberman, S. J. (1973). The analysis of residuals in cross-classified tables. *Biometrics* 29: 205–220.

Haelle, T. (2021). Dozens of major cancer studies can't be replicated. *Science News.* https://www.sciencenews.org/article/cancer-biology-studies-research-replication-reproducibility.

Hald, A. (2003). *A History of Probability and Statistics and Their Applications Before 1750.* Hoboken, NJ: Wiley.

Hall, J.W. (1972). A comparison of Halpin and Croft's organizational climates and Likert and Likert's organizational systems. *Administrative Science Quarterly* 17: 586–590.

Ioannidis, J.P.A. (2005). Why most published research findings are false. *PLoS Medicine* 2 (8): e124. https://doi.org/10.1371/journal.pmed.0020124.

Jackson, E.M. and Howton, A. (2008). Increasing walking college students using a pedometer intervention: differences according to body mass index. *Journal of American College Health* 57 (2): 159–164.

Jackson, J.P. and Weidman, N.M. (2006). *Race, Racism, and Science: Social Impact and Interaction.* New Brunswick, NJ: Rutgers University Press.

Jastrow, R. (1992). *God as the Astronomers.* New York: W. W. Norton and Company, Inc.

Jones, I. (2015). *Research Methods for Sports Studies*, 3e. New York: Routledge.

Jourbert, C.E. (1989). Birth order and narcissism. *Psychological Reports* 64: 721–722.

Katz, V.J. (1993). *A History of Mathematics*. HarperCollins College Publishers.

Kaufman, A.S. (1979). *Intelligent Testing with the WISC-R*. New York: Wiley.

Keppel, G. and Wickens, T.D. (2004). *Design and Analysis: A Researcher's Handbook*, 4e. Upper Saddle River, NJ: Prentice Hall.

Kirschner, P.A. and Karpinski, A.C. (2010). Facebook and academic performance. *Computers in Human Behavior* 26: 1237–1245.

Kish, L. (1990). A CHOICES profile: Rensis Likert: social scientist and entrepreneur. *Choices* 5 (4): 36–38.

Kraus, N., Hornickel, J., Strait, D. et al. (2014). Engagement in community music classes sparks neuroplasticity and language development in children from disadvantaged backgrounds. *Frontiers in Psychology* Retrieved from journal. frontiersin.org. https://doi.org/10.3389/fpsyg.2014.01403/full.

Krauss, I.K. (1980). Between- and within-group comparisons in aging research. In: *Aging in the 1980's Psychological Issues* (ed. L.W. Poon), 542–551. Washington, DC: American Psychological Association.

Kühl, S. (1994). *The Nazi Connection: Eugenics, American Racism, and German National Socialism*. New York: Oxford University Press.

Landauer, T.K. and Whiting, J.W.M. (1964). Infantile stimulation and adult stature of human males. *American Anthropologist* 66: 1007–1028.

Lashley, K.S. (1915). The acquisition of skill in archery. *Carnegie Institutions Publications* 211: 107–128.

Latané, B. and Darley, J. (1970). *The Unresponsive Bystander: Why Doesn't He Help*. New York: Appleton-Century-Crofts.

Leek, J.T. and Jager, L.R. (2017). Is most published research really false? *Annual Review of Statistics and its Applications* 4: 109–122.

Lehman, M., Moelats, D., Varewyck, M. et al. (2013). Activating and relaxing music entrains the speed of beat synchronized walking. *PLoS One* 7: e67932. https://doi.org/10.1371/journal.pone.0067932.

Leonhardt, D. (2000). John Tukey, 85, Statistician; Coined the Word 'Software'. http://www.nytimes.com/2000/07/28/us/john-tukey-85-statistician-coined-the-word-software.html.

Lifton, R.J. (2000). *The Nazi Doctors: Medical Killing and the Psychology of Genocide*. New York: Basic Books.

Lombardi, C.M. and Hurlbert, S.H. (2009). Misprescription and misuse of one-tailed tests. *Austral Ecology* 34: 447–468.

Loschelder, D.D., Swaab, R.I., Trötschel, R., and Galinsky, A.D. (2014). The first-move disadvantage. *Psychological Science* 25: http://journals.sagepub.com/doi/pdf/10.1177/ 0956797613520168.

Makel, M.C., Plucker, J.A., and Hegarty, B. (2012). Replications in psychology research: how often do they really occur? *Perspectives on Psychological Science* 7 (6): 537–542.

Mammarella, N., Russo, R., and Avons, S.E. (2002). Spacing effects in cued-memory tasks for unfamiliar faces and nonwords. *Memory & Cognition* 30: 1238–1251.

Mandelbrot, B. (1967). How long is the coast of Britain? Statistical self-similarity and fractional dimension. *Science* 156 (3775): 636–638.

Marczyk, G., DeMatteo, D., and Festinger, D. (2005). *Essentials of Research Design and Methodology*. Hoboken, NJ: Wiley.

McCullagh, P. (2003). John Wilder Tukey. *Biological Memoirs of Fellows of the Royal Society* 49: 537–555.

McGrath, A. (2015). *The Big question: Why we can't Stop Talking About Science, Faith, and God*. New York: St. Martin's Press.

Mehr, S.A., Schachner, A., Katz, R.C., and Spelke, E.S. (2013). Two randomized trials provide no consistent evidence for nonmusical cognitive benefits of brief preschool music enrichment. *PLoS One* 8: e82007. https://doi.org/10.1371/journal.pone.0082007.

Mirels, H.L. and Dean, J.B. (2006). Right-wing authoritarianism, attitude salience, and beliefs about matters of fact. *Political Psychology* 27 (6): 839–866.

Moore, D.A. and Tenney, E.R. (2012). Time pressure, performance, and productivity. In: *Looking Back, Moving Forward: A Review of Group and Team-Based Research (Research on Managing Groups and Teams, Volume 15)* (ed. M.A. Neale and E.A. Mannix), 305–326. Emerald Group Publishing Limited.

Morling, B. (2020). *Research methods in psychology: Evaluating a world of information*, 4e. New York, NY: WW Norton & Company.

Nichols, A.L. and Maner, J.K. (2008). The good subject effect: investigating participant demand characteristics. *Journal of General Psychology* 135: 151–165.

Oishi, S. and Schimmack, U. (2010). Residential mobility, well-being, and mortality. *Journal of Personality and Social Psychology* 98 (6): 980–994.

Pearson, K. (1894). Contributions to the mathematical theory of evolution. *Philosophical Transactions of the Royal Society of London* 185: 71–110.

Pearson, K. (1896). Mathematical contributions to the theory of evolution III. Regression, heredity and panmixia. *Philosophical Translations of the Royal Society of London, Series A* 187: 253–318.

Pearson, K. (1903). On the inheritance of the mental and moral characters in man, and its comparison with the inheritance of the physical characters. *Journal of the Anthropological Institute* 33: 179–237.

Pearson, K. (1906). Walter Frank Weldon. 1860 – 1906. *Biometrica* 5: 1–52.

Pfungst, O. (1911). *Clever Hans (The Horse of Mr. Von Osten): A Contribution to Experimental Animal and Human Psychology*. New York: Holt and Co.

Pham, M.T., Hung, I.W., and Gorn, G.J. (2011). Relaxation increases monetary valuations. *Journal of Marketing Research* 48 (5): 814–826.

Pittelkow, M., Field, S.M., Isager, P.M. et al. (2023). The process of replication target selection in psychology: what to consider? *The Royal Society* 10: https://doi.org/10.1098/rsos.210586.

Plato, Phaedrus, 265e.

Porter, R.W., Brady, J.V., Conrad, D. et al. (1958). Some experimental observations on gastrointestinal lesions in behaviorally conditioned monkeys. *Psychosomatic Medicine* 20: 379–394.

Post, D.G. and Eisen, M. (2000, January). How long is the coastline of Law? Thoughts on the fractal nature of legal systems. *Journal of Legal Studies* 29 (1): 545–584.

Reeves, J. (2015). The secularization of chance: toward understanding the impact of the probability revolution on Christian belief in divine providence. *Zygon* 50 (3): 604–620.

Robbins, R.A. (1988). *Objective and subjective factors in estimating life expectancy* (Unpublished manuscript). The Pennsylvania State University at Harrisburg, Division of Behavioral Science and Education.

Rohwedder, S. and Willis, R.J. (2010). Mental retirement. *Journal of Economic Perspectives* 24: 119–138. https://doi.org/10.1257/jep.24.1.119.

Rosenthal, R. and Fode, K. (1963). The effect of experimenter bias on performance of the albino rat. *Behavioral Science* 8: 183–189.

Rowatt, W.C., Powers, C., Targhetta, V. et al. (2006). Development and initial validation of an implicit measure of humility relative to arrogance. *The Journal of Positive Psychology* 1 (4): 198–211.

Rubin, M., Paolini, S., and Crisp, R. (2010). A processing fluency explanation of bias against migrants. *Journal of Experimental Social Psychology* 46 (1): 21–28.

Russell, B. (1936/1997). *Religion and Science*. Oxford: Oxford University Press.

Ryan, T.P. (2008). *Modern Regression Methods*, 2e. Hoboken, NJ: Wiley and Sons.

Sande, G. (2001). John Wilder Tukey. *Physics Today* 54 (7): 80. http://physicstoday.scitation.org/doi/10.1063/1.1397408?journalCode=pto.

Sedlmeier, P. and Gigerenzer, G. (1989). Do statistical studies of power have an effect on the power of studies. *Psychological Bulletin* 105 (2): 309–316.

Seery, M.D., Holman, E.A., and Silver, R.C. (2010). Whatever does not kill us: cumulative lifetime adversity, vulnerability, and resilience. *Journal of Personality and Social Psychology* 99: 1025–1041.

Semmel, B. (1958). Karl Pearson: socialist and darwinist. *British Journal of Sociology* 9 (2): 111–125.

Shadish, W.R., Cook, T.D., and Campbell, D.T. (2002). *Experimental and Quasi-Experimental Designs for Generalized Causal Inference*, 2ne. Boston: Houghton Mifflin Co.

Shavelson, R.J. (1996). *Statistical Reasoning for the Behavioral Sciences*, 3e. Boston: Allyn and Bacon.

Siegal, S. and Castellan, N.J. (1988). *Nonparametric Statistics for the Behavioral Sciences*, 2e. New York: McGraw Hill.

Simpson, T. (1755). A letter to the right honourable George Earl of Macclesfield, President of the Royal Society, on the advantage of taking the mean of a number of observations, in practical astronomy. *Philosophical Translations of the Royal Society of London* 46: 82–93.

Simpson-Housley, P. and DeMan, A. (1989). Flood experience and posttraumatic trait anxiety in Appalachia. *Psychological Reports* 64: 896–898.

Sines, J.O. (1959). Selective breeding for development of stomach lesions following stress in the rat. *Journal of Comparative and Physiological Psychology* 52: 615–617.

Sines, J.O., Cleeland, C., and Adkins, J. (1963). The behavior of normal and stomach lesion susceptible rats in several learning situations. *Journal of Genetic Psychology* 102: 91–94.

Snedecor, G.W. (1956). *Statistical methods applied to experiments in agriculture and biology*, 5e. The Iowa State College Press.

Student (1908). The probable error of a mean. *Biometrika* 6: 1–25.

Tankard, J.W. (1984). *The Statistical Pioneers*. Cambridge, MA: Schenkman.

Ting, C.H. and Chen, Y.Y. (2020). Smartphone addiction. In: *Adolescent Addiction: Epidemiology, Assessment, and Treatment* (ed. C.A. Essau and P.H. Delfabbro), 215–240. Elsevier Academic Press https://doi.org/10.1016/B978-0-12-818626-8.00008-6.

Tucker, J. (2016). Does social science have a replication crisis? *Washington Post*. https://www.washingtonpost.com/news/monkey-cage/wp/2016/03/09/does-social-science-have-a-replication-crisis/?utm_term=.1c5d71247aa0.

Turiel, E. (2020). Eugenics, prejudice, and psychological research. *Human Development* 64: 103–107.

UNESCO (1952). The race concept: results of an Inquiry, p. 27.

Walker, H.M. (1934). Abraham De Moivre. *Scripta Mathematica* 2: 316–333.

Walker, H.M. (1940). Degrees of freedom. *Journal of Educational Psychology* 55: 253–269.

Weisberg, H.I. (2014). *Willful Ignorance: The Mismeasure of Uncertainty*. Hoboken, NJ: Wiley.

Welkowitz, J., Ewen, R.B., and Cohen, J. (1988). *Introductory statistics for the behavioral sciences*. New York: Harcourt Brace Jovanovich.

Wilson, W. A. (2016). Scientific regress. *First Things*. Retrieved from https://www.firstthings.com/article/2016/05/scientific-regress.

Woods, A.M. and Rusin, M.J. (1988). Stress management and the elderly. In: *Stress Management for Chronic Disease* (ed. M.L. Russell), 49–62. New York: Pergamon.

Yarhouse, M.A., Dean, J.B., Stratton, S.P., and Lastoria, M. (2018). *Listening to Sexual Minorities: A Study of Faith and Sexual Identity on Christian College Campuses*. Downers Grove, Illinois: Intervarsity Press.

Zenderland, L. (1998). *Measuring Minds: Henry Herbert Goddard and the Origins of American Intelligence Testing*. Cambridge: Cambridge University Press.

Zullow, H.M. and Seligman, M.E.P. (1990). Pessimistic rumination predicts defeat of presidential candidates. *Psychological Inquiry* 1 (1): 52–61.

Glossary

68-95-99.7 rule: If a data-set is normally distributed, this rule roughly approximates score distribution. About 68% of scores fall within ±1 standard deviation of the mean; about 95% of scores fall within ±2 standard deviations of the mean; and about 99.7% of scores will fall within ±3 standard deviations of the mean.

A posteriori approach: Determining the probability of an event empirically by gathering data and then dividing the number of times *A* has occurred by the total number of data points gathered.

A priori approach: Logically determining the probability of an event by dividing the number of ways the event can occur by the total number of possible outcomes.

A priori tests: A category of tests, selected prior to rejecting a null hypothesis in an overall analysis, that are used to locate differences between pairs of means.

Abscissa: The horizontal (*X*) axis of a graph.

Addition rule: Used to determine the probability of occurrence of one or more of many possible events.

Alpha level: The value set by the researcher that specifies the probability of making type I error (rejecting a true null hypothesis).

Alternative hypothesis: The opposite of the null hypothesis. The hypothesis that is automatically accepted when the null hypothesis is rejected. Accepting the alternative hypothesis means that the results of a study are probably not due to chance; there is probably an effect, difference, or relationship between variables.

Analysis of variance (ANOVA): A statistical test designed to determine if there are population differences among several sample means.

Asymptotic: When a graph line approaches but never reaches the *X* axis.

Balancing: A means of controlling an extraneous variable by representing it equally across all conditions.

Bar graph: A frequency distribution for categorical (nominal) data.

Essentials of Behavioral and Social Science Statistics, Third Edition.
K. Paul Nesselroade, Jr. and Laurence G. Grimm.
© 2025 John Wiley & Sons, Inc. Published 2025 by John Wiley & Sons, Inc.
Companion website: https://www.wiley.com/go/Essentials_of_Behavioral_and_Social_Science_Statistics3e

Between-group variation: A measure of the variation among group means in a study with two or more groups.
Between-participants designs: A research design in which the scores of one condition are unrelated to (independent of) the scores in any other condition.
Bimodal: A distribution with two modes.
Bivariate distribution: A distribution of two variables in which scores are paired. A correlation is based on a bivariate distribution.
Categorical data: Nominal data arranged by categories. For example, marital status: single, married, divorced, widowed. See *Nominal scale.*
Cell residual: A ratio measure of observed to expected frequencies within a cell.
Cell: The section of a design identifying one group amidst others. For instance, a 2×2 factorial design has four cells: $\text{Cell}_{1,1}$; $\text{Cell}_{1,2}$; $\text{Cell}_{2,1}$; and $\text{Cell}_{2,2}$.
Central limit theorem: A mathematical claim that (i) a sampling distribution of means approaches a normal curve as the sample size increases, (ii) the mean of the sampling distribution is equal to the mean of the population of raw scores, and (iii) the standard deviation of the sampling distribution is equal to $\sigma/\sqrt{n}$.
Central tendency: A statistic that indicates wherein the distribution scores tend to bunch. The mean, median, and mode are common measures of central tendency.
Chi-square test: One of two significance tests (goodness of fit; test for independence) using frequency count data.
Class intervals: Groups of equal-sized ranges that create tabular simplicity.
Coefficient of determination: A squared correlation coefficient. It measures the amount of variation of *Y* scores accounted for by the variation of *X* scores. A measure of common or shared variance.
Cohen's *d*: A simple, direct, and often used measure of effect size.
Conditional distribution: In regression, the spread of *Y* scores for a given *X* score.
Conditional probability: An expression of likelihood given that another particular event has occurred.
Confidence interval: A range of values within which a researcher can state with a certain degree of confidence that a population parameter will fall.
Confounding variable: An uncontrolled extraneous variable that varies with an independent variable. It offers a plausible alternative explanation for the results of a study.
Contingency table: A table that categorizes observations as frequency counts along a two-factor grid.
Continuous variable: A variable that theoretically has an infinite number of points between any two numbers.
Control group: In experimental research, a condition marked by the absence of the independent variable.

Correlation (or prediction or association): A description of the degree to which two or more variables relate to one another. One of the goals of a researcher.

Correlation coefficient: A measure of the degree of association between two variables. It can range from −1 to +1. The higher the absolute value of the correlation coefficient, the stronger the relationship between the two variables.

Correlational designs: Studies that do not control and manipulate variables. Correlational research examines the covariation among variables.

Counterbalancing: A strategy used in repeated-measures designs to control order effects in which the order of conditions experienced by the participants differs.

Cramér's *V*: A measure of association for nominal variables. In this text, it is used as a measure of effect size for both chi-square tests.

Critical values: The values that separate the rejection regions from those regions of a null distribution that would not lead to a rejection of the null hypothesis.

Cumulative frequency distribution: A frequency distribution, which includes a column that indicates the total frequency of scores up to and including a given class interval.

Curvilinear relationship: A nonlinear relationship between two variables such that a line of best fit between them is curved (rather than straight).

Degrees of freedom: The number of values that are free to vary with certain restrictions placed on all values.

Dependent variable: A measured variable that is presumed to vary due to the independent variable. In correlational research, a *predicted variable* or *criterion variable* is often called a dependent variable, although causality is not implied.

Dependent-samples *t* test: A significance test used with dependent sampling in which scores across conditions are paired.

Description: Defining, identifying, classifying, categorizing, and organizing a topic of interest. This is the initial goal of the researcher.

Descriptive statistics: Statistical techniques designed to describe and summarize data in an abbreviated form.

Deviation score (or error score): The difference between a score and the mean of the distribution.

Dichotomous variable: A variable that takes only two contrary values (e.g. depressed = 0, not depressed = 1).

Discontinuous (or discrete) variable: A variable that can take on only a finite number of values and for which there are no meaningful intermediate values.

Error (or secondary) variance: A measure of the variation between group means or among scores within a group due to uncontrolled, random factors (including individual differences between participants) in the experiment or study.

Estimated standard error: The use of sample mean differences to estimate the standard deviation of the sampling distribution.

Eta-squared: A statistic that estimates the size of an effect, a ratio of primary variance over total variance.

Expected frequencies: The number of observations expected to occur when the null hypothesis is true.

Experiment: A reserved term for a particular research design involving (i) a high degree of control over the independent variable, (ii) careful measurement of the dependent variable, and (iii) complete control of all other variables.

Experimental error: A measure of variance due to unsystematic influences on individual scores. It includes variance due to unreliable measures, inconsistent administration, and disturbances in the research environment.

Experimental group: A group marked by the presence of an independent variable; distinguished from a "control group," which is marked by the absence of an independent variable.

External validity: The extent to which findings can be generalized to different populations, settings, or treatment or measurement variables.

Extraneous variable: Any variable found in an experimental context that is not either being manipulated or carefully measured.

***F* distribution:** A theoretical sampling distribution of *F* values; positively skewed and centered around 1. Specific shape depends on the degrees of freedom.

Factor: A design term indicating a variable with different levels or conditions. In an experimental context, it is an independent variable.

Factorial (or complex) design: A research design that has two or more factors (or independent variables, if used in an experimental context).

Fisher's *LSD*: (Least Significant Difference) A multiple comparison test used to locate the source of significance following a significant *F* test. Also called a *protected t test.*

Frequency polygon: A graph with measured scores on the *X* axis and frequencies on the *Y* axis. Each point above an *X* score represents the frequency of occurrence of that score.

Goodness-of-fit test: A chi-square test that examines the degree to which observed data from a set of categories arranged along a single factor coincides with theoretical or previously empirical-derived expectations.

Grouped frequency distribution: A table possessing equal-sized class intervals with an adjacent column noting the frequency of occurrence for values corresponding to each class interval.

Heteroscedasticity: In regression analysis, a violation of homoscedasticity wherein the variances of the conditional distributions are unequal.

Higher order effect: A relative judgment of different effect types. In an analysis involving multiple effect types, the effect involving the greatest number of factors or independent variables is the highest-order effect, and so on.

Histogram: A graph of vertical bars with shared borders in which the height of each bar corresponds to the frequency of scores for a given class interval.
Holding constant: Controlling extraneous variables by not letting them vary.
Homogeneity of variances: Variances of equivalent values in two or more populations.
Homoscedasticity: An assumption in regression analysis in which all conditional distributions have equal variances.
Hypothesis: A prediction, emerging from a theory, about what data are expected to be found given a particular situation.
Hypothesis testing: A method for testing claims made about population parameters. Hypothesis tests are also called *significance tests.*
Independent observations: The methodological assumption that each score within a *sample* is independent of all other scores.
Independent variable: The manipulated variable believed to be the "cause" part of a cause–effect relationship. In regression, the predictor variable.
Independent-samples design: See *Between-participants design.*
Independent-samples *t* test: An inferential test comparing the means of two samples.
Individual differences: A measure of variance that is due to the unique skills, abilities, and tendencies of individual participants.
Inferential statistics: Statistical techniques using sample data that allow researchers to make inferences about the characteristics of the population from which the sample came.
Interaction: When the effect of one factor on the dependent variable is altered depending on the value of a second factor.
Internal validity: The degree to which one can unambiguously attribute changes in the dependent variable to the manipulation of the independent variable.
Interquartile range (IQR): The width of the middle half of a distribution.
Interval estimate: See *Confidence interval.*
Interval scale: A measurement scale in which the quantitative distance between intervals is held constant. The zero point, however, is not the absence of quantity.
Kurtosis: The peakedness or flatness of a distribution curve.
Least squares criterion: A straight line through bivariate data where the sum of the squared errors is at a minimum.
Least squares method: The method by which the *least squares criterion* is met.
Leptokurtic: A distribution curve that has a narrow and accentuated peak.
Linear relationship: A relationship between two variables in which as the value of one variable changes, the value of a second variable changes by a constant.
Main effect: An effect found among the conditions of one factor, independent of the influence of another factor.

Manipulation: The controlled presentation of an independent variable.

Mann–Whitney *U* test: A nonparametric hypothesis test of two independent samples. The nonparametric counterpart to the independent-samples t test.

Matched-samples design: A research design to control participant variables (e.g. IQ, age, and ethnicity) in which participants are paired on an important participant variable and then randomly split and assigned to different groups.

Mean deviation: A measure of dispersion or variability in the distribution of scores. The mean deviation is the sum of the absolute values of deviation scores divided by the number of scores.

Mean square between: A weighted measure of variation between group means.

Mean square within: A weighted average of within-group variances of two or more samples, also called the pooled variance.

Mean: The sum of scores divided by the number of scores.

Measurement: The assignment of numbers to attributes, objects, or events according to a set of predetermined rules.

Measures of variability (or dispersion): Numerical measures that reflect the degree to which scores of a distribution are spread out.

Median: The value in a distribution which splits the bottom half of the scores, quantitatively speaking, from the top half of the scores.

Midpoint: The balance point of an interval.

Mode: The score in a distribution that occurs most frequently.

Multiple comparisons: Tests of differences between cell or condition means performed after an ANOVA.

Multiple regression: A statistical technique that uses two or more variables to predict the value of another variable.

Multiplication rule: A rule used to determine the probability of the joint occurrence of two or more events.

Mutually exclusive events: When the occurrence of one event precludes the occurrence of another event.

Negatively skewed distribution: A skewed distribution in which the elongated tail points toward the smaller or negative numbers.

Nominal scale: A measurement scale that conveys no quantitative information, but rather merely distinguishes one attribute, object, or event from others.

Nonparametric tests: Statistical tests that do not make assumptions about population parameters and do not require interval or ratio data.

Normal distribution: A symmetrical, bell-shaped curve escalating gradually at first and then more sharply, eventually inflecting, and then tapering to a peak

Normality: The statistical assumption that the population from which the sample is taken is normally distributed.

Null hypothesis: Symbolized as H_0, the null hypothesis is a statement about some population characteristics. It usually states no effect, no difference, or no relationship between variables.

Observation: Careful and systematic measurements of events occurring in the world by using either one of the five senses or scientific tools and instruments.

Observed frequencies: The actual frequency counts recorded in a set of categories.

Omega-squared: A statistic that estimates the size of an effect, an adjusted ratio of primary variance over total variance.

One-tailed (or directional) test: A statistical test in which the rejection region lies only in one tail of the sampling distribution.

One-way ANOVA: A statistical test comparing several sample means to determine if one or more have come from different populations. Used only with between-participants research designs with one independent variable or factor.

Operational definition: A description of the concrete measurement of a concept for the purposes of a given research project.

Order effects: Differences between treatment conditions due to the order of presentation in a repeated-measures design.

Ordinal scale: A scale of measurement registering the relative position between attributes, objects, or events.

Ordinate: The vertical (Y) axis of a graph.

Parameters: Numerical values reflecting properties of a population.

Parametric tests: Significance tests that require interval or ratio scaled data and that make assumptions about the characteristics of populations.

Participant variable: A characteristic of a participant that is fixed at the time of the experiment.

Pearson correlation: A measure of association between two variables created by Karl Pearson. A very powerful and frequently used measure of association.

Percentile rank: A number assigned to a score that indicates the percentage of scores found below that score.

Percentile: A distribution value below which falls a certain percentage of scores.

Platykurtic: A distribution curve that is broad and has a muted peak.

Point estimation: Using a sample statistic to infer the value of a population parameter.

Point-biserial correlation: A correlational analysis used when one variable is continuous and the other variable is dichotomous, symbolized as r_{pb}.

Population: Every member of a given group.

Positively skewed distribution: A skewed distribution in which the elongated tail points toward the larger positive numbers.

Post hoc (or a posteriori) tests: A category of tests, selected prior to the rejection of the null hypothesis in an overall analysis, which are used to locate differences between pairs of means.

Power: The probability of an inferential test to correctly reject a null hypothesis.

Primary variance: See *Treatment variance*.

Probabilistic dependence: When knowledge of the occurrence of one event changes the probability of occurrence of a second event.

Probabilistic independence: When knowledge of the occurrence of one event has no effect on determining the probability of occurrence of a second event.

Probability: A measure of the likelihood that an event will occur. Probability values range from 0 to 1.

Qualitative independent variable: An independent variable that changes by kind or type. For example, different kinds of drugs.

Quantitative independent variable: An independent variable that changes by an amount. For example, different doses of a drug.

Quartile: One-fourth of a distribution of scores.

Quasi-experiment: A research design that has many features of an experiment, but one in which participants are not randomly assigned to conditions.

Random assignment: The placing of participants into study conditions such that each participant is equally likely to be assigned to a given condition.

Random sampling: A means of selecting participants for a study such that each participant in the population has an equal chance of being included in the sample.

Range: A measure of dispersion in a distribution. The highest score of a distribution minus the lowest score.

Ratio scale: A scale of measurement that has all the characteristics of an interval scale plus a "true" zero point.

Raw (original) scores: Quantitative scores obtained in a study.

Real limits: The upper and lower boundaries of an interval of measurement. The upper real limit of the number is one-half the unit of measurement above the number; the lower real limit of a number is one-half the unit of measurement below the number.

Regression equation: A formula used to predict a Y value given a specific X value.

Regression line: A mathematically generated line fitted to a scatter plot by the least squares criterion.

Regression: A set of statistical procedures applied to bivariate data that allows the value of one variable to be used to predict the value of another variable.

Repeated-measures ANOVA: A statistical test used to compare several sample means to determine if one or more come from different populations. Used with within-participants research designs with one independent variable or factor.

Representativeness: The methodological assumption that the participants comprising the sample represent the population in question.

Research hypothesis: A statement (prediction) about the expected outcome of a study; usually derived from theory or previous research findings.

Sample: A subset of participants (or observations) drawn from a population.

Sampling distribution: A theoretical distribution of a statistic based on all possible random samples of size n taken from the same population.

Sampling error: The difference between a sample statistic and a population parameter (e.g. $M - \mu$).

Sampling: The process of selecting participants (or observations) into a study.

Scatterplot: A graphic representation of a bivariate distribution.

Scientific hypothesis: See *Research hypothesis.*

Secondary variance: See *Error variance.*

Semi-interquartile range (SIQR): The interquartile range divided by 2.

Shared (or common) variance: The amount of variance in one variable that can be explained or accounted for by variance in a second variable.

Simple frequency distribution: Scores arranged from highest to lowest, with the frequency of occurrence of each score indicated in a column beside the scores.

Skewed distribution: An asymmetrical distribution in which scores tend to bunch at either the left or the right end of the curve; may be either positive or negative.

Slope: The angle of a straight line. A descriptive feature of a regression line.

Spearman rank correlation: A coefficient for *ranked* data, symbolized as r_s.

Standard deviation: A measure of dispersion or variability. It is the square root of the variance.

Standard error of the estimate: In regression, a measure of prediction error. The average standard deviation of the conditional distributions.

Standard error of the mean: A standard deviation of a sampling distribution of means.

Standard normal distribution: A distribution of z scores with a mean of 0 and a standard deviation of 1.

Standard score: A raw score expressed in standard deviation units (e.g. a z score).

Statistical hypothesis: A numerical claim on the outcome of a study, coming in mutually exclusive and collectively exhaustive pairs (null and alternative).

Statistics: Numerical values reflecting properties of a sample.

***t* Distribution:** A family of various bell-shaped distributions of t values. Also called the *Student t distribution.*

***t* Test:** Any number of inferential tests based on the t distribution.

Test for independence: A chi-square test that examines the degree to which observed data from a set of categories arranged along two factors coincide with theoretical or previously empirical-derived expectations.

Theory: An attempt to explain a topic of interest by referring to general principles and relationships that are independent of the topic to be explained.

Treatment (or primary) variance: The amount of variance among sample means due to the features of the various conditions.

Treatment: In experiments, an analogous term to "independent variable" – most often used when the procedure involves direct interaction with the participant.

Tukey's *HSD*: (Honestly Significant Difference) A conservative multiple comparison test used to locate the source of significance in designs with more than two conditions.

Two-tailed (or nondirectional) test: A statistical test in which the rejection region is divided equally between the two tails of the sampling distribution.

Two-way ANOVA: A statistical test designed to determine if there are population differences within a two-factor research design.

Type I error: A decision error made when rejecting a true null hypothesis.

Type II error: A decision error made when failing to reject a false null hypothesis.

Understanding: The ability to make some cause-and-effect statement regarding a topic of interest and other variables. The highest goal of the researcher.

Unimodal: A distribution with one mode.

Variance: A measure of dispersion or variability. It is the *average squared deviation score*; commonly used in formulas for hypothesis testing.

Wilcoxon signed-ranks test: A nonparametric hypothesis test of two dependent samples. The nonparametric counterpart to the dependent-samples t test.

Within-group variation: A measure of the variation of scores within a group.

Within-participants design: An experimental design in which participants are exposed to all conditions of the study.

***Y* intercept:** The point on the Y axis at which a prediction line intersects.

***z* score:** A transformed raw score that indicates the number of standard deviation units a raw score is from the mean of the distribution.

***z* test:** A significance test used to decide if a sample mean comes from a population with a specified mean.

List of Selected Formulas

Population/ sample mean $$\mu = \frac{\Sigma X}{N} \qquad M = \frac{\Sigma X}{n}$$ (3.1, 3.2)

Definitional formulas

Population variance $$\sigma^2 = \frac{\Sigma(X-\mu)^2}{N}$$ (4.6)

Sample variance $$s^2 = \frac{\sum (X-M)^2}{n-1}$$ (4.7)

Computational formulas

Population variance $$\sigma^2 = \frac{\sum X^2 - [(\sum X)^2/N]}{N}$$ (4.12)

Sample variance $$s^2 = \frac{\sum X^2 - [(\sum X)^2/n]}{n-1}$$ (4.13)

Any formula for the standard deviation is the square root of the variance formula

Population/ sample formulas for *z* $$z = \frac{X-\mu}{\sigma} \quad z = \frac{X-M}{s}$$ (5.2, 5.3)

Population/ sample formulas for *X* given *z* $$X = \mu + z\sigma \quad X = M + zs$$ (5.4, 5.5)

Probability of favorable event $$P = \frac{(\textit{number of favorable events})}{(\textit{total number of events})}$$ (6.1)

Essentials of Behavioral and Social Science Statistics, Third Edition.
K. Paul Nesselroade, Jr. and Laurence G. Grimm.
© 2025 John Wiley & Sons, Inc. Published 2025 by John Wiley & Sons, Inc.
Companion website: https://www.wiley.com/go/Essentials_of_Behavioral_and_Social_Science_Statistics3e

Addition rule formula for two events	$P(A \text{ or } B) = P(A) + P(B) - P(A \text{ and } B)$	(6.4)
Multiplication rule formula for two events	$P(A \text{ and } B) = P(A \mid B)P(B)$	(6.6)
Conditional probability formula	$P(A \mid B) = \dfrac{P(A \text{ and } B)}{P(B)}$	(6.7)
Standard error of the mean	$\sigma_M = \dfrac{\sigma}{\sqrt{n}}$	(7.1)
Estimated standard error of the mean	$s_M = \dfrac{s}{\sqrt{n}}$	(7.2)
Single-sample z and t test	$z_{obt} = \dfrac{M - \mu}{\sigma_M} \quad t_{obt} = \dfrac{M - \mu}{s_M} \quad (df = n - 1)$	(8.1, 8.3)
Cohen's d for single-sample z and t test	$d = \dfrac{M - \mu}{\sigma} \quad d = \dfrac{M - \mu}{s}$	(8.2, 8.4)
Variance formula $s_{M_1 - M_2}$	$s_{M_1 - M_2} = \sqrt{\dfrac{s_1^2(n_1 - 1) + s_2^2(n_2 - 1)}{n_1 + n_2 - 2}\left(\dfrac{1}{n_1} + \dfrac{1}{n_2}\right)}$	(9.1)
The t ratio	$t_{obt} = \dfrac{M_1 - M_2}{s_{M_1 - M_2}} \quad (df = n_1 + n_2 - 2)$	(9.3)
Cohen's d for independent-samples t test	$d = \dfrac{M_1 - M_2}{\sqrt{\dfrac{s_1^2(n_1 - 1) + s_2^2(n_2 - 1)}{n_1 + n_2 - 2}}}$	(9.4)
The estimate of the standard error of the difference, s_D	$s_D = \dfrac{s_D}{\sqrt{n_p}}$	(9.6)
Computational formula for s_D	$s_D = \sqrt{\dfrac{\Sigma D^2 - [(\Sigma D)^2/n_p]}{n_p - 1}}$	(9.7)
Dependent-samples t test	$t_{obt} = \dfrac{(M_X - M_Y)}{s_D} \quad (df = n_p - 1)$	(9.8)

Cohen's *d* for dependent-samples *t* test

$$d = \frac{M_X - M_Y}{s_D} \quad (9.9)$$

Delta

$$\delta = \gamma\sqrt{n} \quad (10.1)$$

Gamma

$$\gamma = \frac{\mu_{alt} - \mu_0}{\sigma} \quad (10.2)$$

Determining sample size for a single-sample *t* test

$$n = \left(\frac{\delta}{\gamma}\right)^2 \quad (10.3)$$

***F* ratio for one-way ANOVA**

$$F = \frac{MS_{BG}}{MS_W}\,(df_{BG} = k - 1;\ df_W = N - k) \quad (11.6)$$

One-way ANOVA Omega-squared, ω^2

$$\varpi^2 = \frac{SS_{BG} - df_{BG}(MS_W)}{SS_T + MS_W} \quad (11.7)$$

One-way ANOVA Eta-squared, η^2

$$\eta^2 = \frac{SS_{BG}}{SS_T} \quad (11.8)$$

One-way ANOVA Tukey's *HSD*

$$HSD = q\sqrt{\frac{MS_W}{n}} \quad (11.9)$$

One-way ANOVA Fisher's *LSD* test

$$t = \frac{M_i - M_j}{\sqrt{MS_W\left(\frac{1}{n_i} + \frac{1}{n_j}\right)}} \quad (11.10)$$

***F* ratio for two-way ANOVA**

$$F_A = \frac{MS_A}{MW_W};\ F_B = \frac{MS_B}{MW_W};\ F_{A\times B} = \frac{MS_{A\times B}}{MW_W} \quad (12.5\text{-}7)$$

$$(df_W = N - k;\ df_A = levels - 1;$$

$$df_B = levels - 1;\ df_{A\times B} = df_A \times df_B)$$

Two-way ANOVA Omega-squared

$$\varpi_A^2 = \frac{SS_A - (df_A)MS_W}{SS_T + MS_W}\ \varpi_B^2 = \frac{SS_B - (df_B)MS_W}{SS_T + MS_W} \quad (12.8\text{–}12.10)$$

$$\varpi_{A\times B}^2 = \frac{SS_{A\times B} - (df_{A\times B})MS_W}{SS_T + MS_W}$$

Two-way ANOVA Eta-squared

$$\eta_A^2 = \frac{SS_A}{SS_A + SS_W} \quad \eta_B^2 = \frac{SS_B}{SS_B + SS_W} \tag{12.11–12.13}$$

$$\eta_{A \times B}^2 = \frac{SS_{A \times B}}{SS_{A \times B} + SS_W}$$

Two-way ANOVA Tukey's *HSD*

$$HSD_A = q_A\sqrt{\frac{MS_W}{n_A}} \quad HSD_B = q_B\sqrt{\frac{MS_W}{n_B}} \tag{12.14–12.16}$$

$$HSD_{A \times B} = q_{A \times B}\sqrt{\frac{MS_W}{n_{A \times B}}}$$

Two-way ANOVA Fisher's *LSD* test

$$t = \frac{M_i - M_j}{\sqrt{MS_W\left(\frac{1}{n_i} + \frac{1}{n_j}\right)}} \tag{12.17}$$

***F* ratio for repeated-measures ANOVA**

$$F = \frac{MS_{BG}}{MS_{error}} \tag{13.1}$$

$$[df_{BG} = k - 1; df_{error} = (N - k) - (n - 1)]$$

Repeated-measures ANOVA omega-squared

$$\varpi^2 = \frac{SS_{BG} - df_{BG}(MS_{error})}{SS_T + MS_{error}} \tag{13.4}$$

Repeated-measures ANOVA eta-squared

$$\eta^2 = \frac{SS_{BG}}{SS_T} \tag{13.5}$$

Repeated-measures ANOVA Tukey's *HSD*

$$HSD = q\sqrt{\frac{MS_{error}}{n}} \tag{13.6}$$

Repeated-measures ANOVA Fisher's *LSD*

$$t = \frac{M_i - M_j}{\sqrt{MS_{error}\left(\frac{1}{n_i} + \frac{1}{n_j}\right)}} \tag{13.7}$$

Computational formula for Pearson *r*

$$r = \frac{n_p(\Sigma XY) - (\Sigma X)(\Sigma Y)}{\sqrt{\left[n_p(\Sigma X^2) - (\Sigma X)^2\right]\left[n_p(\Sigma Y^2) - (\Sigma Y)^2\right]}} \tag{14.1}$$

Linear regression equation	$Y_p = M_Y + b(X - M_X)$	(15.1)
Computational formula for the slope	$b = \frac{n_p(\Sigma XY) - (\Sigma X)(\Sigma Y)}{[n_p(\Sigma X^2)] - (\Sigma X)^2}$	(15.2)
Correlation formula for s_e	$s_e = s_Y\sqrt{(1 - r^2)\left[\frac{n_p}{(n_p - 2)}\right]}$	(15.5)
Chi-square	$\chi^2 = \Sigma\frac{(f_0 - f_e)^2}{f_e}$	(16.1)
Chi-square Frequency expected	$f_e = \frac{f_c f_r}{N}$	(16.2)
Cramér's *V* for Chi-square	$\phi = \sqrt{\frac{\chi^2}{(N)\left(df_{row/column}\right)}}$	(16.4)
Standardized residual for Chi-square	$R = \frac{f_o - f_e}{\sqrt{f_e}}$	(16.5)

List of Symbols

Symbols	
α	Level of significance; probability of Type I error
b	Slope
β	Probability of Type II error
B	Number of scores below a given score X
$1 - \beta$	Power
χ^2	Chi-square
$cum\ f$	Cumulative frequency
d	Cohen's d effect size measure
D	Difference score
$\overline{D}$	Mean of difference scores
df	Degrees of freedom
df_A	Degrees of freedom for Factor A
$df_{A \times B}$	Degrees of freedom for the interaction
df_B	Degrees of freedom for Factor B
df_{BG}	Degrees of freedom between group
df_{error}	Degrees of freedom error
df_W	Degrees of freedom within group
δ	Delta
E	The number of scores equal to X
f	Frequency of occurrence of a score
f_c	Frequency of scores in a column

(*Continued*)

Essentials of Behavioral and Social Science Statistics, Third Edition.
K. Paul Nesselroade, Jr. and Laurence G. Grimm.
© 2025 John Wiley & Sons, Inc. Published 2025 by John Wiley & Sons, Inc.
Companion website: https://www.wiley.com/go/Essentials_of_Behavioral_and_Social_Science_Statistics3e

Symbols	
f_e	Expected frequency
f_o	Observed frequency
f_r	Frequency of scores in a row
F	*F* statistic in ANOVA
γ	gamma
η^2	Eta-squared; effect size measure
H_0	Null hypothesis
H_1	Alternative hypothesis
HSD	Tukey's honestly significant difference value
k	Number of groups; number of the last group
LSD	Fisher's least significant difference value; protected *t* test
LL	Lower limit of an interval
μ	Population mean
μ_M	Mean of a sampling distribution of means
μ_{alt}	Mean of the alternative ("treated") population
μ_0	Population mean when null hypothesis is true
M	Sample mean
MS	Mean square
MS_A	Mean square for Factor *A*
$MS_{A \times B}$	Mean square for *A* × *B* interaction
MS_B	Mean square for Factor *B*
MS_{BG}	Mean square between group
MS_W	Mean square within group
MS_{error}	Mean square error
N	Number of scores in a population
n	Number of scores in a sample
n_k	Number of scores in the last group
n_p	Number of pairs of scores
P	A person in a repeated-measures ANOVA
P	Probability
$P(A)$	Probability of event *A*
$P(B)$	Probability of event *B*
$P(A \text{ and } B)$	Probability of event *A* and event *B* co-occurring
$P(\text{not } A)$	Probability of event *A* not occurring

Symbols	
$P(A \text{ or } B)$	Probability of event A or event B occurring
$P(A\|B)$	Probability of event A occurring given event B has occurred
PR	Percentile rank
q	The studentized range statistic
Q_1	First quartile
Q_3	Third quartile
r	Pearson product moment correlation coefficient for a sample
ρ	Rho
R	Standardized residual
Σ	Sum up the following
s	Sample standard deviation
s^2	Sample variance
s_D	Standard deviation of difference scores
$s_{\overline{D}}$	Standard error of the difference for dependent samples
s_e	Standard error of the estimate
s_Y	Standard deviation of Y scores
s_M	Estimated standard error of the mean
$s_{M_1-M_2}$	Estimated standard error of the difference between sample means
SS	Sum of squares
SS_{BG}	Between-group sum of squares
SS_{BP}	Between-participants sum of squares
SS_{error}	Sum of squares error
SS_W	Within-group sum of squares
σ	Population standard deviation
σ^2	Population variance
σ_M	Standard error of the mean (population)
t	t statistic
t_{obt}	t test
UL	Upper limit of an interval
ϕ	Cramér's V; effect size measure for chi-square
ω^2	omega-squared
$\overline{X}$	Sample mean

(*Continued*)

Symbols	
X_H	Highest score in a distribution
X_L	Lowest score in a distribution
x	Deviation score
Y_p	Y predicted
z	z score
z_{obt}	z test

Index

Essentials of Behavioral and Social Science Statistics, Third Edition.
K. Paul Nesselroade, Jr. and Laurence G. Grimm.
© 2025 John Wiley & Sons, Inc. Published 2025 by John Wiley & Sons, Inc.
Companion website: https://www.wiley.com/go/Essentials_of_Behavioral_and_Social_Science_Statistics3e

r

s